CALIXARENES 50TH ANNIVERSARY:
COMMEMORATIVE ISSUE

Calixarenes 50th Anniversary: Commemorative Issue

Edited by

J. VICENS, Z. ASFARI
*Ecole Européenne des Hautes
Etudes des Industries Chimiques
de Strasbourg (EHICS), France*

and

J. M. HARROWFIELD
*The University of Western
Australia, Perth, Australia*

Reprinted from
Journal of Inclusion Phenomena and Molecular Recognition in Chemistry,
Volume 19, Nos. 1–4 (1994)

KLUWER ACADEMIC PUBLISHERS
DORDRECHT / BOSTON / LONDON

Library of Congress Cataloging-in-Publication Data

```
Calixarenes 50th anniversary : commemorative issue / edited by Jacques
  Vicens and Zouhair Asfari and Jack McB. Harrowfield.
      p.   cm.
   Includes index.
   ISBN 0-7923-3393-4 (HB : acid-free)
   1. Calixarenes.   I. Vicens, J.   II. Asfari, Zouhari.
 III. Harrowfield, Jack McB.   IV. Title: Calixarenes fiftieth
 anniversary.
 QD341.P5C347  1995
 547'.632--dc20                                           95-4009
```

ISBN 0-7923-3393-4

Published by Kluwer Academic Publishers,
P.O. Box 17, 3300 AA Dordrecht, The Netherlands.

Kluwer Academic Publishers incorporates
the publishing programmes of
D. Reidel, Martinus Nijhoff, Dr W. Junk and MTP Press.

Sold and distributed in the U.S.A. and Canada
by Kluwer Academic Publishers,
101 Philip Drive, Norwell, MA 02061, U.S.A.

In all other countries, sold and distributed
by Kluwer Academic Publishers Group,
P.O. Box 322, 3300 AH Dordrecht, The Netherlands.

Printed on acid-free paper

All Rights Reserved
© 1994 Kluwer Academic Publishers
No part of the material protected by this copyright notice may be reproduced or
utilized in any form or by any means, electronic or mechanical,
including photocopying, recording or by any information storage and
retrieval system, without written permission from the copyright owner.

Printed in the Netherlands

Table of Contents

Foreword — 1

Introducing Paper

T. KAPPE / The Early History of Calixarene Chemistry — 3

Review Articles

V. BÖHMER, D. KRAFT, and M. TABATABAI / Inherently Chiral Calixarenes — 17

Y. OKADA and J. NISHIMURA / The Design of Cone-Fixed Calix[4]arene Analogs by Taking *syn*-[2.*n*]Metacyclophanes as a Building Block — 41

P. E. GEORGHIOU and Z. LI / Conformational Properties of Calix[4]naphthalenes — 55

H.-J. SCHNEIDER and U. SCHNEIDER / The Host-Guest Chemistry of Resorcinarenes — 67

F. BOTTINO and S. PAPPALARDO / Synthesis and Properties of Pyridinocalixarenes — 85

D. M. ROUNDHILL, E. GEORGIEV, and A. YORDANOV / Calixarenes with Nitrogen or Phosphorus Substituents on the Lower Rim — 101

G. BRODESSER and F. VÖGTLE / Homocalixarenes and Homocalixpyridines — 111

Z. ASFARI, S. WENGER, and J. VICENS / Calixcrowns and Related Molecules — 137

D. DIAMOND / Calixarene-Based Sensing Agents — 149

General Papers

P. TIMMERMAN, H. BOERRIGTER, W. VERBOOM, G. J. VAN HUMMEL, S. HARKEMA, and D. N. REINHOUDT / Proximally Functionalized Cavitands and Synthesis of a Flexible Hemicarcerand — 167

H. TAKEMURA, T. SHINMYOZU, H. MIURA, I. U. KHAN, and T. UNAZU / Synthesis and Properties of *N*-Substituted Azacalix[*n*]arenes — 193

J. A. WYTKO and J. WEISS / Arranging Coordination Sites around Cyclotriveratrylene — 207

H. IKI, T. KIKUCHI, H. TSUZUKI, and S. SHINKAI / X-Ray Crystallographic Studies of Tricarbonylchromium Complexes of Calix[4]arene Conformers on an Unusual Conformation which Appears in Cone Conformers — 227

O. ALEKSIUK, F. GRYNSZPAN, and S. E. BIALI / Preparation, Structure and Stereodynamics of Phosphorus-Bridged Calixarenes — 237

J. M. HARROWFIELD, W. R. RICHMOND, and A. N. SOBOLEV / Inclusion of Quaternary Ammonium Compounds by Calixarenes — 257

W. XU, R. J. PUDDEPHATT, L. MANOJLOVIC-MUIR, K. W. MUIR, and C. S. FRAMPTON / Calixarenes: Structure of an Acetonitrile Inclusion Complex and Some Metal Rimmed Derivatives — 277

Z. ASFARI, J.-P. ASTIER, C. BRESSOT, J. ESTIENNE, G. PEPE, and J. VICENS / Synthesis, Characterization, and X-Ray Structure of 1,2-Bis-crown-5-calix[4]arene. Modeling of Metal Complexation — 291

T. SAWADA, A. TSUGE, T. THIEMANN, S. MATAKA, and M. TASHIRO / Complexation Properties and Characterization of Four Conformers of a [2.1.2.1]Metacyclophane — 301

T. YAMATO, M. YASUMATSU, Y. SARUWATARI, and L. K. DOAMEKPOR / Synthesis and Ion Selectivity of Macrocyclic Metacyclophanes Analogous to Spherand-Type Calixarenes   315

Y. SHIBUTANI, H. YOSHINAGA, K. YAKABE, T. SHONO, and M. TANAKA / Polymeric Membrane Sodium-Selective Electrodes Based on Calix[4]arene Ionophores   333

P. D. BEER, Z. CHEN, P. A. GALE, J. A. HEATH, R. J. KNUBLEY, M. I. OGDEN, and M. G. B. DREW / Recognition by New Diester and Diamide-Calix[4]arene Diquininones and a Diamide-Benzo-15-Crown-5 Calix[4]arene   343

L. ZHANG, A. MACIAS, R. ISNIN, T. LU, G. W. GOKEL, and A. E. KAIFER / The Complexation of Ferrocene Derivatives by a Water-Soluble Calix[6]arene   361

A. F. DANIL DE NAMOR, J. WANG, I. GOMEZ ORELLANA, F. J. SUEROS VELARDE, and D. A. PACHECO TANAKA / Thermodynamic and Electrochemical Aspects of *p-tert*-butyl-calix(*n*)arenes (*n* = 4, 6, 8) and Their Interactions with Amines   371

Advanced Materials

T. M. SWAGER and B. XU / Liquid Crystalline Calixarenes   389

C. HILL, J.-F. DOZOL, V. LAMARE, H. ROUQUETTE, S. EYMARD, B. TOURNOIS, J. VICENS, Z. ASFARI, C. BRESSOT, R. UNGARO, and A. CASNATI / Nuclear Waste Treatment by Means of Supported Liquid Membranes Containing Calixcrown Compounds   399

Conference Diary   409

Author Index   411

Subject Index   415

Foreword

This issue is to celebrate the 50th anniversary of the proposal of a macrocyclic structure for the calixarenes. In February 1944, Zinke and Ziegler published an article entitled: 'Zur Kenntnis des Härtungsprozesses von Phenol-Formaldehyde-Harzen' in *Chemische Berichte*, volume 77, page 264. Here, they proposed a cyclic tetrameric structure $(C_{11}H_{14}O)_4$ for an oligomer extracted from the condensation product mixture obtained by reacting *p-tert*-butylphenol with formaldehyde in the presence of sodium hydroxide.

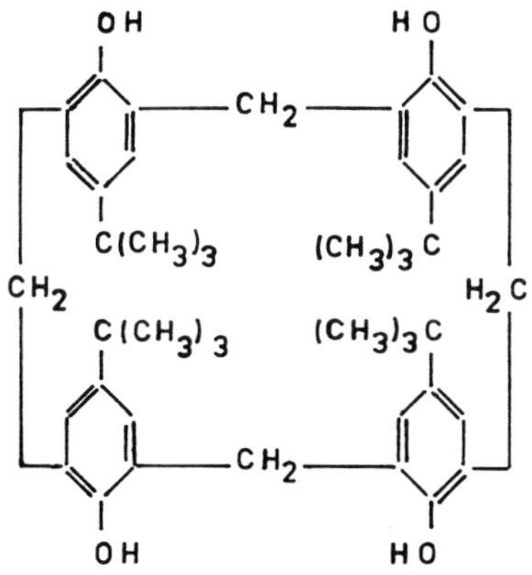

Fifty years on, calixarenes are the basis of many different areas of chemical research occurring at an increasing pace over the past decade in particular. The present volume is not intended to provide an overview of the developments but is rather a celebration of some of the highlights. It is hoped that this selection from the intricate mosaic of diversity that characterizes calixarene chemistry will stimulate and foster further development of this fascinating field.

J. Vicens
Z. Asfari
J. Harrowfield

The Early History of Calixarene Chemistry*

THOMAS KAPPE
Institute of Organic Chemistry, Karl-Franzens-University, A-8010 Graz, Austria

(Received in final form: 25 May 1994)

Abstract. The history of calixarene chemistry from the mid-thirties until 1978 is described.

Key words: History of chemistry, calixarenes, Zinke, Ziegler, Hönel.

1. Introduction

Fifty years ago, in 1944, Alois Zinke and Erich Ziegler presented the constitution (Figure 1) of a cyclic tetramer, obtained in two steps from *p-tert*-butylphenol and formaldehyde and which is called today *p-tert*-butylcalix[4]arene, in *Berichte der Deutschen Chemischen Gesellschaft* [1] and in a review [2]. They had already described the synthesis of this compound (or was it the calix[8]arene?) in 1941 without, however, assigning a structure to it [3]. In 1941 they wrote: "Die erhaltenen Produkte sind offenbar ziemlich hochmolekular, denn vorläufige, mit dem Acetylderivat ausgeführte Molukulargewichtsbestimmungen ergaben einen Durchschnittswert von 1725. Es dürfte sich um Verbindungen (oder ein Gemisch von Verbindungen) handeln, die aus etwa 8 Butylphenolkernen aufgebaut sind. Ätherartige Struktur scheint nicht vorzuliegen, denn Versuche, eine Spaltung mit Bromwasserstoff zu erreichen, verliefen negativ." According to present knowledge [4, 5] it is very likely that Zinke and Ziegler indeed had the acetyl derivative of the octamer at hand in 1941. But why did they propose the calix[4]arene structure in 1944 (without giving further experimental evidence for this conclusion)? Before answering this question we should take a short look at the circumstances of the discovery, the *curricula vitae* of Zinke and Ziegler, and of a third person (whose name has been mentioned in connection with calixarene chemistry so far only in footnotes [6]): Herbert Hönel.

2. People, Time, and Circumstances

Graz, the second-largest city of Austria, is located in the south-eastern part of the country, about 50 km away from the Slovenian and Hungarian borders, respectively. Since Graz is somewhat off the beaten track, the city is not so well known to tourists as, for instance Vienna, Salzburg and Innsbruck. However, its university

---

* This paper is dedicated to the commemorative issue on the 50th anniversary of calixarenes.

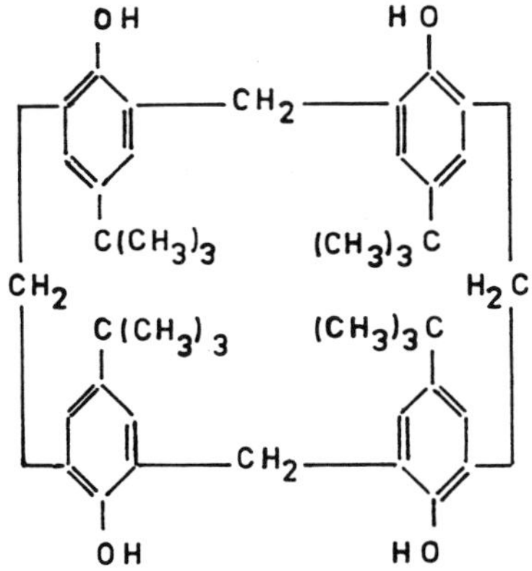

Fig. 1. Calix[4]arene as shown in Ref. [1, 2].

was founded more than 400 years ago and five Nobel prize winners have worked here. The names of Fritz Pregel (who developed the microanalysis of organic compounds), Otto Loewi (the pharmacologist who discovered acetylcholine as a neurotransmitter) and Erwin Schrödinger, the physicist, should be familiar to every chemist.

Alois Zinke was born in 1892 in the small village of Bärnbach, located 20 km west of Graz, where his father was a director of a glass factory. At the age of ten he was sent to high school ('Landesoberrealschule') in Graz, and he stayed in that city for the rest of his life. He continued his studies at the 'Technische Hochschule' of Graz, but after two years he changed to the University where he received his Ph.D. in 1915 under the tutelage of Roland Scholl. He was assistant at the chemical institute from 1914 to 1922. In 1920 he became 'Privat-Dozent' (unpaid lecturer) in organic chemistry, and in 1922 he was appointed Professor at the 'Technische Hochschule' in Graz, but only three years later he returned to the university as 'außerordentlicher Professor'. In 1941 he became full professor and director of the Institute of Organic and Pharmaceutical Chemistry of the University, a position he held until his death in 1963 [7].

Zinke's contribution to organic chemistry include research on natural resins, e.g. the isolation of siaresinol and sumaresinol acids which he showed to be triterpenes. Most of his research, however, was devoted to highly condensed aromatic ring systems (today called PAHs), particularly perylene. He became acquainted with these ring systems during his work on alizarines with R. Scholl, and interestingly, the last three papers he published in the 'sixties also dealt with 'perylene and its derivatives', namely parts 64–66 of this series.

However, for about two decades – from 1936 to 1956 – Zinke devoted a large part of his research to exploring the chemistry of phenol-formaldehyde resin formation. This excursion into a new field of research brought about – among many other most valuable results – the discovery of calix[n]arenes. We will come back to this point.

Alois Zinke, 1892–1963.

Erich Ziegler was born in Marburg at the river Drau in 1912. This city is located 70 km south of Graz, and was incorporated into the territory of Yugoslavia after the First World War in 1919. In 1923 the family moved to Graz. His father had decided that Erich Ziegler should become a druggist, but during his apprenticeship he was so much attracted by chemistry that in 1933 he enrolled for chemistry at the university of Graz. He joined Zinke's group just at the time when the phenol-formaldehyde project started. He obtained his Ph.D. in 1938 with a thesis on the reaction products of cresols with formaldehyde. He worked as an unpaid assistant for Zinke until he obtained a formal appointment in 1940. He was called up to the army from August 1939 to February 1940, and again from January 1942 to March 1944. The last 18 months of this period he spent in various army hospitals, since he had been seriously injured during the war in Russia. I mention these circumstances also as one explanation of the long interval between the first and second publication on cyclic phenol-formaldehyde products by Zinke and Ziegler, referred to in the introduction.

Ziegler's further academic career progressed smoothly: after his dismissal from the army, he completed his habilitation with his inaugural lecture as 'docent' (the experimental work had already been completed before 1942). He worked for some time still in the field of phenolic alcohols, and investigated especially their reactions with aromatic diazonium salts. In the early 'fifties his interest turned to heterocyclic chemistry (he published more than 200 papers in a series 'Synthesis of Heterocycles'), and made Graz a well-known center of heterocyclic chemistry. In 1963, after the death of Alois Zinke, his mentor and finally his good friend, he became full professor and director of the Institute of Organic Chemistry. Zinke and Ziegler had both planned a new building for the institute, but it was not before the spring of 1969 that it was ready for occupancy. Ten years later, Ziegler had to retire for health reasons, but until about 1990 he visited the institute regularly, worked in the laboratory and published papers in his new fields of interest: the reactivity of the C=N double bond, the chemistry of betaines, S- and N-ylids. He died on March 27, 1993 [8].

Erich Ziegler, 1912–1993.

While Zinke had never heard about the future of his cyclic phenol-formaldehyde products, Ziegler was informed about 'calixarenes' by Hermann Kämmerer from Mainz, Germany. Ziegler had given a lecture in Mainz on February 4, 1954, with the title 'Über cyclische Phenol-Formaldehyd-Kondensate' [9]. This was at a time when he had already abandoned research in this area. Nevertheless, Ziegler regularly received reprints of Kämmerer's work in this field, and they exchanged Christmas letters for the rest of Ziegler's life.

As indicated in the introduction, we must mention the name of a third person, who initiated the work on phenol-formaldehyde chemistry in Graz. Herbert Hönel [10] was born in Graz in 1890, where he obtained his Ph.D. in chemistry in 1913. After the First World War, in which he took part as a soldier, he found employment in 1919 at the 'Chemische Fabriken Dr. K. ALBERT' in Biebrich on the river Rhine, Germany, where he first came into contact with the synthesis and production of polymers. It was certainly a lucky chance (for both of them) when in 1926 Otto Reichhold [11], director of the largest Austrian varnish and coatings company, met this most talented chemist, and gave him an adequate position in his company. Hönel's first patents dealt with 'oil-reactive phenolic resins' and with 'terpene phenolic resins'. Within a short time he became a renowned pioneer in the field of plastics and coatings. From 1928 till 1939 he crossed the Atlantic several times to work for REICHHOLD in the United States as well. Hönel's success in research and development were cornerstones for the foundation of the 'BECKACITE-Kunstharzfabriken' in 1934 in Vienna and Hamburg, and factories with similar production lines in France and Great Britain. During the war he worked in Hamburg and Vienna.

After the war, he returned to the city where he was born. Alois Zinke provided him with a small laboratory in the basement of the institute of organic chemistry where he continued his studies on water-soluble resins. At the end of the war, when supplies of raw materials became short, he had already produced a coating material for protecting iron from rusting. This material, however, was only good enough in times of war. Then he became really obsessed with the idea of water-soluble resins. The reasons for this goal were clear: the high price of organic solvents, the danger in handling because of their flammability, health and environmental hazards. In 1948 he was co-founder of a small coatings company in Graz. Though in the beginning the company had to produce the material by conventional methods, Hönel gave the company the name 'VIANOVA-Kunstharz AG', indicating that 'new ways' were on the program. To make the story brief: the breakthrough came at the end of the 'fifties, and in the 'sixties all major car makers in Europe, such as Volkswagen, FIAT, Renault etc., used coatings produced under patents and licenses of VIANOVA. Because of their solubility in water the materials developed at VIANOVA could also be modified to serve in the electrolytic deposition of films on metals, a further big success. For many years now two modern research buildings of VIANOVA (which belongs to the HOECHST group today) have been located just 500 m away from the main building of the University of Graz. Hönel himself received many honors, including the Dr. h.c. rer.nat. from this university. He died in June of 1990, some months after the celebration of his 100th birthday.

Before dealing with the discovery of calixarenes, some remarks should be made on the circumstances under which these investigations took place. The time between the two World Wars was a very hard one for Austria (which means the part that remained from the Austrian Empire). This was especially true for the universities and was felt most in experimental disciplines. From 1917 to 1943 Anton Skrabal was head of the chemistry department in Graz. His main research interests were in the field of physical chemistry, especially kinetics. In 1878, when the new chemistry building was completed, it was the most modern one in Europe [12, 13]. Due to Skrabal's interests and his nature, substantial investments in equipment and modernization of the laboratories did not take place (with the exception of a small additional building for physical chemistry in 1927). Electric light was not introduced to the laboratories until 1943, when it had already become an utmost necessity because (due to shortages of rare earth materials) incandescent Welsbach mantles for gaslight illumination were no longer available during the War. In other words, working was only possible from 9 a.m. to 3 p.m. during the winter months!

For those readers who are not familiar with European history: Austria did not exist between March 1938 and May 1945. When the University of Graz became 'Reichsuniversität' in 1938 about one third of the academics (among them Loewi and Schrödinger), and one third of the students had to leave the university. Though the 'Third Reich' highly esteemed chemistry in general this brought no real and lasting improvement to the chemistry department in Graz. Quite to the contrary: the outbreak of the War robbed Zinke of his talented coworkers (Ziegler, Hanus) in the field of phenol-formaldehyde chemistry and left him with a few female graduate students. These were, in short, the conditions and circumstances under which the calixarenes were born.

3. The Discovery of Calixarenes

In 1936, Herbert Hönel supported by Otto Reichold, asked Alois Zinke, whom he had known from his student days, to act as a consultant for the BECKACITE Co., Hamburg–Vienna, in the field of phenol-formaldehyde resins, a proposal which Zinke obviously could not refuse. He started work immediately, together with his two assistants Franz Hanus and Erich Ziegler. After the war, Hanus [14] left the university and Richard Zigeuner joined Zinke's group. At that time, however, competition was strong in this field. There were already several research groups working in the area: those of H. v. Euler and E. Adler [15], K. Hultsch [16], M. Köbner [17], and N. J. L. Megson [18], to name just a few.

Right from the start Zinke planned a comprehensive study of chemical processes involved in the curing (= hardening) of phenol-formaldehyde resins. Thus, the series he published received the title 'Zur Kenntnis des Härtungsprozesses von Phenol-Formaldehyd-Harzen'. The first paper [19] appeared in 1939 [20], the last contribution [21], number XXVIII in this series, in 1958 [22]. In the industrial production of phenolic resins not only phenol itself but also alkylphenolss, and mixtures thereof, are valuable parts of the recipes, Zinke used *ortho* and/or *para* substituted alkylphenols as model compounds. Also, 2- and 4-chlorophenol and 2,4-dichlorophenol were employed to block reactive sites in order to simplify the first reaction steps. The work performed in Graz has certainly added much to our current knowledge of the constitution of phenol-formaldehyde resins as well as of the intermediates and the mechanisms leading to them (these results may be found in appropriate reviews [2, 18, 23] and monographs [16, 24–26]). The discovery of cyclic oligomers had never caused too much excitement in Graz. They fitted nicely in the general picture they were drawing, created much work (in establishing their molecular weight), but were a kind of a dead end insofar as they would not polymerize further.

Nevertheless, the story leading to the discovery is quite interesting. One of the reasons is that the same story was duplicated some 30 years later in the U.S.A., as we will see. In their first papers Zinke and his coworkers had described that the benzylalcohols formed from phenols lead to dibenzylethers (which had also been found by Euler and others), and that these ethers can decompose on heating to give the corresponding diarylmethane derivatives and one equivalent of formaldehyde is liberated (benzaldehydes and the corresponding methylphenols are other reaction products, especially if chlorinated phenols are used as starting material [3]). In their first paper of 1941 in which the preparation of an cyclic oligomer from *p-tert*-butylphenol was reported [3] they described the fact that the sodium salt of 2-hydroxy-3,5-dimethyl-benzylalcohol (*m*-xylenolalcohol) gave the corresponding diphenylmethane derivative (with loss of formaldehyde and water) within four weeks if kept at room temperature. From this observation they concluded 'that the curing process of phenolalcohols can be influenced by the presence of alkali'.

Zinke had been informed by BECKACITE-Kunstharzfabrik that during the curing process of technical resols [27] clear solutions in dry oils or other inert solvents could only be obtained if scrupulously alkali-clean material was used. The technical handling of a product where sludges are precipitating is of course unpleasant (thinking of pipes and valves), and also the customer might have problems and will complain. So the information from BECKACITE (so to speak, Hönel) was also combined with a request to have a look into this matter. For whatever reason (maybe there was a hint from Hönel?), the resol obtained from *p-tert*-butylphenol was studied first. Anyway, this phenol proved to be an excellent choice. When Erich Ziegler heated 50 g of this resol – prepared in the usual way from the phenol, aqueous formaldehyde (35%), and $3N$ sodium hydroxide solution – in linseed oil he obtained a brownish waxy material which yielded 20 g of crystalline glistening material after washing with ethyl acetate. The product could be recrystallized from tetrachloromethane and from benzene [3] or a chlorobenzene–ethanol mixture [1].

Briefly: apart from the two articles mentioned already in the introduction [1, 3] Zinke published three more papers concerning cyclic oligomers obtained from *p*-substituted phenols and formaldehyde in 1948 [28], 1952 [29], and his final paper in 1958 [21]. The *para* substituents (besides *tert*-butyl) were the methyl, phenyl, benzyl, cyclohexyl, *tert*-pentyl, and 1,1,3,3-tetramethylbutyl group. Cryoscopic molecular weight determinations in camphor or naphthalene with the latter compound (which showed good solubility) gave values in excellent agreement with the tetrameric structure [29]. Since none of the original specimens have been saved in Graz (though many of these compounds were prepared by students in the undergraduate courses of organic synthesis in the 'fifties on a multigram scale, as the author of this article remembers) it is very hard to say today which of these samples were pure tetramers [5]. Moreover, in none of the papers is the exact amount of alkali employed defined. Today it is well known that this is one of the most important factors for the outcome of the reaction [30–32]. Usually it was just stated that the resols [27] were precipitated with dilute hydrochloric (!) or acetic acid and washed. Only in one case [33] is it stated: "das durch schwaches Ansäuern mit verd. Essigsäure abgeschiedene und mit wenig Wasser *nicht alkalifrei gewaschene Produkt wird durch Erhitzen auf 120° vorgehärtet und*..." A successful repetition of Zinke's procedures outside of his laboratory was certainly not easy, as the general part of the papers, remarks and references had to be studied. Therefore, it is not surprising that just a few years after the most detailed paper of 1952 (with the molecular weight determination), John Cornforth (an Australian chemist working in Britain who obtained the Nobel Prize in 1975 for his work on enzyme-catalyzed reactions) isolated a pair of compounds in each case when repeating the preparation of two of Zinke's tetramers [34]. We shall come back to this point shortly in the section on 'The Zinke Paradigm'.

In 1981 Gutsche [31] published a paper for the preparation of *p-tert*-calix[4]arene which he thought to be a 'foolproof' procedure. In 1986 [32] he had to admit that not only others, but also people from his own laboratories had difficulties in duplicating this foolproof procedure. He could pinpoint the frequently observed failure in the instruction given to neutralize the finely ground solid 'precursor' (in former days called resols) with aqueous HCl. The ingenious idea of Gutsche was then [32] to completely omit these washings and to add just the necessary (small) amount of NaOH in the very beginning, and so to run through the reaction in a one pot manner (removing the water after some time, adding the diphenyl ether to allow pyrolysis at 250°C). This procedure has now been published in *Organic Syntheses* [35] together with preparations of *p-tert*-butylcalix[6]arene [36], and -calix[8]arene [37].

The question (mentioned in the introduction) of why Zinke and Ziegler gave their first cyclic oligomer a tetrameric structure (Figure 1) in 1944 (though an exact molecular weight determination was not published before 1952) has not yet been answered. One reason which has been put forward by Gutsche [6, 38], namely that they saw the paper of J. Niederl [39] on calix[4]resorcinols, can be ruled out. Due to the circumstances described above, there were no American chemistry journals or *Chemical Abstracts* available in Graz during the War. Niederl's paper [39] was first quoted by Zinke in his papers of 1948 [28] and 1952 [29], and there he gave a wrong year (1936) in the references. However, before 1958 Zinke must have had a Niederl product at hand since, in his last paper [21], he described its degradation with nitric acid to styphnic acid (2,4,6-trinitroresorcinol). The reason for putting forward the tetrameric structure must have been more intuitive, and caused by the fear that his competitors would present this formula soon. I think that the two main reasons for this assignment were (a) Hans v. Euler's [40] publication in 1941 of a cyclic diether containing four phenolic nuclei, and it could be expected that Euler would soon come up with a carbocyclic tetramer, and (b) the porphins! In the 'thirties Hans Fischer had synthesized several porphins and written two books on the chemistry of pyrroles [41]. At that time it was a generally accepted rule that pyrrole and phenols (especially those with a *para* substituent where only the two *ortho* positions are available) show a similar reactivity towards electrophiles.

4. The Zinke Paradigm

Thomas Kuhn, a philosopher and historian of science wrote: "A scientific paradigm is created by an achievement that is sufficiently unprecedented to attract an enduring group of adherents away from competing modes of scientific activity and is sufficiently open ended to leave all sorts of problems for the redefined group of practitioners to resolve" [42]. In 1989 C. David Gutsche presented a lecture with the title 'Calixarenes: paradoxes and paradigms in molecular baskets' at a symposium; this lecture appeared in print in *Pure and Applied Chemistry* the following year [38]. Since this journal and Gutsche's book of 1989 *Calixarenes* [5] are readily available, the further 'early' history of calixarene chemistry will be dealt with in a nutshell.

In the mid-fifties the 'Zinke cyclic tetramer paradigm' [38] became well known to chemists working in the field of phenol-formaldehyde chemistry. In addition, Hayes and Hunter [43] presented in 1956/1958 a stepwise synthesis of methylcalix[4]arene, and – although no direct comparison with a specimen obtained via the Zinke procedure had been carried out – the identity was assumed. The Hayes and Hunter synthesis was later used and largely extended by H. Kämmerer [44]. But before this, as already mentioned before, Sir John Cornforth [34] had repeated Zinke's experiments with two *p*-substituted phenols and had obtained two pairs of compounds, high melting and lower melting species. Because X-ray experiments did not give unequivocal results Cornforth proposed that the phenolic nuclei could not rotate round the bonds joining them to the methylene groups, thus giving rise to diastereoisomers. H. Kämmerer [45] demonstrated in 1955, using temperature-dependent ^{1}H-NMR, that the cyclic tetramers were much more flexible than was assumed earlier by using space filling molecular models. (But this did not solve the question of structure for the second set of compounds obtained by Cornforth. It is known today that these products were the calix[8]arenes.)

Before the '*Early* History of Calixarene Chemistry' finally ended, the Zinke paradigm worked once more. This event is known as the 'Petrolite Chapter' [5, 38]. The PETROLITE Corporation was founded in 1916 and is located in Webster Groves, Missouri, not far from St. Louis, and thus not far from Washington University, St. Louis, C. D. Gutsche's working place. Its original products were demulsifiers for resolving crude oil emulsions. In the 'fifties, a Petrolite demulsifier was introduced onto the market which was made by reacting *p-tert*-butylphenol and formaldehyde under alkaline catalysis, then further oxyalkylating this intermediate with ethylene oxide. The intermediate was first assumed to be an open chain oligomer. Soon, however, the PETROLITE Co. encountered the same technical troubles with insoluble sludges in their product as had been the case at the BECKACITE Co. in the mid-thirties. The problem was brought to a group of chemists of the PETROLITE research laboratories. One of these chemists, J. H. Munch, around 1972 developed [46] a rather brief method of preparing large amounts of glistening crystals from a *p*-substituted phenol and paraformaldehyde in boiling xylene in the presence of some concentrated aqueous potassium hydroxide. (This method is very similar to the procedure now presented in *Organic Syntheses* [37] for the preparation of *p-tert*-butylcalix[8]arene!). Searching the literature, the PETROLITE chemists found the facts concerning the cyclic tetramers mentioned already, and concluded that their products must be the Zinke substances, although their method of preparation was quite different from the Zinke–Conforth procedure. Patents regarding this preparation of 'calix[4]arenes' were filed in 1976/77.

David Gutsche, having been a consultant to PETROLITE Corporation since 1949 [47] was informed about Munch's method in 1972 [46]. At that time, Gutsche had become interested in bioorganic chemistry and considered that Zinke's tetramers should be interesting candidates as enzyme mimics because of their basket-like shape. R. Ott and Zinke had already published the picture of a space-filling molecular model in a cone form, having the four hydroxyl groups close together, in 1954 [48]. Gutsche presented his results concerning the synthesis of five cyclic tetramers at two symposia in 1975 [49]. At these occasions he also coined the name CALIXARENES, although the name did not appear in print before 1978 [4]. It must have been an unpleasant surprise to Gutsche to see the results (exactly the same five examples) he had presented orally, published by a participant (Timothy Patrick) at one of the meetings two years later [50], obviously without the consent of the group at Washington University, as indicated by a rather short excuse [51]. Furthermore, these compounds were made commercially available by Parish Chemical Co., Provo, Utah [52]. The answer from Gutsche came one year later as a short communication to the editor of the *Journal of Organic Chemistry* [4]. The summary of this paper says: "The products obtained from the base-catalyzed condensation of formaldehyde with several *para*-substituted phenols have been shown to be mixtures of two or more components which appear to be cyclic oligomers with five or more aromatic units in the cyclic array." In this paper Gutsche used ^{1}H- and ^{13}C-NMR, osmotic molecular weight determination and mass spectrometry, as well as derivatization techniques to obtain trimethylsilyl derivatives, and TLC separation. As pointed out by Gutsche much later [38]: "One of the most powerful destroyers of paradigms is a new instrumental technique." So the Zinke paradigm found an end in 1978, and the modern times of calixarene chemistry began.

5. Conclusion

Though Gutsche's paper of 1978 [4] certainly was an incisive work in calixarene chemistry, much remained to be done to clarify the situation. In this paper, Gutsche had claimed that no calix[4]arenes were formed by the direct method. However, he had studied only the products obtained by Munch's procedure and the small modification thereof by Patrick. No comparison was made with a compound prepared according to the Zinke (or Cornforth) procedure or by the stepwise synthesis of Hayes–Hunter [43] or Kämmerer [45, 53]. This missing link was provided in a very extensive contribution in 1981 [31]. In this paper Gutsche showed that he could obtain the *tert*-butylcalix[4]arene in 20–25% yield (accompanied by smaller amounts of cyclic octamer) following the Cornforth modification of Zinke's procedure. It should be mentioned that he also repeated the multistep synthesis of Hayes–Hunter, and Kämmerer, respectively (with only minor modifications), and finally proved the identity of both products in this fine piece of work [31]. From Gutsche's investigations, including the analysis of the product mixtures obtained under different reaction conditions, it is now known that calix[8]arenes are formed by the direct method in the presence of a low concentration of alkali (0.03 equiv. with respect to the phenol), and at the relatively low temperature of boiling xylene; calix[6]arenes are obtained as the major product at the same temperature, but using nearly a tenfold amount of alkali (0.3 equiv.); and finally calix[4]arenes are obtained with low alkali concentration (0.045 equiv.) but at rather high temperature, boiling diphenyl ether (b.p. 250°) being the preferred solvent. Under these conditions *p-tert*-butyl[4]calixarene can now be prepared in 61% yield, according to *Organic Syntheses* [35] "pure enough for further use". From these pieces of evidence it is obvious that the Zinke products must have been in general the tetramers (low alkali concentration, rather high temperature), maybe contaminated with small amounts of the octamers. Therefore "to Zinke should go the true parentage" of calixarene chemistry, as noted by Gutsche [54].

* *Note added in proof:* The 'term ZINKE ZIEGLER Synthesis of Calixarenes' has recently been added to the vocabulary of Organic Name reactions [55].

References and Notes

1. A. Zinke and E. Ziegler: *Ber. Dtsch. Chem. Ges.* **77**, 264 (1944).
2. A. Zinke and E. Ziegler: *Wiener Chem. Ztg.* **47**(13/14), 151 (1944), review.
3. A. Zinke and E. Ziegler: *Ber. Dtsch. Chem. Ges.* **74**, 1729 (1941).
4. G. D. Gutsche and R. Muthukrishnan: *J. Org. Chem.* **43**, 4905 (1978).
5. C. D. Gutsche: *Calixarenes*, Monographs in Supramolecular Chemistry, J. F. Stoddard (Ed.), Royal Society of Chemistry (1989).
6. See Ref. [5], p. 7.
7. For Obituaries concerning A. Zinke see: N. J. L. Megson: *Chemistry and Industry* 1076 (1963); E. Ziegler: *Österr. Chem. Z.* **64**, 147 (1963).
8. For more detailed biographies of Ziegler see: H. Wittmann-Zinke: *Österr. Chem. Z.* 131 (1982/5); 123 (1993/5).

9. A second lecture was given by Ziegler in Mainz in 1966, but about a heterocyclic topic.
10. For a detailed biography of H. Hönel see: H. Wittmann-Zinke: *Österr. Chem. Z.* **57** (1990/2).
11. Otto Reichhold met an untimely death at the age of 40 when the Zeppelin airship *Hindenburg* caught fire during the landing operation on May 6, 1937, in Lakehurst, New Jersey.
12. H. Wittmann and E. Ziegler: *Die Entwicklung der chemischen Wissenschaften an der Universität Graz 1850–1982 – Eine Leistungsbericht*, Akad. Druck- u. Verlagsanstalt, Graz, 1985.
13. A. Kernbauer: *Das Fach Chemie an der Philosophischen Fakultät der Universität Graz*, Akad. Druck- u. Verlagsanstalt, Graz, 1985.
14. Frans Hanus left the university after the War. In 1950 he was hired by Herbert Hönel for the VIANOVA Co. (according to Hönel one of the most valuable decisions in his life!). After Hönel's retirement Frans Hanus became president of VIANOVA (to 1979).
15. The work of Euler and Adler until 1943 is nicely reviewed in Ref. [2].
16. K. Hultsch: *Ber. Dtsch. Chem. Ges.* **74**, 898, 1533, 1539 (1941); K. Hultsch: *Chemie der Phenolharze*, Springer-Verlag, Berlin (1950).
17. M. Koebner: *Chem. Ztg.* **54**, 619 (1930); *Angew. Chem.* **46**, 251 (1933); *Brit. Plast.* **11**, 9 (1939).
18. N. J. L. Megson and A. A. Drummond: *J. Soc. Chem. Ind.* **49**, 251T (1930); N. J. L. Megson: *Österr. Chem. Z.* **54**, 317 (1953), historical review after a lecture given in Graz, Oct. 9, 1952.
19. A. Zinke, F. Hanus, and E. Ziegler: *J. Prakt. Chem.* **152**, 126 (1939).
20. However, H. Hönel had already disclosed some of Zinke's preliminary results in a lecture given at a meeting of the 'Oil and Colour Chemists Association' on February 17, 1938 in London.
21. A. Zinke, R. Ott, and F. H. Garrana: *Monatsh. Chem.* **89**, 135 (1958).
22. The total number of papers coming from Zinke's laboratory is much higher since Ziegler and Zigeuner later published their own research in this field independently.
23. E. Ziegler: *Österr. Chem. Ztg.* **49**(5/6), 92 (1948).
24. R. W. Martin: *The Chemistry of Phenolic Resins*, John Wiley, New York (1956).
25. E. Müller: *HOUBEN-WEYL: Methoden der Organischen Chemie*, Vol. XIV/2 *Makromolekulare Stoffe* Part 2, Georg Thieme Verlag, Stuttgart (1963).
26. A. Knop and L. A. Pilato: *Phenolic Resins*, Springer-Verlag, Berlin (1985).
27. Resols are solid materials which are obtained from the primary oily condensation products of phenols and alkylphenols with formaldehyde (catalyzed by alkali) by heating to about 120°C. Further heating leads to 'Resitols', the product of final curing is called 'Resit'. 'Novolaks' are the primary condensation products if acids are used as catalyst.
28. A. Zinke, G. Zigeuner, K. Hössinger, and G. Hoffmann: *Monatsh. Chem.* **79**, 438 (1948).
29. A. Zinke, R. Kretz, E. Leggewie, K. Hössinger, G. Hoffman, and P. Weber v. Ostwalden: *Monatsh. Chem.* **83**, 1213 (1952).
30. See Ref. [5], p. 27–31.
31. C. D. Gutsche, B. Dhawan, K. H. No, and R. Muthukrishnan: *J. Am. Chem. Soc.* **103**, 3782 (1981).
32. G. D. Gutsche, M. Iqbal, and D. Stewart: *J. Org. Chem.* **51**, 742 (1986).
33. See Ref. [29], p. 1226.
34. J. W. Cornforth, P. D'Arcy Hart, G. A. Nicholls, R. J. W. Rees, and J. A. Stock: *Br. J. Pharmacol.* **10**, 73 (1955); J. W. Cornforth, E. D. Morgan, K. T. Potts, and R. J. W. Rees: *Tetrahedron* **29**, 1659 (1973).
35. C. D. Gutsche and M. Iqbal: *Org. Synth.* **68**, 234 (1989).
36. C. D. Gutsche, B. Dhawan, M. Leonis, and D. Stewart: *Org. Synth.* **68**, 238 (1989).
37. J. H. Munch and C. D. Gutsche: *Org. Synth.* **68**, 243 (1989).
38. C. D. Gutsche, J. S. Rogers, D. Stewart, and K.-A. See: *Pure Appl. Chem.* **62**, 485 (1990).
39. J. B. Niederl and H. J. Vogel: *J. Am. Chem. Soc.* **62**, 2512 (1940).
40. H. v. Euler, E. Adler, and B. Bergström: *Ark. Kem. Mineral. Geol.* **14B**, No. 30, 1 (1941).
41. Fischer-Orth: *Die Chemie des Pyrrols II*, Leipzig (1937).
42. T. S. Kuhn: *The Structure of Scientific Revolutions*, 2nd ed., The University Press of Chicago, Chicago (1962) (quoted from Ref. [38]).
43. B. T. Hayes and R. F. Hunter: *Chem. Ind. (London)* 193 (1956); *J. Appl. Chem.* **8**, 743 (1958).
44. V. Böhmer and J. Vicens (Eds.): *Calixarenes: A Versatile Class of Macrocyclic Compounds*, Kluwer Academic Publishers, Dordrecht, the Netherlands, pp. 39–62 (1991).

45. G. Happel, B. Mathaisch, and H. Kämmerer: *Makromol. Chem.* **176**, 3317 (1975).
46. See Ref. [4], footnote 8.
47. See Ref. [5], p. 18.
48. R. Ott and A. Zinke: *Österr. Chem. Ztg.* **55**, 156 (1954).
49. See Ref. [5], p. 21.
50. T. B. Patrick and P. A. Egan: *J. Org. Chem.* **42**, 382 (1977).
51. T. B. Patrick and P. A. Egan: *J. Org. Chem.* **42**, 4280 (1977).
52. See Ref. [4], footnote 2.
53. H. Kämmerer, G. Happel, and F. Caesar: *Makromol. Chem.* **162**, 179 (1972).
54. See Ref. [5], p. 10.
55. A. Hassner and C. Stumer: *Organic Synthesis Based on Name Reactions and Unnamed Reactions*, Pergamon-Elsevier Science Ltd., Oxford, 1994, p. 436.

Inherently Chiral Calixarenes*

VOLKER BÖHMER**, DAGMAR KRAFT and
MONIRALSADAT TABATABAI
Institut für Organische Chemie, Johannes-Gutenberg-Universität, Johann-Joachim-Becher-Weg 34, SB1, D-55099 Mainz, Germany

(Received: 3 March 1994; in final form: 11 July 1994)

Abstract. Due to the nonplanarity of the basic 1_n-metacyclophane system, calixarenes and resorcarenes can be transformed into molecules with inherent chirality. Various attempts to achieve this goal are reviewed. Special emphasis is given to derivatives with C_n-symmetry, including derivatives of spherand calixarenes and other calixarene-like macrocycles.

Key words: Inherent chirality, asymmetric and dissymmetric calixarenes, resorcarenes.

1. Introduction

One of the main reasons for the still increasing interest in calixarenes is their ability to act as host molecules, which is even more pronounced in suitable derivatives that are readily available using calixarenes as starting materials. One of the main features of naturally occurring host molecules is their capacity for enantioselective recognition. Various attempts have therefore been made to obtain chiral host molecules based on calixarenes.

As with any molecule, a calixarene may be converted into chiral derivatives simply by attaching chiral substituents. In principle this can be done at the 'lower rim' (at the phenolic oxygens) or at the 'upper rim' (*p*-positions). Thus, compounds **1** and **2** were obtained by etherification of the corresponding calixarenes with 2-methylbutyl bromide or by Friedel–Crafts acylation with 2-methylbutanoyl chloride [1]. Aminomethylation with chiral amines like proline (**3**) may be mentioned as an example in the resorcarene series [2].

If one uses enantiomerically pure reagents derivatives are obtained in this way directly as pure enantiomers, provided the derivatization reaction proceeds without racemization. Larger quantities thus become available in a quite straightforward way, which is advantageous in comparison to the compounds discussed below.

---

* This paper is dedicated to the commemorative issue on the 50th anniversary of calixarenes.
** Author for correspondence.

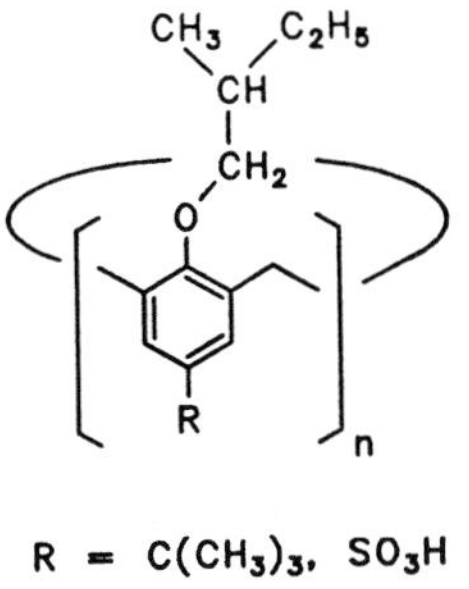

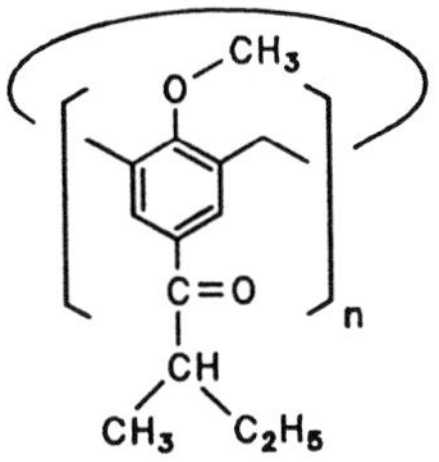

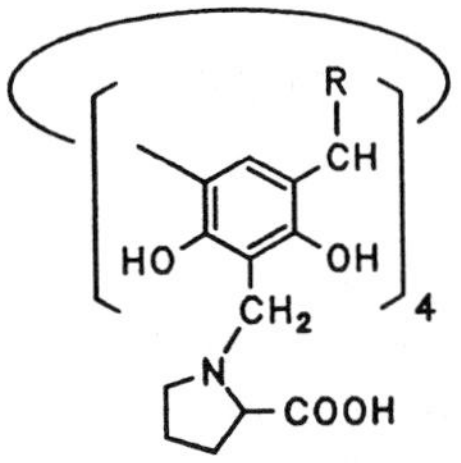

The chirality of compounds like **1–3** is entirely based on the chirality of the derivatizing reagents. However, due to their nonplanar shape, calixarenes offer numerous additional possibilities for producing chiral host molecules, which are not based on a chiral subunit but on the absence of a plane of symmetry or an inversion center in the molecule as a whole. In other words, opening of the macrocyclic structure would lead to an achiral linear molecule. Below we attempt to present a systematic overview of such 'inherently chiral' calixarenes.

2. Asymmetric Calix[4]arenes with Different Phenolic Units

Early attempts in our group were directed at the preparation of calix[4]arenes with three (order AABC) or four different *p*-substituted phenolic units [3]. Molecules of these types [4] can be obtained in a rational way [3, 5] by fragment condensation of suitable trimers with bisbromomethylated phenols (3+1) or suitable dimers with bisbromomethylated dimers (2+2) [6].

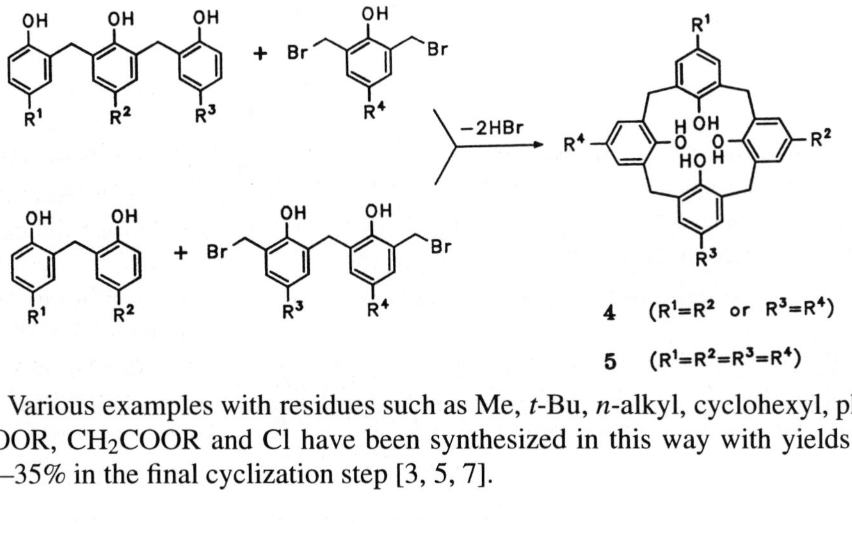

4 ($R^1=R^2$ or $R^3=R^4$)

5 ($R^1=R^2=R^3=R^4$)

(1)

Various examples with residues such as Me, *t*-Bu, *n*-alkyl, cyclohexyl, phenyl, COOR, CH$_2$COOR and Cl have been synthesized in this way with yields up to 25–35% in the final cyclization step [3, 5, 7].

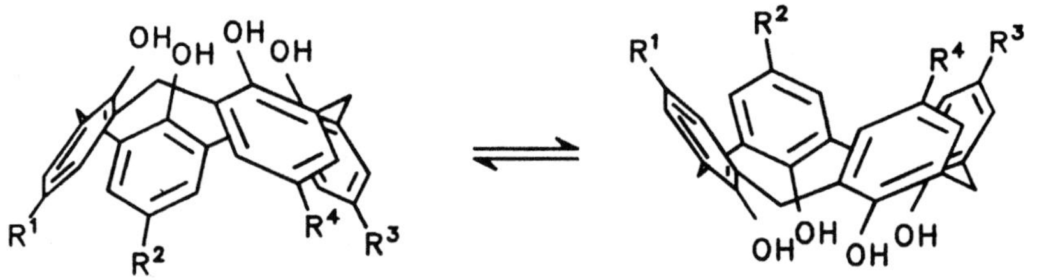

INHERENTLY CHIRAL CALIXARENES

$$\text{(structures shown)} \quad \rightleftharpoons \quad \text{(structures shown)} \qquad (2)$$

To obtain stable enantiomers, the well-known ring inversion of calix[4]arenes, which is in this case synonymous with racemization, must be made impossible. This can be done, for instance, by the introduction of sufficiently large residues on the phenolic oxygen atoms. Due to the asymmetry of such a calixarene complete conversion of all OH groups to a *cone*-derivative must be achieved in a clean reaction [8]. This conversion, which is standard for many symmetrically substituted calix[4]arenes, caused some problems with compounds like **4** and **5**, probably due to their asymmetry [5]. Thus, the tetraester derivative **6**, fixed in the *cone* conformation, obviously assumes a rather distorted shape. This follows from its ^{1}H-NMR spectrum showing, for instance, two singlets for *t*-butyl groups separated by 0.29 ppm, a difference which completely disappears in its Na$^+$ complex. Similar distortions may be present already in partially *O*-alkylated intermediates, favoring the formation of conformations different from *cone* in the final tetraether.

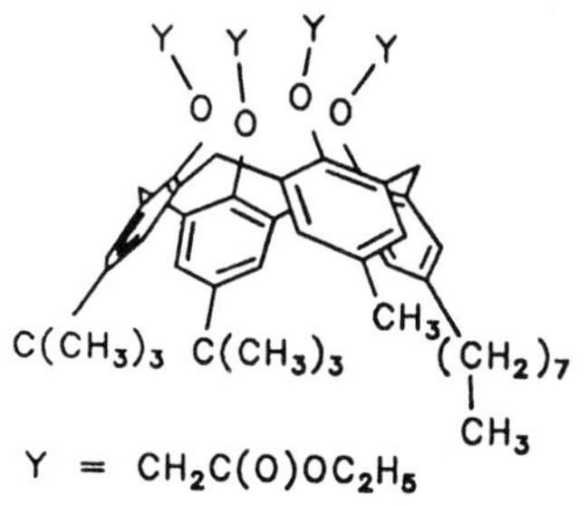

$Y = CH_2C(O)OC_2H_5$

6

The ring inversion (Equation 2) may also be suppressed by suitable bridges between *p*-positions. An example is given by the annelated calix[4]arene **7** which is asymmetric due to the difference of the residues (Me, *t*-Bu) in the *p*-positions [9]. The existence of stable enantiomers was shown here by splitting of ^{1}H-NMR signals in the presence of Pirkle's reagent. A single crystal X-ray analysis suggests a *cone*-conformation for both calix[4]arene substructures, as indicated in the formula. In solution both calix[4]arene parts may be more or less flexible, undergoing the conformational changes *cone* ⇌ *partial cone* ⇌ *1,2-alternate*, but the molecular skeleton does not allow their conversion into the opposite *cone*-conformations.

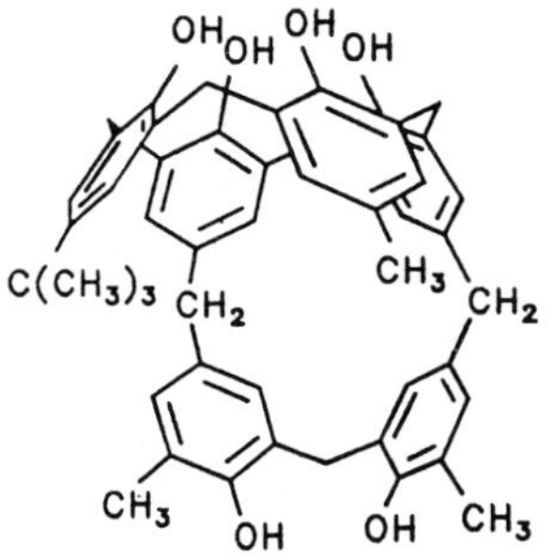

7

Directional bridges between adjacent *p*-positions represent another possibility for introducing molecular asymmetry or dissymmetry. As a recent example, the macrocyclic ether derivative **8** should be mentioned in this connection [10]. This is obtained from the tetrachloromethylated calix[4]arene-tetrapropylether (already fixed in the *cone* conformation) with salicylic acid. Its two salicylic acid residues cause it to have C_2 symmetry.

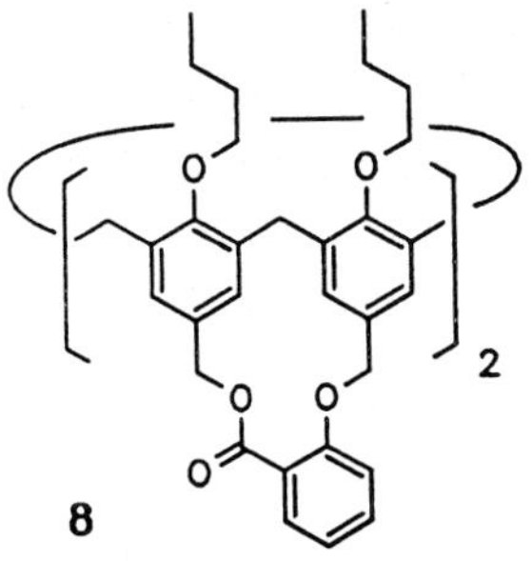

8

3. Calix[4]arenes with a Single *m*-Substituted Phenolic Unit

Asymmetric calix[4]arenes also result from the incorporation of a single *meta*-substituted phenolic unit (**9**), an idea first realized by Vicens *et al.* [11] and subsequently picked up by Shinkai *et al.* [12]. Again, these compounds have been synthesized by '3+1' fragment condensation.

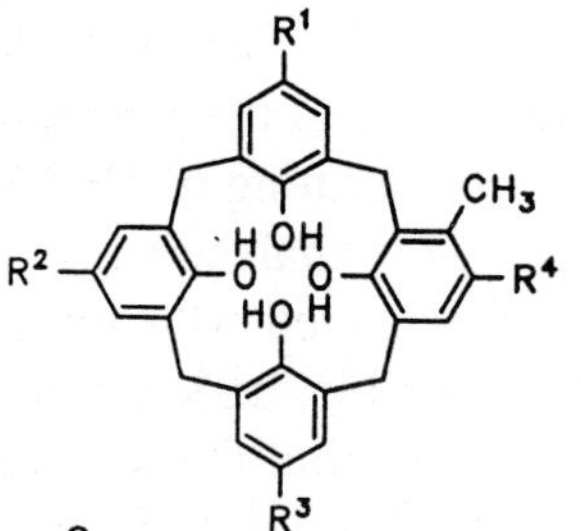

	R^1	R^2	R^3	R^4
a	Me	Me	Me	Me
b	iPr	iPr	iPr	iPr
c	iPr	Phe	iPr	iPr
d	tBu	tBu	tBu	iPr

9

From **9d** the tetrapropyl ether could be prepared in the *cone* conformation and resolved by chromatography on chiral stationary phases [12]. Thus, the first single enantiomer of an inherently chiral calixarene was obtained.

Examples of this type also comprise compounds with a *meta*-hydroxyl group, i.e. with a single resorcinol unit, incorporated in the 2,6-position [13]. Very recently Gutsche synthesized calix[4]arenes with *meta*-substituted phenolic units [14] from the corresponding monoquinone derivatives by addition of various reagents, some examples being shown in Equation 3.

[Structure 10] ← [Structure] → [Structure 11]   (3)

10

R=H, R'=S-C$_6$H$_4$-CH$_3$
R=Ac, R'=O-Ac

11

X=S, CHR, SCH$_2$

4. Asymmetric Calix[4]arenes by *O*-Alkylation with Achiral Residues

In principle the same asymmetric pattern found in **4** or **5** can also be obtained by *O*-alkylation (or *O*-acylation), adding different residues [15–17] to the phenolic oxygen. This strategy has the additional advantage that (with residues larger than ethyl) simultaneously the conformation is fixed and racemization becomes impossible.

For an *all-syn* arrangement of the *O*-alkyl groups, at least two different residues are necessary (**12**), but additional possibilities exist if the *O*-alkyl groups are in the *anti* position. Thus, compounds **13** and **14** with one kind of *O*-alkyl groups are asymmetric (**13**) or dissymmetric (**14** has effectively C_2 symmetry). Examples of these types have been prepared mainly by Shinkai *et al.* [15] and Pappalardo *et al.* [16], and some of them have been resolved by chromatographic techniques [15, 16b, 16c].

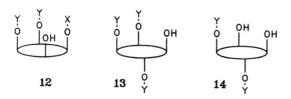

12   13   14

When describing these various *O*-alkylation products the terms *cone, partial cone, 1,2-* and *1,3-alternate* should be used *only* for the tetraethers, where the mutual arrangement of the *O*-alkyl groups unambiguously defines one of these basic conformations. Since the OH group can pass through the annulus a partially *O*-alkylated calixarene can still assume different conformations (at least in principle). This has been demonstrated by the formation of different conformational isomers in subsequent *O*-alkylation steps. For instance for **12** *cone* and *partial* *cone* conformations are possible, for **13** *partial cone* and *1,2-alternate*, and for **14** all conformations except the *cone* conformation. Thus the expressions *cone*, etc., should be used here only to assign these *conformations*, while the mutual situation of the *O*-alkyl groups can be described using *syn* and *anti* or the superscripts α and β. For instance **14** (Y = Pr) could be named *anti*-1,2-dipropylether, abbreviated by Pr^{α}, Pr^{β}, H, H or symbolized by

Using different *O*-alkyl residues 34 types of tetraalkylethers, 13 types of trialkylethers and 8 types of dialkylethers are possible, among which 17, 9 and 3 are chiral [18].

Cavitands are more-or-less rigid, bowl shaped molecules with an enforced cavity. Many examples have been prepared from resorcarenes by intramolecular connection of the hydroxyl functions of adjacent resorcinol units [19]. Asymmetric cavitands [20] have been obtained using a similar principle as described above by introducing just two different bridges (A, B) as shown in **15**.

5. *O*-Alkylation or *O*-Acylation Products of Larger Calixarenes

The number of possible *O*-alkylation products of calix[5]arenes is greater than those of calix[4]arenes and consequently there are also more possibilities for constructing asymmetrical substituted derivatives by *O*-alkylation. In general, selective functionalization of calix[5]arenes is not yet very advanced. The crown ether derivatives which we recently obtained are the first examples of selective *O*-alkylation [21a]. Such a 1,3-crown ether (which has C_S symmetry as indicated) can be made asymmetric by further monoalkylation or monoacylation at one of the remaining *proximal* OH groups [21b].

[Structure diagrams showing σ and 16]   (4)

While the structure of a monoester can be easily established by NMR and mass spectrometry, five singlets for *t*-butyl groups and 10 doublets for Ar–H protons (partly superimposed) prove the asymmetry of **16** (Figure 1). This clearly shows that one of the adjacent OH groups has reacted, but it gives no definite information on the orientation (*syn* or *anti*) of the ester group with respect to the crown ether chain. A high field shift of one *t*-butyl singlet indicates at least a rather distorted conformation.

The recent rapid development of selectively functionalized calix[6]arenes also has a large, yet unexplored potential for obtaining inherently chiral derivatives. Two different diastereomeric 1,2-dibenzylethers of *t*-butylcalix[6]arene (*syn* and *anti*) have been obtained, for instance, from which the *anti* isomer (**17**) is chiral (C_2 symmetry) [22]. The free energy barrier for their mutual interconversion (about 27 kcal mol^{-1}) should also be sufficient to isolate the enantiomers of **17**. 1,2,4-Tri-*O*-alkyl or -acylderivatives [23] should be separable into the enantiomers as long as neither the *O*-acyl nor the *p*-substituent can penetrate the annulus. The macrobicyclic 2,5-esters **18** obtained from 1,4-di-*p*-methylbenzylethers are also chiral, having C_2-symmetry in the *all-syn* isomer [24].

[Structures 17 and 18]

Y = CH$_2$C$_6$H$_4$CH$_3$
X = –CO–R–CO–

6. Chirality in Symmetrically Bridged Calixarenes

It is interesting to note that rigid bridges like the phthaloyl residue, which in principle has a symmetry plane, can impose an asymmetric conformation on the

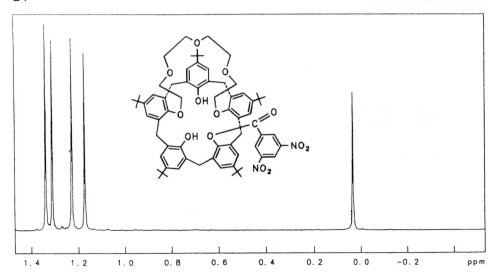

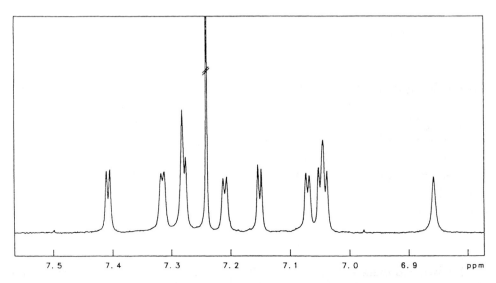

Fig. 1. Sections of the ^{1}H-NMR spectrum of the calix[5]arene derivative **16**.

calixarene skeleton. At room temperature, for instance, the 1,3-phthaloyl-bridged calix[5]arene **19** shows five singlets for the *t*-butyl groups (1.38, 1.30, 1.24, 1.14 and 1.11 ppm), proof of the absence of any symmetry element. At 120°C, however, three singlets (1.26, 1.10 and 1.21 ppm, ratio 2 : 2 : 1) are observed, as with the more flexible crown ether derivatives. A similar observation was made for the 1,2-phthaloyl bridged *t*-butylcalix[6]arene (five *t*-butyl singlets at room temperature and three singlets of equal intensity at 120°C), where the asymmetry was shown

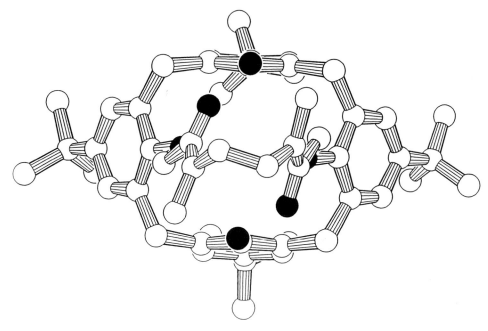

Fig. 2. Single crystal X-ray structure of the macrobicyclic amide **20a**.

also by X-ray structural analysis [25].

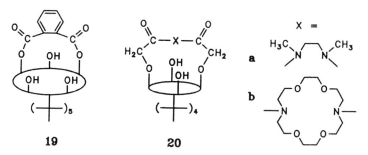

1,3-Bridged calix[4]arenes (e.g. calixcrowns) normally have C_{2V} symmetry. This is also observed for the macrocyclic amides **20** obtained by reaction of *t*-butylcalix[4]arene-1,3-diacidchlorides with various diamines, including ethylenediamine [26]. *N,N'*-Dimethylethylenediamine, however, leads to a cyclic amide with C_2 symmetry, which is obviously caused by the steric demands of this rather rigid bridge.

The structure found in the crystalline state (Figure 2) is also maintained in solution, as shown by ^{1}H-NMR spectroscopy in the presence of Pirkle's reagent. The four doublets (*meta*-coupling) observed for the aromatic protons in the corresponding amide **20b** obtained with diaza-18-crown-6 must be explained in a similar way.

A calix[4]arene bridged via the four phenolic oxygen functions by a pyrophosphate group (**21**) assumes a dissymmetric conformation at low temperature [27]. While the room-temperature ^{1}H-NMR spectrum shows one signal for the *t*-butyl groups and two different aromatic protons, the low-temperature (203 K) NMR spectrum shows two *t*-butyl groups and four different aromatic protons, which is in agreement with a chiral molecule having C_2-symmetry.

A similar observation was made for the calix[6]arene **22** which is bridged by two phosphate groups where, even at room temperature, the ^{1}H-NMR spectrum shows a dissymmetric conformation with C_2 symmetry.

Inherently chiral derivatives are also known from calix[8]arenes. Dimetalla complexes were obtained with titanium (as well as with Zr, V, Sn) in which the two titanium atoms have a pseudooctahedral environment bonded to all of the eight calixarene oxygens, two of which are bridging, and two isopropoxide oxygens. The C_2 symmetry was proved by single crystal X-ray analysis and in solution by two-dimensional ^{1}H-NMR spectroscopy [28]. Various complexes of *t*-butylcalix[8]arene with lanthanide ions have a similar structure [29].

7. 'Symmetry Breaking' by Di-*O*-Alkylation

Calix[4]arenes consisting of two different *p*-substituted phenolic units are readily prepared by fragment condensation ('2+2' for AABB, '3+1' for ABAB). In the *cone* conformation their molecules possess C_S and C_{2V} symmetry. 1,3-Diethers

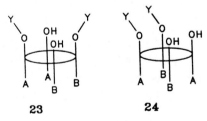

in *syn*-arrangement are generally readily available from calix[4]arenes [30]. They can also be prepared in high yields from calix[4]arenes of the type AABB, and therefore they represent an attractive type of inherently chiral derivative [31]. The asymmetry here is due to the fact that in **23** the symmetry plane of the calix[4]arene substructure does not coincide with the symmetry planes of the arrangement of the

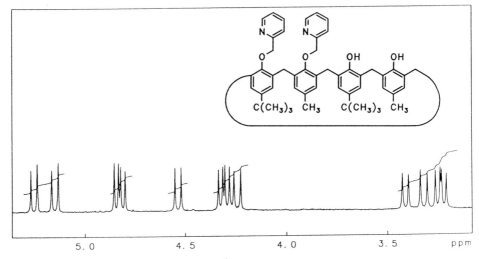

Fig. 3. Methylene protons section of the ^{1}H-NMR spectrum of a 1,2-diether derivative of type **24**.

alkyl groups. In a certain sense this situation is comparable to that encountered in atropisomers of biphenyls, where each aromatic ring usually represents a symmetry plane.

The same symmetry considerations are valid for 1,2-diether derivatives of calix[4]arenes of the type ABAB (**24**). Their synthesis, however, is a rather difficult approach, since 1,2-di-*O*-alkylation is not nearly as selective as 1,3-di-*O*-alkylation [30]. Since two different mono-, 1,3-di- and triether derivatives are possible (neglecting even derivatives with *anti*-arrangement of the *O*-alkyl groups), a rather complex reaction mixture exists, even if these compounds are formed only in small amounts. As one example [32] the section of the methylene protons is shown in Figure 3. Twelve doublets of equal intensity with geminal coupling prove that all six Ar–CH$_2$–Ar and O–CH$_2$–Py groups are different, each having two diastereotopic protons [32].

With calix[4]arene-like macrocycles of type **25** having two different bridges X and Y [33–35] *all* mono- and triether derivatives are asymmetric. Their 1,3-diethers have C_2 symmetry, while the two possible 1,2-diethers have a symmetry plane. The first examples of such compounds were mentioned by Nishimura [35].

8. Further Asymmetric Derivatives

t-Butylcalix[4]arene may be converted by mild oxidation (Me$_3$PhN$^+$Br$_3^-$/NaHCO$_3$ in CH$_2$Cl$_2$) into spirodienones [36] which are extremely interesting examples of chiral derivatives. The structure of the monospirodienone **26** has been established by single crystal X-ray analysis. Two signals for the different OH groups and four AX systems for the methylene bridges are observed in the ^{1}H-NMR spectrum. On the one hand, the chirality of **26** is mainly due to the asymmetrically substituted spirocarbon atom and therefore this compound is not inherently chiral. However, there is also some similarity to a calix[4]arene having three different phenolic units or three different oxygen functions (two hydroxyl groups in proximal positions, a carbonyl and an ether group) in the order AABC (cf. Section 4).

Three of six possible bis-spirodienone derivatives have been obtained in a similar way from *p-tert*-butylcalix[4]arene, one of them (**27**) showing C_2 symmetry. The other two are achiral (C_1 and C_S symmetry). Tris-spirodienones have been obtained meanwhile from *t*-butylcalix[6]arene [36e].

26   **27**

Calix[4]arene ethers, like the tetrapropylethers, can be converted into mono-Cr(CO)$_3$ complexes [37]. Starting with the *cone* or the *1,3-alternate* isomer, these derivatives still have a symmetry plane, while the corresponding derivative of the isomer in the *1,2-alternate* conformation becomes asymmetric. Starting with a *partial cone* conformer, an asymmetric molecule can be obtained only by introducing the Cr(CO)$_3$ at one of the aromatic rings next to the inverted phenol ring.

9. Dissymmetric Calixarenes with C_n Symmetry

Dissymmetric molecules are chiral but still have symmetry elements: a single *n*-fold axis (C_n symmetry) or, in addition, *n* two-fold axes perpendicular to it (D_n symmetry). Some derivatives with C_2 symmetry have been already mentioned above in connection with asymmetric compounds of the same type.

Most attractive, and not just from an aesthetic point of view, are dissymmetric calix[*n*]arenes having an *n*-fold symmetry axis. Several calix[4]arenes with C_4

INHERENTLY CHIRAL CALIXARENES   29

symmetry have been obtained from 3,4-disubstituted phenols as shown in Equation (5) with yields up to 30% in the cyclization step [38].

	R^1	R^2
a	CH$_3$	CH$_3$
b	CH(CH$_3$)$_2$	CH$_3$
c	Cl	CH$_3$
d	(CH$_2$)$_3$	
e	(CH$_2$)$_4$	
f	CH=CH–CH=CH	

(5)

The regular incorporation of the phenolic units has been demonstrated not only by their ^{1}H-NMR spectra, but also, for one example, by X-ray analysis. It shows the molecule in a 'pinched' *cone* conformation with (essentially) C_2 symmetry. In solution no deviation from the (average) C_4 symmetry is observed down to temperatures of -100°C. In comparison to calix[4]arenes with *p*-substituted phenolic units a slightly lower energy barrier (13.4 vs. 14.6 kcal mol^{-1}) is found for the ring inversion.

Due to the equivalence of the phenolic units, it is now possible to construct in an unambiguous way all kinds of *O*-alkylation products (from mono- to tetraethers), as with achiral calix[4]arenes. Of course, the symmetry is reduced in partially *O*-alkylated derivatives. Mono- and triether derivatives are asymmetric (C_1) while 1,3-diether derivatives with the usual *syn*-arrangement of the *O*-alkyl groups have C_2 symmetry. An *anti*-1,3-diether (the formation of which is not observed under usual reaction conditions) would have an inversion center in its *1,2-alternate* conformation, hence being achiral.

As an example, Figure 4 shows the ^{1}H-NMR spectrum of a mono-*p*-nitrobenzyl-ether. Eight singlets for the methyl groups or four singlets for the aromatic protons demonstrate that all phenolic units are different. Especially noteworthy is the AB system (two doublets with geminal coupling) for the diastereotopic O–CH$_2$–Ar protons which has nothing to do with restricted rotation around these σ-bonds but is entirely due to the fact that the benzyl group is attached to a chiral skeleton.

Various *syn*-1,3-diether derivatives were obtained in good yields. Some of them were resolved into pure enantiomeric forms [39] by chromatography with chiral stationary phases. Figure 5 demonstrates that their CD spectra are similar in principle, but show subtle differences, which must be due to small conformational differences. As special examples for 1,3-derivatives the crown ethers **29** should be mentioned ($R_1 = R_2 = Me$). These were obtained as usual by reaction with tetra- and pentaethyleneglycol ditosylates, respectively, in about 20% yield [40].

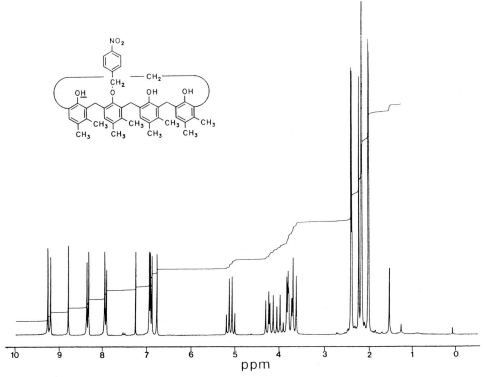

Fig. 4. ¹H-NMR spectrum of the mono-*p*-nitrobenzylether of **28a**.

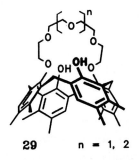

29   n = 1, 2

¹H-NMR spectral studies show that the tetraether derivatives of **28** assume a 'pinched' *cone* conformation (as in the crystalline state) at moderately low temperatures. This is clearly due to the steric demands of the *meta*-methyl groups. The C_4-symmetry suggested by the structure is observed in solution only at higher temperatures, as the time average of two conformations with C_2 symmetry.

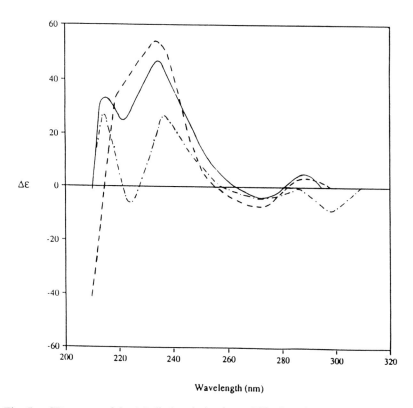

Fig. 5. CD spectra of the 1,3-diether derivatives of **28a**, **b** and **c**.

$$C_2 \;\rightleftarrows\; [C_4]^{\ddagger} \;\rightleftarrows\; C_2 \tag{6}$$

The energy barrier for this 'pseudorotation' process, which most probably has a C_4 symmetrical transition state, is in the range of $\Delta G^{\#} = 13\text{–}14\,\text{kcal mol}^{-1}$. A similar molecular motion (with a much lower energy barrier) was already assumed from relaxation time studies for the corresponding derivatives of t-butylcalix[4]arene [41].

The regular incorporation of the β-naphthol units in **28** is determined by the synthetic strategy. The condensation of α-naphthol with formaldehyde has recently allowed calix[4]arenes to be synthesized in which the naphthol units are incorporated via their 2,4-positions. In this way not only the isomer with C_4 symmetry is formed, but simultaneously also the isomers with C_1 and C_S symmetry. It was possible, however, to isolate the C_4 symmetric compound in 9–10% yield, while a fourth isomer with C_{2V} symmetry was not found [42]. Due to the absence of intraannular substituents these molecules are quite flexible.

Calix[n]arenes with two *meta*-substituted phenolic units in opposite positions may be chiral (C_2 symmetry) or achiral (C_S symmetry). Addition reactions to 1,3-di-quinone derivatives, in analogy to Equation 3, give a mixture of both isomers, from which for one example the C_2 isomer **30** was isolated. The C_2 isomer is formed in a more rational way according to Equation 7 with 20% yield in the cyclization step [38]. However the five-step synthesis of the dinuclear precursor means that this, too, is not a very attractive alternative.

(7)

10. Compounds with Resorcinol Units

Acid catalyzed condensation of 2,4-dihydroxy-3-hydroxymethylbenzophenone in dioxane also leads to a calixarene with regular incorporation of the phenolic units [43]. This was easily proved by the ^{1}H-NMR spectrum which showed just one set of signals for the repeating phenolic unit (e.g. two singlets for OH and one singlet for the Ar–H of the resorcinol units and an AB system for the Ar–CH$_2$–Ar protons). Mass spectrometry and finally also a single crystal X-ray structure revealed that, surprisingly in this case, the calix[5]arene **31** was formed. This is not only a rare example of a molecule with intrinsic C_5 symmetry, it is also the first example of a calix[5]arene consisting entirely of resorcinol units. In contrast to the 'usual' resorc[4]arenes, however, the resorcinol units are not incorporated via the 4,6-positions but via the 2,6-positions. This shows also that calixarenes (which in a narrow sense are derived from phenols) and resorcarenes should be considered together. In a more general sense the name 'calixarene' may be used for *all* 1_n-*metacyclophanes*.

31

Compound **31** shows another surprising result. While the energy barrier for the cone-to-cone ring inversion in calix[5]arenes is usually lower than in calix[4]arenes (the values for *p*-methylcalix[4]- and -[5]arene are $\Delta G^{\neq} = 14.6$ and 12.7 kcal mol^{-1}) dynamic NMR gave a comparatively high barrier of $\Delta G^{\neq} = 17.3$ kcal mol^{-1} for **31**. The reason for this is not yet entirely understood.

Two interesting approaches to C_4 symmetrical derivatives have recently been described in the resorcarene family. Esterification with Cl–P(O)–(OR)$_2$ usually gives the corresponding octaester. For R = Et, *i*-Pr, however, a tetraester was isolated in good yields, to which the authors assign structure **32a** [44]. Two singlets found in the ^{1}H-NMR spectrum for each of the aromatic protons (in 2- and 5-position of the resorcinol units) are explained by a distorted conformation also found for octaesters. However, this would afford two signals for the OH groups and two different phosphorus ester groups, where only a single set of signals is found. Thus the NMR data are in even better agreement with structure **32b**.

32a   X = P(O)(OR)$_2$ **32b**   **33**

Aminomethylation of resorcarenes with several primary amines leads to compounds for which the formula **33** suggests C_4 symmetry [2a]. However, no spectroscopic data have been published for this type of compound, nor do the authors mention this potential chirality. We recently prepared various compounds of this type for which a single set of signals for all protons of the resorcinol unit was observed, which is comparable only with the C_4 symmetrical formula **33**. Meanwhile this was unambiguously proved for one example (R=p–C$_6$H$_4$–NO$_2$) by single crystal X-ray analysis [45]. C_4 symmetry was found in the crystalline state for product **34** due to the network of intramolecular hydrogen bonds, all of which proceed in the same direction [46a] and a tetralactone with C_4-symmetry derived from a resorcarene has been described by Cram et al. [46b]. Finally, in this connection a calixarene-like macrocyclic compound should be mentioned, in which four uracil units are connected in a regular fashion by (en)Pt bridges between their nitrogen atoms. The molecule has C_4 symmetry, but a 1,3-alternate conformation with C_2 symmetry is found in the crystalline state [47].

34

11. Spherand-Type Calixarenes

Condensation of 2,2'-dihydroxy-5,5'-di-*tert*-butyldiphenol with formaldehyde leads to macrocyclic molecules **35** which combine structural features of a calixarene and a spherand [48, 49]. They have three or four methylene groups: less than a calix[6]- or -[8]arene and more than the corresponding spherand. Their ^{1}H-NMR spectrum which shows one singlet for *t*-butyl, methylene and hydroxyl protons and a pair of doublets for the aromatic protons, indicate that these parent macrocycles are either highly symmetrical or rather flexible molecules. The splitting observed at lower temperature for the methylene protons is in favour of the latter explanation, although the exact minimum energy conformation is not yet known.

O-alkylation of the parent spherand calixarenes leads to derivatives in which the biphenyl subunits are fixed in a certain configuration, as in usual atropisomers. Although such a molecule then contains chiral subunits it may nevertheless be regarded as an inherently chiral molecule. Opening of the macrocycle would lead to a linear molecule in which (at least for small residues R) racemization would be

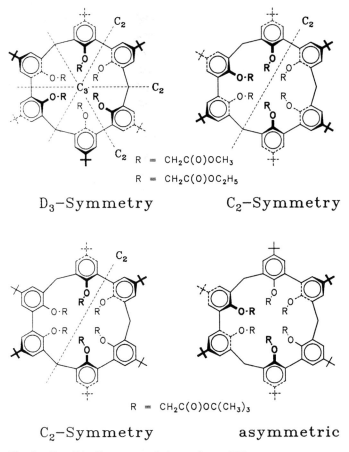

Fig. 6. Possible diastereomeric hexaethers of **35a**.

possible via a transoid transition state, while in the macrocycle the cisoid transition state is the only possibility.

35a n=3   **35b** n=4

Two diastereoisomers are possible for the trimer, both of which are chiral. *R,R,R* or *S,S,S* configuration of the biphenyl units leads to D_3 symmetry and *R,R,S/S,S,R*

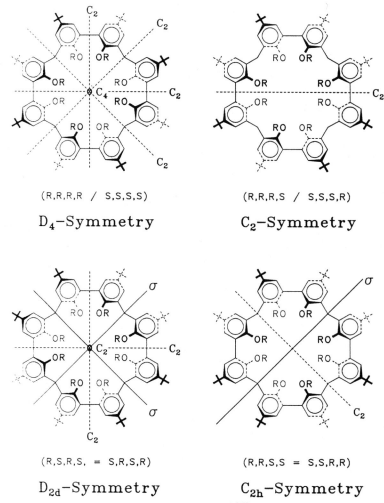

Fig. 7. Possible diastereomeric octaethers of **35b**.

configuration results in C_2 symmetry.

While Yamato *et al.* were able to isolate both hexamethylethers [48] (the D_3 isomer only in low yield and in inpure form), our own attempts at alkylation with ethylbromoacetate gave only the C_2 isomer (in yields up to 92%). In addition to the ^{1}H-NMR spectroscopic evidence, its structure was also determined by single crystal X-ray analysis [49]. Further chemical modification by hydrolysis or transesterification is possible, without changing the configuration.

If ethylbromoacetate is replaced in the alkylation reaction by t-butylbromoacetate, the residues R are bulky enough to hinder not only the rotation around the Ar–Ar bonds but also around the Ar–CH$_2$–Ar bonds. This means the diphenylmethane subunits become asymmetric (atropisomeric), too. In addition to a C_2 symmetric derivative now also an asymmetric (C_1) derivative exists for the $R,R,S/S,S,R$ configuration of the biphenyl units (see Figure 6). Both diastereomers could be isolated in pure form and converted by transesterification into the same (C_2 symmetrical) hexamethylester which was also obtained from the hexaethylester.

Four different octaether derivatives should be available from the cyclic tetramer (Figure 7), two of which are chiral while the other two are meso-forms. However, up to now all our attempts at alkylation have led only to mixtures of these diastereomers, in accordance with the results of the O-methylation [48].

12. Conclusion

This survey demonstrates the huge potential which is available in the calixarene family for obtaining inherently chiral host molecules or chiral building blocks for the construction of even larger molecular systems. Clearly the possibilities are not exhausted by the examples discussed above and further strategies will be developed in the future. Although convincing examples have not yet been described, chiral recognition and discrimination will be one of the main topics in calixarene chemistry in the future.

References and Notes

1. (a) T. Arimura, H. Kawabata, T. Matsuda, T. Muramatsu, H. Satoh, K. Fujio, O. Manabe, and S. Shinkai: *J. Org. Chem.* **56**, 301 (1991). (b) A. Ikeda, T. Nagasaki, and S. Shinkai: *J. Phys. Org. Chem.* **5**, 699 (1992).
2. (a) Y. Matsushita and T. Matsui: *Tetrahedron Lett.* **46**, 7433 (1993). (b) H.-J. Schneider and U. Schneider: *J. Incl. Phenom.* **19**, 67–83 (1994) (this volume).
3. (a) V. Böhmer, L. Merkel, and U. Kunz: *J. Chem. Soc., Chem. Commun.* 896 (1987). (b) V. Böhmer, F. Marschollek, and L. Zetta: *J. Org. Chem.* **52**, 3200 (1987).
4. The first example of type AABC was obtained by chance from AABB as a side product during its conversion to AACC. K. H. No and C. D. Gutsche: *J. Org. Chem.* **47**, 2713 (1982).
5. L. Zetta, A. Wolff, W. Vogt, K.-L. Platt, and V. Böhmer: *Tetrahedron* **47**, 1911 (1991).
6. Either $R^1 = R^2$ or $R^3 = R^4$ is necessary to obtain a definite product in the 2+2 approach.
7. Unfortunately, convincing conditions have not yet been found for preparing calix[5]- or calix[6]arenes in a similar way.
8. Otherwise an untreatable mixture of isomeric or similar compounds will be obtained. For instance, four different mono- or trialkylated derivatives in *syn* arrangements, or four different tetra-O-alkylated derivatives in the *partial cone* conformation are possible.
9. R. Dörrenbächer, E. Paulus, W. Vogt, and V. Böhmer: unpublished results. For first examples of achiral annelated calixarenes see V. Böhmer, R. Dörrenbächer, W. Vogt, and L. Zetta: *Tetrahedron Lett.* **33**, 769 (1992).
10. A. Ikeda and S. Shinkai: *J. Chem. Soc., Perkin Trans. 1*, 2671 (1993).

11. (a) H. Casabianca, J. Royer, A. Satrallah, A. Taty-C, and J. Vicens: *Tetrahedron Lett.* **28**, 6595 (1987). For a recent X-ray structure see: (b) Y. Ueda, T. Fujiwara, K.-I. Tomita, Z. Asfari, and J. Vicens: *J. Incl. Phenom.* **15**, 341 (1993).
12. (a) S. Shinaki, T. Arimura, H. Kawabata, H. Murakami, K. Araki, K. Iwamoto, and T. Matsuda: *J. Chem. Soc., Chem. Commun.* 1734 (1990). (b) S. Shinkai, T. Arimura, H. Kawabata, H. Murakami, and K. Iwamoto: *J. Chem. Soc., Perkin Trans. 1*, 2429 (1991).
13. M. Tabatabai, W. Vogt, and V. Böhmer: *Tetrahedron Lett.* **31**, 3295 (1990).
14. P. A. Reddy and C. D. Gutsche: *J. Org. Chem.* **58**, 3245 (1993).
15. (a) K. Iwamoto, A. Yanagi, T. Arimura, T. Matsuda, and S. Shinkai: *Chem. Lett.* 1901 (1990). (b) K. Iwamoto, H. Shimizu, K. Araki, and S. Shinkai: *J. Am. Chem. Soc.* **115**, 3997 (1993).
16. (a) S. Pappalardo, L. Giunta, M. Foti, G. Ferguson, J. F. Gallagher, and B. Kaitner: *J. Org. Chem.* **57**, 2611 (1992). (b) S. Pappalardo, S. Caccamese, and L. Giunta: *Tetrahedron Lett.* **32**, 7747 (1991). (c) S. Caccamese and S. Pappalardo: *Chirality* **5**, 159 (1993). (d) G. Ferguson, J. F. Gallagher, L. Giunta, P. Neri, S. Pappalardo, and M. Parisi: *J. Org. Chem.* **59**, 42 (1994). (e) F. Bottino and S. Pappalardo: *J. Incl. Phenom.* **19**, 85 (1994) (this volume).
17. A recently described phosphonate derivative corresponds to the general formula **12**. L. T. Burne, J. M. Harrowfield, D. C. R. Hockless, B. J. Peachey, B. W. Skelton, and A. H. White: *Aust. J. Chem.* **46**, 1673 (1993).
18. Unfortunately a recent classification (see [15b]) of all possible inherently chiral calix[4]arenes which can be obtained by O-alkylation is ambiguous and potentially misleading. V. Böhmer, D. Kraft, and W. Vogt: *Supramol. Chem.* **3**, 299 (1994).
19. D. J. Cram, S. Karbach, H.-E. Kim, C. B. Knobler, E. F. Maverick, J. L. Ericson, and R. C. Helgeson: *J. Am. Chem. Soc.* **110**, 2229 (1988).
20. D. J. Cram, L. M. Tunstad, and C. B. Knobler: *J. Org. Chem.* **57**, 528 (1992).
21. (a) D. Kraft, R. Arnecke, V. Böhmer, and W. Vogt: *Tetrahedron* **49**, 6019 (1993). (b) R. Arnecke, V. Böhmer, and W. Vogt: unpublished results.
22. P. Neri, C. Rocco, G. M. L. Consoli, and M. Piattelli: *J. Org. Chem.* **58**, 6535 (1993).
23. R. G. Janssen, W. Verboom, S. Harkema, G. J. van Hummel, D. N. Reinhoudt, A. Pochini, R. Ungaro, P. Prados, and J. de Mendoza: *J. Chem. Soc., Chem. Commun.* 506 (1993).
24. S. Kanamathareddy and C. D. Gutsche: *J. Am. Chem. Soc.* **115**, 6572 (1993).
25. D. Kraft, V. Böhmer, W. Vogt, G. Ferguson, and J. F. Gallagher: *J. Chem. Soc., Perkin Trans. 1*, 1221 (1994).
26. V. Böhmer, G. Ferguson, J. F. Gallagher, A. J. Lough, M. A. McKervey, E. Madigan, M. B. Moran, J. Phillips, and G. Williams: *J. Chem. Soc., Perkin Trans. 1*, 1521 (1993).
27. (a) F. Grynszpan, O. Aleksiuk, and S. E. Biali: *J. Chem. Soc., Chem. Commun.* 13 (1993). (b) O. Aleksiuk, F. Grynszpan, S. E. Biali: *J. Incl. Phenom.* **19**, 237 (1994) (this volume).
28. (a) G. E. Hofmeister, F. E. Hahn, and S. F. Pedersen: *J. Am. Chem. Soc.* **111**, 2318 (1989). (b) G. E. Hofmeister, E. Alvarado, J. A. Leary, D. I. Yoon, and S. F. Pedersen: *J. Am. Chem. Soc.* **112**, 8843 (1990).
29. (a) B. M. Furphy, J. M. Harrowfield, D. L. Kepert, B. W. Skelton, A. H. White, and F. R. Wilner: *Inorg. Chem.* **26**, 4231 (1987). (b) J. M. Harrowfield, M. I. Ogden, and A. H. White: *Aust. J. Chem.* **44**, 1237 and 1249 (1991).
30. J.-D. van Loon, W. Verboom, and D. N. Reinhoudt: *Org. Prep. Proced.* **24**, 437 (1992).
31. V. Böhmer, A. Wolff, and W. Vogt: *J. Chem. Soc., Chem. Commun.* 968 (1990).
32. F. Marschollek, V. Böhmer, and W. Vogt: unpublished results.
33. B. Dhawan and C. D. Gutsche: *J. Org. Chem.* **48**, 1536 (1983) and references therein.
34. (a) T. Yamato, Y. Saruwatari, S. Nagayama, K. Maeda, and M. Tashiro: *J. Chem. Soc., Chem. Commun.* 861 (1992). (b) T. Yamato, M. Yasumatsu, H. Ota, Y. Saruwarati, and L.K. Doamekpor: *J. Incl. Phenom.* **19**, 315 (1994) (this volume).
35. (a) Y. Okada, F. Ishii, Y. Kasai, and J. Nishimura: *Chem. Lett.* 755 (1992). (b) Y. Owada and J. Nishimura: *J. Incl. Phenom.* **19**, 41 (1994) (this volume).
36. (a) O. Aleksiuk, F. Grynszpan, and S. E. Biali: *J. Chem. Soc., Chem. Commun.* 11 (1993). (b) A. M. Litwak and S. E. Biali: *J. Org. Chem.* **57**, 1943 (1992). (c) A. M. Litwak, F. Grynszpan, O. Aleksiuk, S. Cohen, and S. E. Biali: *J. Org. Chem.* **58**, 393 (1993). (d) F. Grynszpan, O. Aleksiuk,

and S. E. Biali: *J. Org. Chem.* **59**, 2070 (1994). (e) F. Grynszpan, and S. Biali: *J. Chem. Soc., Chem. Commun.*, 2545 (1994).
37. H. Iki, T. Kikuchi, and S. Shinkai: *J. Chem. Soc., Perkin Trans. 1*, 205 (1993).
38. (a) A. Wolff, V. Böhmer, W. Vogt, F. Ugozzoli, and G. D. Andreetti: *J. Org. Chem.* **55**, 5665 (1990). (b) G. D. Andreetti, V. Böhmer, J. G. Jordon, M. Tabatabai, F. Ugozzoli, W. Vogt, and A. Wolff: *J. Org. Chem.* **58**, 4023 (1993).
39. S. T. Pickard, W. H. Pirkle, M. Tabatabai, W. Vogt, and V. Böhmer: *Chirality* **5**, 310 (1993).
40. M. Tabatabai, V. Böhmer, and W. Vogt: unpublished results.
41. A. Yamada, T. Murase, K. Kikukawa, T. Arimura, and S. Shinkai: *J. Chem. Soc., Perkin Trans. 2*, 793 (1991).
42. (a) P. E. Georghiou and Z. Li: *Tetrahedron Lett.* **34**, 2887 (1993). (b) P. E. Georghiou and Z. Li: *J. Incl. Phenom.* **19**, 55 (1994) (this volume).
43. M. Tabatabai, W. Vogt, V. Böhmer, G. Ferguson, and E. F. Paulus: *Supramol. Chem.* in press.
44. L. N. Markovsky, V. I. Kal'chenko, D. M. Rudkevich, and A. N. Shivanyuk: *Mendeleev Commun.* 106 (1992).
45. R. Arnecke, V. Böhmer, E.F. Paulus, and W. Vogt: *J. Am. Chem. Soc.* submitted.
46. (a) D. A. Leigh, P. Linnane, R. G. Pritchard, and G. Jackson: *J. Chem. Soc., Chem. Commun.* 389 (1994). (b) H.-J. Choi, M. L. C. Quan, C. B. Knobler, and D. C. Cram: *J. Chem. Soc., Chem. Commun.*, 1733 (1992).
47. H. Rauter, E. C. Hillgeris, A. Erxleben, and B. Lippert: *J. Am. Chem. Soc.* **116**, 616 (1993).
48. T. Yamato, K. Hasegawa, Y. Saruwatari, and L. K. Doamekpor: *Chem. Ber.* **126**, 1435 (1993).
49. P. O'Sullivan, V. Böhmer, W. Vogt, E. F. Paulus, and R. A. Jakobi: *Chem. Ber.* **127**, 427 (1994).

The Design of Cone-fixed Calix[4]arene Analogs by Taking syn-[2.n]Metacyclophanes as a Building Block

YUKIHIRO OKADA and JUN NISHIMURA
Department of Chemistry, Gunma University, Kiryu 376, Japan

(Received: 3 March 1994; in final form: 16 November 1994)

Abstract. Rigidified calix[4]arene analogs were synthesized from *syn*-[2.n]metacyclophanes as a building block. Their structure was firmly locked in the cone conformation. An enlarged calix[4]arene analog was obtained after the cyclobutane ring cleavage of the parent analog by Birch reduction. Several ionophores have been derived from the analogs and been found to select the larger ions during the extraction of alkali metals, transition metals, and lanthanoids. The ionophore having oligoethylene glycol units showed an effective catalytic activity for S_N2 reactions such as ester synthesis, Williamson ether synthesis, and Finkelstein reaction in several media.

Key words: Syn-conformer, cyclophane, metal ion, extraction, catalytic activity.

1. Introduction

The chemistry of calix[n]arene attracts much attention in many interesting applications such as the binding of organic molecules and inorganic ions [1]. Accordingly, the modifications of calixarene and its derivatives are widely investigated to explore the new aspects in this chemistry. Rigidification is one of the most important modifications which usually results in arranging binding sites favorably for various guests and giving specific selectivities in affinity. Many successful examples in this respect are disclosed in the synthesis for the families of calix[4]arenes by using bisphenol derivatives as a building block [2]. In order to make calixarenes conformationally rigid, alkylidene bridges and/or bulky substituents were also introduced and it resulted in better selectivity and efficiency for binding metal ions [3]. Recently, we reported the photochemical synthesis of *syn*-dimethoxy[2.n]metacyclophanes, whose stereochemistry was controlled by the steric effect of the methoxyl group [4]. These cyclophanes with a *syn* conformation were used as a building block for the construction of a more rigid calix[4]arene skeleton [5], because many macrocyclic compounds composed of a metacyclophane skeleton are reported as artificial receptors [6]. In this review, we report the synthetic method using *syn*-dimethoxy[2.n]metacyclophane by [2 + 2] photocycloaddition and the first

---

* This paper is dedicated to the commemorative issue on the 50th anniversary of calixarenes.

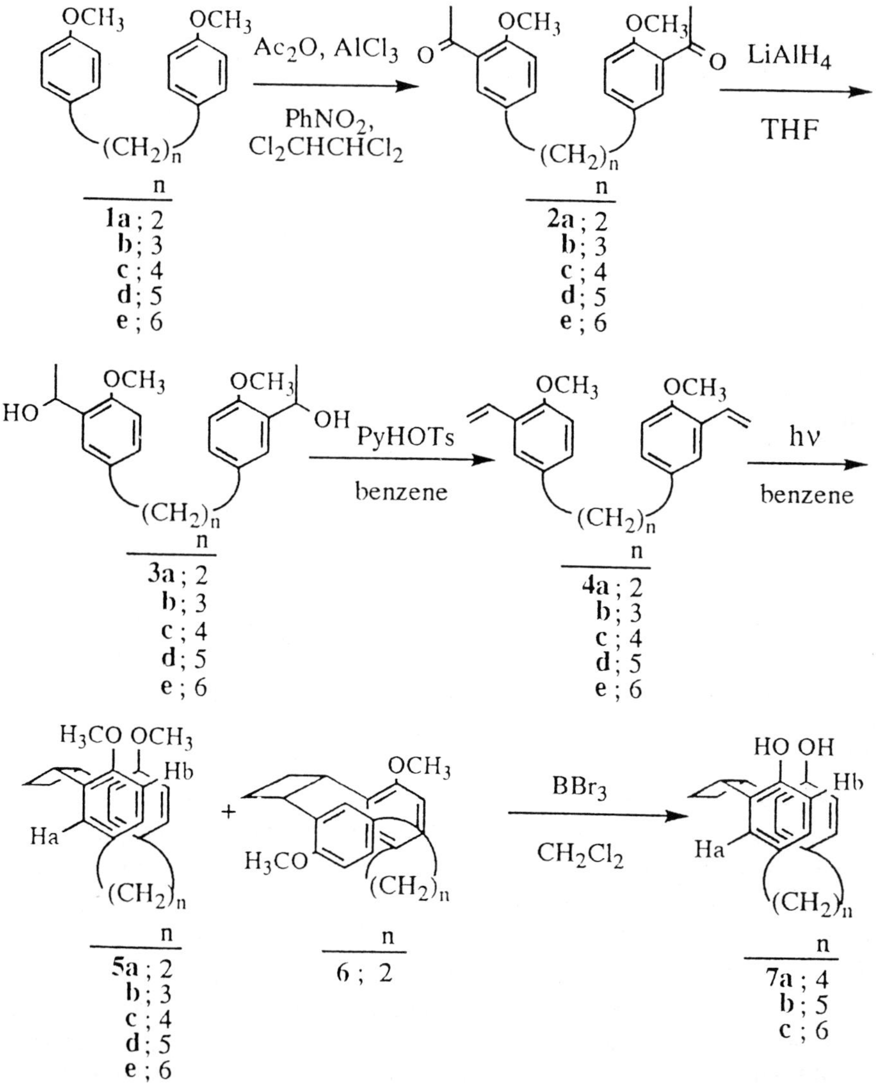

Scheme 1.

synthesis and characterization of calix[4]arene analogs firmly locked in the cone conformation.

2. Synthesis of *syn*-[2.*n*]Metacyclophanes

The synthetic route of dimethoxy[2.*n*]metacyclophanes is shown in Scheme 1. α,ω-Bis(*p*-methoxyphenyl)alkanes **1** were used as starting materials [7]. Diketones

TABLE I. Conformational Analysis of Cyclophanes **5**, **6** and **7**

Compound	Observed			Corrected	Assignment
	Ha	Hb	$\Delta\delta^a$	$\Delta\delta^{b,c}$	
5a	7.03	6.08	0.95	0.72	*syn*
6	4.38, 5.18	6.83, 6.89	−2.45 – −1.71	−2.68 – −1.94	*anti*
5b	7.05	6.24	0.81	0.58	*syn*
5c	7.04	6.32	0.72	0.49	*syn*
5d	7.04	6.43	0.61	0.38	*syn*
5e	6.98	6.49	0.49	0.26	*syn*
7a	6.95	6.36	0.59	0.33	*syn*
7b	6.95	6.46	0.49	0.23	*syn*
7c	6.91	6.50	0.41	0.15	*syn*

[a] $\Delta\delta = \delta_{Ha} - \delta_{Hb}$.
[b] Corrected by –0.23 ppm for **5** and **6**, since 2,4-dimethylanisole gives the chemical shifts difference between Ha (3–) and Hb (6–) positions.
[c] Corrected by –0.26 ppm for **7**.

2 were obtained in 58–93% yields by treatment with acetic anhydride and AlCl$_3$ in nitrobenzene and 1,1,2,2-tetrachloroethane at room temperature for 12 h [8]. Diols **3** were obtained in quantitative yields by the reduction with LiAlH$_4$ in THF at room temperature for 1 h. Diolefins **4** were obtained in 72–92% yields by dehydration with pyridinium p-toluenesulfonate in benzene under reflux for 5 days. [2 + 2] Photocycloaddition of diolefins **4** was carried out by irradiation with a 400 W high-pressure Hg lamp (Pyrex filter) in benzene for 26–92 h under N$_2$ [9–11]. After evaporation, [2.n]metacyclophanes **5b–e** were isolated in 61–87% yields by column chromatography. [2.2]Metacyclophanes **5a** and **6**, however, were found to be an equilibrium mixture, so that they could not be separated with either HPLC or TLC. But the ^{1}H NMR peaks for each isomer were detected separately.

Structural determination was carried out by NMR spectroscopy in CDCl$_3$, including COSY, NOESY, ^{13}C, and DEPT experiments. The cyclobutane ring of metacyclophanes **5** and **6** was assigned to be of *cis* configuration by the ^{1}H NMR chemical shifts ($\delta 3.72 - 4.74$) of the cyclobutane methine protons [12]. The direction of the cyclobutane ring to the methoxyl group was easily confirmed by NOESY experiments; i.e., the methylene protons of the cyclobutane ring clearly show an NOE interaction with the Ha aromatic protons (see Scheme 1). The methoxyl groups possess NOE interactions with not only the methine protons of the cyclobutane ring but also the Hb aromatic protons. Accordingly, the cyclobutane ring is concluded to face to the opposite direction of the methoxyl groups as shown in Scheme 1. The ^{1}H NMR chemical shifts of the Ha and Hb aromatic protons are listed in Table I. According to the molecular framework examination, **5** and **6** are apt to take a *syn* conformation [13, 14], because the steric interaction between the methoxyl group and the cyclobutane methylene protons seems to be severe, if they take an *anti* conformation.

The conformation was experimentally determined by the $\Delta\delta$ value [15] as shown in Table I [16–18]. It was also confirmed by ^{1}H NMR spectra, since the *syn* conformer showed a symmetrical spectral pattern of C_s symmetry, while the *anti* conformer displayed an unsymmetrical one due to C_1 symmetry. The anisotropic shielding effect of the CH_3O group on the chemical shift of Hb was estimated by using 2,4-dimethylanisole as a model, whose proton chemical shifts corresponding to Ha and Hb are $\delta 6.95$ and 6.72, respectively. Hence, the chemical shift deviation due to the effect of the CH_3O group is calculated as 0.23 ppm. Dimethoxy[2.*n*]metacyclophanes **5b–e** are concluded to be of *syn* conformation because the corrected $\Delta\delta$ value is positive with small values from 0.26 to 0.58. According to ^{1}H NMR and COSY spectra, [2.2]metacyclophanes **5a** and **6** formed a mixture of *syn*- and *anti*-isomers in the ratio of 4 : 3. *syn*-Dimethoxy[2.2]metacyclophane is highly strained, so that the repulsion between benzene rings is considered to overcome the steric hindrance between the methoxyl groups and the ethano bridge.

The synthetic route to dihydroxy[2.*n*]metacyclophanes **7** is shown in Scheme 1. Anisole derivatives **5** and **6** were treated with excess of boron tribromide in dry CH_2Cl_2 at r.t. for 12 h [19]. Phenol derivatives **7b** and **c** were obtained in 95 and 93% yields, respectively. On the other hand, **5c** did not give any desired products under the same conditions. Since the reaction gave a complex product mixture, we chose milder conditions. Thus, **5c** was carefully treated with an equimolar amount of BBr_3 at 0°C for 2 h and then gave **7a** in 78% yield. Unfortunately, **5a**, **5b**, and **6** did not give any phenol products.

The configuration of the cyclobutane ring for dihydroxy[2.*n*]metacyclophanes **7** was assigned to be *cis* by the chemical shift of its methine protons ($\delta 4.31 - 4.46$). The conformation of **7a–c** was determined by the corrected $\Delta\delta$ value as shown in Table I. The $\Delta\delta$ values are small and range from 0.15 to 0.33. Accordingly, we concluded that **7a–c** take a *syn* conformation. So the structure of **7a–c** is the same as the corresponding dimethoxy[2.*n*]metacyclophanes **5** even after the cleavage of methoxyl groups.

3. Synthesis of Cone-fixed Calix[4]arene Analogs

We examined the synthesis of a three-bridged calix[4]arene by using *syn*-dihydroxy-[2.5]metacyclophane **7b** as a building block (see Scheme 2). This is because we thought that it is not only an easy preparation but we could also take advantage of the *syn* conformation. Thus, cyclophane **7b** (2.0 g, 6.5 mmol) was treated with LiOH (0.31 g, 13 mmol) and paraformaldehyde (2.0 g, 65 mmol) in 20 mL of diglyme at 140–150°C for 12 h under N_2 to afford desired product **8a** [5]. Interestingly, a remarkable template effect was observed on this condensation reaction; i.e., lithium hydroxide gave **8a** in excellent 89% yield. When other larger metal

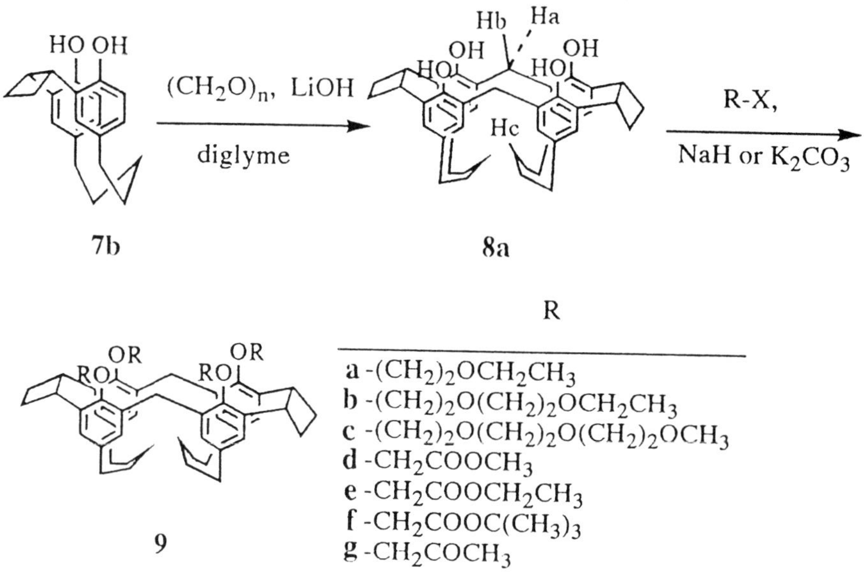
HO OH
$(CH_2O)_n$, LiOH
diglyme
Hb Ha
OH HO
OH HO
Hc
R-X,
NaH or K_2CO_3
7b
8a
OR RO
OR RO
9
R
a -$(CH_2)_2OCH_2CH_3$
b -$(CH_2)_2O(CH_2)_2OCH_2CH_3$
c -$(CH_2)_2O(CH_2)_2O(CH_2)_2OCH_3$
d -CH_2COOCH_3
e -$CH_2COOCH_2CH_3$
f -$CH_2COOC(CH_3)_3$
g -CH_2COCH_3

Scheme 2.

ions were used, the yield was gradually decreased in the order of Na^+ (42%), K^+ (15%), and Rb^+ (5%). Then finally cesium hydroxide did not give **8a** at all under the same reaction conditions.

The structure of **8a** was mainly elucidated by ^1H NMR spectroscopy in $CDCl_3$, including COSY, NOESY, and ^{13}C NMR. Typical findings are summarized as follows: (1) the methylene bridge shows on AB type coupling (Ha at δ3.28 with J=14 Hz and Hb at δ3.97 with J=14 Hz), which is the same as those ascribed to the calixarene cone-form. Moreover, this same coupling constant is maintained even in pyridine-d_5 (Ha at δ3.39 with J=14 Hz and Hb at δ4.32 with J=14 Hz). (2) The Hc proton resonance (δ–0.22) shifts to a higher field by ca. 0.4 ppm from that of the starting material **7b**, due to the additional shielding effect coming from another cyclophane system. (3) The hydroxy protons of **7b**, whose OH–OH distance is estimated as 4.4 Å from a CPK model, resonate at δ5.04, just the typical chemical shift value for simple monomeric phenols, suggesting the lack of a hydrogen bond in the molecule. On the contrary, the hydroxy protons of **8a** resonate at a much lower field, δ7.78. This large down-field shift clearly suggests the presence of intramolecular hydrogen bonding between two adjacent hydroxy groups, attached to each of two metacyclophane units, whose OH–OH distance is estimated to be 2.4 Å from a CPK model.

As mentioned above, we successfully developed a new methodology to obtain cone-fixed calixarene analog **8a**. This method can be applied to make another analog **8b** by using *syn*-dihydroxy[2.6]metacyclophane **7c** as shown in Scheme 3 [20]. In fact, the cesium hydroxide catalyzed reaction gave **8b** in an excellent 78%

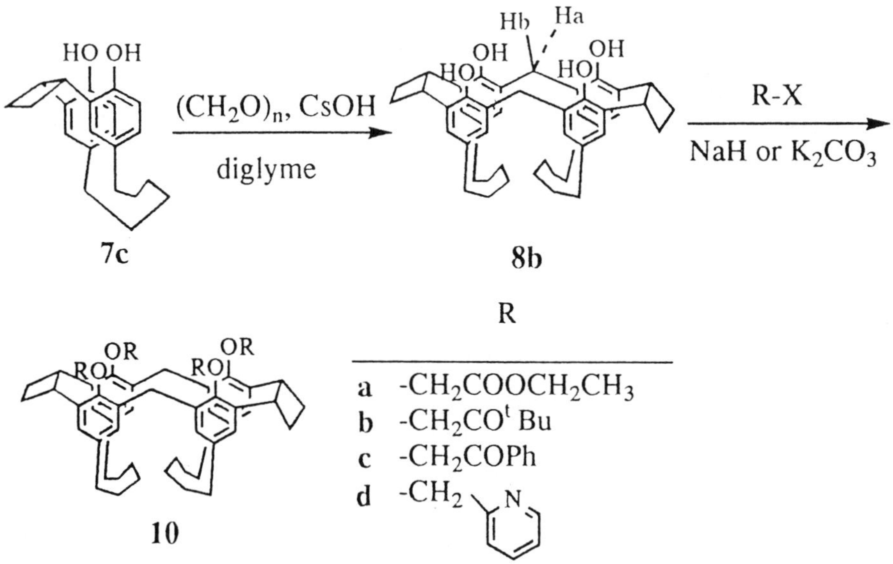

Scheme 3.

	R
a	-CH$_2$COOCH$_2$CH$_3$
b	-CH$_2$CO^{t}Bu
c	-CH$_2$COPh
d	-CH$_2$-(2-pyridyl)

yield. When other smaller metal ions were used, the yield was gradually decreased in the order of K$^+$ (21%) and Li$^+$ (19%). This template effect is opposite to that using **7b**.

The typical NMR spectroscopic features are summarized as follows: (1) the aromatic protons of **8b** are located at nearly the same position of δ6.71 and 6.85 as those of **8a**. (2) The cyclobutane methine protons of **8b** resonate at δ4.32, shifting to higher field than that of **8a** at δ4.56. (3) Its AB type coupling of the methylene bridge, which demonstrates the cone-form structure, appears in a higher field region from δ3.17 (Ha) and 3.84 (Hb), due to the movement of the methylene protons to the more shielding region, than that of **8a** (Ha at δ3.28 and Hb at δ3.97). (4) The hydroxy protons of **8b** resonate at δ6.71 a moderately higher field than that of **8a** at δ7.78, because the neighboring OH–OH distance for **8b** is estimated about 0.4 Å longer than that for **8a** by the molecular framework examination and also the distance of the opposite hydrogen bond sites of **8b** is approximately estimated ca. 5.0 Å. These results suggest that its hydrogen bonding of two neighboring hydroxy groups is weaker than that of **8a**.

Calix[4]arene analog **8a** can be easily modified by Birch reduction under the reported conditions shown in Scheme 4 [21]. We examined the direct reduction of **8a**, but the desired product **11** was not produced and only the starting material was recovered. This result suggests that phenoxide ion generated in the media prevented the desired radical anion from forming, due to the electronic repulsion between solvated electrons and phenoxides. Accordingly, the phenolic OH group was protected by etherification before Birch reduction. The methoxymethylation

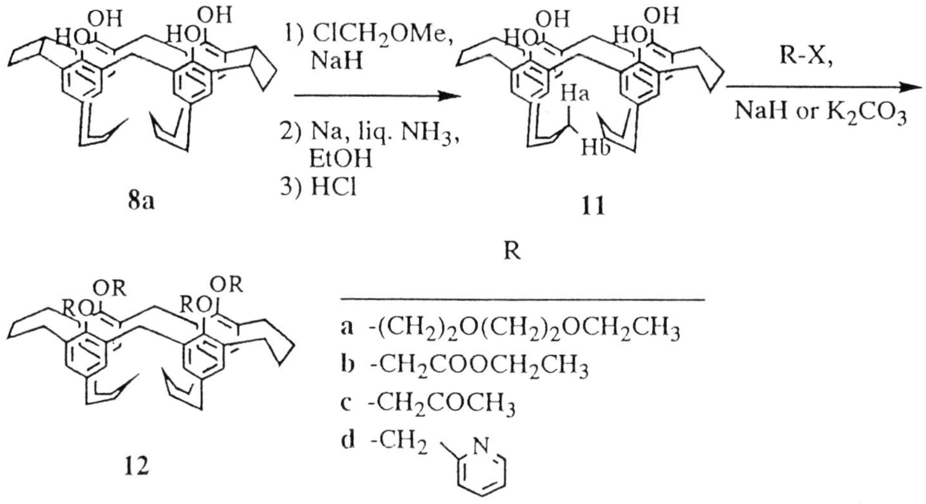
OH
HO
1) ClCH2OMe, NaH
2) Na, liq. NH3, EtOH
3) HCl
8a
Ha
Hb
11
R-X,
NaH or K2CO3
OR
RO
12
R
a -(CH2)2O(CH2)2OCH2CH3
b -CH2COOCH2CH3
c -CH2COCH3
d -CH2
N

Scheme 4.

was performed with **8a** (31 mM), chloromethyl methyl ether (10 equiv), and NaH (2 equiv) in dry THF/DMF (9/1) at 40–45°C for 12 h under N_2. Birch reduction was done with the ether (4.3 mM), Na (150 equiv), and EtOH (4 equiv) in liq. NH_3/dry THF (1/1) at –60°C for 4 h under N_2. The deprotection was carried out with the crude product (1 mM) in THF/aq. HCl (1/1) at 50°C for 12 h. Analog **11** was obtained in overall 86% yield. The structural determination is summarized as follows: (1) the cyclobutane ring methine protons (δ4.56) of **8a** have disappeared. (2) The AB type coupling of methanobridges is shifted from δ3.28 and δ3.97 for **8a** to δ3.33 and δ4.16 for **11**, due to the release of strain. (3) The inner methine protons (Ha) of **11** resonate at a normal chemical shift of δ0.59 in contrast with that of **8a** at δ–0.22. This result suggests that the distance between benzene nuclei is increased to decrease the shielding effect to these protons. Furthermore, Ha and Hb lie in the unequal environments so that they resonate at different positions, δ0.59 and δ0.84. (4) The hydroxy proton chemical shift of **11** (δ6.14) reveals that it has weaker hydrogen bonding than that of cyclophane **8a** (δ7.78), and therefore the benzene rings of **11** have been pushed apart by reduction of the cyclobutane ring.

Thus, we have successfully obtained rigidified calix[4]arene analogs **8a**, **8b**, and **11**, which kept the cone conformation. Moreover, these analogs are proved to maintain the cone structure from r.t. to 150°C in DMSO-d_6 or from r.t. to 100°C in pyridine-d_5 by VT NMR experiments. Analogs **8b** and **11** are rather limited in their inner movements, although they are more flexible than **8a**.

TABLE II. Extraction (%) of alkali metal, transition metal, and lanthanoid picrates in CH_2Cl_2 [a]

Compd	Li^+	Na^+	K^+	Rb^+	Cs^+	NH_4^+	Cr^{3+}	Mn^{2+}	Ag^+	Hg^{2+}	La^{3+}	Sm^{3+}	Yb^{3+}
9a	2.6	1.8	14.0	15.5	18.9	2.4	<1	9.3	6.9	6.0	–	–	–
9b	1.7	3.0	15.1	23.2	38.4	2.7	5.0	<1	<1	12.6	–	–	–
9c	1.9	8.4	25.3	27.5	13.8	2.3	5.0	1.4	8.1	12.1	–	–	–
9d	11.1	10.6	35.0	28.5	27.0	9.9	7.8	7.9	6.7	14.9	–	–	–
9e	15.9	32.8	90.2	95.7	88.3	33.0	16.5	6.3	12.5	26.7	22.1	11.1	8.8
9f	12.1	44.8	50.7	38.8	46.1	36.3	42.5	69.2	60.2	49.6	48.3	33.6	15.4
9g	7.9	13.1	36.3	39.1	38.0	9.9	7.5	7.9	10.2	14.1	–	–	–
10a	7.0	19.5	40.4	42.3	35.5	8.6	11.2	3.0	18.9	15.9	4.8	4.9	2.8
10b	49.1	55.6	57.9	63.6	48.0	44.6	69.1	74.5	96.3	89.3	59.9	44.8	13.8
10c	14.1	17.4	24.1	27.8	21.6	20.1	53.3	34.3	66.6	39.1	49.5	38.4	19.6
10d	4.2	6.1	60.6	46.2	10.9	6.3	96.8	84.6	97.9	97.1	98.0	88.5	47.6
12a	<1	4.9	5.1	6.2	6.8	<1	4.2	<1	<1	1.7	–	–	–
12b	<1	4.7	5.4	6.3	6.7	<1	11.8	2.7	2.9	17.6	–	–	–
12c	<1	<1	<1	<1	<1	<1	8.2	<1	<1	12.3	–	–	–
12d	<1	<1	<1	<1	<1	<1	77.6	26.6	98.4	97.6	44.9	39.1	18.6

[a] Extraction conditions: 2.5×10^{-4} M of receptor in CH_2Cl_2; 2.5×10^{-4} M of picric acid in 0.1 M of MOH for alkali metals or 2.5×10^{-5} M of picric acid in 1×10^{-3} M of metal nitrate for transition metals and lanthanoids at $22°C$. Receptor solution (5.0 mL) was shaken (10 min) with picrate solution (5.0 mL) and % extraction was measured by the absorbance of picrate in CH_2Cl_2. Experimental error was ±2%.

4. Ionophoric Behaviour of Rigidified Calix[4]arene Analogs

Functionalized calixarene analogs **9**, **10**, and **12** were obtained in 63–97% yields from their parents **8a**, **8b**, and **11**, respectively, with R–X and NaH or K_2CO_3 [5]. The cone-fixed conformation of calixarene analogs should affect their ion binding properties. In fact, the methylene protons of **9e** altered their chemical shifts upon titration with metal thiocyanate in $CDCl_3$: i.e., a 1 : 1 mixture of **9e** and KSCN showed broad peaks at $\delta 3.14$ and $\delta 3.76$ [22]. Hence, it is concluded that the K^+ ion was strongly bound on **9e**. In fact, the binding constant for K^+ ion has the largest value among the alkali metal ions and ammonium ion. The order is K^+ (log Ka=4.56) > Rb^+ (4.31) > Cs^+ (4.12) > NH_4^+ (4.01) > Na^+ (3.84) > Li^+ (3.34). This result suggests that larger alkali metals ions (K^+, Rb^+, and Cs^+) interact more strongly with **9e** than smaller ones (Li^+ and Na^+).

Based on these observations, we determined the extractability of alkali metal, transition metal, and lanthanoid ions from the aqueous phase to the organic phase [3, 23]. The results for metal picrates are summarized in Table II. The effects of the restricted ring conformation clearly appears in the ion selectivity; i.e., **9** extracted large ions like K^+, Rb^+, and Cs^+ more efficiently than small ones like Li^+ and Na^+. This ion selectivity of **9** resembles that of calix[6]arene derivative **14** rather than calix[4]arene one **13**, because the cavity of cylindrical **9** is as large as that of flexible calix[6]arene **14**. It is interesting to note that the cavity size of **9** is larger than that of calix[4]arene cone conformer **13**, although **9** and **13** have similar structural elements. These results indicate that the cavity size governing ion selectivity is determined by the aromatic ring frameworks rather than the kinds of binding sites. Introduction of bulky *tert*-butyl group **9f** instead of methyl or ethyl groups remarkably increases the extractability of the small Na^+ ion, probably due to the formation of a large hydrophobic cavity stabilizing the picrate anion [3]. The ion selectivity, however, apparently decreases except for the case of Li^+ ion.

Ionophore **9** apparently exhibited high selectivities and extractabilities for Hg^{2+} and Ag^+ ions in this transition metal extraction, because the rigidified ionophore **9** prefers the large metal ions. Lanthanoids have similar physical properties except the ion radius and the largest ion La^{3+} has nearly the same ion radius as K^+ ion. Ionophores **9e** and **9f** selectively separated La^{3+} ion from all other lanthanoids. In general, the extraction gradually decreased with decreasing ionic size.

13 n=4
14 n=6

All of **10** are excellent ionophores of alkali, transition metals, and lanthanoids (see Table II). The ion selectivity of **10** is recognized for K^+ and Rb^+ ions among alkali metals, Ag^+ among transition metals, and La^{3+} among lanthanoids. Interestingly, **10d** having picolyl ligands is an excellent ionophore in respect to both selectivity and extractability for alkali metals, transition metals, and lanthanoids. These results show that **10** prefers large ions to small ions. The ion selectivity of **10** is also more marked than that of **9**, because the rotation of binding sites of **10** is more free than that of **9**, due to the formation of the large cavity. The stoichiometry for extraction is determined by the distribution ratio as a function of ionophore concentration ($[M]=2.5 \times 10^{-4}$ to 2.5×10^{-5}). The slopes of log D vs. log [M] plots was unity. This result clearly suggests that the metal ions formed 1 : 1 complexes with ionophore **10**.

Ionophore **12** showed moderate extraction for alkali metals as depicted in Table II. The ion selectivity of **12** is recognized for Cs^+ ion. This selectivity for the large Cs^+ ion shows that the binding sites of **12** exist largely apart from each other. It has moderate extractability for transition metals with high selectivity. In particular, **12d** having picolyl ligands is an excellent ionophore in respect to both selectivity and extractability for Ag^+ and Hg^{2+}. And also the largest La^{3+} ion in lanthanoids was most efficiently extracted to the organic phase. The extractability of **12d** gradually decreased from large La^{3+} to small Yb^{3+} ion.

5. The Catalytic Activity of Calix[4]arene Analog

The results of ion extraction experiments suggest a possibility that **9** might display a catalytic activity for a nucleophilic substitution reaction with inorganic reagents in organic media. Accordingly, we investigated the catalytic activity of **9c** on

$$\text{PhOH} + \text{PhCH}_2\text{Br} \xrightarrow{a} \text{PhOCH}_2\text{Ph} \quad (1)$$

$$\text{CH}_3(\text{CH}_2)_7\text{-X} \xrightarrow{b} \text{CH}_3(\text{CH}_2)_7\text{-Y} \quad (2)$$

a) MOH / Cat. / CCl_4 or CD_2Cl_2

b) MY / Cat. / CD_3COCD_3 or CD_3CN

some S_N2 reactions, which were ester synthesis, Williamson ether synthesis, and the Finkelstein reaction, and compared the results with those of calix[n]arene derivatives ($n=4$ and 6 for **15** and **16**) [24]. A 2-[2-(2-methoxyethoxy)ethoxy]ethyl unit was chosen as a binding site, because it is a stable substituent under neutral and basic conditions [25]. The reaction was followed by ^{1}H NMR and the rate was estimated as a pseudo-first-order rate constant k (s^{-1}) for the increase of product.

15 n=4
16 n=6

Firstly, we examined the esterification with metal acetate, benzyl bromide, and catalyst **9c**, **15**, or **16**. The reaction proceeded in the presence of these catalysts, but not in the absence of them. Therefore, their catalytic activity is apparent. The rate constants remained in the same range for all runs. Moreover, the reaction was slow (k=$10^{-6} - 10^{-7}$ s^{-1}) because of the low nucleophilicity of acetate ion. Since the substitution by acetate did not give much information on their catalytic activity, we chose the Williamson ether synthesis with phenol, benzyl bromide, metal hydroxide, and catalyst (see Equation 1), because phenolate has high nucleophilicity and hydrophobicity [24]. In fact, the difference of their catalytic activities clearly appeared in this case as shown in Table III. Catalysts made the reaction markedly faster than that without them. In carbon tetrachloride as a nonpolar solvent, the catalytic activity of **9c** increased remarkably when larger ions were used and the maximum rate constant was recorded in the RbOH system. This behavior of **9c** resembles that of calix[6]arene derivative **16** because both **9c** and **16** have the same affinity for large ions. The other experiments were performed in

TABLE III. Rate constant of the reaction between phenoxide and benzyl bromide[a]

base (MOH)	solvent	k (10^{-7} s^{-1})			
		none	9c	15	16
NaOH	CCl$_4$	4.2	62	50	67
KOH	CCl$_4$	6.1	79	56	70
RbOH	CCl$_4$	5.6	160	60	180
NaOH	CD$_2$Cl$_2$	8.4	120	90	140
NaOH	sat CD$_2$Cl$_2$[b]	17	590	120	390
KOH	CD$_2$Cl$_2$	15	520	480	600
RbOH	CD$_2$Cl$_2$	18	760	420	990
CsOH	CD$_2$Cl$_2$	28	320	220	710

[a] Phenol: benzyl bromide: base: catalyst=1: 1: 3.5: 0.029 (molar ratio); phenol, 0.43 mol/l; temp., 32 ± 1°C. Experimental error was ±10%.
[b] Saturated with D$_2$O.

dichloromethane as a low polar solvent. The rate constant in this solvent increased by around 2–8 times as compared to that in CCl$_4$, probably due to the increment of the solubility of ion-catalyst complex and the activity of the nucleophile. The increasing order of rate constant, when **9c** was used, can be explained by its selectivity for binding the alkali metal ion and solubilizing the metal hydroxide; K$^+$ and Rb$^+$ are suitable for complexation with **9c**, but the Cs$^+$ ion is too large to fit the binding site of **9c** effectively [5]. By the addition of a little water to make a two-phase system, the rate became larger, but the order of enhancement by the catalysts did not change. This experimental result shows that the present catalysts can be used as phase transfer catalysts. The order of catalytic activity for this ether synthesis is **15 < 9c ≤ 16**.

We also examined the Finkelstein reaction of octyl halide involving a conversion from bromide to iodide or from iodide to bromide (see Equation 2) [26]. The results are summarized in Table IV. The Finkelstein reaction proceeded without any catalysts, because halide salts are soluble in various solvents. But, when catalyst was combined in this system, the rate was enhanced clearly by 1.5–2 times (see Table IV). In the case involving a conversion of octyl bromide to iodide, **9c** showed the maximum rate for KI and RbI in acetone or acetonitrile, indicating that its complexation with the alkali metal ion is an important factor to accelerate this reaction. Note that **9c** has the largest rate constant among the catalysts for all metal iodides examined in this reaction.

In the other case involving a conversion of octyl iodide to bromide, **9c** also showed the maximum rate for K$^+$, Rb$^+$ and C$_s$ salts, although the reaction of iodide to bromide is difficult. Furthermore, **9c** is again the best catalyst for all

TABLE IV. Rate constant of Finkelstein reaction[a]

substrate	reagent (MY)	solvent[b]	temp.[c] (°C)	k (10^{-7} s^{-1}) none	9c	15	16
n-C$_8$H$_{17}$Br	KI	A	50	27	42	28	31
	RbI	A	50	22	41	24	24
	CsI	A	50	14	24	18	20
	KI	B	50	–	39	20	28
	RbI	B	50	–	36	17	25
	CsI	B	50	–	31	14	20
n-C$_8$H$_{17}$I	KBr	B	75	9.2	15	12	13
	RbBr	B	75	–	14	10	11
	CsBr	B	75	–	12	8.0	9.7

[a] Substrate: reagent: catalyst=1: 5: 0.05 (molar ratio); substrate, 0.24 mol/l. Experimental error was ±10%.
[b] A: CD$_3$COCD$_3$, B: CD$_3$CN.
[c] ±2°C.

metal bromides. These results suggest that the cylindrical structure of **9c** gives a favorable environment for this Finkelstein reaction.

6. Conclusion

We successfully synthesized calix[4]arene analogs **8a** and **8b** firmly locked in the cone conformation in excellent yield. An enlarged calix[4]arene analog **11** was also obtained from **8a** by Birch reduction. All ionophores **9, 10,** and **12** derived from these analogs have been found to select the larger ions in the extraction of alkali metals, transition metals, and lanthanoids. Ionophore **9c** having oligoethylene glycol units showed an effective catalytic activity for some S$_N$2 reactions in several media.

Acknowledgement

This work was supported in part by grants from the Ministry of Education, Science, and Culture, Japan.

References

1. C.D. Gutsche: *Acc. Chem. Res.* **16**, 161 (1983); S. Shinkai: *J. Incl. Phenom.* **7**, 193 (1989).
2. D.W. Chasar: *J. Org. Chem.* **50**, 545 (1985); V. Böhmer, H. Goldmann, and W. Vogt: *J. Chem. Soc., Chem. Commun.* **1985**, 667.
3. F. Arnaud-Neu,, E.M. Collins, M. Deasy, G. Ferguson, S.J. Harris, B. Kaitner, A.J. Lough, M.A. McKervey, E. Marques, B.L. Ruhl, M.J. Schwing-Weill, and E.M. Seward: *J. Am. Chem. Soc.* **111**, 8681 (1989); V. Böhmer, W. Vogt, and H. Goldmann: *J. Org. Chem.* **55**, 2569 (1990); E. Ghidini, F. Ugozzoli, R. Ungaro, S. Harkema, A.A. El-Fadl, and D.N. Reinhoudt: *J. Am. Chem. Soc.* **112**, 6979 (1990).

4. Y. Okada, K. Sugiyama, Y. Wada, and J. Nishimura: *Tetrahedron Lett.* **31**, 107 (1990); Y. Okada, K. Sugiyama, M. Kurahayashi, and J. Nishimura: *ibid.* **32**, 2367 (1991); Y. Okada, S. Mabuchi, M. Kurahayashi, and J. Nishimura: *Chem. Lett.* **1991**, 1345.
5. Y. Okada, F. Ishii, Y. Kasai, and J. Nishimura: *Chem. Lett.* **1992**, 755; Y. Okada, F. Ishii, Y. Kasai, and J. Nishimura: *Tetrahedron Lett.* **34**, 1971 (1993); Y. Okada, Y. Kasai, F. Ishii, and J. Nishimura: *J. Chem. Soc., Chem. Commun.* **1993**, 976.
6. D.J. Cram: *Science* **219**, 1177 (1983); *Synthesis of Macrocycles*, ed. by R.M. Izatt and J.J. Christensen, Wiley, New York, 1987; C.D. Gutsche: *Calixarenes*, Royal Society of Chemistry, London, 1989; K. Koga and K. Odashima: *J. Incl. Phenom.* **7**, 53 (1989).
7. J. Nishimura, N. Yamada, Y. Horiuchi, E. Ueda, A. Ohbayashi, and A. Oku: *Bull. Chem. Soc. Jpn.* **59**, 2035 (1986).
8. *Organic Reactions Vol. V*, ed. by R. Adams, Wiley, New York, 1949, Chap. 5.
9. J. Nishimura, H. Doi, E. Ueda, A. Ohbayashi, and A. Oku: *J. Am. Chem. Soc.* **109**, 5293 (1987).
10. M. Yamamoto, T. Asanuma, and Y. Nishijima: *J. Chem. Soc., Chem. Commun.* **1975**, 53; T. Asanuma, M. Yamamoto, and Y. Nishijima: *ibid.* **1975**, 56.
11. H. Nozaki, I. Otani, R. Noyori, and M. Kawanisi: *Tetrahedron* **24**, 2183 (1968).
12. J. Nishimura, A. Ohbayashi, H. Doi, K. Nishimura, and A. Oku: *Chem. Ber.* **121**, 2019 (1988).
13. T. Kamada and O. Yamamoto: *Bull. Chem. Soc. Jpn.* **52**, 1159 (1979).
14. T. Kamada and O. Yamamoto: *Bull. Chem. Soc. Jpn.* **53**, 994 (1980).
15. D. Krois and H. Lehner: *Tetrahedron* **38**, 3319 (1982).
16. R.H. Mitchell, T.K. Vinod, and G.W. Bushnell: *J. Am. Chem. Soc.* **107**, 3340 (1985); R.H. Mitchell, G.J. Bodwell, T.K. Vinod, and K.S. Weerawarna: *Tetrahedron Lett.* **29**, 3287 (1988).
17. Y.-H. Lai and S.-M. Lee: *J. Org. Chem.* **53**, 4472 (1988).
18. H. Meier, E. Praß, and K. Noller: *Chem. Ber.* **121**, 1637 (1988).
19. T.W. Greene: *Protective Groups in Organic Synthesis*, Wiley, New York, 1981, p. 91.
20. Y. Okada, Y. Kasai, and J. Nishimura: *Synlett.* in press.
21. A. Greenberg and J.F. Liebman: *Strained Organic Molecules*, Academic Press, New York (1978), p. 58; J. Nishimura, A. Ohbayashi, E. Ueda, and A. Oku: *Chem. Ber.* **121**, 2025 (1988); B.P. Mundy and M.G. Ellerd: *Named Reactions and Reagents in Organic Synthesis*, Wiley, New York, 1988, p. 31.
22. A. Arduini, A. Pochini, S. Reverberi, and R. Ungaro: *Tetrahedron* **42**, 2089 (1986).
23. S.-K. Chang and I. Cho: *Chem. Lett.* **1984**, 477; M.A. McKervey, E.M. Seward, G. Ferguson, B. Ruhl, and S.J. Harris: *J. Chem. Soc., Chem. Commun.* **1985**, 388; T. Nagasaki and S. Shinkai: *Bull. Chem. Soc. Jpn.* **65**, 471 (1992).
24. H. Taniguchi and E. Nomura: *Chem. Lett.* **1988**, 1773.
25. Y. Okada, Y. Sugitani, Y. Kasai, and J. Nishimura: *Bull. Chem. Soc. Jpn.* **67**, 586 (1994).
26. D. Landini and F. Montanari: *J. Chem. Soc., Chem. Commun.* **1974**, 879; M. Cinquini and F. Montanari: *ibid.* **1975**, 393.

Conformational Properties of the Calix[4]naphthalenes*

PARIS E. GEORGHIOU** and ZHAOPENG LI
Department of Chemistry, Memorial University of Newfoundland, St. John's, Newfoundland, Canada A1B 3X7

(Received: 3 March 1994; in final form: 28 June 1994)

Abstract. The base-catalyzed condensation of 1-naphthol and formaldehyde in refluxing dimethylformamide affords three isomeric cyclic tetramers which are conformationally flexible over a wide temperature range. Their tetrabenzoate esters however show restricted flexibility. Variable-temperature NMR and low-temperature COSY is used to analyze the conformational preferences of these calix[4]naphthalenes.

Key words: 1-Naphthol, formaldehyde, calix[4]arenes, calix[4]naphthalenes, NMR, conformational analysis.

1. Introduction

It has been 50 years since Zincke and Ziegler assigned cyclic tetrameric structures to substances obtained from the base-induced reaction of *p*-substituted phenols with formaldehyde [1], substances which we have since come to know as the calixarenes. However, it has taken almost 50 years since their assignment for the first report to appear, in 1993, of cyclic tetrameric structures from the base-induced reaction of formaldehyde with the naphthalene analogue of phenol, 1-naphthol [2].

The naphthols are more reactive than phenols and resemble resorcinol rather than phenol in many of their reactions [3]. The complexity of the reaction of 1-naphthol with formaldehyde is well known [4] and it has been assumed that cross-linked polymers are formed since reaction can occur at both C-2 and C-4, the positions which are respectively *ortho* and *para* to the hydroxyl group. In 1907 Breslauer and Pictet [5] reported obtaining an amorphous product having empirical formula $C_{23}H_{16}O_3$ from the reaction of 1-naphthol with formaldehyde in the presence of potassium carbonate. Abel [6] reported obtaining a "brown, brittle, alkali-soluble resin" on heating 1-naphthol with formaldehyde in 50% acetic acid containing a small quantity of hydrochloric acid. By way of contrast, 2-naphthol reacts mainly at C-1, and as a result, condenses with formaldehyde under either

---

* This paper is dedicated to the commemorative isssue on the 50th anniversary of calixarenes.
** Author for correspondence.

base [7] or acid [8] catalysis to readily afford a well-defined, single product, *bis*-(2-hydroxy-1-naphthyl)-methane, **I**.

When we reinvestigated the base-induced reaction of 1-naphthol with formaldehyde we isolated and identified three isomeric tetrameric compounds [2] which we named "calix[4]naphthalenes" by analogy with the calix[4]arenes and calix[4]resorcinarenes. However, unlike the latter which are derived from *p*-substituted phenols and resorcinol respectively, several different isomers can theoretically exist for the calix[4]naphthalenes. Additionally, the conformations that are possible for some of these isomers are further complicated due to the dissymmetry that is introduced by the naphthalene rings. In this contribution we will describe some properties of this new class of compounds.

2. Results and Discussion

When a solution containing purified 1-naphthol, formaldehyde and potassium carbonate was heated in DMF under reflux for 30 h, a crude product was obtained that thin-layer chromatography indicated to be a complex mixture. The ^{1}H-NMR spectrum of this crude product however had surprisingly clearly-defined features. In particular, the signals in the 4.0–4.8 ppm region resembled those usually seen for the methylene bridge protons of the various conformers of calixarene derivatives [9, 10]. The limited solubility of the crude product in the usual organic solvents prohibited purification by conventional chromatographic techniques. However, fractional crystallization of the crude reaction mixture afforded three crystalline products. Mass spectra indicated that these are isomeric tetramers, each having a molecular ion peak at $m/z = 624$.

If ring A of the naphthalene ring only is considered, there are four isomers that are possible for cyclic tetramers which can be formed from the condensation of 1-naphthol with formaldehyde. These are depicted as **II–V**. Using symmetry considerations alone, the number of ^{13}C-NMR resonance signals that would be expected for **II** would be eleven, for **III** the number would be twelve, and for **IV** the number would be twenty three. Isomer **V**, the least symmetrical of the isomers, would be expected to show forty four carbon signals.

I

II : R = H
IX : R = Bz

III : R = H　　　　IV : R = H　　　　V : R = H
　　　　　　　　　X : R = Bz　　　　XI : R = Bz

The first substance to precipitate from the reaction mixture consisted of at least two compounds as ascertained by thin layer chromatography. Crystallization from acetone yielded a homogeneous product whose ^{13}C-NMR (DMSO-d_6) spectrum shows twelve signals consisting of five quaternary aromatic carbon signals, five methine aromatic carbon signals, a single aliphatic methylene carbon signal and an aliphatic methyl carbon signal. The methyl carbon signal could be shown to be due to acetone which could not be removed by overnight drying under vacuum. The structure for this compound was therefore assigned to be **II**. It is uncertain at this stage whether or not this compound forms an inclusion complex with the acetone and we are striving to obtain suitable crystals for X-ray diffraction analysis. The ^{1}H-NMR spectrum of **II** (Figure 1) includes a relatively high-field aromatic signal which is a four-proton singlet at 6.62 ppm due to the four intra-annular naphthalene protons (H-41, H-42, H-43 and H-44). The methylene protons (on C-10, C-20, C-30 and C-40) appear as an eight-proton singlet at 4.29 ppm. These data together with the HETCOR, NOED spectra and MS data (for a more detailed assignment, see [2]) are consistent for structure **II** which possesses C_4 symmetry. The relatively higher-field aromatic signal in this and the other isomers can be accounted for by examination of molecular models which reveal that these intra-annular protons can be situated directly above a shielding aromatic ring current. That the methylene protons appear as a singlet at ambient temperature indicates that the compound has a flexible structure and that the positions of these methylene protons are rapidly interchanging.

The second compound obtained from the crude reaction product was crystallized from ethyl acetate. This compound was the same substance which initially co-crystallized with **II**. The ^{13}C-NMR (DMSO-d_6) spectrum of the pure product reveals only twenty one clearly defined signals. However, the APT-^{13}C-NMR spectrum shows that there are ten quaternary aromatic carbon signals, ten methine aromatic carbon signals and three aliphatic methylene carbon signals. One pair of quaternary carbon signals and a pair of aromatic methine signals clearly overlap. In addition, the height of one of the aliphatic methylene carbon signals is double that of each of the other two. The ^{1}H-NMR spectrum (Figure 2) shows the higher-

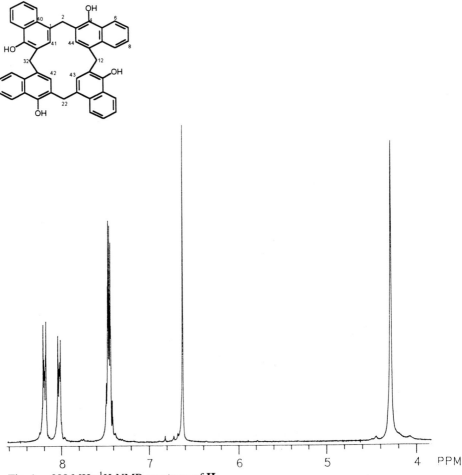

Fig. 1. 300 MHz ^{1}H-NMR spectrum of **II**.

field aromatic signals as two (two-proton) singlets of equal intensity at 6.83 and 6.72 ppm for the intra-annular protons. The methylene protons appear as three singlets, at 4.40, 4.29 and 4.08 ppm with relative intensities in the ratio of 1 : 2 : 1, respectively. In addition to these data, the HETCOR, NOED spectra and MS data (for a more detailed assignment, see [2]) are consistent for structure **IV** which possesses a plane of symmetry through the C-20 and C-40 methylene groups and which is perpendicular to the macrocyclic ring. This isomer is also conformationally flexible at ambient temperature.

The third isomer, and the one which was the most difficult to isolate, was recrystallized from diethyl ether. Its ^{13}C-NMR (DMSO-d_6) spectrum shows only forty two signals, with some obvious overlapping in the group of signals which are centred around 124.5 ppm. The APT-^{13}C-NMR spectrum clearly indicates twenty

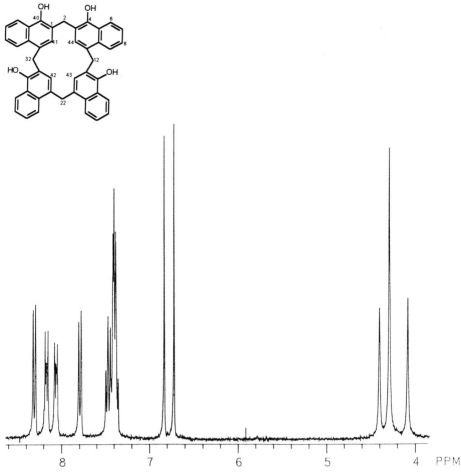

Fig. 2. 300 MHz ^{1}H-NMR spectrum of **IV**.

quaternary aromatic carbon signals; twenty methine aromatic carbon signals, of which the same group of signals centred at 124.45 ppm was not clearly resolved; and, four aliphatic methylene carbon signals. The ^{1}H-NMR spectrum (Figure 3) shows four relatively high-field aromatic (one-proton) singlets of equal intensities, at 6.80, 6.70, 6.66 and 6.64 ppm for the four intra-annular protons. The methylene protons appear as four two-proton singlets of equal intensities, at 4.45, 4.32, 4.21 and 4.09 ppm. These data together with the HETCOR, NOED spectra and MS data (see [2]) were also consistent for structure **V** which does not possess any symmetry. This isomer is also conformationally flexible at ambient temperature.

When the ^{1}H-NMR spectra of these three isomers are superimposed, it is obvious that they are the major components of the crude reaction product. In the crude reaction product the ratio of the three isomers can be estimated from integration of

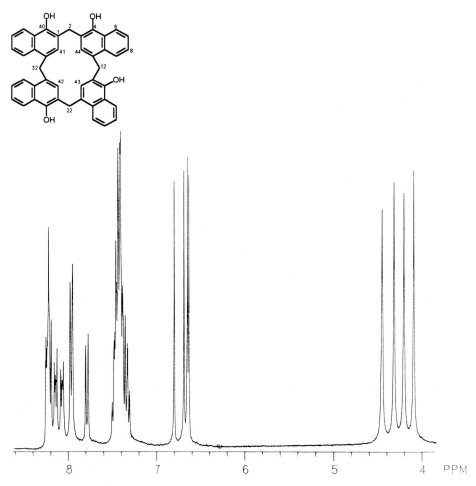

Fig. 3. 300 MHz ^{1}H-NMR spectrum of **V**.

the intra-annular aromatic signals in the range 6.5–6.9 ppm. The ratio of isomers **II** : **IV** : **V** can be estimated to be approximately 1.0 : 2.2 : 3.0. The total isolated yield (unoptimized) of the three isomers was only 25%, with the ratios of the respective isomers being 1.0 : 1.6 : 0.5. We have been unable to isolate, nor is there any evidence from the ^{1}H-NMR spectra for any significant amount of the fourth potential isomer, **III**.

There are three potential intermediate *bis*-(1-hydroxynaphthyl)methanes which could be produced that could lead to the formation of cyclic tetramers. They are the *ortho,ortho*-coupled product **VI**, the *para,para*-coupled product **VII**, and the *ortho,para*-coupled product, **VIII**. The following analysis assumes that dimerization of some of these intermediates can account for the formation of the calix[4]naphthalenes that were observed. Dimerization of **VIII** with two equivalents of formaldehyde would lead to the formation of both **II** and **IV**, but not of **III** or **V**. Dimerization of **VI**, dimerization of **VII** and cross-condensation of **VI** and **VII** with formaldehyde would all lead to the formation of the tetramer **III** which is not observed. However, the formation of **V** indicates that **VIII** could have condensed with formaldehyde and either **VI** or **VII**. Interestingly, the only calix[4]naphthalene derived from a similar reaction of 1,8-naphthalenesulfone with formaldehyde shows twelve carbon signals in its ^{13}C-NMR spectrum [9] and is analogous to **III**.

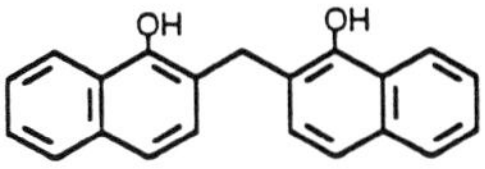

VI 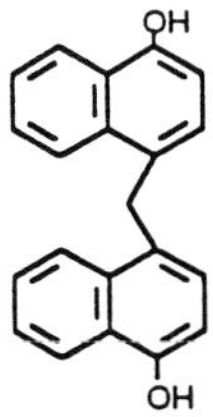   **VII** 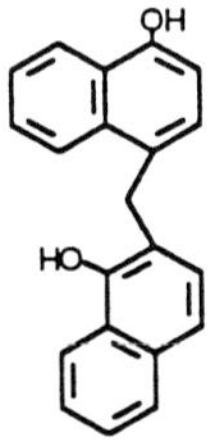   **VIII**

All three calix[4]arenes are conformationally flexible from ambient temperature down to -50°C, as evidenced by the sharp signals observed for the all of the methylene bridge and intra-annular protons (Figures 1–3). This is not the case when the corresponding tetrabenzoates were similarly analyzed. Using standard procedures, **IX**, **X** and **XI** were prepared from **II**, **IV** and **V** respectively. Each tetrabenzoate showed a molecular ion peak at $m/z = 1040$ by positive FAB mass spectrometry. The ambient ^{1}H-NMR spectra of each of the tetrabenzoates are characterized by the apparent almost total absence of the intra-annular signals in the 6.0–7.0 ppm region. The methylene protons appear as broad signals. Variable-temperature NMR experiments performed on each of the tetrabenzoates reveal many well-defined signals at the lower temperatures (see Figures 4d–6d). Only at approximately 100°C do the ^{1}H-NMR spectra resemble the ambient spectra of **II**, **IV** and **V** depicted above, with the intra-annular and methylene protons appearing as singlets.

Figure 4 shows the ^{1}H-NMR (CD$_2$Cl$_2$) spectra of **IX** taken at (a) ambient temperature, (b) 0°C, (c) -20°C, and (d) -50°C. At -50°C conformational freezing is apparent by the presence of well-defined signals in the 4.0–5.0 ppm and 5.8–8.4 ppm ranges.

Figure 5 shows the ^{1}H-NMR (CD$_3$Cl) spectra of **X** taken at (a) 50°C, (b) ambient temperature, (c) 0°C, (d) -20°C, and (e) -50°C. Conformational freezing is also apparent at -50°C by the presence of well-defined signals.

In order to interpret the low-temperature spectra, COSY determinations were conducted at -50°C. COSY of **IX** confirmed that the (singlet) signals at 7.14, 7.08, 5.86 and 5.83 ppm (see Figure 4d) are due to the intra-annular protons on C41-44. A conformer which would result in the intra-annular protons being observed as four singlets is a 'partial cone' conformer which is analogous to those

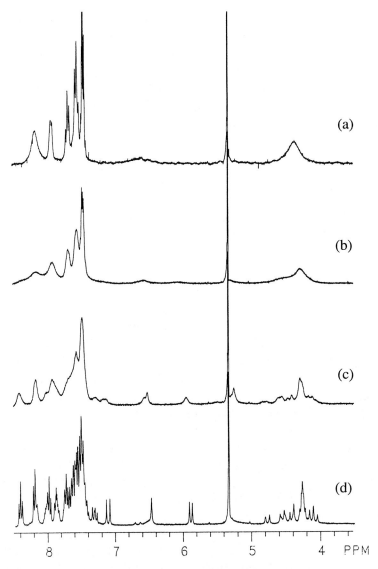

Fig. 4. 300 MHz ^{1}H-NMR spectra of **IX** at: (a) +20°C, (b) 0°C, (c) −20°C, and (d) −50°C.

partial cone conformers observed with the calix[4]arenes [10, 11], rather than a 'flattened partial-cone' type of conformer seen with the calix[4]resorcinarenes [12]. This is confirmed by the COSY spectrum shown in Figure 6 which shows eight pairs of doublets due to the methylene geminal protons. However, at least one other conformer (25–30%) is present as evidenced by the large unresolved signal (singlet?) centred at 4.25 ppm and the corresponding aromatic intra-annular signal which appears as a singlet at 6.45 ppm. At this stage it is not possible to assign these latter signals specifically to any of the other three possible conformers

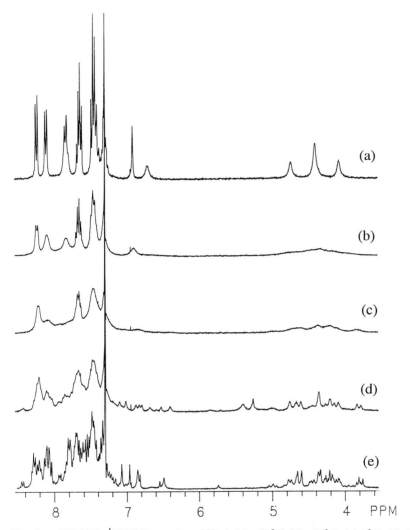

Fig. 5. 300 MHz ^{1}H-NMR spectra of **X** at: (a) +50°C (b) +20°C, (c) 0°C, (d) -20°C, and (e) -50°C.

(cone, 1,3-alternate or 1,2-alternate, all of which are analogous to those observed with the calix[4]arenes) as all could be predicted to have a singlet for all four intra-annular protons. Molecular modeling calculations using Alchemy III [13] confirm that the partial cone is the lowest energy conformer and suggest that the 1,2-alternate would be the next higher energy conformer, followed by the cone and 1,3-alternate conformers. However, the energy differences between these conformations that these calculations reveal do not appear to be meaningful as they are much higher than would be expected considering the relative amounts of the two major conformers that are evident at -50°C.

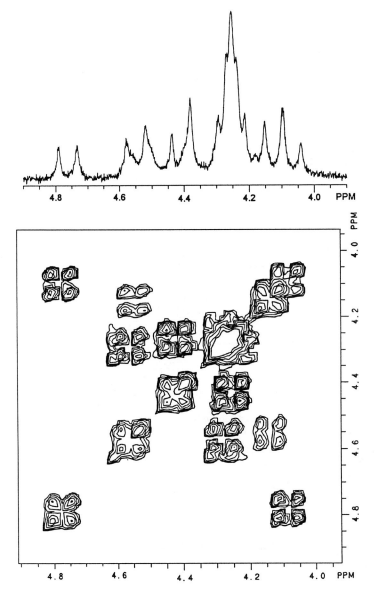

Fig. 6. 300 MHz COSY spectrum (-50°C) of methylene protons in **IX**.

The -50°C COSY spectrum of **X** is considerably more complex than that of **IX**. At least eighteen pairs of doublets and up to four singlets can be discerned in the methylene proton region (Figure 7). At least six signals (singlets) due to the intra-annular protons can also be discerned, albeit with difficulty. As with **IX** a partial-cone conformer could account for eight pairs of doublets; a 1,2-alternate (unsymmetrical) conformer could account for an additional six pairs of doublets; and a 1,2-alternate (symmetrical) conformer could account for an additional four

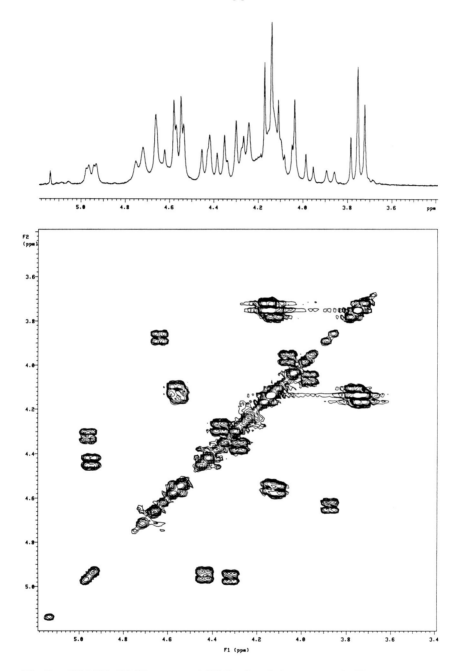

Fig. 7. 500 MHz COSY spectrum (-50°C) of methylene protons in **X**.

pairs of doublets and two singlets. Similar molecular modeling calculations also indicate that as with **IX**, the partial cone is the lowest energy conformer for **X** as well. The -50°C COSY spectrum of **XI** is the most complex of the three and has not yet been fully assigned.

The well-resolved ^{1}H-NMR signals observed for both the naphthalene ring and the methylene bridge protons of the three calix[4]naphthalenes **II**, **IV** and **V** indicate that there is rapid conformational interconversion in these molecules at ambient temperatures. The corresponding ^{1}H-NMR signals of the respective tetrabenzoates **IX**, **X** and **XI** indicate restricted conformational mobility as a result of steric crowding due to the presence of the bulky benzoate groups. However, at ambient temperatures there is no evidence for single conformer formation as noted for example by Jaime et al. [14] for their calixarene derivatives.

Acknowledgements

We thank Dr. C. R. Jablonski and Ms. N. Brunet for the 300 MHz ^{1}H-NMR spectra and Dr. T. Nakashima of the Department of Chemistry, University of Alberta, Edmonton, Alberta for the 500 MHz ^{1}H NMR and COSY spectra. Memorial University of Newfoundland and N.S.E.R.C. are thanked for financial support for one of us (Z.L.).

References

1. A. Zincke and E. Ziegler: *Ber.* **77**, 264 (1944).
2. P. E. Georghiou and Z. Li: *Tetrahedron Lett.* **34**, 2887 (1993).
3. N. Donaldson: *The Chemistry and Technology of Naphthalene Compounds*, Edward Arnold Publishers Ltd., London (1958).
4. J. F. Walker: *Formaldehyde*, American Chemical Society Monograph Series, No. 159, Reinhold Publishing Corp., New York (1964).
5. J. Breslauer and A. Pictet: *Ber.* **40**, 3785 (1907).
6. J. Abel: *Ber.* **25**, 3477 (1892).
7. O. Manasse: *Ber.* **27**, 2409 (1894).
8. H. Hosaeus: *Ber.* **25**, 3213 (1892).
9. P. E. Georghiou and Z. Li: *unpublished results*.
10. C. D. Gutsche: in *Calixarenes*, J. F. Stoddard (Ed.), Monographs in Supramolecular Chemistry Vol. 1, Royal Society of Chemistry, Cambridge (1989).
11. C. D. Gutsche: in *Calixarenes: A Versatile Class of Macrocyclic Compounds*, J. Vicens, V. Böhmer (Eds.), Topics in Inclusion Science, Kluwer Academic Publishers, Dordrecht (1991).
12. A. G. S. Högberg: *J. Am. Chem. Soc.* **102**, 6046 (1980).
13. *Alchemy III. 3D Molecular Modeling Software*, Tripos Associates Inc., St. Louis, Missouri (1992).
14. C. Jaime, J. de Mendoza, P. Prados, P. M. Nieto, and C. Sanchez: *J. Org. Chem.* **56**, 3372 (1991).

The Host-Guest Chemistry of Resorcinarenes [1]*

H.-J. SCHNEIDER* and U. SCHNEIDER
FR Organische Chemie der Universität des Saarlandes, D66041 Saarbrücken, Germany

(Received: 11 May 1994; in final form: 24 September 1994)

Abstract. Conformations, acid-base and supramolecular properties of phenolic metacyclophanes obtained from the condensation of resorcinol with aldehydes are discussed, including the mechanisms involved in the formation of these macrocycles. The strong binding of choline-type compounds and the inhibition of acetylcholine hydrolysis with the **rccc** stereoisomers is mechanistically evaluated; a **rctt** isomer shows strong conformational coupling for, e.g., choline binding and simultaneous proton release. The presence of larger alkyl residues at the bottom of the **rccc** macrocycle leads to an additional binding site for small lipophilic substrates, which is independent of the upper complexation center for positively charged substrates. Substitution at the upper rim by carboxylic groups at the 2-position of the phenyl rings yields receptors for, e.g., α, ω-diammonium ions with alternate equatorial and axial arylunits. Positively charged substituents at the upper rim, introduced by aminoalkylation, lead to little change of complexation as a result from their orientation away from the binding center. Aminoacid substituents, for the same reason, do not lead to enantioselective complexation, but allow particularly for strong binding of transition metal ions. Preliminary studies show that resorcinarenes bearing a wide array of positive charges are potent groove binders to ds-DNA without intercalative contributions.

Key words: Acidities, amino acids, binding mechanisms, calixarenes, choline, conformations, copper complexes, NMR, polyphenolates, resorcinarenes, supramolecular complexes.

1. Introduction

Although Adolf von Baeyer obtained crystalline material from the condensation of resorcinol with aldehydes [2] in 1872 it took almost a century and the efforts of several groups to secure the structure of these products. In 1884 further derivatives were isolated [3], and in 1884 Michael [3a] suggested that they might have a cyclic structure. This was on the basis of molecular weight determinations a supposition supported by Niederl and Vogel in 1940 [4]. Erdtman *et al.* finally proved the metacyclophane structure **1** of derivatives [5] by X-ray analysis. In 1980 Högberg published his NMR studies on the conformational changes of the corresponding esters in chloroform solution [6a] and also clarified some of the acid-catalyzed rearrangememts of the stereoisomers [6b], which later were all characterized, in particular by Mann *et al.* [7].

Today, these macrocycles have opened access to promising supramolecular systems and are discussed in recent monographs on cyclophanes [8]. We prefer to call them resorcinarenes and not calixarenes for two reasons: first, the name calixarenes

---
* This paper is dedicated to the commemorative issue on the 50th anniversary of calixarenes.

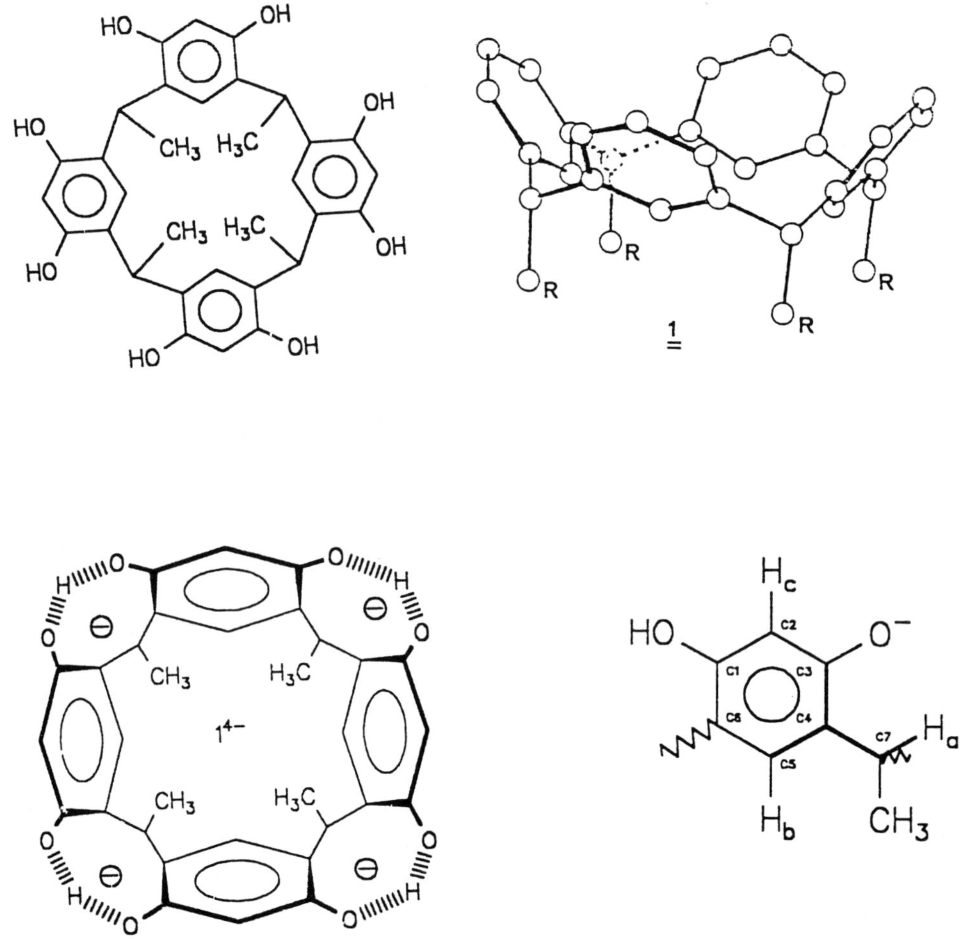

Scheme 1. The structure of the **rccc** product **1**.

was given to the cyclophanes obtained from 4-alkylphenol condensations; second, the cyclic polyphenol systems from resorcinol adopt a vase-like structure only with *one* of the four isomers, and even then only under special conditions (see below). Cram et al. have used resorcinarenes as a building block for their cavitands and carcerands [9], which are based on covalent links between the neighbouring phenolic groups of **1** [10]. Aoyama and collaborators have developed efficient sugar binding hosts by condensation of resorcinol with long chain alkanals leading to chloroform-soluble receptors (e.g. **1**, R=$(CH_2)_{10}CH_3$) with 8 well ordered phenolic groups for hydrogen bonding with carbohydrates [11]. Other modifications of the basic resorcin[4]arene skeleton involve binding elements for metals and their complexes [12] and the formation of liquid crystals [13].

The use of the basic resorcin[4]arene skeleton itself as host compound for positively charged substrates was initiated in 1985 in Saarbrücken [14]. The devel-

THE HOST-GUEST CHEMISTRY OF RESORCINARENES 69

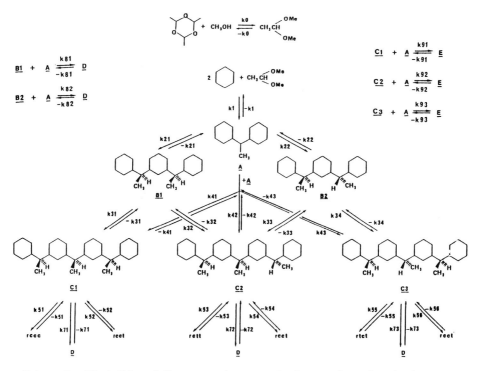

Scheme 2. The buildup of oligomers and macrocycles (see text for explanations).

opment of these systems and of those obtained by substitution both at the top of the macrocycle (at position C-2) and at the bottom (at C-7) will be the focus of the present review. This also reflects our primary interest in the physical characterization of supramolecular complexes in solution, both with respect to the binding mechanisms and energies involved, as well as to the conformations [15]. NMR spectroscopy provides the most important method for this. We also want to point out the possibilities of developing receptors with polytopic and allosteric binding sites from these resorcinarenes.

2. The Mechanisms of Macrocyclization

Before we discuss the supramolecular features of the resorcin[4]arene systems we address the questions of *why and how* the macrocycle containing four resorcinol units forms in such high yields without high dilution and without template effects; these are known to play an important role in calixarene synthesis based on the reaction of 4-alkylphenoles with formaldehyde. Baeyer must have wondered about this, since he isolated crystalline – although at that time unidentified – products only from resorcinol [2], in contrast to the alkylphenol reactions. The conditions which can make such macrocycles available on a large scale, with potential industrial uses [16], are not only of great practical importance, but also fascinating enough

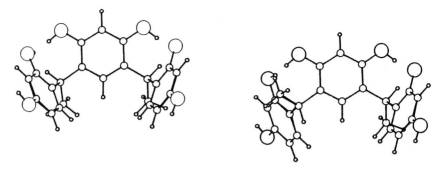

Scheme 3. An illustration of the most stable "folded" (left) and less stable "unfolded" (right) conformations with trimers. The folded structure shows the terminal phenols oriented towards the front side; in the unfolded one phenol is oriented backwards, causing an unfavourable interaction between the hydroxy and methyl groups as well as the absence of a hydrogen bond.

for a mechanistic study. Weinelt, during a stay in Saarbrücken, was in fact able to rationalize on a quantitative basis the peculiarities of such a macrocyclization [17].

The buildup sequence of the macrocycles from acetaldehyde (used as paraldehyde) and resorcinol leads to one dimer **A** and from there to two diastereomeric trimers **B**, further to three tetramers **C**, and then either by ring closure to the four possible stereoisomeric cyclophanes (**rccc,rcct, rctt, rtct**) or further on to pentamers **D**, etc., with more than 50 different rate constants (Scheme 2). Seven of these structures could be identified and their interconversion kinetics followed by NMR; a computerized fitting procedure by numerical integration enabled some rate constants to be obtained and to define the essential conditions for the effective ring closure: (a) under the reaction conditions the higher oligomers **D** etc. degrade back to the tetramers relatively quickly, (b) the tetramers react even faster to form rings, which, (c) open slowly compared to the chain propagation steps. In line with factor (a), there are more polymers under non-homogeneous reaction conditions, e.g. in water instead of methanol as solvent, as the oligomers become less soluble. The reason for the fast cyclization – even in relatively concentrated solutions – became obvious by the conformational preorganization of the open chain precursors: molecular mechanics calculations indicate that they favour folded structures (Scheme 3) in which the terminal phenyl substituents approach each other. There are two factors responsible for this: in the nonfolded conformers hydrogen bonds between phenolic groups are impossible, *and* there is a repulsive 1,5-interaction between the OH and CH_3-substituents which is relieved in the folded form.

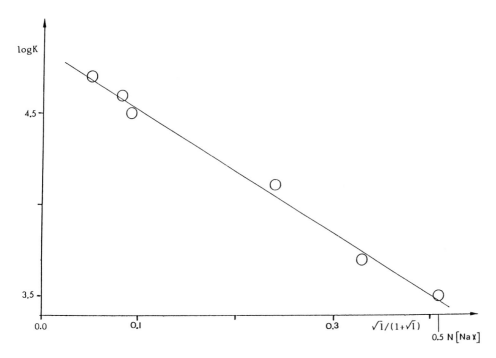

Fig. 1. Equilibrium constants (log K) for **1** + Et$_4$NBr as function of ionic strength [21].

3. The Basic Skeleton: Conformations and Complexation; An Element of an Allosteric Proton Pump; Choline Binding and Acetlycholine Hydrolysis

In basic solutions **1** forms a stable, calixarene-like "cone" conformation which is still one of the strongest synthetic receptors for choline-type substrates [14, 19], surpassing the natural receptor [19c]. The cavity is quite shallow and the binding free energy ΔG_{cplx} is to 80% due to the formation of four salt bridges; a comparison with a large number of other ion pairs in water [20] showed that this leads to a Coulomb-stabilization of $4 \times 5 = 20$ kJ/mol at low salt concentrations. A comparison with the binding of electroneutral substrates of similar shape such as *tert*-butanol demonstrate that other interactions contribute only about 5 kJ/mol [14]. The ΔG_{cplx} values correlate surprisingly well with Debye–Hückel coefficients of ionic strength (Figure 1) [21]. If the charges are separated by a distance r – as one can study using tetralkylammonium substrates R$_4$N$^+$ with different chain length – we see that ΔG_{cplx} decreases as a function of r^{-1} [18]. This is further experimental evidence for dominating Coulomb forces, supported by a realistic dielectric constant obtained from the slope (Figure 2).

A study of the temperature dependence of ΔG shows the binding of choline chloride to be dominated by ΔH (33 kJ/mol), with negligible entropy contributions [22]. The association and dissociation rate constants are, with k = 2.5×10^8

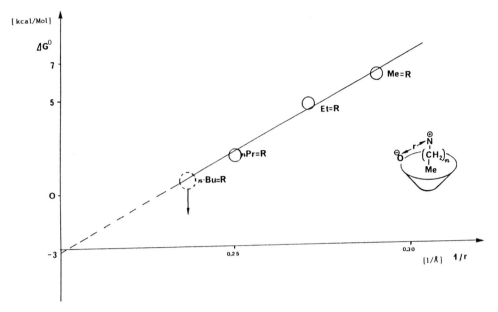

Fig. 2. The dependence of the association free energies between **1** and R$_4$NCl with R = Me to R = n-Bu and the resulting variation of distances r between the charges [18].

mol^{-1}s^{-1}, controlled by diffusion, and, with k = 5 × 10^3 s^{-1}, controlled by the complex stability, respectively, as found by dynamic NMR measurements [18]. *Solvent effects* [21] on the equilibria are characterized by an *increase* of ΔG_{cplx} with increasing water content. This effect, somewhat surprising for an ion pair, is related to the smaller desolvation energy per charge for large organic ions such as **1**$^{4-}$ compared to small ions such as bromide, and possibly to still important hydrophobic contributions. Nevertheless, the observed linear correlations with hydrophobicity parameters of binary alcohol-water mixtures exhibit the lowest slope or sensitivity of all host-guest equilibria studied so far, which is in line with particularly strong polar binding contributions here.

The acidities of the phenolic groups in the octol **1** (R = CH$_3$) reveal the reasons for the particular stabilization of the cone conformation as a consequence of a cyclic hydrogen bonded system (see Scheme 1) with four delocalized negative charges: four protons are dissociating at a pK value *two units below* resorcinol itself, whereas the remaining four cannot be abstracted even by sodium methoxide in methanol solution [18]. The picture changes if phenyl instead of methyl substituents are present (**1**, R = C$_6$H$_5$); here one observes slow decomposition in strongly alkaline solutions [22].

The **rctt** stereoisomer **2** (R = CH$_3$) shows, in neutral as well as in strongly basic solution, two sets of equally intense ^{1}H and ^{13}C NMR signals for the phenyl rings, demonstrating slowly interconverting C$_{2v}$ conformations **A** and **C** (Scheme 4). Abstraction of two protons from **A** leads to conformer **B**, which is retained in the

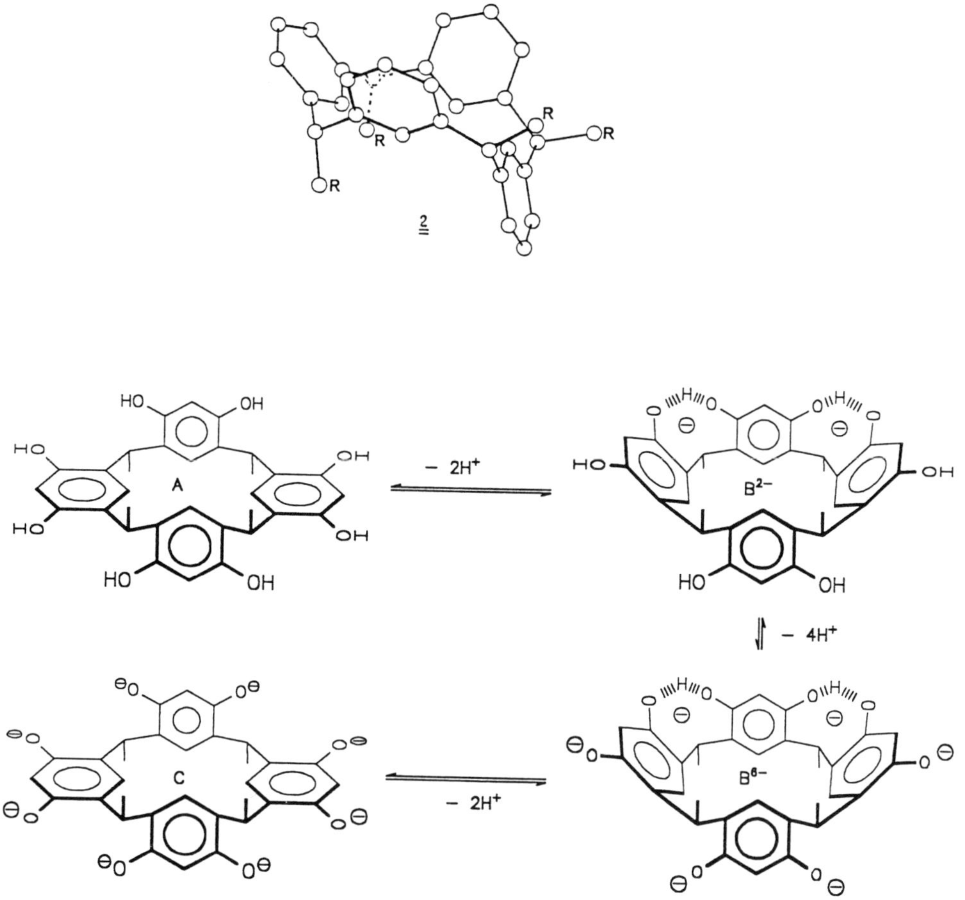

Scheme 4. The **rctt** isomer **2** and its interconversions.

hexaphenolate in more basic solution, before converting to the octaphenolate form **C** after adding more base. The corresponding interconversion from **C** to **B** occurs if instead of sodium hydroxide a methylammonium substrate such as choline is added.

The system represents a very simple allosteric element of a proton pump. The mechanistic origin for the strong conformational coupling between the uptake of protons and of positively charged substrates is the repulsion between the pseudo-axial methyl substituents and either -OH or O$^-$ groups of the two pseudoaxial phenylrings in **B**, which becomes acceptable only if either two hydrogen bonds are formed between the three axial phenol units, or if the system gains energy by the complexation of the $^+NR_4$ substrates. Therefore, this complexation leads to the uptake of protons and *vice versa*. Only the "half-cone" structure **B** can efficiently bind choline; on the other hand the thus enforced vicinity of the neighbouring phenolic groups is stabilized by proton dissociation leading to hydrogen bonds.

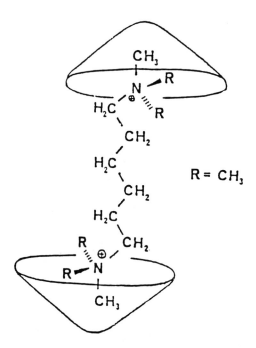

Scheme 5. An illustration of a ditopic complexation of 1,5-bis(trimethylammonium)pentane by the **rccc** isomer **1**.

In contrast to the **rccc** isomer **1**, the addition of sufficient base to the **rctt** isomer **2** leads to the abstraction of *all* eight protons, because again the unfavourable interaction of the axial methyl groups in **2** would be enhanced two times if the last phenyl ring in this "partial cone" conformer **B** turned upwards for the sake of additional hydrogen bonds.

The "cone" conformations for the tetraphenolates **1** were confirmed in solution for R = CH_3 and R = C_6H_5 by an analysis of the vicinal coupling between C-4 *or* C-5 and H-a. The value of (4 + 0.5) Hz for *both* couplings and *both* derivatives indicate angles which are similar and around 25° or 150°, in agreement with force field modelled geometries [22]. The tetraphenolate **1** binds not only ammonium compounds but also, e.g., phosphonium derivatives with the same high affinity of about 25 kJ/mol as long as there is at least one methyl group on the onium center.

Ditopic onium derivatives are, as expected, complexed twice at the two ends (Scheme 5), with negligible weakening of ΔG at the second binding site. The complexation of choline is so strong that significant retardation is observed for the hydrolysis of acetylcholine. The kinetically derived inhibition constant is close to the spectrocopically determined thermodynamic equilibrium value [23]. This rarely

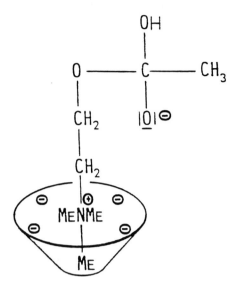

Scheme 6. A transition state model for the hydrolysis of acetylcholine; the normal stabilization by attraction between the N^+ and the O^- charges is diminished by the four negative charges of receptor 1^{4-} with a subsequent lower hydrolysis rate.

observed reaction retardation, not to be confused with the much more frequent inhibition of a catalyst, is based on the diminution of the activating effect of the $^+NMe_4$ charge on the ester group by the four-fold negative charge of the receptor 1^{4-} (Scheme 6).

4. Substitution at the Bottom: A Second, Lipophilic Binding Site

The condensation of resorcinol with "longer" aldehydes than acetaldehyde provides a convenient entry to the construction of narrow lipophilic binding centers at the bottom of the **rccc** isomer **1**. Benzaldehyde furnishes only a weak complexer even for small substrates like acetonitrile [24] or diethylether, and an intracavity-inclusion is not proven here. Phenyl rings directly attached at the bottom of **1** do not leave enough space, as is evident also from molecular models; even with the 4-phenyl-benzaldehyde product (R = biphenyl) we detect no measurable NMR shift changes upon the addition of e.g. benzylalcohol [22]. In contrast, more flexible bottom substituents R leave enough room (Scheme 7) and several lipophilic substrates are efficiently bound (Scheme 8).

As expected from the conformationally mobile sidechains and the predominantly lipophilic and/or hydrophobic binding mechanism there is relatively little selectivity, apart from an increase of ΔG with increasing lipophilicity of the substrate. Nevertheless, this is probably the *first* host system for which the inclusion of open chain substrates like diethylether has been proven by NMR shielding variations [22]. NMR measurements at host concentrations between 5×10^{-2} M^{-1} and

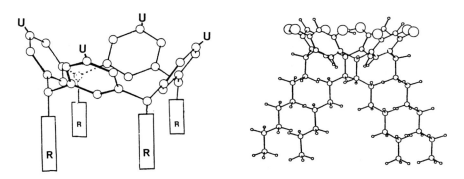

Scheme 7. The structures of the **rccc** isomer with substituents R at the bottom (QUANTA /CHARMm simulation for R = C_7H_{15}).

	R = →	$-(CH_2)_5CH_3$	$-(CH_2)_{11}CH_3$	$-CH_2-\phi$
$(CH_3CH_2)_2O$		9.9	10.1	9.7
$\phi-CH_2OH$		8.4	9.7	5.7
$\phi-(CH_2)_3OH$		15.7		9.7

Scheme 8. Complexation free energies ΔG [kJ mol^{-1}] for the bottom-substituted **rccc** isomers with different substituents R and different substrates.

8×10^{-5} M^{-1} showed no evidence for micelle formation. The upper binding site of host **1** is not involved in the complexation, as is evident from a separate experiment in which the binding of diethylether in the presence of tetramethylammonium salt was measured and found to be the *same* as without the occupied upper site. This shows that there is no cooperativity in these ditopic receptors.

5. Substitution at C-2: the Upper Rim

5.1. ALTERNATING TETRACARBOXYLATE CONFORMATIONS/SALT BRIDGES

The **rccc** isomer with four carboxylic acid functional groups at C-2 is unable to form a cone-like tetraphenolate **1** as all phenolic groups are involved in hydrogen bonding to the disssociated carboxylates. This is anologous to the behaviour of salicylic acid, and the pK value measured for **3** is indeed similar to that of salicylic acid. The NMR spectrum of **3**, which above pH 5 is present as a tetraanion, shows

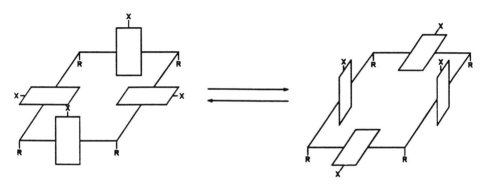

Scheme 9. The **rccc** isomer with four carboxylic substituents at C-2.

at, e.g. 270 K, two sets of equally intense phenyl protons, which coalesce at 293 K (400 MHz ^{1}H-NMR). This indicates an equilibrium of pseudorotating conformers with alternate equatorial and axial phenyl rings (Scheme 9), interconverting with a barrier of $\Delta G^* = 72$ kJ/mol. This value, measured in water, comes close to the pseudorotation barrier of $\Delta G = 60$ kJ/mol observed by Högberg [6a] for the butyrates of **1** in chloroform.

The tetrabenzoate host **3** can bind ammonium ions either by a double salt bridge from one onium center to two opposing axial benzoate units (with tetramethylammonium salts), or with α, ω-diammonium-alkanes, to the pseudoequatorial benzoate functionalities with one salt bridge at each end. A maximum ΔG is found if the alkane chain length matches approximately the distance between the two eq. COO$^-$ groups (Scheme 10). However, whereas the complexation with the small cation $^+$NMe$_4$ corresponds exactly to the general [20] increment of 5 kJ/mol per salt bridge, the α, ω-alkyldications show additional stabilization (Scheme 10). Molecular modelling indicates that this may be the result of a contact between the

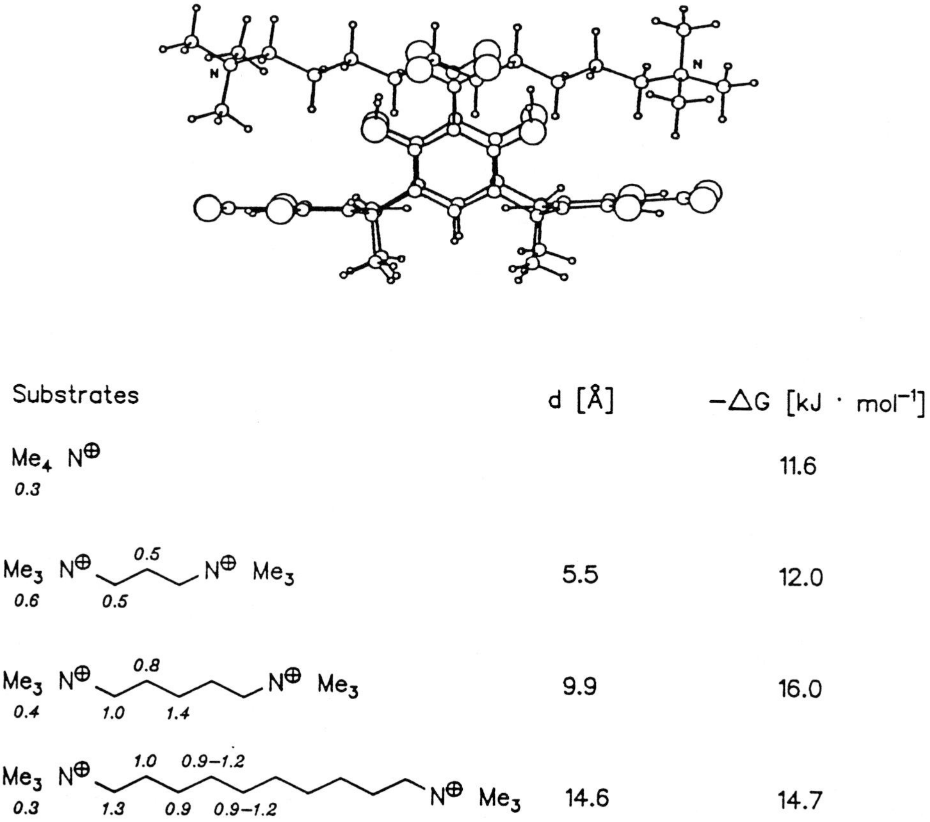

Substrates	d [Å]	$-\Delta G$ [kJ · mol^{-1}]
Me$_4$ N$^\oplus$ (0.3)		11.6
Me$_3$ N$^\oplus$ (0.6) —0.5— (0.5) N$^\oplus$ Me$_3$	5.5	12.0
Me$_3$ N$^\oplus$ (0.4) —0.8— 1.0 1.4 N$^\oplus$ Me$_3$	9.9	16.0
Me$_3$ N$^\oplus$ (0.3) 1.3 1.0 0.9 0.9–1.2 0.9–1.2 N$^\oplus$ Me$_3$	14.6	14.7

Scheme 10. QUANTA/CHARMm simulation of the complex **3** + Me$_3$N-(CH$_2$)$_{10}$NMe$_3$ and complexation free energies [kJ/mol] of **3** with substrates of different chainlength. Complexation induced NMR shifts (all upfield CIS, in [ppm]) are in *italics*.

pseudoaxial phenylrings and the alkane chain; the observed complexation-induced NMR shifts support this (Scheme 10).

5.2. CATIONIC SUBSTITUENTS AT C-2/AMINOACIDS AND METAL COMPLEXES

A large number of amino-methylsubstituents were introduced by the Mannich reaction which works as smoothly with the resorcinarene **1** as it had been reported for calixarenes [25]. From the eight aminoderivatives thus prepared and investigated [22] we restrict ourselves here to the L-proline compound **4** which was obtained without any racemization. The system can adopt four different protonation states (Scheme 11) which were all characterized. Computer-aided potentiometric titrations yielded the relevant pK values with an average error of +0.2 units, with excellent fits assuming the same value for the different independently treated phenyl units in the macrocycle. ^{1}H-NMR titrations allowed the different steps as shown in Scheme 12 to be identified. Due to the betaine formation the pK of the

THE HOST-GUEST CHEMISTRY OF RESORCINARENES

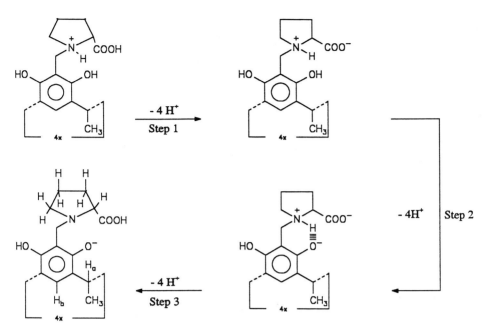

Scheme 11. The protonation steps for the proline-substituted macrocycle **4**.

pK$_S$ - Values

Dissociating Group	4	Resorcinol	Proline
COOH	1.8	--	1.9
OH	6.3	9.9	--
NH	13.0	--	10.8

Scheme 12. Comparison of pK values of **4**, resorcinol, and proline.

carboxylic group is as low as in proline itself; the basicity of the nitrogen, however, is even higher as a result of the extra-stabilization by the negative phenolate charge. Acidity increases by the presence of additional proton donors, such as for **1** or **4** in comparison to resorcinol alone, have been reported also for structurally related calixarenes [26].

Although the macrocycle **4** contains four chiral units in a well ordered arrangement, all attempts to discriminate between enantiomers of racemic substrates bearing suitable charges, such as D,L-carnitine, ephidrinium or 1-phenylethylammonium bromides failed: even at 400 MHz the ^{1}H-NMR spectra showed no line splitting of the quite stable (usually K>10^3 M^{-1}) complexes with **4** in water. This is consistent with the complexation free energies observed with many alkylamino-analogs to **4**, where the presence of additional charges such as in the piperazine derivative **5** showed little changes compared to the unsubstituted ring **1**. In general accordance

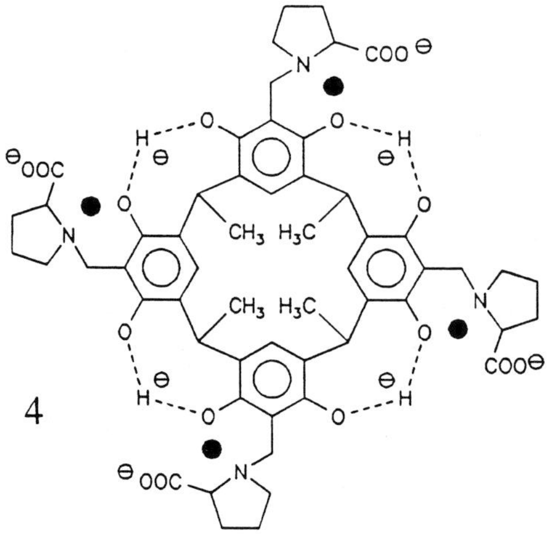

4

●: Cu^{2+}

pK values for Cu^{2+}-complexes:

with **4**:	15.0
with proline:	9.0

Scheme 13. Structure and binding pK values for the complex **4** with copper(II) ions.

with the observed NMR shifts the substituents point away from the cavity and the substrate binding center; in particular, if they can expose in this way an additional charge to the aqueous environment.

One of the attractive features of the aminoacid-substituted macrocycles is the possibility to provide strong and chiral binding elements for transition metal ions. Potentiometric titrations with **4** showed a stability increase from Zn^{2+} to Cu^{2+} to Fe^{3+}, similar to the parent aminoacid complexes [27]. However, the stabilities themselves are considerably increased, e.g. with Cu^{2+} by six units (Scheme 13). This is the result of the additional Coulomb attraction by the phenolate (half)-anions in the macrocycle, leading to four independent strong tridentate binding sites. Attempts to use transition metal ions implemented in the macrocycle for the catalysis of acetylcholine hydrolysis showed moderate success, consisting essentially in the removal of the inhibition discussed above for bound acetylcholine.

Finally, we discuss briefly the first attempts to use resorcinarenes as ligands for double-stranded DNA. The piperazinium-derivative **5** seems to be a particular attractive species in view of the extended array of positive charges for groove binding and the aryl units which may lend themselves to intercalation. Preliminary investigations (B. Palm, [28]) show only slight increases of viscosity and *no* NMR upfield shifts of any ligand signals, ruling out intercalative mechanisms. Both fluorescence binding assays and the melting point increases are similar to those observed with several azoniacyclophanes which also bear four ammonium groups in a macrocycle. The latter have been shown to fit into a regular affinity scheme of polyamine groove binders [29], characterized again by a salt bridge increment value of 4–6 kJ/mol.

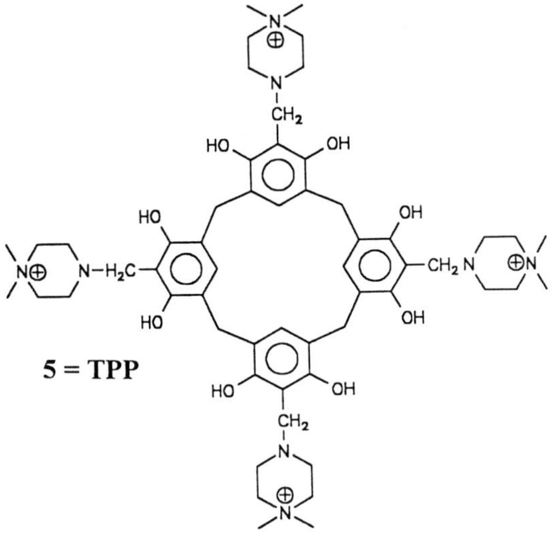

THE HOST-GUEST CHEMISTRY OF RESORCINARENES

5 = TPP

Experiments with Calf Thymus DNA; MWav $9*10^6$

A: Affinity (assay with Ethidiumbromide); $1/C_{50} = A$

ΔT_M: Melting point changes (with 10^{-6} M Ligand)

Ligand	$A*10^{-6}$	ΔT_M
TPPi	**0.8**	**1.6**
CP66	0.7	3.0
CP55	0.11	1.4
CP44	3.7	5.1
CP33	0.14	2.2
+N4-open chain-av.	0.5	5-10

No intercalation: a) NMR $\Delta \vartheta < 0.1$ ppm

b) Viscosity increase too small (L/Lo = 0.95 to 1.05 with intercalator ≥ 1.15)

Scheme 14. Binding of the resorcarene **5** to **DNA**.

References

1. Supramolecular Chemistry, Part **53**; part **52**: T. Ikeda, K. Yoshida, H.-J. Schneider: *J. Am. Chem. Soc.*, in print.
2. A. v. Baeyer: *Ber. Dtsch. Chem. Ges.* **5**, 280, 1094 (1872).

3. (a) A. Michael: *Ber. Dtsch. Chem. Ges.* **17** (Vol. 3), 20 (1884); (b) A. Michael and J.P. Ryder: *ibid.* **19**, 1388 (1886); (c) R. Möhlau and R. Koch: *ibid.* **27**, 2887 (1894); (d) C. Liebermann and S. Lindenbaum: *ibid.* **37**, 1171 (1904).
4. J.B. Niederl and H.J. Vogel: *J. Am. Chem. Soc.* **62**, 2512 (1940).
5. H. Erdtman, S. Högberg, S. Abrahamsson and B. Nilsson: *Tetrahedron Lett.* 1679 (1968); B. Nilsson: *Acta Chem. Scand.* **22**, 732 (1968).
6. (a) A.G.S. Högberg: *J. Am. Chem. Soc.* **102**, 6046 (1980); (b) *J. Org. Chem.* **45**, 4498 (1980).
7. G. Mann, F. Weinelt, and S. Hauptmann: *J. Phys. Org. Chem.* **2**, 531 (1989); for stereoisomers of condensation products from long chain alkanals, see: L. Abis, E. Dalcanale, A. DuVosel, S. Spera: *J. Org. Chem.* **53**, 5475 (1988).
8. (a) F. Diederich: *Cyclophanes* (Monographs in Supramolecular Chemistry) J.F. Stoddart (Ed.) (1991), Royal Soc. Chem., Cambridge; (b) F. Vögtle: *Cyclophan-Chemie*, Teubner, Stuttgart (1990).
9. D.J. Cram: *Angew. Chem.* **100**, 1041 (1988); *Angew. Chem., Int. Ed. Engl.* **27**, 1009 (1988); D.J. Cram, M.T. Blanda, K. Paek and C.B. Knobler: *J. Am. Chem. Soc.* **114**, 1765 (1992), and references cited therein.
10. For related substitution reactions at the phenolic groups, see (a) P. Soncini, S. Bonsignore, E. Dalcanale, and F. Ugozzoli: *J. Org. Chem.* **57**, 4608 (1992); (b) T. N. Sorrell and J. L. Richards: *Synlett.* 155 (1992); (c) P. Timmerman, M.G.A. van Mook, W. Verboom, G.J. van Hummel, S. Harkema, and D.N. Reinhoudt: *Tetrahedron Lett.* 3377 (1992); other resorcinol-based hosts: (d) O. Manabe, K. Asakura, T. Nishi, and S. Shinkai: *Chem. Lett.* **1219** (1990); (e) E. Weber, H.J. Koehler, K. Panneerselvam, and K. K. Chacko: *J. Chem. Soc., Perkin Trans.* 2, 1599 (1990); (f) E. Chapoteau, B. P Czech, A. Kumar, A. Pose, R. A. Bartsch, R. A. Holwerda, N.K. Dalley, B.E. Wilson, and W. Jiang: *J. Org. Chem.* **54**, 861 (1989).
11. (a) Y. Aoyama: *Adv. Supramol. Chem.* **2**, 65 (1992); (b) K. Kobayashi, Y. Asakawa, Y. Kikuchi, H. Toi and Y. Aoyama: *J. Am. Chem. Soc.* **115**, 2648 (1993), and references cited therein.
12. P.D. Beer and E.L. Tite: *J. Chem. Soc., Dalton Trans.* 2543 (1990).
13. G. Cometti, E. Dalcanale, A. DuVosel and A. M. Levelut: *Liquid Crystals* **11**, 93 (1992), and earlier references.
14. H. J. Schneider, D. Güttes and U. Schneider: *Angew. Chem.* **98**, 635 (1986), *Angew. Chem. Int. Ed. Engl.* **25**, 647 (1986).
15. (a) H. J. Schneider: *Angew. Chem.* **103**, 1419 (1991); *Angew. Chem., Int. Ed. Engl.* **30**, 1417 (1991); (b) H. J. Schneider, T. Blatter, U. Cuber, R. Juneja, T. Schiestel, U. Schneider, and P. Zimmermann in: *Frontiers in Supramolecular Chemistry and Photochemistry*, H. J. Schneider and H. Dürr, Eds; VCH, Weinheim, pp. 29–56 (1991); (c) H. J. Schneider: *Rec. Trav. Chim. Pays-Bas* **112**, 412 (1993).
16. *German Patent* DD 290 412 (29.5.1991); DDP 287, 158 (21.1.1991).
17. F. Weinelt and H. J. Schneider: *J. Org. Chem.* **56**, 5527 (1991).
18. H. J. Schneider, D. Güttes, U. Schneider: *J. Am. Chem. Soc.* **110**, 6449–6454 (1988).
19. For other potent choline receptors see: (a) D.A. Dougherty and D. A. Stauffer: *Science (Washington)* **250**, 1558 (1990); (b) R. Meric, J. P. Vigneron, and J. M. Lehn: *J. Chem. Soc., Chem. Commun.*, 129 (1993); and references cited therein. For the natural receptor, see (c) F.B. Hasan, S.G. Cohen, and J.B. Cohen: *Biol. Chem.* **255**, 3898 (1980).
20. (a) H. J. Schneider and I. Theis: *Angew. Chem.* **101**, 757 (1989) *Angew. Chem., Int. Ed. Engl.* **28**, 753 (1989); (b) H. J. Schneider, T. Schiestel, and P. Zimmermann: *J. Am. Chem. Soc.* **114**, 7698 (1992).
21. H. J. Schneider, R. Kramer, S. Simova and U. Schneider: *J. Am. Chem. Soc.* **110**, 6442 (1988); see also H. J. Schneider and I. Theis: *J. Org. Chem.* **57**, 3066 (1992).
22. (a) U. Schneider: Dissertation, Universität des Saarlandes 1992; (b) U. Schneider and H. J. Schneider: unpublished results; *Chem. Ber.* **127**, 2455 (1994).
23. H. J. Schneider and U. Schneider: *J. Org. Chem.* **52**, 1613 (1987).
24. J.A. Tucker, C.B. Knobler, K.N. Trueblood and D.J. Cram: *J. Am. Chem. Soc.* **111**, 3688 (1989).
25. See e.g., C.D. Gutsche and K.C. Nam: *J. Am. Chem. Soc.* **110**, 6153 (1988).
26. O. Manabe, K. Asakura, T. Nishi, and S. Shinkai: *Chem. Lett.* 1219 (1990); and earlier references.

27. A.E. Martell and R.M. Smith: *Critical Stability Constants*, Vol. 3, Plenum Press, New York (1974).
28. B. Palm: Diplomarbeit, Universität des Saarlandes (1993).
29. H. J. Schneider and T. Blatter: *Angew. Chem.* **104**, 1244 (1992); *Angew. Chem., Int. Ed. Engl.* **31**, 1207 (1992).

Synthesis and Properties of Pyridinocalixarenes[*]

FRANCESCO BOTTINO and SEBASTIANO PAPPALARDO[**]
Dipartimento di Scienze Chimiche, Università di Catania, Viale A. Doria 8, 95125 Catania, Italy

(Received in final form: 3 March 1994)

Abstract. Several facets of pyridinocalixarene chemistry have been investigated including reaction pathways for their formation from the base-catalyzed alkylation of the parent calixarenes with PicCl·HCl, effect of the nature and identity of the base on regio- and stereoselective O-alkylations, creation of molecular asymmetry in calix[4]arenes and enantiomeric resolution, conformation and conformational mobility, and complexation.

Key words: Regio- and stereoselective O-alkylations, conformation and conformational mobility, inherently chiral calix[4]arenes and enantiomeric HPLC resolution, N-oxide ligands and complexes.

1. Introduction

Readily available calixarenes are currently enjoying considerable interest in host–guest or supramolecular chemistry as three-dimensional building blocks for the design of new lipophilic receptors and carriers with specific properties [1, 2]. The architecture of this class of compounds is such that they simultaneously possess a potential hydrophobic cavity, suitable for the inclusion of small organic molecules, and a hydrophilic site (hydroxyl groups) very attractive for chemical modification. The chemistry of calix[4]arenes, i.e. the smallest members of this family, has been disclosed in the last few years, and general procedures have been developed for regio– [3–13] and stereoselective [14–19] functionalizations at the lower rim. The larger calix[n]arenes (n = 5–8) have received less attention, mainly because of a higher degree of functionality and flexibility which complicate their chemistry, and only very recently useful guidelines are emerging for the selective functionalization of these substrata [20–25].

Functionalization of calixarenes by the base-catalyzed O-alkylation with α-halomethyl-N-heterocyclic reagents has been recently introduced in order to extend the coordination chemistry of calixarenes to transition metals [7, 26, 27]. This review deals with the synthesis and properties of calixarenes endowed with pyridine pendant groups at the lower rim, derivable from p-*tert*-butylcalix[n]arenes (n = 4, 6, 8) by direct O-alkylation with 2-(chloromethyl)pyridine hydrochloride (PicCl·HCl) in the presence of base. The influence of the molar ratios between the reactants and the nature of the base applied on the product distribution and

---
[*] This paper is dedicated to the commemorative issue on the 50th anniversary of calixarenes.
[**] Author for correspondence.

conformational outcome is emphasized. Furthermore, the discovery of regioselective *syn-proximal* (1,2)-di-*O*-alkylation of calix[4]arenes [7–13] has opened new perspectives toward the synthesis of atropisomeric inherently chiral calix[4]arenes, owing to the nonplanar structure of these compounds. Most of the chiral derivatives here described can be optically resolved into their antipods by enantioselective HPLC methods. The locked conformation of pyridinocalix[4]arenes and fluxional properties of the larger calixarene homologues is discussed. Further chemical transformations of the title compounds and their potentialities are briefly outlined.

2. Pyridinocalix[4]arenes

A schematic representation of the twelve pyridino homologues derivable from *p-tert*-butylcalix[4]arene **1** and possible reaction pathways for their formation are shown in Scheme 1. Of these compounds, only *anti*-(1,2) and *anti*-(1,3)-di-*O*-alkylated regioisomers **3c** and **3d** have not been reported yet. The remainder have been obtained either by direct *O*-alkylation of the parent calix[4]arene **1** or by alkylation of an appropriate precursor. The mechanism of alkylation of calix[4]arene **1** with PicCl·HCl is based on product analysis of stepwise alkylation experiments, and MM2 calculations of the involved intermediates and their anions [18].

2.1. EXHAUSTIVE *O*-ALKYLATION AND CONFORMER DISTRIBUTION

Tetrakis[(2-pyridylmethyl)oxy]calix[4]arene conformers **5a–d** are obtained by subjecting the parent calix[4]arene **1** to a large excess of PicCl·HCl in anhydrous dimethylformamide (DMF) in the presence of a base. The conformational outcome of the reaction strongly depends upon the identity and strength of the base applied. The reaction with NaH is stereoselective and produces only the cone conformer **5a** (80%) [7, 18]. By using weaker bases, such as alkali metal carbonates, mixtures of conformers are obtained, and with Cs_2CO_3 it is possible to isolate after careful chromatography the four possible cone **5a** (9%), partial cone **5b** (54%), 1,2-alternate **5c** (1–2%) and 1,3-alternate **5d** (18%) conformers [18, 28].

Stereochemical assignments follow from distinctive proton and carbon NMR spectral patterns arising from each conformation. Selected ^{1}H-NMR regions of the four extreme conformations of tetra-substituted calix[4]arenes are illustrated in Figure 1, and are in agreement with the spectral patterns reported by Gutsche for calix[4]arene conformers [29]. The ^{13}C-NMR resonances of both $ArCH_2Ar$ and OCH_2Py groups provide a further diagnostic tool for distinguishing among the various conformers. Consistently with the single rule proposed by de Mendoza *et al.* for the determination of the conformation of calix[4]arenes [30], the signals of the methylene groups connecting two adjacent phenyl moieties in a *syn* orientation (e.g., in the cone conformation) appear at 30.7 ± 0.8 ppm, and those of the OCH_2Py groups linked to them at 76.4 ± 2.2 ppm; on the other hand, they show up at

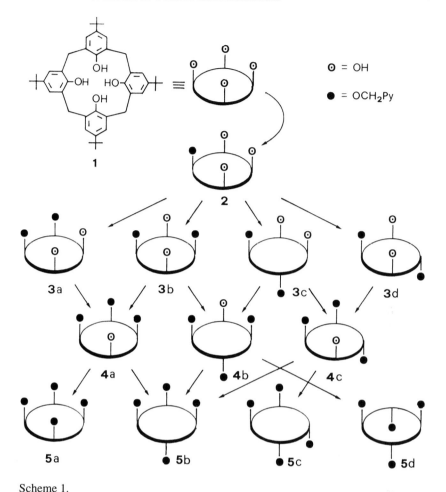

Scheme 1.

38.2 ± 1.0 ppm and 70.8 ± 1.8 ppm, respectively, when both phenyl moieties are *anti* oriented (e.g., in the 1,3-alternate conformation) [18].

Cone and partial cone structures **5a** and **5b** have been further confirmed by single-crystal X-ray analyses [18]. In both compounds a methanol of solvation is hydrogen bonded to one pyridine N atom and is *exo* to the calix cavity. In the partial cone conformer **5b** the conformation adopted is such that the pendant OCH$_2$Py group of the 'inverted' aryl ring lies in, and effectively fills, the calix cavity produced by the remaining three aryl rings. Self-inclusion phenomena (i.e., inclusion of flexible substituents within the calix cavity) are not uncommon in calixarene chemistry, especially in the larger analogues [31]. Studies in solution (1D and 2D NMR) and *in vacuo* (MM2 calculations) are in agreement with the conformation found in the solid state [18].

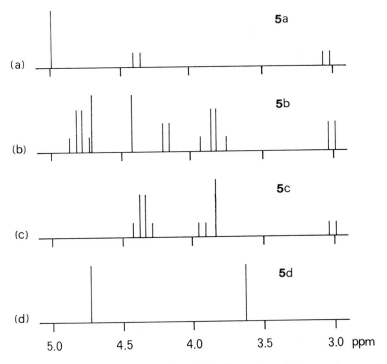

Fig. 1. Schematic representation of the observed methylene and oxymethylene ^{1}H-NMR patterns for tetrakis[2-pyridylmethyl)oxy]calix[4]arene cone (a), partial cone (b), 1,2-alternate (c), and 1,3-alternate (d) conformers **5a–d**.

2.2. PARTIAL O-ALKYLATION. REGIO- AND STEREOSELECTIVITIES

Increased calix[4]arene/PicCl·HCl molar ratios and diminished reaction times give mixtures of products representing various stages of alkylation; nevertheless, under appropriate reaction conditions (solvent, base, temperature, reaction time) selective functionalization can be realized. Accordingly, monoalkylated derivative **2** is generated in 59% yield by reaction of **1** with PicCl·HCl in toluene at 70°C for 20 h in the presence of NaH [32]. When alkylation of **1** is conducted with 2.2 equiv. of PicCl·HCl in dry DMF in the presence of NaH (10 equiv.), *syn-proximal* di[(2-pyridylmethyl)oxy]calix[4]arene regioisomer **3a** is obtained in up to 85–90% yield [7, 33]. Conversely, substitution of alkali metal carbonates (Na$_2$CO$_3$ or K$_2$CO$_3$) for NaH affords regioselectively the *syn-distal* (1,3)-regioisomer **3b** in 50–60% yield, even in the presence of an excess of electrophile [18, 26]. Similarly, alkylation of de-*tert*-butylated calix[4]arene with 2-bromomethyl-6-hydroxymethylpyridine (2 equiv.) and K$_2$CO$_3$ (1 equiv.) in refluxing acetonitrile gives the *syn-distal* di-O-alkylated derivative as an intermediate in the synthesis of a calix[4]arene cryptand [27].

The structure of **3a** has been elucidated by single-crystal X-ray analysis. Compound **3a** adopts a relatively distorted cone conformation in the solid state, creating a hydrophobic cavity which is large enough to accommodate an ethanol molecule with the hydroxyl end of the molecule directed out of the calix cavity [34].

Although theoretically four different disubstituted regioisomers **3a–d** can exist (Scheme 1), direct substitution on the parent calix[4]arene **1** in the presence of the above bases regioselectively affords **3a** or **3b**, with no trace of the other two possible *anti* regioisomers. However, the isolation of tetra-*O*-alkylated 1,2-alternate conformer **5c** in the Cs_2CO_3-catalyzed exhaustive alkylation of **1** suggests that the obvious *anti*-(1,2)- or *anti*-(1,3)-di-*O*-alkylated precursors are transient species under the experimental conditions employed [35].

In order to gain an insight into the origin of conformational isomers in the base-catalyzed exhaustive alkylation of **1** with PicCl·HCl, stepwise alkylation reactions with regioisomers **3a** and **3b** have been carried out in DMF by varying the electrophile molar ratio, reaction time and base (alkali metal carbonates) [18]. Alkylation of **3a** with one equivalent of PicCl·HCl affords stereoselectively tri-*O*-alkylated cone conformer **4a** in high yield, irrespective of the nature of the base used. *Syn-distal* regioisomer **3b** was proved to be less reactive under analogous reaction conditions, and alkylation with Cs_2CO_3 gave tri-*O*-alkylated partial cone conformer **4c** (34%, based on unreacted **1**) as the main product, along with minor amounts of cone **4a** (11%). Alkylation of each of the two di-*O*-alkylated regioisomers with an excess of electrophile and base has shown that in the final substitution step the template effect of the base plays an important role in determining the conformational outcome of the reaction: the conformation of the tri-*O*-alkylated precursor(s) (cone and/or partial cone) is completely retained with Na^+ cation in the base, mainly retained with K^+, and completely inverted (cone to partial cone) and/or partial cone to 1,3-alternate) with Cs^+. Since *syn*-1,2 and *syn*-1,3-di-*O*-alkylated intermediates **3a, b** can be generated *in situ* with excellent regioselectivity during the one-pot exhaustive alkylation of the parent calix[4]arene **1** with excess PicCl·HCl, the reaction can be easily driven to the desired conformer(s) by a proper choice of the base [18].

2.3. BIS(SYN-PROXIMALLY) AND BIS(SYN-DISTALLY) FUNCTIONALIZED CALIX[4]ARENES

Di-*O*-alkylated regioisomers are a potential source of calix[4]arenes with mixed ligating groups at the lower rim. Reinhoudt has reported that functionalization of the free hydroxyl groups of *syn-proximal* di-*O*-alkylated calix[4]arenes affords a variety of cone bis(*syn-proximally*) functionalized calix[4]arenes [13]. Cation complexation studies on these derivatives have shown that subtle changes in regioselective functionalization influences the selectivity for Na^+ considerably.

Alkylation of (1,2)-di[(2-pyridylmethyl)oxy]calix[4]arenes in THF in the presence of NaH produces the *achiral* bis(*syn-proximally*) functionalized calix[4]arenes (cone conformers) in good yield [18]. On the other hand, when alkylation of **3a** is conducted in DMF in the presence of Cs_2CO_3, *chiral* partial cone structures are

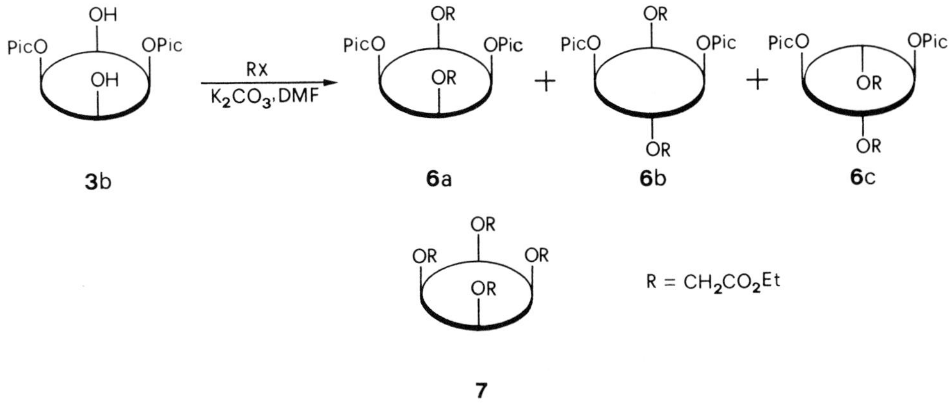

Scheme 2.

formed preferentially, owing to the absence of symmetry elements [33, 36]. This class of compounds is discussed in detail in the next section.

The alkylation of *syn-distal* regioisomer **3b** with ethyl bromoacetate and K_2CO_3 in DMF produces a mixture of bis (*syn-distally*) tetra-*O*-alkylated cone **6a** (32%), partial cone **6b** (7.7%) and 1,3-alternate **6c** (1.1%) conformers [37, 38], as shown in Scheme 2. Solvent extraction data have indicated that cone **6a** shows a strong metal affinity, comparable to that of cone tetrakis[(ethoxycarbonylmethyl)oxy]calix[4]-arene **7**, and binds not only Na^+ but Li^+, partial cone **6b** shows a poor metal affinity (enhanced K^+ selectivity but lower Ex% values), while 1,3-alternate **6c** has the highest Ex% for K^+ among the three conformational isomers. Thus, metal recognition with calix[4]arene ionophores can be exploited not only on the basis of the nature of binding sites and ring size [39], but also on the basis of conformational changes [37, 38].

2.4. MOLECULAR ASYMMETRY AND ENANTIOSELECTIVE HPLC RESOLUTION

Although chiral calix[4]arenes can be generated by simply attaching chiral residues at the upper [40] or lower [41–43] rim of the calixarene skeleton, recent interest has been focused on the possibility of synthesizing 'inherently' chiral calix[4]arenes, which are build up of nonchiral subunits and consequently owe their chirality to the fact that the calixarene molecule is nonplanar. Molecular asymmetry can arise from the substitution pattern at the lower rim and/or conformation. In this respect, Shinkai has recently reported a systematic classification of all possible chiral isomers derivable from calix[4]arene, and delineated some basic concepts for the design and synthesis of chiral derivatives [44].

Two strategies have been used for the preparation of inherently chiral calix[4]arenes: the fragment condensation (convergent stepwise synthesis of asymmetric calix[4]arenes from appropriate linear precursors) [45–52] and the lower-rim functionalization of a preformed calix[4]arene [32, 33, 36, 44]. The second strategy

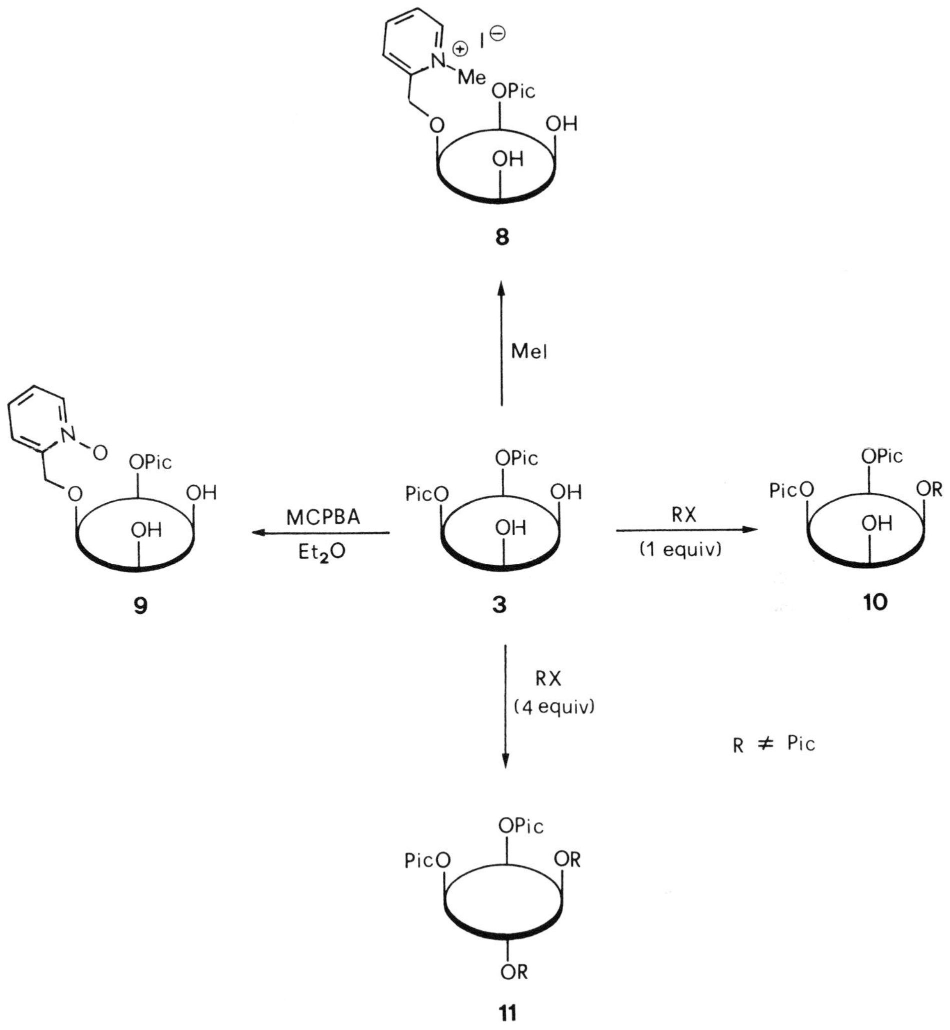

Scheme 3.

is more attractive for practical reasons, and is based on the regio- and stereoselective functionalization of conventional calix[4]arenes at the lower rim.

Syn-proximal di[(2-pyridylmethyl)oxy]calix[4]arene **3a** has proved to be a useful achiral intermediate for the production of inherently chiral derivatives, and some chemical modifications are shown in Scheme 3. Dissolution of **3a** in MeI at room temperature produces the chiral *N*-methylpyridinium derivative **8** in a nearly quantitative yield [18], while MCPBA oxidation in diethyl ether has led to the isolation of the chiral mono-*N*-oxide **9** [53]. The reluctance of **8** to undergo further *N*-alkylation suggests the hypothesis that the lone pair of the residual ring nitrogen is

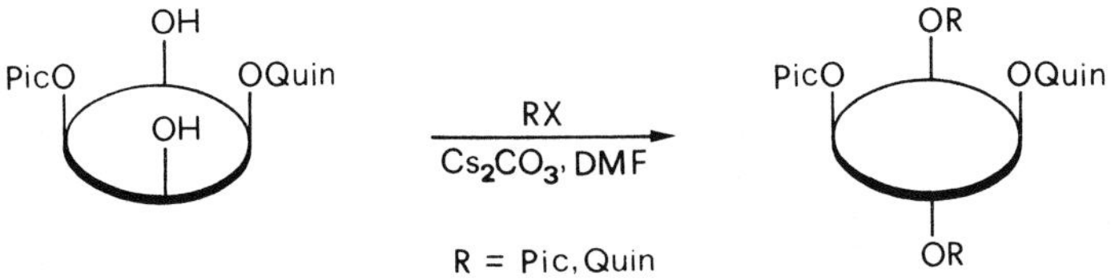

Scheme 4.

directly involved in a sort of 'self-complex' structure, with the N-methylpyridinium cation surrounded by oxygen and nitrogen donor atoms [18].

When **3a** is treated with an electrophile RX (1 equiv.) in anhydrous DMF in the presence of Cs_2CO_3 (1 equiv.) at 60°C for a few hours, racemic *cone* calix[4]arenes **10**, endowed with mixed ligating functionalities in the sequence $A^\alpha A^\alpha B^\alpha C^\alpha$ [54] at the lower rim are obtained in 42–93% yield with excellent stereoselectivity [33]. The conformational outcome of this reaction is not affected by metal template effects, but rather is determined by the strong hydrogen bonding stabilization of the phenolate intermediate in the cone conformation. The reaction appears to be general, as demonstrated by the wide variety of binding functionalities (including alkenic, alcoholic, ether, amino, ester, amide, keto, aromatic and N-heteroaromatic groups) which can be easily introduced at the lower rim via ether formation [33]. The first chiral calix[4]arene possessing the $A^\alpha A^\alpha B^\alpha C^\alpha$ pattern of substituents at the lower rim was obtained by Shinkai by using a different strategy, i.e. a BaO/Ba(OH)$_2$ assisted bis-O-propylation of monopyridinocalix[4]arene **2** [32].

Conversely, alkylation of **3a** with an excess of electrophile RX under similar conditions gave the chiral tetra-O-alkylated *partial cone* derivatives **11**, possessing an $A^\alpha A^\alpha B^\alpha B^\beta$ sequence of substituents at the lower rim [33, 36]. In a similar way, exhaustive alkylation of mixed *syn-distal* **12**, generated from **2** by treatment with 2-chloromethylquinoline hydrochloride (QuinCl·HCl), with either PicCl·HCl or QuinCl·HCl furnished the corresponding chiral partial cone derivatives **13**, having the $A^\alpha A^\alpha A^\beta B^\alpha$ pattern of substituents at the lower rim [33], as shown in Scheme 4.

Apart from NMR spectral patterns showing molecular asymmetry, evidence of chirality for cone **10** and partial cone structures **11** and **13** was provided by the addition of Pirkle's reagent (S)-(+)-(9-anthryl)-2,2,2-trifluoroethanol to a chloroform solution of each calixarene, which caused doubling of (in principle) all signals.

The direct HPLC separation of most chiral tri-O-alkylated calix[4]arenes **10** has been achieved using Chiralcel OD phase, while it was ineffective for partial cone products **11** and **13** [55]. Nevertheless, compound **11** (R = CH$_2$Quin) could be separated into its entantiomers by using a Chiralpak OP(+) HPLC column [36]. Sufficient amounts of each pair of enantiomers from racemic **10** (R = benzyl) and **11** (R = CH$_2$Quin) could be obtained to measure their CD spectra. These are almost mirror images of each other, indicating that the eluates from the two chromatographic peaks are optical isomers.

Among the various factors influencing enantioselection, hydrogen bonding between the residual hydroxyl group of tri-*O*-alkylated compounds and the Chiralcel OD phase seems to play an important role. As a matter of fact, the separation factor of compound **10** (R = benzyl) (α = 3.45 under optimum conditions) dramatically drops to 1.24 when the OH group is replaced by a propoxy group [55].

2.5. *N*-OXIDE DERIVATIVES

Regio- and conformational isomers of calix[4]arenes bearing pyridine-*N*-oxide pendant groups at the lower rim have been recently synthesized by MCPBA oxidation of appropriate pyridinyl precursors in order to provide different atropisomeric (cone, partial cone, 1,2- and 1,3-alternate) *N*-oxide ligands [53, 56].

Extraction data of alkali metals with the picrate method have shown that pyridinyl compounds are better ionophores than their *N*-oxide counterparts (deleterious role of hydrogen bonding between *N*-oxide functionalities and water molecules). The highest phase-transfer values are observed for cone conformers, where selectivity follows the order $Na^+ > K^+ > Rb^+ > Cs^+ > Li^+$. In aprotic solvents *N*-oxide ligands are much stronger complexers, and alkali and lanthanide metal complexes have been prepared and characterized. The Eu(III) and Tb(III) complexes of tetra-*N*-oxide cone conformer are fluorescent upon UV light excitation at 312 nm. Unfortunately, these complexes totally lose their luminescence upon addition of water, indicating their modest stability in aqueous medium [56, 57].

3. Pyridinocalix[6]arenes

A schematic representation of the twelve pyridino homologues derivable from *p-tert*-butylcalix[6]arene **14** and possible reaction pathways are shown in Scheme 5. Of these compounds, only diether **16b** and tetraether **18b** have not been isolated yet. The remainder have been obtained either by direct alkylation of **14** or by stepwise synthesis [23, 26, 38, 58]. In contrast to pyridinocalix[4]arenes and calix[6]arenes bearing very bulky substituents at the lower rim [59], no conformational isomerism is observed for these compounds, indicating that the calix[6]arene annulus is large enough to allow the oxygen-through-the annulus rotation of the pyridylmethyloxy groups.

Alkylation of the parent calix[6]arene **14** with PicCl·HCl (0.5–30 equiv.) was investigated in anhydrous DMF at 60°C (and in one case in refluxing CH_3CN) in the presence of base [NaH, K_2CO_3, or BaO/Ba(OH)$_2$]. Exhaustive alkylation reactions are clean, while reactions with limiting amounts of electrophile produce complex reaction mixtures, which can be eventually separated into the puric components by extensive chromatographic means.

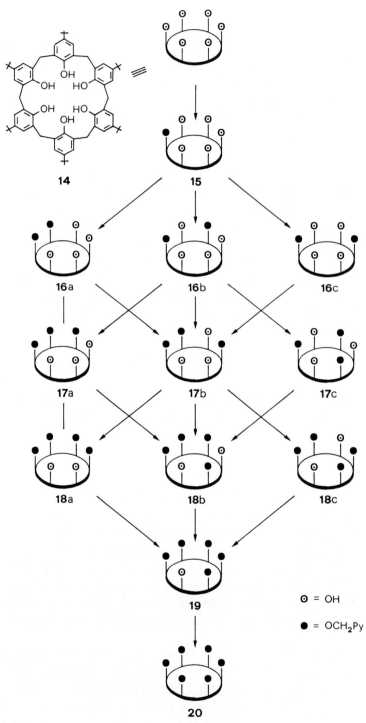

Scheme 5.

3.1. SYNTHESES

Hexaether **20** is obtained in 81% yield by subjecting **14** to a large excess of PicCl·HCl in the presence of K_2CO_3 [23, 26]. The use of NaH instead of K_2CO_3 affords regioselectively 1,2,4,5-tetraether **18c** (61%), along with pentaether **19** (5%) and hexaether **20** (8%) [23, 38, 58]. The selective formation of **18c** parallels the results reported by Gutsche for the NaH-catalyzed aroylation and arylmethylation of **14** [21, 22].

Conditions for partial alkylation have been found by tuning the amount of electrophile and the identity of the base. Thus, monoether **15** is generated in 55% yield by reacting **14** with PicCl·HCl (0.5 equiv.) and K_2CO_3. Reactions of **14** with increasing amounts of PicCl·HCl (1–4 equiv.) and K_2CO_3 in DMF produce invariably complex mixtures with 1,2,3-triether **17a** as the major product (up to 51%), along with variable amounts of diethers **16a** and **16c** and 1,2,4-triether **17b** as by products. When alkylation of **14** is carried out in CH_3CN, the product distribution is similar to that described in DMF, but in addition a small amount of 1,3,5-triether **17c** could be isolated, suggesting that the obvious precursor, i.e. 1,3-diether **16b**, is a transient species under the experimental conditions employed. The 1,2,3,4-tetraether **18a**, scantly present in the K_2CO_3-catalyzed alkylation products, can be synthesized as the sole product (35%) by treating 1,2,3-triether **17a** with PicCl·HCl (1 equiv.) in the presence of NaH [23].

Alkylation of **14** with PicCl·HCl (2 equiv.) in the presence of $BaO/Ba(OH)_2$ affords regioselectivity the barium complex of 1,4-diether **16a** in 80% yield [23]. The formation of a labile K^+ complex has been hypothesized by Gutsche to account for the selective 1,4-di-*O*-alkylation of **14** when using Me_3SiOK, Me_3COK, or KH as the base [22].

3.2. FLUXIONAL PROPERTIES

p-tert-Butylcalix[6]arene **14** is a conformationally flexible molecule, as deduced from the broad singlet for the $ArCH_2Ar$ protons in the ^{1}H-NMR spectrum at room temperature, and dynamic measurements have demonstrated a facile interconversion among various conformers with a coalescence temperature (T_c) of 11°C and an inversion barrier of 13.3 kcal/mol in $CDCl_3$ [60]. Conversion of **14** to monoether **15** reduces the conformational mobility, resulting in a set of three pairs of doublets at room temperature for the CH_2 protons, which remain invariant up to 345 K in $CDCl_3$, and coalesce at temperatures higher than 360 K, with an estimated $\Delta G > 18$ kcal/mol. If the rule found for pyridinocalix[4]arenes can be extended to the larger calix[6]arene homologues, a signal at 77.91 ppm for the oxymethylene carbon of **15** may be suggestive for a cone conformation [23].

The methylene region of 1,2-diether **16a** is characterized by a well-resolved 20-line pattern (five pairs of doublets in the ratio 2 : 1 : 2 : 2 : 1) coalescing at 360 K, while that of the 1,4- regioisomer **16c** displays a 5-line pattern (a pair of broad doublets and a singlet in the ratio 2 : 1) with a T_c of 320 K. In 1,2,3-triether

17a the bridging methylenes show up as a 9-line pattern (two pairs of doublets and a singlet in the ratio 1 : 1 : 1) in CD_2Cl_2, and as three broad singlets of equal intensity in DMSO-d_6 at 355 K. On the other hand, tri-O-alkylated regioisomers **17b** and **17c**, as well as their higher homologues **18–20** display broadened ^{1}H-NMR spectra which sharpen at higher temperatures. This behaviour suggests that T_c for pyridinocalix[6]arenes decreases by increasing progressively the degree of substitution at the lower rim; besides, within each series of partially alkylated regioisomers, T_c increases with increasing number of adjacent unalkylated phenol units (e.g. in tri-O-alkylated compounds $T_{c1,2,3} > T_{c1,2,4} > T_{c,1,3,5}$), suggesting that hydrogen bonding among hydroxyl groups plays a major role in raising the energy barrier to conformational inversion [23].

The conformational behaviour of 1,2,4,5-tetraether **18c** has been investigated in detail by different techniques that include dynamic NMR spectroscopy, MM2 molecular mechanics calculations and X-ray analysis [58]. From VT-NMR analysis in the range 220–345 K, two coalescence temperatures at 227 ($\Delta G = 11.1$ kcal/mol) and 315 K ($\Delta G = 14.2$ kcal/mol) have been ascertained in $CDCl_3$. Theoretically, 14 different relative orientations of the aromatic rings with respect to the best plane containing the bridging methylenes are possible. The symmetry of NMR signals and NOESY data restricted the possible conformations of **18c** in solution, at temperature below 315 K, to those indicated as A and C in Figure 2. Whereas conformers of type C could be ruled out on the basis of MM2 results showing conformer A as the most stable, conclusive evidence in favour of the existence of conformer A in solution was provided by ^{1}H-NMR and ROESY spectra of **18c** in CD_2Cl_2 at 183 K. The conformation of the lowest energy conformer A can be described as 1,2,4,5-alternate with the OH-bearing phenyl rings in anti orientation with each other. In the solid state a similar conformation A1 is found for **18c**, the only difference being the *syn* orientation of OH···O intramolecular hydrogen bonds.

On the basis of the above results, the fluxional behaviour of **18c** can be summarized as shown in Figure 3. At temperatures below 220 K conformer A, having a C_2 axis of symmetry, is predominant and in slow exchange with A1, as illustrated in Figure 2. Above 220 K this equilibrium becomes fast on the NMR time scale, and averaged spectra are obtained consistent with two C_2 axes of symmetry. At temperatures below 315 K conformer D (the third one in the MM2 energy scale), begins to appear as a consequence of the new equilibrium between conformers A and D in the slow exchange rate (Figure 3). The equilibrium between A and D becomes fast, on the NMR time scale, at temperatures above 315 accounting for the second coalescence in the VT-NMR studies. Conformers of type D are believed to be pivotal intermediates in the conformational pathways leading to the other conformers [58].

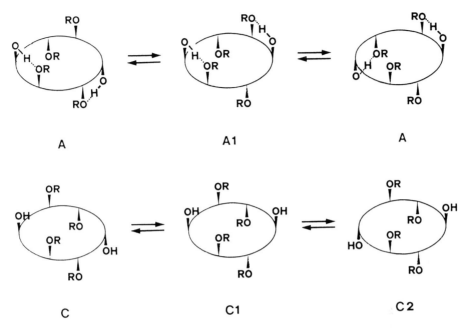

Fig. 2. Schematic representation of the two possible conformations compatible with the NMR spectral properties of **18c** in the temperature range 227–315 K. Reprinted with permission from *J. Am. Chem. Soc.* **114**, 7814 (1992). Copyright 1992 American Chemical Society.

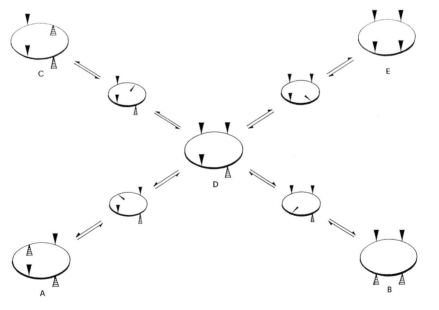

Fig. 3. Proposed mechanism for the conformational interconversions of **18c**. Conformational intermediates are represented by smaller circles where the thinner triangle indicates a pyridine or *tert*-butyl group inside the cavity. Reprinted with permission from *J. Am. Chem. Soc.* **114**, 7814 (1992). Copyright 1992 American Chemical Society.

4. Pyridinocalix[8]arenes

Due to the increased number of OH functionalities, 29 pyridino homologues of *p-tert*-butylcalix[8]arene **21** can exist: one mono-, hepta-, and octa-substituted derivatives, four di-, five tri-, eight tetra-, five penta- and four hexasubstituted regioisomers. The products of the partial alkylation of **21** with PicCl·HCl are hitherto unknown, while fully alkylated octakis[(2-pyridylmethyl)oxy]calix[8]arene **22** has been obtained in 24% yield by using conditions similar to those described for calix[6]arene **20**. From solvent extraction data, Shinkai has shown that compounds **20** and **22** display a high affinity for UO_2^{2+} at elevated temperature (100°C, *o*-dichlorobenzene) [26].

21 R = H
22 R = OCH$_2$Py

5. Concluding Remarks

The base-catalyzed alkylation of *p-tert*-butylcalix[*n*]arenes (*n* = 4, 6, 8) with PicCl·HCl has been investigated. Regio- and stereoselective *O*-alkylations have paved the way to atropisomeric inherently chiral calix[4]arenes endowed with mixed ligating functionalities at the lower rim. Tri-*O*-alkylated racemates (compounds **10**) can be optically resolved by enantioselective HPLC, but in order to pursue further studies (chiral recognition, enantioselection) larger quantities of the pure enantiomers are desirable. These may become available by converting **10** into diastereomers upon further alkylation of the residual OH group with suitable optically active derivatizing agents [44].

The properties of calixarene-based host molecules are strongly influenced by the conformation of the calixarene moiety. Pyridinocalix[4]arenes are conformationally inflexible molecules (the introduction of just one pyridylmethyl group at the lower rim suffices to suppress conformational changes), while the larger calix[6]arene homologues show fluxional properties.

Finally, calixarenes endowed with pyridine pendant groups at the lower rim are well suited for the design of receptors containing positively or negatively charged recognition sites by chemical transformations at the pyridine ring nitrogen, and recently a quaternised pyridinocalix[4]arene has been shown to possess molecular recognition for anionic guest species [61].

Acknowledgment

The authors wish to thank the Italian Ministry of Education for partial support of this work.

References and Notes

1. C. D. Gutsche: *Calixarenes*. Monographs in Supramolecular Chemistry, Vol. 1, J. F. Stoddard (Ed.), The Royal Society of Chemistry, Cambridge (1989)
2. J. Vicens and V. Böhmer (Eds.): *Calixarenes: A Versatile Class of Macrocyclic Compounds*, Kluwer Academic Publishers, Dordrecht (1991).
3. R. Ungaro, A. Pochini, and G. D. Andreetti: *J. Incl. Phenom.* **2**, 199 (1984).
4. P. J. Dijkstra, J. A. J. Brunink, K.-E. Bugge, D. N. Reinhoudt, S. Harkema, R. Ungaro, F. Ugozzoli, and E. Ghedini: *J. Am. Chem. Soc.* **111**, 7565 (1989).
5. J.-D. van Loon, A. Arduini, W. Verboom, R. Ungaro, G. J. van Hummel, S. Harkema, and D. N. Reinhoudt: *Tetrahedron Lett.* **30**, 2681 (1989).
6. J.-D. van Loon, A. Arduini, L. Coppi, W. Verboom, A. Pochini, R. Ungaro, S. Arkema, and D. N. Reinhoudt: *J. Org. Chem.* **55**, 5639 (1990).
7. F. Bottino, L. Giunta, and S. Pappalardo: *J. Org. Chem.* **54**, 5407 (1989).
8. J.-D. van Loon, D. Kraft, M. J. K. Ankoné, S. Arkema, K. Vogt, V. Böhmer, and D. N. Reinhoudt: *J. Org. Chem.* **55**, 5176 (1990).
9. A. Arduini, A. Casnati, L. Dodi, A. Pochini, and R. Ungaro: *J. Chem. Soc., Chem. Commun.* 1597 (1990).
10. L. C. Groenen, B. H. M. Ruel, A. Casnati, P. Timmerman, W. Verboom, S. Harkema, A. Pochini, R. Ungaro, and D. N. Reinhoudt: *Tetrahedron Lett.* **32**, 2675 (1991).
11. K. A. See, F. R. Fronczek, W. H. Watson, R. P. Kashyap, and C. D. Gutsche: *J. Org. Chem.* **56**, 7256 (1991).
12. K. Iwamoto, K. Araki, and S. Shinkai: *Tetrahedron* **47**, 4325 (1991).
13. J. A. J. Brunink, W. Verboom, J. F. J. Engbersen, S. Harkema, and D. N. Reinhoudt: *Recl. Trav. Chim. Pays-Bas* **111**, 511 (1992).
14. M. Iqbal, T. Mangiafico, and C. D. Gutsche: *Tetrahedron* **43**, 4917 (1987).
15. K. Iwamoto, K. Fujimoto, T. Matsuda, and S. Shinkai: *Tetrahedron Lett.* **31**, 7169 (1990).
16. C. D. Gutsche and P. A. Reddy: *J. Org. Chem.* **56**, 4783 (1991).
17. K. Iwamoto, K. Araki, and S. Shinkai: *J. Org. Chem.* **56**, 4955 (1991).
18. S. Pappalardo, L. Giunta, M. Foti, G. Ferguson, J. F. Gallagher, and B. Kaitner: *J. Org. Chem.* **57**, 2611 (1992).
19. K. Iwamoto and S. Shinkai: *J. Org. Chem.* **57**, 7066 (1992).
20. A. Casnati, P. Minari, A. Pochini, and R. Ungaro: *J. Chem. Soc., Chem. Commun.* 1413 (1991).
21. J. S. Rogers and C. D. Gutsche: *J. Org. Chem.* **57**, 3152 (1992).
22. S. Kanamathareddy and C. D. Gutsche: *J. Org. Chem.* **57**, 3160 (1992).
23. P. Neri and S. Pappalardo: *J. Org. Chem.* **58**, 1048 (1993).
24. R. G. Janssen, W. Verboom, S. Harkema, G. J. van Hummel, D. N. Reinhoudt, A. Pochini, R. Ungaro, P. Prados, and J. de Mendoza: *J. Chem. Soc., Chem. Commun.* 506 (1993).
25. P. Neri, C. Geraci, and M. Piattelli: *Tetrahedron Lett.* **34**, 3319 (1993).
26. S. Shinkai, T. Otsuka, K. Araki, and T. Matsuda: *Bull. Chem. Soc. Jpn.* **62**, 4055 (1989).
27. P. D. Beer, J. P. Martin, and M. G. B. Drew: *Tetrahedron* **48**, 9917 (1992).

28. Spectral data (CDCl$_3$) for **5c**: ^{1}H-NMR δ 1.13 [s, C(CH$_3$)$_3$, 36H], 3.01, 3.94 (ABq, J = 12.4 Hz, ArCH$_2$Ar, 4H), 3.84 (s, ArCH$_2$Ar, 4H), 4.31, 4.41 (ABq, J = 13.2 Hz, OCH$_2$Py, 8H), 6.05 (d, J = 7.8 Hz, 3-PyH, 4H), 6.91, 7.25 (ABq, J = 2.2 Hz, ArH, 8H), 6.92 (m, 5-PyH, 4H), 7.06 (td, J = 7.7, 1.7 Hz, 4-PyH, 4H), and 8.25 (d, J = 4.0 Hz, 6-PyH, 4H); ^{13}C-NMR δ 29.88, 39.21 (t, ArCH$_2$Ar), 31.36 [q, C(CH$_3$)$_3$], 33.97 [s, C(CH$_3$)$_3$], 74.14 (t, OCH$_2$Py), 121.49, 122.68 (d, 3,5-Py), 125.28, 125.71 (d, Ar), 132.65, 134.09 (s, bridgehead-C), 136.07 (d,4-Py), 145.12 [s, C$_{Ar}$—(CH$_3$)$_3$], 147.66 (d, 6-Py), 153.28 (s, C$_{Ar}$—OCH$_2$Py), and 157.65 (s, 2-Py).
29. C. D. Gutsche, B. Dhawan, J. A. Levine, K. H. No, and L. J. Bauer: *Tetrahedron* **39**, 409 (1983).
30. C. Jaime, J. de Mendoza, P. Prados, P. M. Nieto, and C. Sanchez: *J. Org. Chem.* **56**, 3372 (1991).
31. G. D. Andreetti and F. Ugozzoli: Ref. [2], pp. 87–123.
32. K. Iwamoto, A. Yanagi, T. Arimura, T. Matsuda, and S. Shinkai: *Chem. Lett.* 1901 (1990).
33. G. Ferguson, J. F. Gallagher, L. Giunta, P. Neri, S. Pappalardo, and M. Parisi: *J. Org. Chem.* **59**, 42 (1994).
34. G. Ferguson, J. F. Gallagher, and S. Pappalardo: *J. Incl. Phenom.* **14**, 349 (1992).
35. K. Iwamoto, K. Araki, and S. Shinkai: *J. Chem. Soc., Perkin Trans. 1* 1611 (1991).
36. S. Pappalardo, S. Caccamese, and L. Giunta: *Tetrahedron Lett.* **32**, 7747 (1991).
37. S. Shinkai, T. Otsuka, K. Fujimoto, and T. Matsuda: *Chem. Lett.* 835 (1990).
38. S. Shinkai, K. Fujimoto, T. Otsuka, and H. L. Ammon: *J. Org. Chem.* **57**, 1516 (1992).
39. M. J. Schwing-Weill, F. Arnaud-Neu, and M. A. McKervey: *J. Phys. Org. Chem.* **5**, 496 (1992).
40. C. D. Gutsche and K. C. Nam: *J. Am. Chem. Soc.* **110**, 6153 (1988).
41. R. Muthukrishnan and C. D. Gutsche: *J. Org. Chem.* **44**, 3962 (1979).
42. T. Arimura, S. Edamitsu, S. Shinkai, O. Manabe, T. Muramatsu, and M. Tashiro: *Chem. Lett.* 2269 (1987).
43. S. Shinkai, T. Arimura, H. Satoh, and O. Manabe: *J. Chem. Soc., Chem. Commun.* 1495 (1993).
44. K. Iwamoto, H. Shimizu, K. Araki, and S. Shinkai: *J. Am. Chem. Soc.* **115**, 3997 (1993).
45. V. Böhmer, L. Merkel, and U. Kunz: *J. Chem. Soc., Chem. Commun.* 896 (1987).
46. V. Böhmer, F. Marschollek, and L. Zetta: *J. Org. Chem.* **52**, 3200 (1987).
47. H. Casabianca, J. Royer, A. Satrallah, A. Taty-c, and J. Vicens: *Tetrahedron Lett.* **28**, 6595 (1987).
48. S. Shinkai, T. Arimura, H. Kawabata, H. Murakami, K. Araki, K. Iwamoto, and T. Matsuda: *J. Chem. Soc., Chem. Commun.* 1734 (1990).
49. S. Shinkai, T. Arimura, H. Kawabata, H. Murakami, and K. Iwamoto: *J. Chem. Soc., Perkin Trans. 1* 2429 (1991).
50. L. Zetta, A. Wolff, W. Vogt, K.-L. Platt, and V. Böhmer: *Tetrahedron* **47**, 1911 (1991).
51. A. Wolff, V. Böhmer, W. Vogt, F. Ugozolli, and G. D. Andreetti: *J. Org. Chem.* **55**, 5665 (1990).
52. G. D. Andreetti, V. Böhmer, J. G. Gordon, M. Tabatabai, F. Ugozzoli, W. Vogt, and A. Wolff: *J. Org. Chem.* **58**, 4023 (1993).
53. S. Pappalardo, L. Giunta, P. Neri, and C. Rocco: *Proceedings 2nd Workshop on Calixarenes and Related Compounds*, Kurume (Japan), 2–4 June 1993, OP-5.
54. Notations α and β have been introduced by Shinkai (see Ref. [12, 19, 44]) to define the stereochemistry of atropisomeric calix[4]arenes.
55. S. Caccamese and S. Pappalardo: *Chirality* **5**, 159 (1993).
56. S. Pappalardo, F. Bottino, L. Giunta, M. Pietraszkiewicz, and J. Karpiuk: *J. Incl. Phenom.* **10**, 387 (1991).
57. M. Pietraszkiewicz, J. Karpiuk, and A. K. Rout: *Pure Appl. Chem.* **65**, 563 (1993).
58. P. Neri, M. Foti, G. Ferguson, J. F. Gallagher, B. Kaitner, M. Pons, M. A. Molins, L. Giunta, and S. Pappalardo: *J. Am. Chem. Soc.* **114**, 7814 (1992).
59. P. Neri, C. Rocco, G. M. L. Consoli, and M. Piattelli: *J. Org. Chem.* **58**, 6535 (1993).
60. C. D. Gutsche and L. J. Bauer: *J. Am. Chem. Soc.* **107**, 6052 (1985).
61. P. D. Beer, C. A. P. Dickson, N. Fletcher, A. J. Goulden, A. Grieve, J. Hodacova, and T. Wear: *J. Chem. Soc., Chem. Commun.* 828 (1993).

Calixarenes with Nitrogen or Phosphorus Substituents on the Lower Rim[*]

D. M. ROUNDHILL[**], E. GEORGIEV and A. YORDANOV
Department of Chemistry, Tulane University, New Orleans, LA 70118, U.S.A.

(Received: 3 March 1994; in final form: 27 July 1994)

Abstract. Calix[n]arenes ($n = 4, 6$) with diphenylphosphinite groups appended to their lower rim have been synthesized by reaction first with base, followed by chlorodiphenylphosphine. The reaction has also been carried out with the partially methoxylated calix[n]arenes. Calix[6]arenes with phosphate groups selectively bridging adjacent pairs of oxygens have been synthesized by reaction first with base, followed by ethyl dichlorophosphate. Calix[n]arenes ($n = 4,6$) with 2-aminoethyloxy groups appended to the lower rim have been synthesized both by the reduction of an amide or nitrile group. Calixarenes with 2-hydroxyethyloxy and 2-bromoxyethyloxy groups appended to the lower rim have also been prepared. A route to preparing calixarene-functionalized polymers by the alkylation of polyethyleneimine is also described.

Key words: Calixarene, diphenylphosphinite, phosphate, amine, polymer.

1. Introduction

Although numerous examples exist of calixarenes with nitrogen or phosphorus substituents on the upper rim, relatively few such compounds have been prepared where these substituents are attached to the lower rim [1, 2]. Such compounds are, however, of importance because they can be potentially used as encapsulating ligands for metal ions. In order for these compounds to be useful as ligands, however, it is necessary that the nitrogen or phosphorus heteroatom either be in the tricoordinate form, or that an atom bonded to this heteroatom can act as a two-electron donor to a metal ion. In this article we describe some calixarenes that have nitrogen or phosphorus heteroatoms attached to their lower rim, and show how they can have applications to metal ion complexation chemistry.

2. Results and Discussion

Calixarenes with diphenylphosphinite groups attended to the lower rim can be synthesized by treating the *tert*-butylcalix[4]arene first with sodium hydride, then with chlorodiphenylphosphine (Equation 1) [3].

---

[*] This paper is dedicated to the commemorative issue on the 50th anniversary of calixarenes.
[**] Author for correspondence.

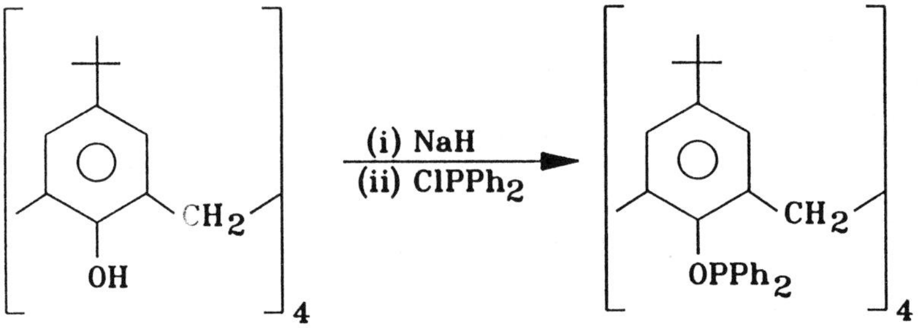

(1)

The product **1** has diphenylphosphinite groups attached to each of the lower rim positions. Compound **1** coordinates to Cu(I) via phosphorus to give a polymetallic complex that has a 1 : 1 ratio of Cu : P atoms. By contrast, **1** forms a bimetallic complex with two Fe(CO)$_3$ units [4]. The number of diphenylphosphinite groups on the lower calixarene rim can be controlled by introducing methoxy groups at selected positions on this rim. For the case of *tert*-butylcalix[4]arene and *tert*-butylcalix[6]arene, reaction with potassium carbonate followed by methyl tosylate leads to the conversion of alternating hydroxy groups into methoxy groups. Treatment of these compounds with sodium hydride followed by chlorodiphenylphosphine results in the formation of calix[4]arene **2** and calix[6]arene **3**, respectively, having alternating methoxy and diphenylphosphite substituents on the lower rim (Equation 2) [5].

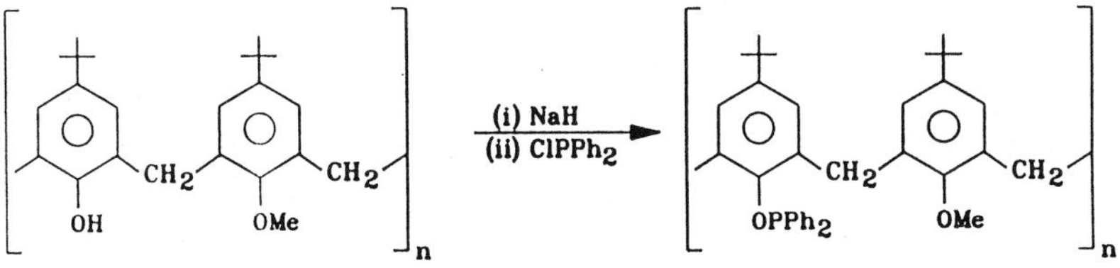

(n = 2, **2**; n = 3, **3**)

(2)

Compound **2** is conformationally rigid in solution at 25°C. The ^{31}P{^{1}H}-NMR spectrum of **2** shows a singlet resonance at δ 114.2. The ^{1}H-NMR spectrum shows two singlets for the methoxy groups at δ 2.38 and 3.07, along with four sets of doublet resonances for the methylene groups at δ 3.13 (^2J(HH) = 12.0 Hz), 3.62 (^2J(HH) = 15.4 Hz), 4.05 (^2J(HH) = 15.4 Hz) and 4.29 (^2J(HH) = 12.0 Hz). These data identify the conformation as the partial cone shown in Figure 1. For the calix[6]arene **3** the ^{31}P{^{1}H}-NMR spectrum shows a single resonance at δ 117.2, with the ^{1}H-NMR spectrum showing a single resonance for the OCH_3 protons at δ 2.34. These data confirm the structure of the compound, but do not indicate any particular conformational preference. Similar di- and tri-substituted phosphinites have been prepared from the di- and tri-substituted calix[4]aryl and calix[6]aryl

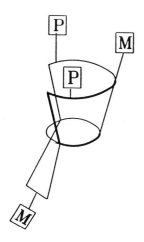

Fig. 1. Conformer of **2** where 'M' refers to OMe and 'P' refers to OPPh$_2$.

acetate [6]. In this synthetic procedure, however, LDA was used in place of sodium hydride as the base.

Phosphate groups can also be appended to the lower rim of *tert*-butylcalix[4]-arene and -calix[4]arene. The method of choice involves treating the calixarene with diethoxyphosphoryl chloride in basic solution. By changing the reaction conditions that are used it is possible to prepare compounds having different numbers of phosphate groups appended to the lower rim [7–10].

Phosphate groups can also be used to bridge and cap the oxygen atoms on the lower calix[4]arene rim. An example of a capped calix[4]arene where all four oxygens on the lower rim bind to a single phosphorus heteroatom is found in the reaction of *tert*-butylcalix[4]arene with P(NMe$_2$)$_3$ (Equation 3). The product **4** has a hypervalent phosphorus capping the lower rim of the calix[4]arene [11].

$$\text{(3)}$$

Phosphorus heteroatoms can also be incorporated onto the lower rim of a calixarene in a pairwise manner. Such bridged compounds have been prepared by treating *tert*-butylcalix[6]arene with base followed by ethyl dichlorophosphate [12]. Thus the addition of cesium fluoride followed by 1 equivalent of ethyl dichlorophosphate results in the incorporation of a single ethyl phosphate group into the compound, in a manner where it spans two adjacent positions on the lower calix[6]arene rim (Equation 4).

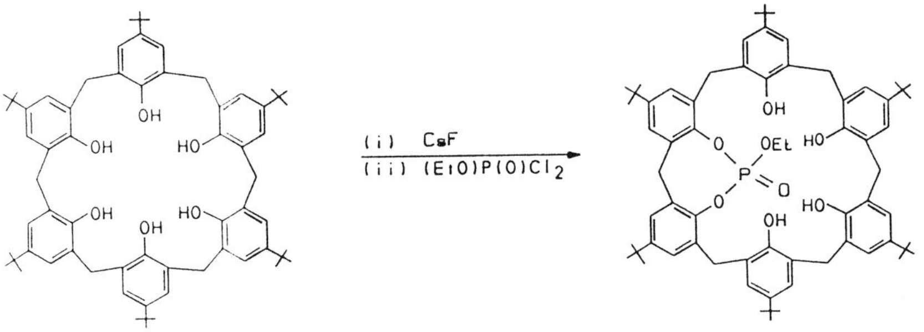

(4)

The $^{13}C\{^{1}H\}$-NMR spectrum of this compound **5** shows three resonances for the *tert*-butyl groups at δ 31.26, 31.57 and 31.62. The presence of these resonances supports the structure shown for the compound. When potassium hydride is used as base in place of cesium fluoride, *tert*-butylcalix[6]arene can be reacted with 2 equivalents of ethyl dichlorophosphate to give the compound **6**, where now two sets of adjacent lower rim oxygens are spanned by ethyl groups (Equation 5).

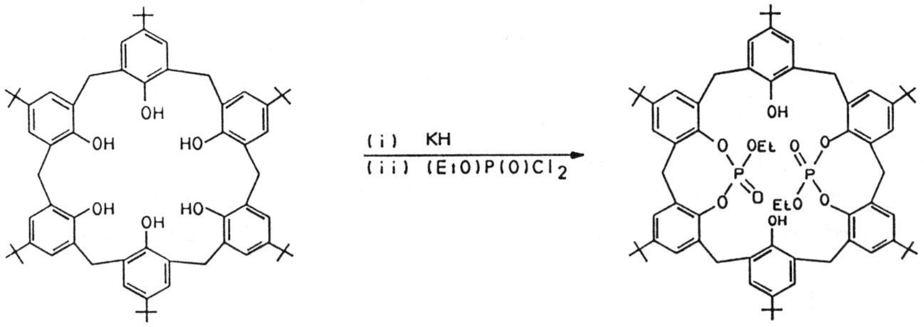

(5)

The ^{1}H-NMR spectrum of **6** shows resonances at δ 1.03 and 1.38 for the *tert*-butyl groups in the ratio of 2 : 1, and a single multiplet for the methylenes at δ 4.64. These data support the structure shown where pairs of oxygens across the lower rim are bridged, and do not support a structure where adjacent pairs of phenolic groups around the lower calix[6]arene rim are sequentially spanned by ethyl phosphate groups.

Just as phosphorus heteroatoms can be appended to the lower calixarene rim, so can functional groups that contain a nitrogen atom. Among the nitrogen containing substituents that can be introduced onto the lower rim of a calixarene are the amide and amine moieties. Compounds having an amide substituent can be prepared by the sequence of reactions shown in Scheme 1 [13]. A similar calix[4]arene with amide substituents on the lower rim has been used as a ligand for the encapsulation of

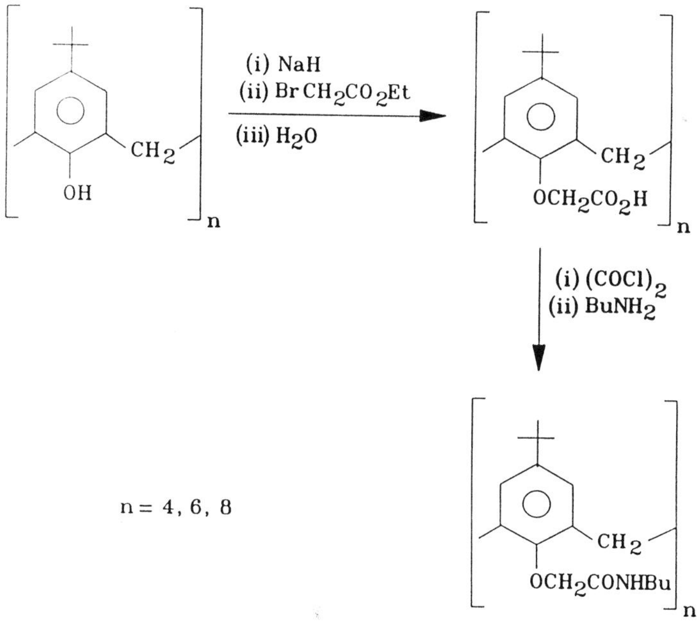

Scheme 1.

lanthanide ions (Figure 2) [14]. The Tb^{3+} complex that is obtained with this ligand has both a high luminescence quantum yield and a long luminescence lifetime in aqueous solution at ambient temperature, a feature that makes it a potentially useful complex for time-resolved fluorimmunoassay.

We have prepared calix[4]arenes and calix[6]arenes that have a 2-aminoethyloxy group attached to the lower rim. The synthetic procedure involves reduction of either an amide or a nitrile group with borane to the amine. These two routes are shown in Equations 6 and 7.

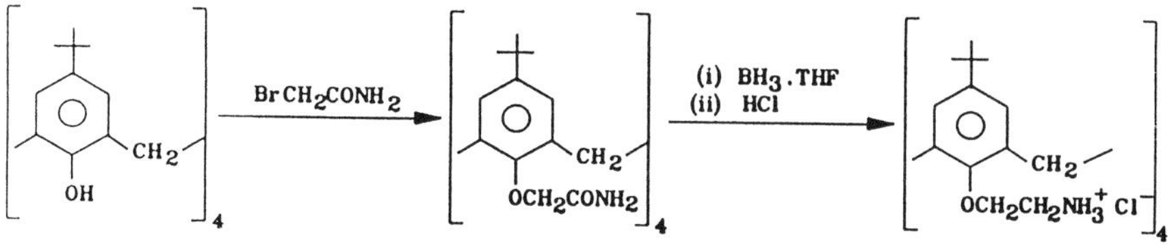

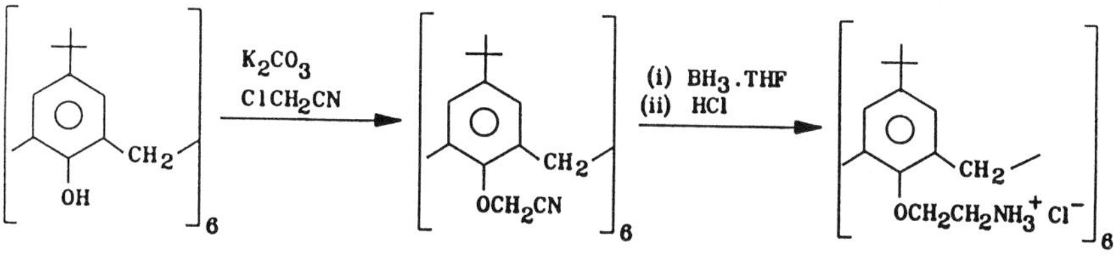

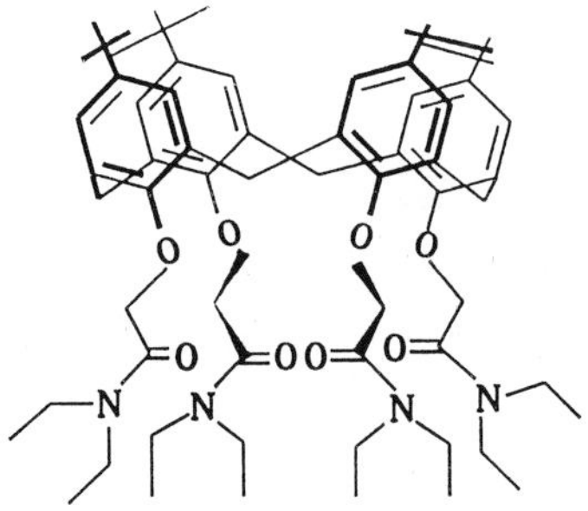

Eu^{3+}, Tb^{3+} and Gd^{3+} are encapsulated

Lifetime of LMCT (nsec)

Complex	300K (H$_2$O)	300K (D$_2$O)	77K (D$_2$O)	Φ 300K (H$_2$O)
Eu^{3+}	0.65	1.9	1.8	2 x 10^{-4}
Tb^{3+}	1.5	2.6	1.6	2 x 10^{-1}

Fig. 2. Lifetime and quantum yield data for encapsulated lanthanide ions.

The intermediate compound **7** is identified in the ^{1}H-NMR spectrum by resonances at δ 4.18 (CH_2) and 7.35 (NH_2). Compound **8** has the amine resonance at δ 8.43, which is identifiable by its exchange reaction with D$_2$O. The latter method via the cyanomethoxy intermediate can be used to prepare partially substituted calixarenes, since the addition of chloroacetonitrile to calix[4]arene can be used to selectively substitute the two alternate positions on the lower rim (Equation 8).

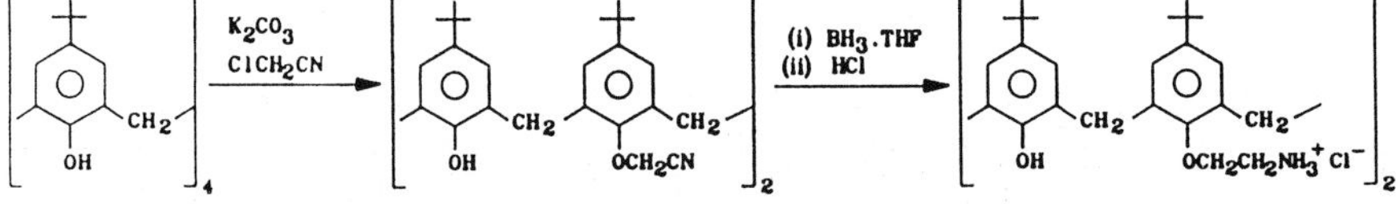

$$\tag{8}$$

9

Subsequent reduction with BH$_3$·THF, followed by quaternization of the amine with HCl, can be used to obtain the alkylammonium salt **9**. The compound shows a resonance at δ 8.57 (8H) due to the OH and NH groups which undergoes exchange with added D$_2$O.

We have also prepared a calix[4]arene with a 2-bromoethyloxy substituent appended onto the lower rim. The compound has been prepared by the sequence of reactions shown in Scheme 2. The sequence involves first converting the calixarene into the acetate derivative by reaction with ethyl bromoacetate, followed by reduction of this intermediate ester to the 2-hydroxyethyloxy derivative **10**. This com-

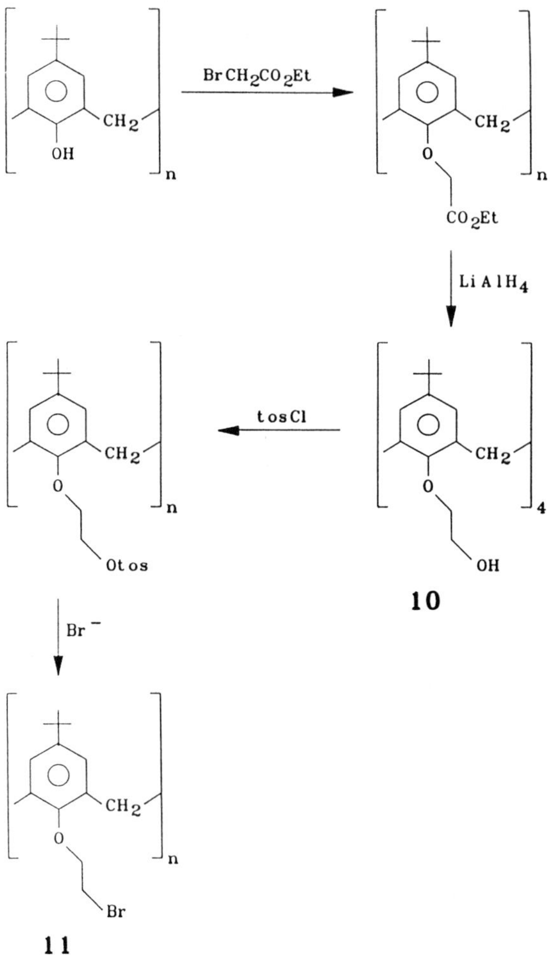

Scheme 2.

pound is characterized by resonances in the ^{1}H-NMR spectrum at δ 4.03 (OCH_2) and δ 5.6 (OH). This intermediate **10** can be converted into the 2-bromoethyloxy compound **11** by replacement of the tosylate group with bromide ion. Compound **11** shows characteristic resonances in the ^{1}H-NMR spectrum for the OCH_2 and CH_2Br groups at δ 4.24 and 3.86, respectively. This new 2-bromoethyloxy calix[4]arene is a potentially useful synthon for the introduction of a wide range of heteroatom substituents onto the lower rim.

A relatively new development in calixarene chemistry is their immobilization on polymeric supports. For certain applications such as the separation of heavy metal ions from aqueous media, it is advantageous for these calixarene-impregnated polymers to be available in an insoluble form. This goal is relatively easy to achieve since insoluble materials can be obtained by cross-linking the calixarene to the polymer via several attachment points. Since reactions leading to the chemical modification of calixarenes usually lead to functionalization at multiple sites, precursor calixarenes for such uses are readily available. Several examples of such cross-linked calixarene modified polymers are reported in the literature. An example by Harris involves the synthesis of silicone bound calixarenes where the attachment is via multiple ester functionalities between the polymer and the calixarene [15, 16]. A further cross-linked calixarene modified polymer has been prepared by Shinkai from treating polyethyleneimine with a chloro-sulfonated calixarene [17]. These insoluble modified polymers have been used by Shinkai for the extraction of uranium ion from seawater [18, 19]. Recently Harris has reported the synthesis of a new calixarene bearing a single methacrylate functionality that is potentially suitable for use as a reagent to give living polymers and copolymers. Attempted homopolymerization of this compound, however, yielded an oligomer that had only approximately six calixarene units in the chain [20]. More recently we have prepared a cross-linked material by treating a 50% aqueous solution of polyethyleneimine with 5,11,17,23-tetrachloromethyl-25,26,27,28-tetrahydroxycalix[4]arene [21].

For other applications such as in electrokinetics or the synthesis of radiopharmaceuticals it is desirable that the calixarene-functionalized polymer be soluble in aqueous media. In order to achieve such a goal it is necessary to avoid the formation of cross-linking within the polymer. Such a polymer **12** that is soluble in aqueous or organic solvents is formed when a calixarene with a single haloalkyl substituent on the upper rim is reacted with the polyethyleneimine solution (Equation 9).

(9)

This availability of two separate rims that can be independently modified for either selective ligation or for the attachment to polymers makes the calixarenes an attractive class of compounds for a wide range of applications involving metal ions.

Acknowledgment

We thank the Center for Bioenvironmental Research at Tulane University for financial support.

References

1. C. D. Gutsche: *Calixarenes*, Royal Society of Chemistry, Cambridge, U.K. (1989).
2. J. Vicens and V. Böhmer: *Calixarenes: A Versatile Class of Macrocyclic Compounds*, Kluwer, Dordrecht, the Netherlands (1991).
3. C. Floriani, D. Jacoby, A. Chiesi-Villa, and C. Guastini: *Angew. Chem., Int. Ed. Engl.* **28**, 1376 (1989).
4. D. Jacoby, C. Floriani, A. Chiesi-Villa, and C. Rizzoli: *J. Chem. Soc. Dalton Trans.* 813 (1993).
5. J. K. Moran and D. M. Roundhill: *Inorg. Chem.* **31**, 4213 (1992).
6. D. Matt, C. Loeber, J. Vicens, and Z. Asfari: *J. Chem. Soc., Chem. Commun.* 604 (1993).
7. L. N. Markovskii, V. I. Kal'chenko, and N. A. Parkhomenko: *Zh. Obschch. Khim.* **60**, 2811 (1990).
8. Y. Ting, W. Verboom, L. C. Groenen, J.-D. van Loon, and D. N. Reinhoudt: *J. Chem. Soc., Chem. Commun.* 1432 (1990).
9. F. Grynszpan, Z. Goren, and S. E. Biali: *J. Org. Chem.* **56**, 532 (1991).
10. R. G. Janssen, W. Verboom, S. Harkema, G. J. van Hummel, D. N. Reinhoudt, A. Pochini, R. Ungaro, P. Prados, and J. de Mendoza: *J. Chem. Soc., Chem. Commun.* 506 (1993).
11. D. V. Khasnis, M. Lattman, and C. D. Gutsche: *J. Am. Chem. Soc.* **112**, 9422 (1990).
12. J. K. Moran and D. M. Roundhill: *Phosphorus, Sulfur, and Silicon* **71**, 7 (1992).
13. S.-K. Chang, S.-K. Kwan, and I. Cho: *Chem. Lett.* 947 (1987).
14. N. Sabbatini, M. Guardigli, A. Mecati, V. Balzani, R. Ungaro, E. Ghidini, A. Casnati, and A. Pochini: *J. Chem. Soc., Chem. Commun.* 878 (1990).
15. S. J. Harris, M. A. McKervey, D. P. Melody, J. G. Woods, and J. M. Rooney: *Eur. Patent Appl.*, 151, 527 (1985); *Chem. Abstr.* **103**, 216, 392x (1985).
16. B. Kneafsey, J. M. Rooney, and S. J. Harris: *U.S. Patent*, 4,912,183 (1990); *Chem. Abstr.* **114**, 123, 273 (1991).
17. S. Shinkai, H. Kawaguchi, and O. Manabe: *J. Polym. Sci.* **C26**, 391 (1988).
18. Y. Kondo, T. Yamamoto, O. Manabe, and S. Shinkai: *Jpn. Kokai Tokkyo Koho* JP 62/210055 A2 [87/210055] (1987); *Chem. Abstr.* **108**, 116, 380b (1988).
19. Y. Kondo, T. Yamamoto, O. Manabe, and S. Shinkai: *Jpn. Kokai Tokkyo Koho* JP 63/7837 A2 [88/7837] (1988); *Chem. Abstr.* **109**, 137, 280b (1988).
20. S. J. Harris, G. Barrett, and M. A. McKervey: *J. Chem. Soc., Chem. Commun.* 1224 (1991).
21. E. M. Georgiev, K. Troev, and D. M. Roundhill: *Supramol. Chem.* **2**, 61 (1993).

… 111

Homocalixarenes and Homocalixpyridines

G. BRODESSER* and F. VÖGTLE
Institut für Organische Chemie und Biochemie der Universität Bonn, Gerhard-Domagkstrasse 1, D-53121 Bonn, Germany

(Received: 11 May 1994; in final form: 24 June 1994)

Abstract. A new type of host is introduced: *all*-homocalixarenes. Phane syntheses leading to molecules which may be termed, in the most general sense, homocalixarenes, are outlined in a brief overview. The design, synthesis, conformations and host properties of *all*-homocalixarenes and *all*-homocalixpyridines are described in detail.

Key words: Phane syntheses, *all*-homocalixarenes, *all*-homocalixpyridines, host tailoring, large functionalized cavities, host properties, liquid-liquid-extraction experiments, crystal structures.

1. Introduction

Molecular recognition is a fundamental aspect of biological processes and particular interest is focused on its understanding. One pathway to this understanding is the study of 'model' compounds, the products of 'supermolecular chemistry', wherein specific properties of biomolecules may be considered in isolation. The control of both functional groups and molecular stereochemistry in these synthetic molecules provides numerous insights into the relationship between structure and function, insights which are often obscured within the full complexity of the biological systems.

The calixarenes [1, 2] are a versatile group of 'host' molecules which have indeed been used to probe various aspects of enzyme functions [3]. Nonetheless, it is not to be expected that a single molecular structure could provide a basis for the mimicry of all enzymic behaviour, and thus new molecules with controllable architecture are valuable. This work describes such a group of molecules, closely related to the calixarenes obtained through the syntheses which combine aspects of cyclophane and host/guest chemistry. It defines the group of '*all*-homocalixarenes'.

2. Phane Chemistry as a Tool for Host Tailoring

The field of cyclophane chemistry encompasses the chemistry of a remarkable variety of molecules, so that it is of general importance to survey its current state of development [4]. Molecular stereochemistry has long been a fundamental issue, and it is the wide, known range of partial structures, fixed or conformationally

---

* This paper is dedicated to the commemorative issue on the 50th anniversary of calixarenes.

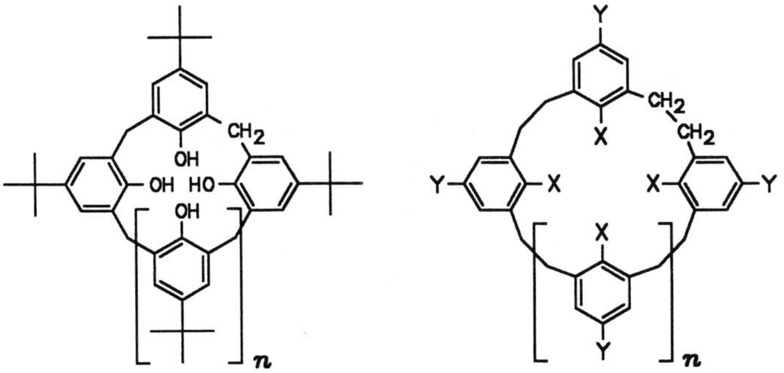

Fig. 1. Structural features of calixarenes (left side) versus homocalixarenes (right side).

flexible, that affords such a rich field for research into molecular design, synthesis, structural analysis, physical properties and, chemical reactions, all the way to supramolecular chemistry! Phanes are ideal model compounds in which benzene and other aromatic rings can be placed relative to one another in a tailor-made fashion. One can introduce into the rings of many cyclophanes substituents which act as ligands towards both cations and neutral molecules. One can achieve a particular arrangement of functional groups and thereby trigger their chemical interaction. The possibility of locating groups precisely in space allows cyclophane chemistry to be used to provide the essential units for specialised structures such as 'nests', hollow cavities, 'multi-floor structures', helices, macropolycycles and long hollow tubes, as well as major ligand systems like the calixarenes and, now, the *all*-homocalixarenes.

Whereas calixarenes contain a $[1_n]$metacyclophane skeleton, the *all*-homocalixarenes are $[2_n]$metacyclophanes in which an additional CH_2 group is present in all their bridges (Figure 1).

2.1. NOMENCLATURE

The close and obvious relationship between these new molecules and the calixarenes justifies the use of a closely related nomenclature. '*all*-Homocalix[n]arene' (*all*-HC) is used as a generic term for those calixarene analogues which contain symmetrically one additional CH_2 group in all their bridges. '*all*-Homocalix[n]pyridine' denotes the group of homocalixarenes with pyridines as the aromatic units, an example being 6,13,20,27-tetramethoxy-*all*-homocalix[4]pyridine (**3**). Whereas the prefix 'homo' indicates the enlargement of the cavity, the prefix '*all*' is intended to indicate that the aliphatic bridges in the macrocycle are enlarged symmetrically, so that it is not necessary to explicitly state the number of additional carbon atoms as in 'di-, tri- ... homocalixarene'

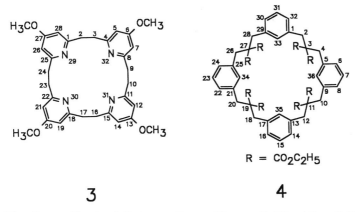

3   **4**

Fig. 2. Structural features of an *all*-homocalix[4]pyridine (left side) and an *all*-bishomocalix[4]arene (right side).

(e.g., **2**: Y = H, X = OH is 29,30,31,32-tetrahydroxy-*all*-homocalix[4]arene). With two additional CH$_2$ groups in all aliphatic bridges, the molecules are designated '*all*-bishomocalix[n]arenes' (e.g., **4** in Figure 2).

We have chosen the metacyclophane skeleton as a versatile and stable platform for functional units. A critical appraisal of the efficiency of the syntheses of cyclophanes incorporating more than two aromatic units is given below. The specific focus is upon those metacyclophanes which can be considered, in the broadest sense, *all*-homocalixarenes.

2.2. METACYCLOPHANES

2.2.1. *Müller–Röscheisen Cyclisation*

The first [2.2]metacyclophane was obtained by Pellegrin using the Wurtz reaction [6]. Using a modified Wurtz reaction, Jenny *et al.* were able to obtain higher, 'oligomeric' [2$_n$]metacyclophanes from 1,3-bis(bromomethyl)benzene (**5**) [7], actually by employing Müller–Röscheisen conditions of sodium tetraphenylethene in THF at -80 °C [8]. The reaction mixture yields the full series of oligomeric [2$_n$]metacyclophanes up to the 50-membered [2$_{10}$]metacyclophane. Over a period of three years, all were isolated by various procedures. The first isolated cyclo-oligomer of this series, the first isolated *all*-homocalixarene, was the [2$_4$]metacyclophane (**6**), obtained in 1.7% yield by Burri and Jenny in 1966 [9]. Shortly thereafter they reported the isolation of the [2$_3$]metacyclophane (7.5%) and the other cyclo-oligomers up to the [2$_{10}$]metacyclophane (<1%) [10–13] except the [2$_7$]metacyclophane and the [2$_9$]metacyclophane, which were isolated two years later (<1%) [14].

[Structure 5] → Na/TPE, −80°C → [Structure with n = 2–10]

Also, using sodium tetraphenylethene in THF, Tashiro *et al.* prepared disubstituted [2.2]metacyclophanes in 1981 via intermolecular ring closure of the dimeric building block 1,2-bis[5-*tert*-butyl-3-(iodomethyl)-2-methoxyphenyl]-ethane [15]. As a side product of intramolecular reaction, they obtained a methoxy-substituted tetra-*tert*-butyl-[2$_4$]metacyclophane in 34% yield. Demethylation with BBr$_3$ gave **7**.

6   **7**

2.2.2. *Samarium (II) Diiodide Coupling*

Coupling of 1,3-bis(bromomethyl)benzenes in the presence of SmI$_2$ in THF gives cyclic and acyclic oligomers, the distribution depending upon the amount of SmI$_2$ used. Five equivalents of SmI$_2$ gave the unsubstituted [2$_3$]metacyclophane in 10.2% yield and the tetramer **4** in 7.5% yield but no higher oligomers were observed [16].

2.2.3. *Sulfone Extrusion Method*

In 1969, Vögtle [17] first outlined the sulfone route for the preparation of [2.2.0]metacyclophane as one of the rare metacyclophanes containing three aromatic units. In 1981, Vögtle *et al.* prepared the [2.2.2.2]biphenylophane **10** in 48% yield by sulfone pyrolysis at 600 °C/10^{-6} torr [18].

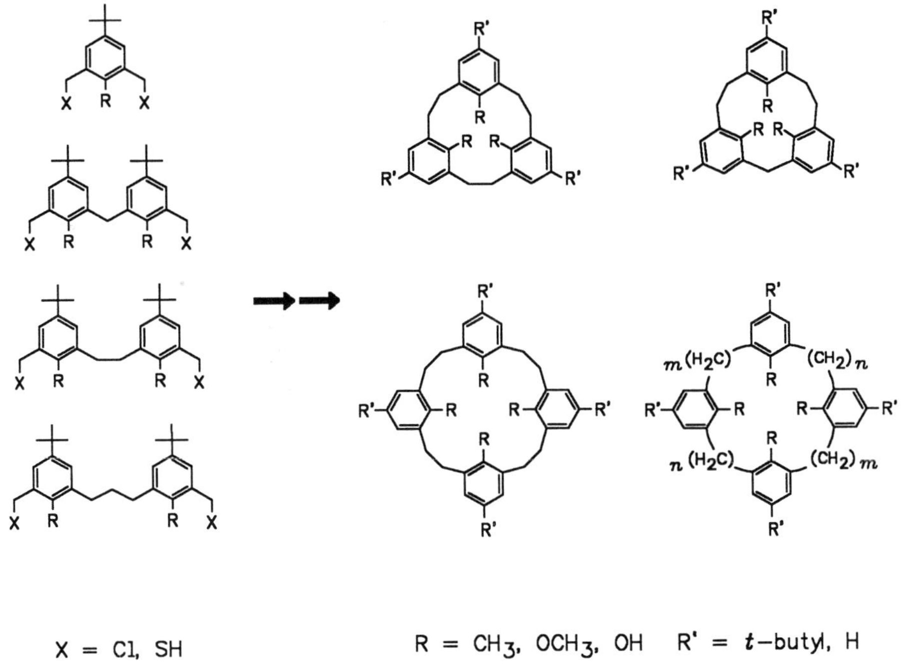

Fig. 3. Conventional sulphur method for synthesis of [2_n]- and [$m.n$]metacyclophanes.

X = Cl, SH   R = CH$_3$, OCH$_3$, OH   R' = t-butyl, H

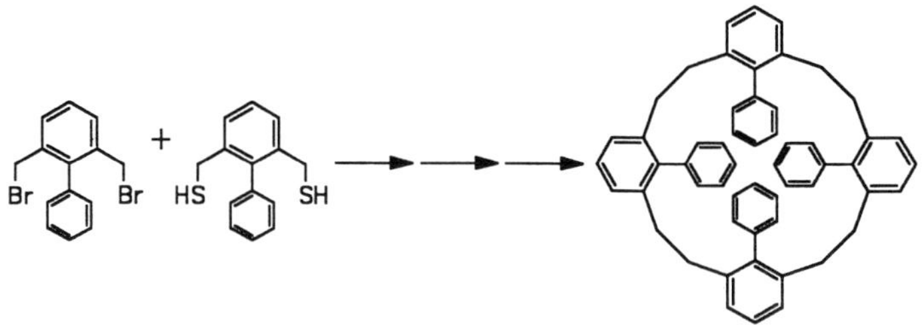

8   9   10

Whereas **7** isolated as a side product, in 1989 Tashiro *et al.* deliberately prepared trimeric and tetrameric metacyclophanes by the conventional sulfone method. Cyclisation of monomeric or dimeric thiol- and chloride-functionalised precursors linked by (CH$_2$)$_n$ bridges in various combinations, yielded, after ring contraction via sulfone pyrolysis, a multitude of [2_n]- and [m.n]-metacyclophanes as shown in Figure 3 [19–24].

2.2.4. Condensation with Aldehydes

The base-catalysed condensation of **11** with formaldehyde in p-xylene yields, in three steps, the tetrahydroxy[3.1.4.1]metacyclophane **12** [25]. Recently, Yamato et al. [26] reported the synthesis of hexa- and octa-hydroxy$[2.1]_n$metacyclophanes **14** and **15** from 1,2-bis(5-*tert*-butyl-2-hydroxy-phenyl)ethane **13** via the same method in 70–90% yield. Interestingly, no dimer was formed.

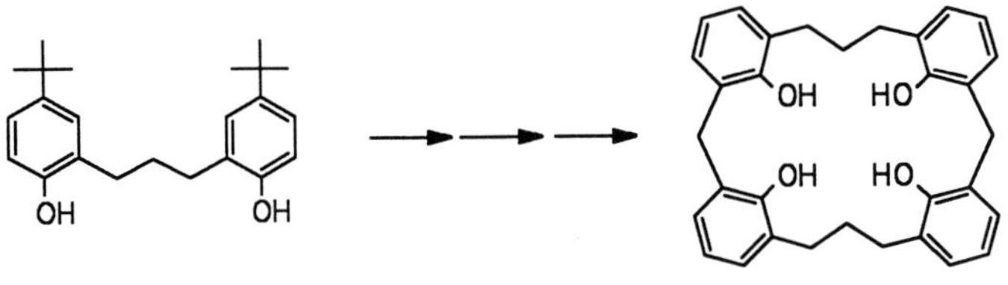

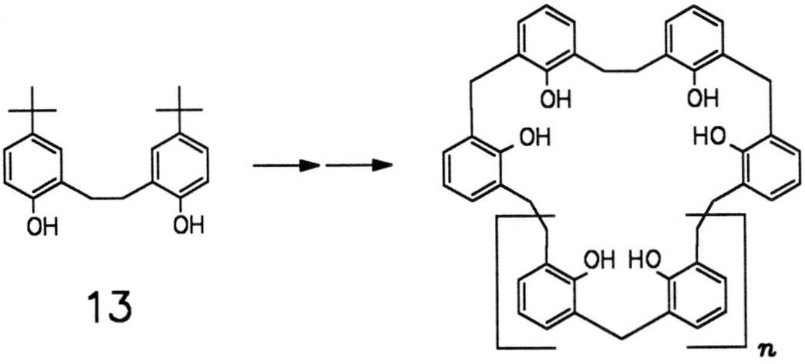

14: $n = 1$

15: $n = 2$

2.2.5. Malonate Cyclisation

In 1972, Vögtle et al. reported a synthesis for $[3_n]$metacyclophanes via condensation of 1,3-bis(bromomethyl)benzene **5** with malonate **16** [27].

2.2.6. Cycloalkylation of Benzene by Diols

Addition of a benzene solution containing 3,3'-bis(hydroxymethyl)bibenzyl **18** to concentrated sulfuric acid yields the [2.1.1]metacyclophane **19** in 17% yield [28].

2.2.7. Via Dianions

A [5.5.5]metacyclophane **22** was obtained in 1% yield by reacting the dianion of 2,6-dimethylanisole **20** with the α, ω-dihalide **21** [29].

As an alternative to sulfone extrusion, Burns et al. [30] recently described a five-step synthesis via the dianion of **24** for the preparation of [3$_4$]metacyclophane **25** in 21% yield.

2.3. HETERAPHANES

As well as those thiaphanes mentioned in Section 2.2 as intermediates in [2$_n$]- and [m.n]phane syntheses, there are oxa- and azaphanes like **26**, **27** and **28**, which are named homooxa- and homoazacalixarenes because they were first obtained through the 'Petrolite' procedure of calixarene synthesis in 1962 [31, 32]. The compounds were first characterised in 1979, and in 1983 Gutsche [35] described detailed syntheses based on inter- or intra-annular thermally-induced dehydration of bis(hydroxymethyl)aromatic building blocks.

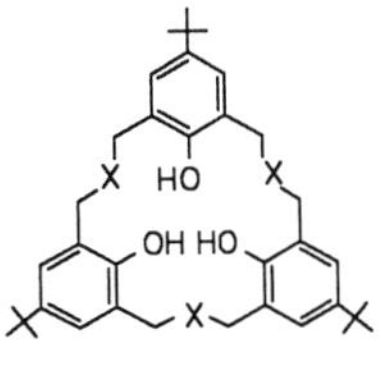

26 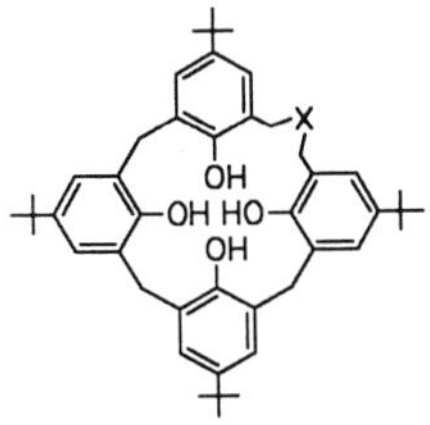   27 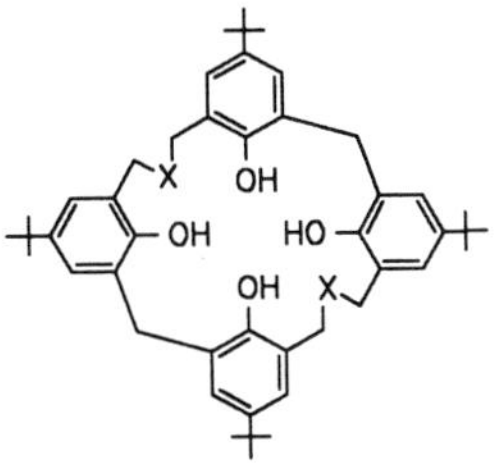   28

X = O, NCH$_2$C$_6$H$_5$

More recently, syntheses of oxacalixarene analogues (azacalixarenes) based on condensation between benzylamine and 2,6-bis(hydroxymethyl)-4-alkylphenol have been described [36].

2.4. HETEROPHANES

[2.2.2](2,5)Thiophenophane has been obtained as a by-product of the synthesis of [2.2](9,10)naphthaleno(2,5)thiophenophane by crossed Hofmann degradation of the corresponding quaternary ammonium hydroxides [37].

Whereas acid-catalysed condensations of furans [38, 40], thiophenes [41–44] or pyrroles [45, 46] provide heterocalixarenes **29** in one step, the introduction of pyridine rings is more difficult. Newkome *et al.* were successful in preparing the triketone **30** and the tetraketone **31** [47, 48].

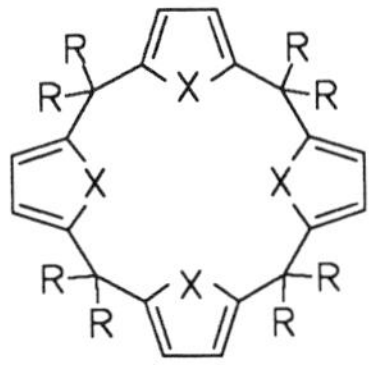

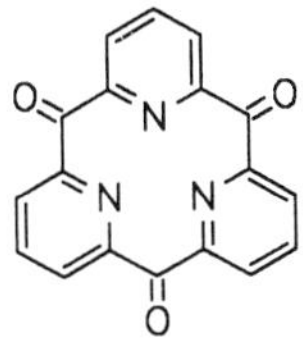

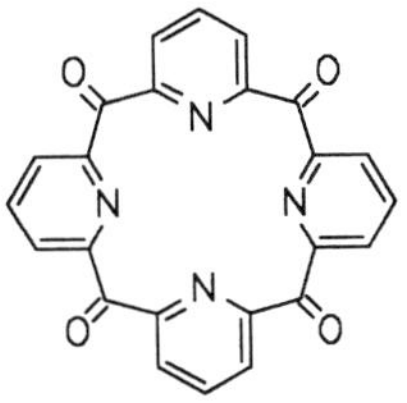

29   X = O
        X = S
        X = NH   **30**   **31**

R = H, CH$_3$, C$_2$H$_5$, C$_6$H$_5$

Pyridinophanes with ethano bridges have been synthesised by Jenny *et al.* via Müller–Röscheisen cyclisation of 3,5-bis(bromomethyl)pyridine [49, 50] and 2,6-bis(bromomethyl)pyridine [51, 52], leading to [2$_n$](2,6)pyridinophanes involving up to six pyridine units. Also, the tetrameric cyclooligomer **36** can be derived from intermolecular reaction of the dimeric copper compound **34** [53].

3. From Phanes to *all*-Homocalixarenes and -calixpyridines

3.1. CONCEPT

In order to meet the objectives mentioned in the introduction, we have developed a versatile host architecture [54], which is expected to comply with the requirements for an effective host listed below:

- adjustable cavity size – for steric host/guest fit;
- conformational flexibility straight into time with preorganisation;
- various easily accessible binding sites providing high selectivity through structural modifications so as to control both the electronic fit to charged or neutral but acidic/basic guests and parameters such as charge, dipole moment, lipophilicity and H-bonding ability;
- good solubility in organic solvents, PVC membranes;
- stability in the reaction environment;
- anchoring groups for immobilising and fixing signal-giving units.

Knowledge gained through the study of calixarene systems should allow these new molecules to be used more effectively and efficiently to achieve even more demanding goals.

A near-planar ring skeleton, intended to enhance solubility relative to that of the calixarenes, should result from the introduction of an alternating arrangement of rigid aromatic and flexible aliphatic moieties. Consequently, this should have the result of avoiding the existence of rigid conformers perhaps unable to enclose

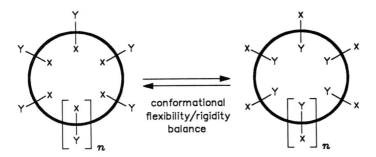

Fig. 4. Schematic carbocyclic *all*-homocalixarenes with variable ring size and functionalities X, Y inside and outside.

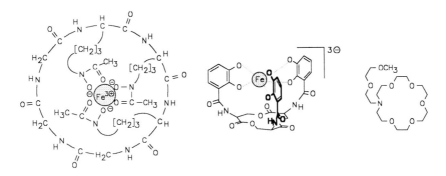

Fig. 5. Ferrichrome (left side) and enterobactin (middle) $Fe^{3\oplus}$ complexes as examples from nature for lariat type complexones (right side).

guests and of stable conformers as mixtures which would have to be separated and characterised. A chemically stable large ring skeleton of adjustable size can be built up and modified by the introduction of functional groups even under vigorous conditions. For tailoring, the host molecule can be endowed with binding sites both inside and outside the cavity (Figure 4).

Although cavity size can be controlled over a wide range by varying the number of aromatic units, ligand arms fixed inside the cavity can be used to assist guest binding by adjusting the steric and electronic host/guest fit, as in natural ionophores (cf. Figures 5, 6).

3.2. SYNTHESIS

Of the multitude of cyclisation methods (Sections 2.2–2.4) possible, the Müller–Röscheisen procedure has the versatility to make it the method of choice. As indicated previously, its application allowed both $[2_n]$metacyclophanes and $[2_n]$pyridinophanes to be obtained for the first time [55].

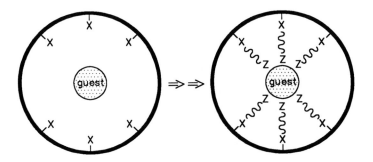

Fig. 6. *all*-Homocalixarenes allow us to adjust the steric and electronic host/guest complementarity.

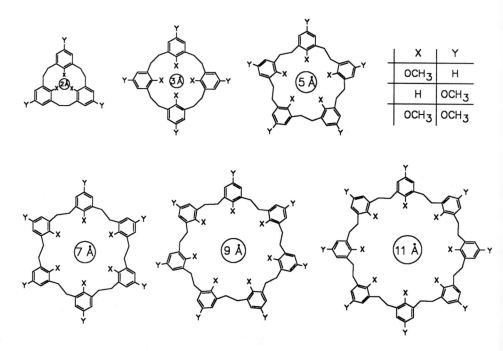

Fig. 7. A spectrum of eligible cavity sizes (longest O-O distances estimated for X = OCH$_3$, Y = H) feasible by Müller–Röscheisen cyclization.

In order to improve the yield of cyclooligomers, we modified the Müller-Röscheisen method; high dilution conditions can be achieved by using perfusers to add the dibromo compound very gradually. In the case of pyridines, the reaction was conducted at -90 °C to minimise by-product formation. A chromatographic procedure was developed to isolate the complete spectrum of cyclooligomers (Figure 7) obtained in the one-step synthesis.

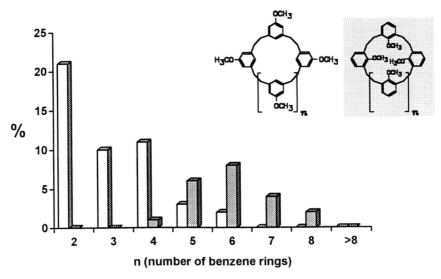

Fig. 8.   Steering of oligomer formation in *all*-homocalixarenes synthesis by intraannular substituents.

3.2.1. *Methoxy-Substituted all-Homocalixarenes as Host Precursors*

From different bis(bromomethyl)benzenes, various methoxy-substituted *all*-homocalixarenes were easily obtained (Figure 7) [55]. The intention was to use them as starting materials for the introduction of additional appropriate heteroatoms as individual coordination sites for guest complexation. The control of cavity size was based upon variation in the *endo* and *exo*-annular substituents X and Y (Figure 8).

3.2.2. *Refunctionalisation*

The homocalixarenes can be prepared not only in a wide range of ring sizes (10–80 members cf.: Figure 7) but also with almost any variation in functional groups. Cleavage of the methoxy protective groups in the macrocycles leads to a series of oligophenols, where the hydroxyl groups provide one useful type of binding site. These groups can be alkylated to provide hosts of high denticity with ligand arms such as those of oxapropionic acid, esters, amides and thioamides (Figure 9).

Ether cleavage with BBr_3 proceeds in 67–99% yield [56]. The oligophenols exhibit greatly differing solubilities, e.g., pentamer **44** and octamer **43** (Figure 18) show high solubility in di- or trichloromethane, while the hexamer is only soluble in DMSO and the heptamer in acetone. Refunctionalisation to oxapropionic methyl ester derivatives by reaction with methyl bromoacetate proceeds in 68–87% yields for the exoannular macrocycles and in 63% yield in the case of the endoannular

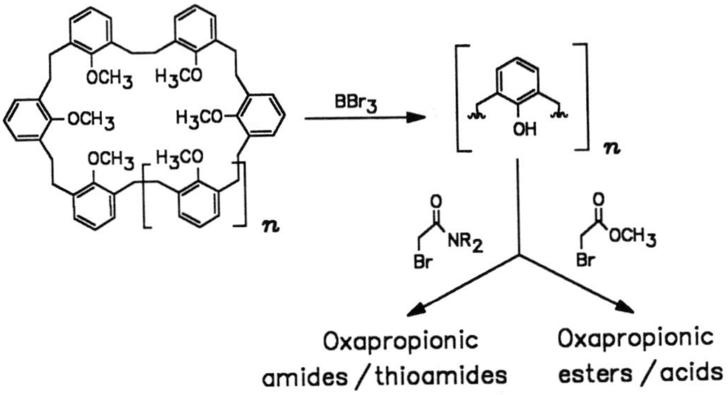

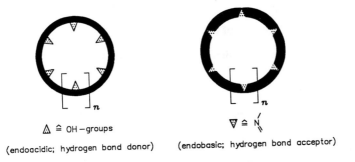

Fig. 9. Introduction of various binding sites into *all*-homocalixarenes.

Fig. 10. Schematic diagram of endoreceptors fitted with convergent acidic or basic functional groups/donor sites.

hexaphenol. The exoannular esters can be converted to the acids in 92–96% yields [56]. The hexaethylamide **42** was obtained in 80% yield [57].

3.2.3. *all*-Homocalixpyridines

In contrast to calixarenes, *all*-homocalixarenes may contain either endoacidic or endobasic cavities (Figure 10), because cyclisation of pyridine units is possible, too. The guest-binding strength of pyridine-containing hosts is readily controlled by the addition of OR groups at the 4-position of the pyridine rings.

The distribution of products in the case of the cyclisation of 2,6-bis(bromomethyl)-4-methoxypyridine depends mainly on dilution conditions and the temperature. At temperatures of about -100 to -90 °C, formation of open-chain products is decreased. Use of high-dilution conditions leads to odd-numbered main products like the trimer **37** (12%) and pentamer **38** (2%) [58]. Increase of concentration produces a broader spectrum of cyclooligomers. Although even-numbered conventional calixarenes are readily obtained, improvements in the synthetic yields for odd-numbered calixarenes have been slowly achieved [59, 60].

Recently, Gutsche *et al.* did report a one-step synthesis giving a quite acceptable yield for *p-tert*-butylcalix[5]arene [61].

37   **38**

3.3. PROPERTIES

3.3.1. *Melting Points*

In contrast to calixarenes, which generally melt well above 250 °C, *all*-homocalixarenes (Figure 7) melt below 250 °C and show a characteristic melting point pattern (Figure 11).

- odd-numbered cyclooligomers melt more easily than adjacent even-numbered ones;
- melting points of even-numbered cyclooligomers decrease with increase in ring size;
- there is a minimum for the heptamer explained by its conformational characteristics.

Jenny, among others, concluded that the compounds with an odd number of rings are not completely strain-free [14], a conclusion that is consistent with the yields.

3.3.2. *NMR Spectra*

The aliphatic protons, which appear as an AA'BB' system in [2.2]metacyclophanes, become a singlet in higher oligomers of homocalixarenes and -calixpyridines (δ = 2.75–2.95). This suggests that they are highly mobile. ^{1}H-NMR singlets are observed at δ 2.78 (ethano protons) and δ 3.60 (methylene protons) even in [2.1.1]metacyclophane [62]. The high-field shift of inner protons in [2.2]metacyclophanes decreases regularly with an increase in ring size (δ = 4.08–6.60). In

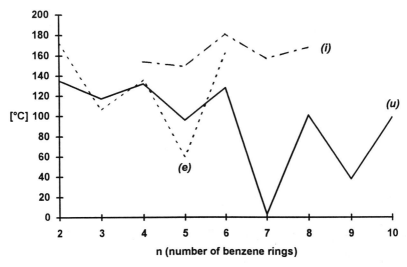

Fig. 11. Melting points of *all*-homocalixarenes (**u**: unsubstituted, **i**: with intraanular OCH$_3$ groups, **e**: with extraanular OCH$_3$ groups) in dependence of the number of benzene units.

methoxy-substituted homocalixarenes, the intra-annular methyl proton resonances are also shifted to higher field (trimer: 3.46; tetramer: 3.12).

3.4. CONFORMATIONS

3.4.1. *Flexibility/Rigidity Balance*

Calixarenes have an inherent rigidity which limits the range over which they can be shaped to a particular host. For this reason, attention has gradually focused on homocalixarenes. Shinkai *et al.*, for example, have reported the ion binding selectivity of hexahomotrioxacalix[3]arene [63, 64]. On the one hand, the multitude of calixarene conformations has advantages regarding preorganisation and selectivity (additional shapes for molecular recognition), but on the other hand their separation causes problems. Due to the expanded aliphatic bridges in *all*-homocalixarenes, the host skeleton is flexible and undergoes a fast ring flip process, so that conformational isomerism is markedly simplified (cf. Figure 4). The benefit of this simplification has been demonstrated by Shinkai *et al.* in the case of a hexahomotrioxacalix[3]arene containing ester groups as ionophores, where only two conformers, cone and partial cone, are detected and the molecule shows selectivity towards Na$^+$. The influence of bulky groups inside the cavity (cf. **10**) of *all*-homocalix[4]arenes on their conformations has been studied by Tashiro *et al.* [65].

3.4.1.1. *Capped all-Homocalixarenes*   Although some level of conformational flexibility is desirable, especially for ligands to be used for guest transport where the

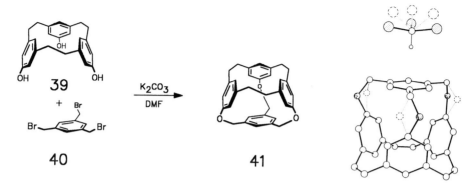

Fig. 12. Preliminary crystal structure of **41**.

kinetics of complexation-decomplexation should be fast, it is possible to envisage situations where greater rigidity of an *all*-homocalixarene could be desirable. One way to achieve this is to form a 'cap' involving ether linkages to a benzene ring. In fact, the capped *all*-homocalixarene **41** can be obtained by reaction under high dilution conditions of the triphenol **39** with the tribromo compound **40** [66]. Preliminary results of an X-ray structural analysis are shown in Figure 12. A trichloromethane molecule is oriented over one of the benzene rings at a distance of 338 pm in the crystal.

3.4.2. Crystallographic Studies

X-ray structure analyses are not available for unsubstituted $[2_n]$metacyclophanes and $[2_n]$pyridinophanes. Only NMR data are available as an index of conformational characteristics. For *all*-homocalixarenes and *all*-homoalixpyridines of various ring sizes, however, it has been possible to obtain single crystals suitable for crystallographic studies. In the trimeric exoannular *all*-homocalixarene and -calixpyridine, the aromatic rings do not lie in a single plane (Figure 13). Two methoxy groups are oriented in the same direction, with the third one opposed ('partial cone' conformation). The distances of the nitrogen atoms projecting into the 15-membered ring of **37** are: 408.4 pm (N8,...N16), 356.1 pm (N8...N24) and 395.3 pm (N16...N24).

With increasing ring size, *all*-homocalixarenes become flatter. Methoxy groups of the tetramer (Figure 13) are directed outwards. Facing benzene rings are almost parallel, as is also found in the methyl tetra-ester. In the endo- and exoannular methoxy-substituted *all*-homocalix[5]arene, the five benzene rings are situated in different planes (Figure 14). All endoannular substituents are oriented in the same direction but opposite to the exoannular methoxy groups. In contrast, the pentameric, methoxy-substituted *all*-homocalixpyridine **38** is nearly planar. Five water molecules and one trichloromethane molecule are included inside the cavity. Four of the water molecules form hydrogen bonds with two nitrogen atoms

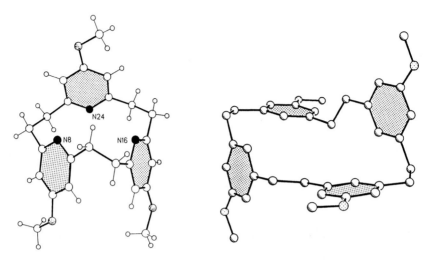

Fig. 13. Crystal structures of **37** (left side), and the *all*-homocalix[4]arene substituted with OCH$_3$ groups (right side).

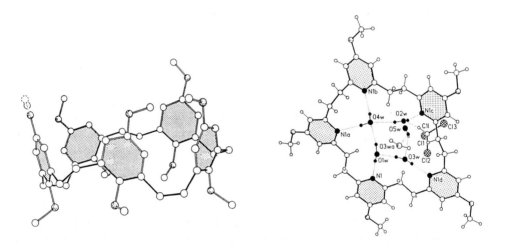

Fig. 14. Crystal structure of the intra- and extraannular methoxy substituted *all*-homocalix[5]arene (left side) compared to **38** (right side).

(N1A...O4w; 290.8 pm; N1B...O4w: 285.2 pm). Other N...O distances range from 284.3 pm (N1...O1w) to 306.7 pm (N1D...O3w). The distance N1A...N1D, 856 pm, is twice as long as the largest distance in trimer **37**.

The near planarity of the macrocyclic ring skeleton, which is not possible in the calixarene series, is also demonstrated by the example of a cyclohexamer, which X-ray structural analysis reveals its acts as a host towards cyclohexane. An endoannularly substituted hexaester, however, shows a very different 'stepped'

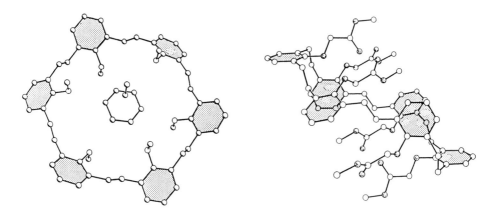

Fig. 15. Crystal structures of a near to planar *all*-homocalix[6]arene and the stepped ester derivative.

structure in which the benzene rings lie in four different planes and only four of the ester groups project in to the cavity.

In the hexaamide **42**, a three-dimensional cavity is formed in which each benzene ring can be considered to be flanked by two similarly-oriented benzene rings. Presumably because of steric hindrance, the amido groups do not project into the cavity but are bent out of the macrocycle, indicating that the ring is quite flexible. All carbonyl groups of the amido units point in the same direction. The orientations of the amido groups relative to the macroring mean plane can be described as: above, below, within, below, above, within. The homocalixarene molecules are connected to embedded water molecules by hydrogen bonds, forming long chains within the crystal.

Hitherto, no X-ray structural analysis of any heptameric *all*-homocalixarene has been obtained. The highest cyclooligomer for which crystallographic information is now available is the octamer **43**. All hydroxy groups are oriented inwards (Figure 17). Each *all*-homocalixarene molecule encapsulates one ethanol and one cyclohexane molecule. Four of the hydroxyl groups form hydrogen bonds to a disordered ethanol molecule. In the case of the *all*-homocalixpyridines, it is remarkable that all the even-numbered oligomers crystallise as needles unsuitable for X-ray crystallography.

3.5. HOST PROPERTIES

Whereas the unsubstituted parent compounds prepared by Jenny *et al.* (Section 2.2.1) were not considered likely to be able to act as ligands or hosts, some indications of the capacity of *all*-homocalixarenes and -calixpyridines to bind to guests are given by the results of X-ray structure analyses. The cluster type inclusion of water by the pentameric pyridine macrocycle **38** via hydrogen bonds seems

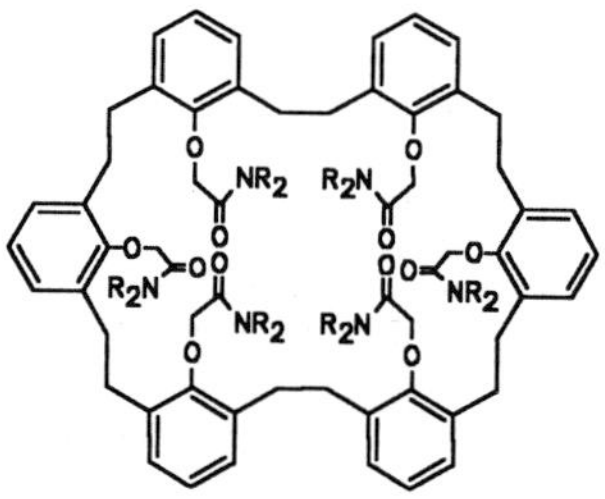

R = Ethyl

42

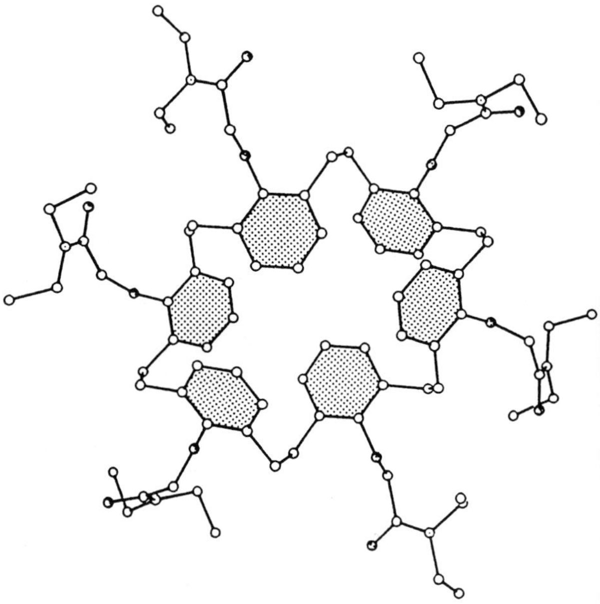

Fig. 16. Crystal structure of **42**.

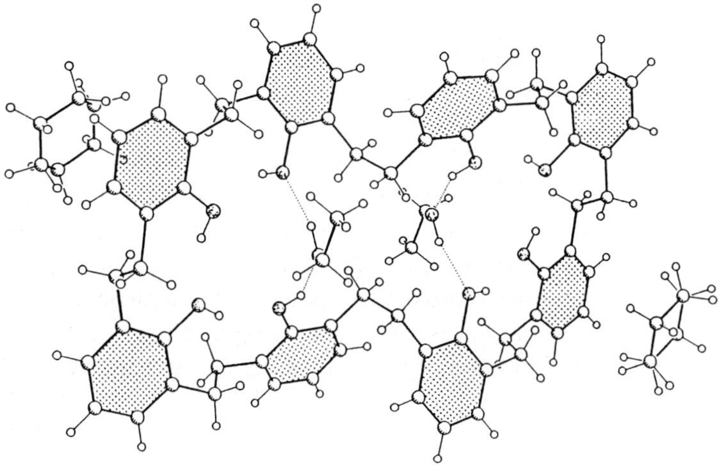

Fig. 17. Crystal structure of **43**.

to be rare and may be an exceptional case. Oligophenol **43** displays guest binding towards ethanol and cyclohexane. Host sensitivities/selectivities and liquid-liquid partition experiments [67–69] have been investigated with regard to the use of

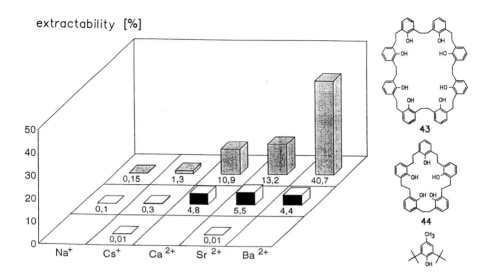

Fig. 18. Extraction properties of **43** and **44**. [M(NO$_3$)$_n$] = 1 × 10^{-4} M; [NaNO$_3$] = 0.1 M; [NaOH] = 0.1 M; [ligand] = 5 × 10^{-3} in trichloromethane.

the hosts in ion-selective electrodes and transport processes [70–72]. The most relevant results are:

- Pentahydroxy-*all*-homocalix[5]arene **44** discriminates strongly in favour of the alkaline earth metal ions Ca^{2+}, Sr^{2+} and Ba^{2+} over the alkali metal ions, even when the latter are present in as much as a 1000-fold excess [56] (Figure 18).
- Extractibility improves at high pH due to the ionisation of the hydroxyl groups.
- Octaphenol **43** is also selective towards the alkaline earths and more efficient in extraction due to its greater number of hydroxyl groups.
- From calcium, strontium and barium, **43** extracts barium selectively as a 1 : 1 complex.
- Complex formation constants of selected hosts have been measured in water [74] (Figure 19). Some values are extremely high, even in comparison with those of crown ethers.

To favour both selectivity and extractability, amido groups were introduced, eliminating the dependence on hydroxyl group ionisation. With this macrocycle, Ba^{2+} is extracted with a pronounced peak selectivity in contrast to other singly and doubly charged cations [57] (Figure 20). Host properties were examined in different lipophilic phases. Higher extractabilities were obtained in toluene than

	Cation:	$Li^{\oplus}$	$Na^{\oplus}$	$K^{\oplus}$	$Sr^{2\oplus}$	$Ba^{2\oplus}$
42	Log K:	2.03	1.66	1.97	2.70	2.81
44			1.96	1.38	1.07	2.47

Fig. 19. Complex formation constants of some *all*-homocalixarenes in water, determined UV-spectroscopically as an increase of solubility of the host in water as a function of rising $M^{n+} + Cl_n^-$ -concentrations.

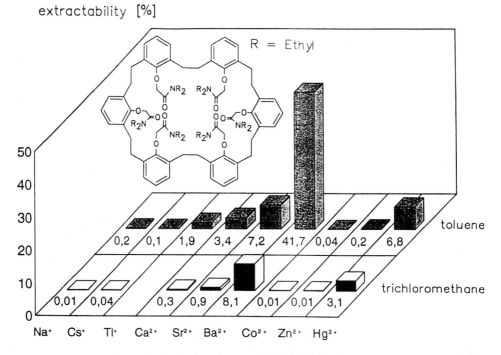

Fig. 20. Extractability of metal ions with *all*-homocalixarene **42**. $[M(NO_3)_n] = 1 \times 10^{-4}$ M; [picric acid] $= 5 \times 10^{-3}$ M; [ligand] $= 1 \times 10^{-3}$ in trichloromethane or toluene.

in trichloromethane because the amido oxygen is blocked as a binding site in the latter solvent, as revealed by infra-red spectroscopy.

It is well known from calixarene chemistry that the introduction of sulfur as a binding site shifts the selectivity from the alkali and alkaline earth metal ions towards the transition metal ions [75]. Recent results have provided evidence of selectivity towards Pd(II) and Hg(II) for a thiomorpholine-substituted hexaamido *all*-homocalixarene.

all-Homocalixpyridines have also been investigated in regard to their host properties. With pyridine nitrogen atoms as preorganised binding sites [76], *all*-homocalixpyridines show an appreciable selectivity towards transition metals, in particular Ag(I), Hg(II), Cu(II) and Pd(II), dependent upon cavity size.

4. Conclusions

A synthetic advance associated with the synthesis of *all*-homocalixarenes is that almost any binding site or functional group can be readily incorporated. These multiple possibilities for structural modification enable the host structure to be optimised for metal cations, anions, and both neutral and charged organic guests. Many applications as analytical sensors can then be envisaged.

Particular prospects are to optimise *all*-homocalixarenes towards environmentally relevant cations with, for example, the objective of removing radioactive and toxic metals from drinking water, and to design new hosts for medical applications such as magnetic resonance imaging based on gadolinium complexes.

Mass spectrometry has revealed [77] that cyclooligomers containing up to 21 aromatic units (i.e., a 105-membered ring host or 'gigantocycle' [78]) can be present in *all*-homocalixarene reaction mixtures. These are possible receptor molecules for larger biologically relevant molecules. Endowed with coordination sites like side arms containing hydrogen-bonding units, sugars, amino acids, peptides and nucleotides might well then be guests. The incorporation of homocalixarenes into nanoscale structures may well be possible by reactions between oligomers endowed with complementary binding sites.

To produce enlarged *all*-homocalixarenes soluble in water and exhibiting very high affinity constants, we have in mind the concept of *all*-bishomocalixarenes in which the aliphatic bridge is functionalised with carboxy substituents.

References

1. C. D. Gutsche: *Calixarenes*, Monographs in Supramolecular Chemistry, Vol. 1, Ed. J. F. Stoddard, The Royal Society of Chemistry, Cambridge (1989).
2. J. Vicens and V. Böhmer (Eds.): *Calixarenes: A Versatile Class of Macrocyclic Compounds*, Kluwer Academic Publishers, Dordrecht (1991).
3. C. D. Gutsche: *Acc. Chem. Res.* **16**, 161 (1983).
4. F. Vögtle: *Cyclophane Chemistry*, Wiley, Chichester (1993).
5. J. L. Sessler and A. K. Burrell: *Top Curr. Chem.* **161**, 177 (1991).
6. M. M. Pellegrin: *Recl. Trav. Chim. Pays-Bas* **18**, 457 (1989).
7. K. Burri and W. Jenny: *Helv. Chim. Acta* **50**, 1978 (1967).
8. E. Müller and G. Röscheisen: *Chem. Ber.* **90**, 543 (1957).
9. K. Burri and W. Jenny: *Chimia* **20**, 403 (1966).
10. W. Jenny and K. Burri: *Chimia* **20**, 436 (1966).
11. W. Jenny and K. Burri: *Chimia* **21**, 186 (1967).
12. W. Jenny and K. Burri: *Chimia* **21**, 472 (1967).
13. W. Jenny and R. Paioni: *Chimia* **22**, 142 (1968).
14. R. Paioni and W. Jenny: *Helv. Chim. Acta* **52**, 2041 (1969).
15. M. Tashiro and T. Yamato: *J. Org. Chem.* **46**, 1543 (1981).
16. S. Takahashi and N. Mori: *J. Chem. Soc., Perkin Trans., 1* 2029 (1991).

17. F. Vögtle: *Liebigs Ann. Chem.* **728**, 17 (1969).
18. K. Böckmann and F. Vögtle: *Chem. Ber.* **114**, 1048 (1981).
19. M. Tashiro and T. Yamato: *J. Org. Chem.* **46**. 1543 (1981).
20. M. Tashiro, T. Watanabe, A. Tsuge, T. Sawada, and S. Mataka: *J. Org. Chem.* **54**, 2632 (1989).
21. A. Tsuge, T. Sawada, S. Mataka, N. Nishiyama, H. Sakashita, and M. Tashiro: *J. Chem. Soc., Chem. Commun.* 1066 (1990).
22. M. Tashiro, A. Tsuge, T. Sawada, T. Makishima, S. Horie, T. Arimura, and S. Mataka, T. Yamato: *J. Org. Chem.* **55**, 2404 (1990).
23. A. Tsuge, T. Sawada, S. Mataka, N. Nishiyama, H. Sakashita, and M. Tashiro: *J. Chem. Soc., Perkin Trans. 1* 1489 (1992).
24. T. Yamato, Y. Saruwatari, S. Nagayama, K. Maeda, and M. Tashiro: *J. Chem. Soc., Chem. Commun.* 861 (1992).
25. T. Yamato, Y. Saruwatari, S. Nagayama, K. Maeda, and M. Tashiro: *J. Chem. Soc., Chem. Commun.* 861 (1992).
26. T. Yamato, Y. Saruwatari, L. K. Doamekpor, K.-i. Hasegawa, and M. Koike: *Chem. Ber.* **126**, 2501 (1993).
27. F. Vögtle and M. Zuber: *Synthesis* 543 (1972).
28. T. Sato, M. Wakabayashi, and K. Hata: *Bull. Chem. Soc. Jpn.* **43**, 3632 (1970).
29. R. B. Bates, S. Gangwar, V. V. Kane, K. Suvannachut, and S. R. Taylor: *J. Org. Chem.* **56**, 1696 (1991).
30. D. H. Burns, J. D. Miller, and J. Santana: *J. Org. Chem.* **58**, 6526 (1993).
31. C. D. Gutsche, B. Dhawan, K. H. No, and R. Muzhukrishnan: *J. Am. Chem. Soc.* **103**, 3782 (1981).
32. K. Hultzsch: *Kunstoffe* **52**, 19 (1962).
33. B. Dhawan and C. D. Gutsche: *J. Org. Chem.* **48**, 1536 (1983).
34. T. Tanno and Y. Mukoyama: *Netsu. Kokasei Jushi* **2**, 132 (1981); *Chem. Abstr.* **96**, 52791k (1982).
35. Y. Mukoyama and T. Tanno: *Org. Coat. Plast. Chem.* **40**, 894 (1979).
36. H. Takemura, K. Yoshimura, I. U. Khan, T. Shinmyozu, and T. Inazu: *Tetrahedron Lett.* **33**, 5775 (1992).
37. S. Mizogami, N. Osaka, T. Otsubo, Y. Sakata, and S. Misumi: *Tetrahedron Lett.* 799 (1974).
38. M. Chastrette and F. Chastrette: *J. Chem. Soc., Chem. Commun.* 534 (1973).
39. A. G. S. Högberg and M. Weber: *Acta Chem. Scand. Ser. B* **37**, 55 (1983).
40. R. M. Musau and A. Whiting: *J. Chem. Soc., Chem. Commun.* 1029 (1993).
41. M. Ahmed and O. Meth-Cohn: *Tetrahedron Lett.* 1493 (1969).
42. M. Ahmed and O. Meth-Cohn: *J. Chem. Soc. C* 2104 (1971).
43. T. Thiemann, Y. Lee, Y. Nagano, M. Tashiro: *Book of Abstr.* 'Second Workshop on Calixarenes and Related Compounds' in Krume, Fukuoka, Japan, **PS/A28** (1993).
44. E. Vogel, P. Röhrig, M. Sicken, B. Knipp, A. Herrmann, M. Pohl, and H. Schmickler, J. Lex: *Angew. Chem.* **101**, 1683 (1989); *Angew. Chem. Int. Ed. Engl.* **28**, 1651 (1989).
45. P. Rothemund and C. L. Gage: *J. Am. Chem. Soc.* **77**, 3340 (1955).
46. J. S. Lindsay, I. C. Schreiman, H. C. Hsu, P. C. Kearney, and A. M. Marguerettaz: *J. Org. Chem.* **52**, 827 (1987).
47. G. R. Newkome, Y. J. Joo, K. J. Theriot, and F. R. Fronczek: *J. Am. Chem. Soc.* **108**, 6074–6075 (1986).
48. G. R. Newkome, Y. J. Joo, and F. R. Fronczek: *J. Chem. Soc., Chem. Commun.* 854 (1987).
49. W. Jenny and H. Holzrichter: *Chimia* **21**, 509 (1967).
50. W. Jenny and H. Holzrichter: *Chimia* **22**, 139 (1968).
51. W. Jenny and H. Holzrichter: *Chimia* **22**, 306 (1968).
52. W. Jenny and H. Holzrichter: *Chimia* **23**, 158 (1968).
53. Th. Kauffmann, G. Beisser, W. Sahm, and A. Woltermann: *Angew. Chem.* **82**, 815 (1970); *Angew. Chem. Int. Ed. Engl.* **9**, 808 (1970).
54. Cf.: G. Brodesser, R. Güther, R. Hoss, S. Meier, S. Ottens-Hildebrandt, J. Schmitz, and F. Vögtle: *Pure Appl. Chem.* **65**, 2325 (1993); G. Brodesser, J. Schmitz, F. Vögtle, K. Gloe, O. Heitzsch, and H. Stephan: *Book of Abstr.* 'Second Workshop on Calixarenes and Related Compounds' in Kurume, Fukuoka, Japan, **IL-8** (1993); G. Brodesser, R. Güther, S. Meier, S. Ottens-Hildebrandt,

J. Schmitz, and F. Vögtle: *Book of Abstr.:* 'XVIII. Int. Symposium on Macrocyclic Chemistry' in Enschede, the Netherlands, **P-12** (1993); F. Vögtle: *Ann. Quim.* **89**, 29 (1993).
55. F. Vögtle, J. Schmitz, and M. Nieger: *Chem. Ber.* **125**, 2523 (1992).
56. J. Schmitz, F. Vögtle, M. Nieger, K. Gloe, H. Stephan, O. Heitzsch, H.-J. Buschmann, W. Hasse, and K. Cammann: *Chem. Ber.* **126**, 2483 (1993).
57. J. Schmitz, F. Vögtle, M. Nieger, K. Gloe, H. Stephan, O. Heitzsch, H.-J. Buschmann: *Supramol. Chem.* **3** (1994), in press.
58. F. Vögtle, G. Brodesser, M. Nieger, K. Rissanen: *Recl. Trav. Chim. Pays-Bas* **112**, 325 (1993).
59. A. Ninagawa and H. Matsuda: *Makromol. Chem., Rapid Commun.* **3**, 65 (1982).
60. M. A. Markowitz, V. Janout, D. G. Castner, and S. L. Regen: *J. Am. Chem. Soc.* **111**, 8192 (1989).
61. D. R. Stewart, C. D. Gutsche: *Org. Prep. Proced. Int.* **25**, 137 (1993).
62. T. Sato, M. Wakabayshi, K. Hata, and M. Kainosho: *Tetrahedron* **27**, 2737 (1971).
63. K. Araki, N. Hashimoto, H. Otsuka, and S. Shinkai: *J. Org. Chem.* **56**, 5958 (1993).
64. K. Araki, K. Inada, H. Otsuka, and S. Shinkai: *Tetrahedron* **49**, 9465 (1993).
65. A. Tsuge, S. Sonoda, S. Mataka, and M. Tashiro: *Chem. Lett.* 1173 (1990).
66. H.-B. Mekelburger, J. Gross, J. Schmitz, M. Nieger, and F. Vögtle: *Chem. Ber.* **126**, 1713 (1993).
67. Cooperation with Prof. Dr. K. Gloe, Dresden.
68. K. Gloe, H. Stephan, O. Heitzsch, J. Schmitz, and F. Vögtle: *Book of Abstr.:* 'XVIII. Int. Symposium on Macrocyclic Chemistry' in Enschede, the Netherlands, **A-70** (1993).
69. H. Stephan, K. Gloe, O. Heitzsch, G. Brodesser, and F. Vögtle: *Book of Abstr.:* '24. GDCH-Hauptversammlung', in Hamburg, Germany, **SUP 9** (1993).
70. Cooperation with Prof. Dr. K. Cammann, Münster.
71. J. Reinbold, B. Ahlers, W. Hasse, K. Cammann, G. Brodesser, and F. Vögtle: *Book of Abstr.:* 'XVIII. Int. Symposium on Macrocyclic Chemistry' in Enschede, the Netherlands, **B-55, 56** (1993).
72. W. Hasse, K. Cammann, F. Vögtle, and G. Brodesser: *Book of Abstr.:* 'Eurosensor VII', in Budapest, Hungary, **AP-24, 4430-1** (1993).
73. Cf.: R. M. Izatt, J. D. Lamb, R. T. Hawkins, P. R. Brown, S. R. Izatt, and J. J. Christensen: *J. Am. Chem. Soc.* **105**, 1782 (1983); S. R. Izatt, R. T. Hawkins, J. J. Christensen, and R. M. Izatt: *J. Am. Chem. Soc.* **107**, 63 (1985).
74. Cooperation with Dr. H.-J. Buschmann, Krefeld.
75. R. Perrin and S. Harris: in *Calixarenes: A Versatile Class of macrocyclic Compounds*, J. Vicens and V. Böhmer (Eds.), Kluwer Academic Publishers, Dordrecht, p. 241 (1991).
76. Cf.: S. Shinkai, T. Otsuka, K. Araki, and T. Matsuda: *Bull. Chem. Soc. Jpn.* **62**, 4055 (1989).
77. Cooperation with Prof. Dr. M. Przybylski, Konstanz.
78. F. M. Menger, S. Broccini, and X. Chen: *Angew. Chem.* **104**, 1542 (1992); *Angew. Chem. Int. Ed. Engl.* **31**, 1492 (1992).

Calixcrowns and Related Molecules*

Z. ASFARI, S. WENGER and J. VICENS**
E.H.I.C.S., URA 405 du C.N.R.S., 1 rue Blaise Pascal, F-67008 Strasbourg, France

(Received in final form: 6 June 1994)

Abstract. The synthesis of 1,2- and 1,3-calix[4]-*bis*-crowns, double calix[4]arenes and double calixcrowns have been shown to depend on the reaction conditions (nature of the base, structure of the ditosylates, and the stoichiometry of the reactants). The 1,3-alternate conformation of the 1,3-calix[4]-*bis*-crowns was shown to be favourable to the selective complexation of cesium cation. The observed Na^+/Cs^+ selectivity was exploited in separation processes using them as carriers in transport through supported liquid membranes (SLMs). The best Na^+/Cs^+ selectivity (1/45 000) was observed for the naphthyl derivative **7**. Calix(aza)crowns and 1,3-calix[4]-*bis*-(aza)-crowns were also produced through the preliminary formation of the Schiff base-calixarenes, which were further hydrogenated. The syntheses consisted of the 1,3-selective alkylation of calixarenes followed by cyclization into a 1,3-bridged calixarene or by the direct 1,3-capping of the calixarene with appropriate ditosylates. Soft metal complexation by these ligands is also presented.

Key words: 1,2- and 1,3-Calix[4]-*bis*-crowns, double calix[4]arenes, double calixcrowns, calix(aza)crowns, Na^+/Cs^+ selectivity, soft metal complexation.

1. Introduction

In the 25 years since Pedersen [1] reported on the synthesis and metal–cation complexing properties of crown ethers the number of crown ether compounds has been continuously increasing. Nowadays the synthesis of a crown ether molecule is directed toward a desired application (molecular and enantiomer recognition, asymmetric catalysis, redox properties, allosteric effect, replication, molecular assembly processes, etc.) [2]. To create these application-directed macrocycles chemists have developed the synthesis of macropolycycles or cage molecules [3] with a molecular framework which combines simple, already existing molecular elements, the functions of which are known. Hegelson *et al.* [4] prepared multistranded systems containing binaphthyl and related compounds with polyether chains for chiral recognition. Weber [5] described a series of multiloop ligands made from hexaphenol and crown ethers. These molecules, which had previously been synthesized as polytopic cation receptors, were observed to be surfactants having a novel construction principle. Sijbesma and Nolte [6] prepared a *bis*-aza-tetra-oxa-crown ether which acts as a molecular clip with allosteric binding properties. Nakano *et al.* [7] proposed a tri-crown ether oligomer as a mimic of the nature and coiling of gramicidin. Rumney and Kool [8] constructed hybrid circular molecules which

---
* This paper is dedicated to the commemorative issue on the 50th anniversary of calixarenes.
** Author for correspondence.

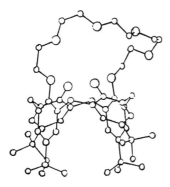

Fig. 1. The first calixcrown (Alfieri *et al.* [11]).

contain two oligonucleotide domains bridged by two oligoethylene glycol chains. These molecules bind with high affinity to complementary strands of RNA and DNA and display exceptional resistance to degradation by nucleases. Kobuke *et al.* [9] designed macrocycles containing both catecholate functions and polyether groups with metal-assisted organization and cooperative metal binding.

Since 1991 we have developed the synthesis of macropolycycles containing in their molecular structure the monocyclic structure of calixarenes and crown ether elements. This combination gives a close coupling of the hydrophobic cavity of the calixarenes able to include organic substrates and the metal cation complexing sites of the crown ether, with potential interactions between them. We have already demonstrated evidence of such cation–substrate contact during a triple inclusion by a calixarene [10]. The crystal structure of the Eu(III) complex of *bis*-(homooxa)-*p*-*tert*-butylcalix[4]arene showed the Eu(III) to be coordinated to a DMSO molecule included in the hydrophobic cavity of the calixarene [10].

Calixcrowns refer to the family of macropolycyclic or cage molecules in which the monocyclic structures of calixarenes and crown ethers are combined through the bridging of phenolic oxygens of a calixarene by a polyether chain. The first member of this family was produced by Alfieri *et al.* [11], who reacted pentaethylene glycol ditosylate with *p-tert*-butylcalix[4]arene under basic conditions to produce 1,3-*p-tert*-butylcalix[4]crown-6, see Figure 1.

2. Calixcrowns

We have synthesized calixcrowns by systematic reactions of calix[4]arenes with various ditosylates. Depending on the reaction conditions (nature of the base, structure of the ditosylates, and the stoichiometry of the reactants) products with different topologies were isolated (Scheme 1).

The reaction of calix[4]arene with 4–6 equivalents of various ditosylates (containing 5, 6 or 7 oxygen atoms) in the presence of K_2CO_3 in refluxing acetonitrile produced 1,3-calix[4]-*bis*-crowns **1–7** in 60–80% yield [12] (Figure 2). The for-

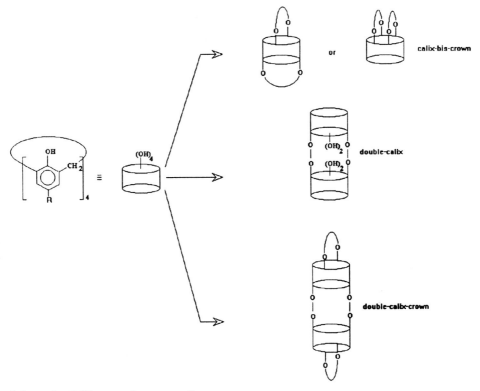

Scheme 1. Different pathways to calixcrowns.

mation of **1–7** implies a preliminary *distal* 1,3-capping of the calix[4]arene by one glycolic chain. The second capping forces the calixarene to adopt the 1,3-alternate conformation which is effectively observed in the ^{1}H-NMR spectra. Similar reactions afforded 1,3-*p-iso*-propylcalix[4]crown-5 **8** (which shows the reaction to be general) and 1,3-*p-tert*-butylcalix[4]crown-5 **9** (for comparison with the published molecule [13]). When K_2CO_3 was replaced by Cs_2CO_3 we also isolated the 1,2-calix[4]-*bis*-crown-5 **10** isomeric to **1**, in 10% yield in which the capping by the glycolic chains is *vicinal* [14]. ^{1}H-NMR indicated that the calix[4]arene moiety is in the cone conformation [14]. 1,2-Dialkylations have been observed by other groups during the use of Cs_2CO_3 [15] and CsF [16].

The conformations of the 1,2- and 1,3-calix[4]-*bis*-crowns were ascertained by the determination of the X-ray structure of 1,3-*p-tert*-butylcalix[4]crown-5 **9** [17] and 1,2-calix[4]-*bis*-crown-5 **10** [14] (Figure 3).

When *shorter* ditosylates were used *lower rim–lower rim* double calixarenes were obtained [18]. For instance, the reaction of calix[4]arene, *p-tert*-butyl- and 1,3-dimethoxy-*p-tert*-butylcalix[4]arene with 2 equivalents of triethylene glycol ditosylate lead to double calixarenes **11**, **12**, and **13** in very good yields (Figure 4). They were shown to consist of two calix[4]arene units linked by two distal glycolic

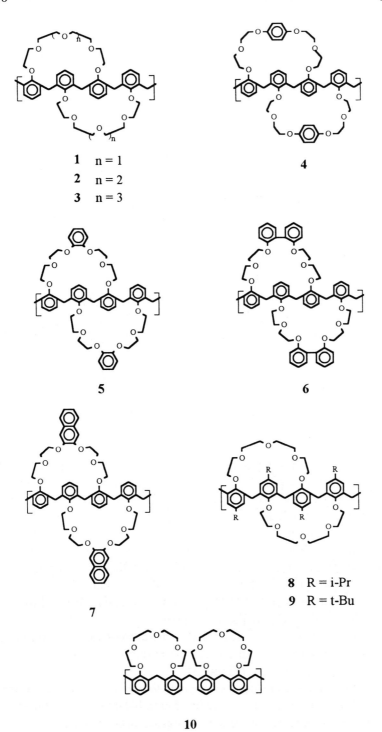

Fig. 2. 1,3-Calix[4]-*bis*-crowns.

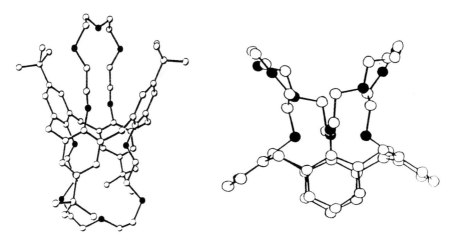

Fig. 3. 1,3-*p-tert*-Butylcalix[4]crown-5 **9**, and 1,2-calix[4]-*bis*-crown-5 **10**.

chains. ^{1}H-NMR showed the calixarene moieties to be in the cone conformation [18]. Recently a double calix[4]arene, very similar to compounds **11–13**, linked with two tetraglycolic chains (R = monoglycol; R' = *t*-Bu) was shown to be a ditopic ionophore in which Na$^+$ or K$^+$ vibrates between the two metal-binding sites in the NMR time scale [19]. The treatment of *p-tert*-butylcalix[4]arene with diethylene glycol ditosylate afforded double-calix[4]arene **14** triply bonded via the oxygen atoms by glycolic chains [20]. ^{1}H-NMR and ^{13}C-NMR displayed complex spectrometric data due to several conformations of the *p-tert*-butylcalix[4]arene subunits and/or *anti* or *syn* linkages [12].

Subsequent to these results, we discovered a striking example of the formation of a double calixcrown by changing the stoichiometry of the reactants. *p-tert*-Butylcalix[4]arene was treated with a 15 equivalent excess of tetraethylene glycol ditosylate to afford double *p-tert*-butylcalix[4]-*bis*-crown-5 **15** in which each calixarene unit is in the 1,3-alternate conformation and 1,3-capped by a tetraethylene glycolic chain [21].

The preparation of 1,3-calix[4]-*bis*-crowns and double calixcrowns was also performed with the mesitylene derived calix[4]arene or calix[4]mesitylene which exists only in the 1,3-alternate conformation [22], allowing us to conclude that *1,3-calix[4]-bis-crowns are formed with longer glycolic chains while double calixcrowns are formed with shorter and more rigid ones* (Figure 5).

3. Metal Complexation Properties

Our interest was in the recovery of cesium from waste waters by solvent extraction using an extractant able to selectively bind the cesium in the presence of large amounts of sodium [25]. Only those calixcrowns in the 1,3-alternate conformation were tested for complexation of alkali cations due to the spherical geometry they

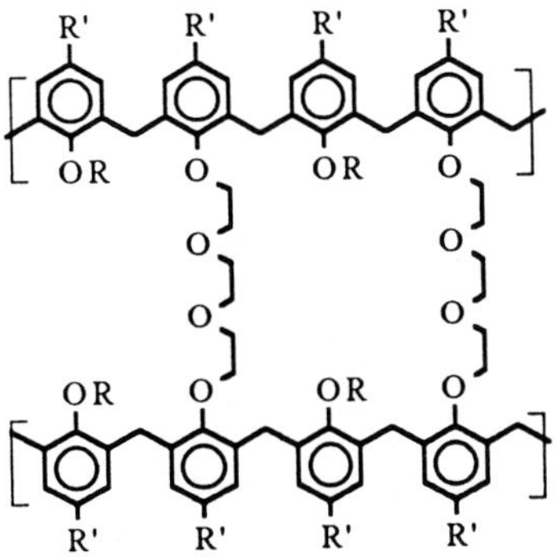

Fig. 4. Double calixarenes.

11 R = R' = H
12 R = H   R' = t-Bu
13 R = Me R' = t-Bu

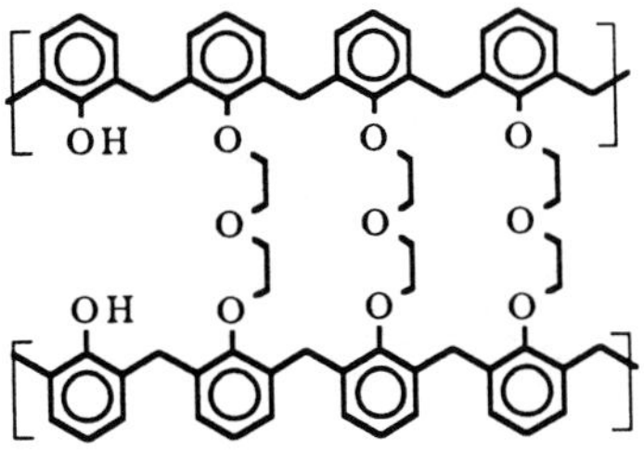

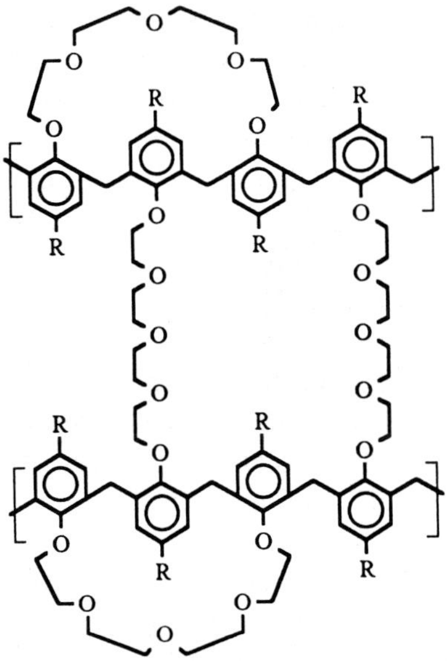

15 R = t-Bu

offer, comparable to the spheric cryptands of Graf and Lehn [26]. 1,3-Calix[4]-*bis*-crowns **1–3**, **5** and **7** were first shown to extract alkali ions, with the exception of Li^+, from the corresponding solid picrates (Me^+Pic^-) in excess in a chloroform phase. The 1 : 1 (Me^+Pic^-)·**1–3**, **5** and **7** complexes were isolated as yellow solids. The stoichiometries were deduced from ^{1}H-NMR. The presence of original and

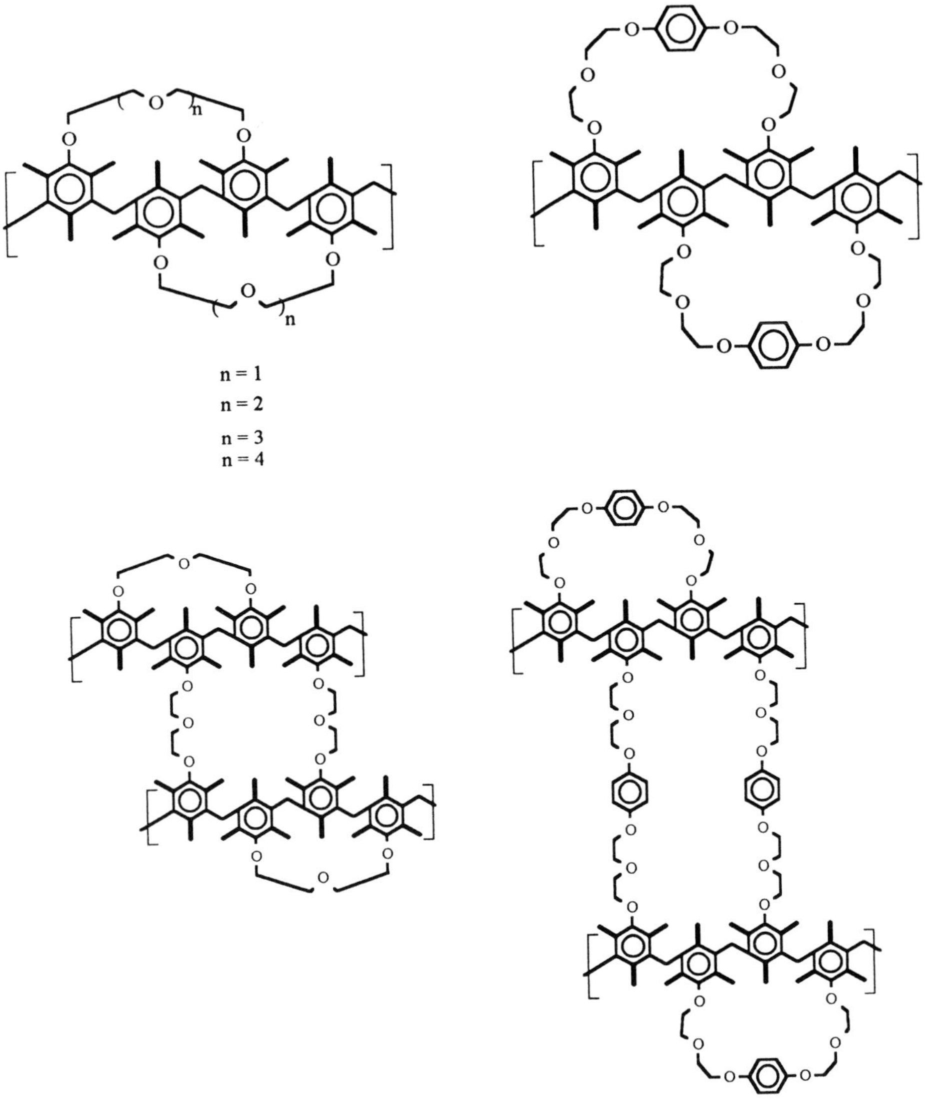

Fig. 5. 1,3-Calix[4]-*bis*-crowns and double calixcrowns derived from calix[4]arene or calix[4]mesitylene.

shifted signals of the protons of the polyether chain in a 1 : 1 integration ratio lends support to a location of the cation in one polyether loop. From the X-ray structure of **9** [17] the approximate radius of one polyether loop was determined to be ∼ 1.5 Å, a size approximately complementary to K^+ and Rb^+, leading us to assume that a calixcrown containing six oxygens may well fit with Cs^+, as observed for **1–3, 5** and **7**. 1,3-Calix[4]-*bis*-crowns **4** and **6** did not extract, probably due to a larger and more rigid glycolic chain. *p-iso*-Propyl and *p-tert*-butyl analogues **8** and **9** were also unable to extract the cations. This was attributed to a shielding of the crown unit by the bulky *iso*-propyl and *p-tert*-butyl groups preventing the cation from complexation, as deduced from the crystal structure of **9** [17].

Additional solvent extraction of alkali picrates from the aqueous phase to dichloromethane solutions and the UV determination of the constants of complex formation in methanol showed the selectivity of complexation to be in the order $Cs^+ > Rb^+ \sim K^+ > Na^+ > Li^+$ for those 1,3-calix[4]-*bis*-crowns **2, 5**, and **7** containing six oxygens in the glycolic chain [27].

By a similar method double *p-tert*-butylcalix[4]-*bis*-crown-5 **15** was shown to be selective for K^+ and Rb^+ with a location of the cation in the central cavity [21]. Ligand **15** extracted Li^+ (3%), Na^+ (2%), K^+ (35%), Rb^+ (45%), Cs^+ (7%) with the highest K values of 4.9 and 5.2 log units, respectively, for K^+ and Rb^+ [21].

4. Industrial Application

The observed Na^+/Cs^+ selectivity was exploited in separation processes using 1,3-calix[4]-*bis*-crowns **2, 5**, and **7** as carriers in transport through supported liquid membranes (SLMs). They transported Cs^+ through a microporous propylene support (NPOE) from an acidic phase (HNO_3, 1N, $NaNO_3$, 3N). The best Na^+/Cs^+ selectivity (1/45 000) was observed for the naphthyl derivative **7** [28]. The selectivity was attributed to π-metal interactions favorable to complexation of a larger Cs^+ cation, which is polarisable and poorly hydrated as compared to sodium cation.

5. Calix(aza)crowns

We have previously described the synthesis of calixcrowns by one-pot reactions. In a different approach the calix(aza)crowns were constructed by stepwise synthesis. The synthesis began with the condensation of cone-1,3-dialdehyde derivative **16** with various primary diamines to afford Schiff base–*p-tert*-butylcalix[4]arenes **17–20** in the cone conformation [29, 30]. Hydrogenation of **18** and **19** with $NaBH_4$ produced chlorohydrates **21**·HCl and **22**·HCl and subsequent deprotonation with NaOH lead to the di-aza-benzo-crown-ether-*p-tert*-butylcalix[4]arenes **21** and **22** in almost quantitative yield [31] (Figure 6).

In a similar manner we achieved the synthesis of *bis*-Schiff base-calix[4]mesitylenes **25** and **26** from the corresponding tetraaldehydes **23** and **24**. We could not hydrogenate **25** and **26** due to their low solubility in the usual solvents for the reaction (Figure 7).

We were unable to prepare the tetraaldehyde precursor corresponding to the calix[4]arene and the synthesis of the hydrogenated *bis*-Schiff base-calix[4]arene **27** was achieved by stepwise synthesis. Aldehyde **28** was prepared from salicylaldehyde and 2-(2-chloroethoxy)ethanol. Condensation of **28** with 1,4-diaminobutane lead to the Schiff base **29** which was hydrogenated with $NaBH_4$ to give the chlorhy-

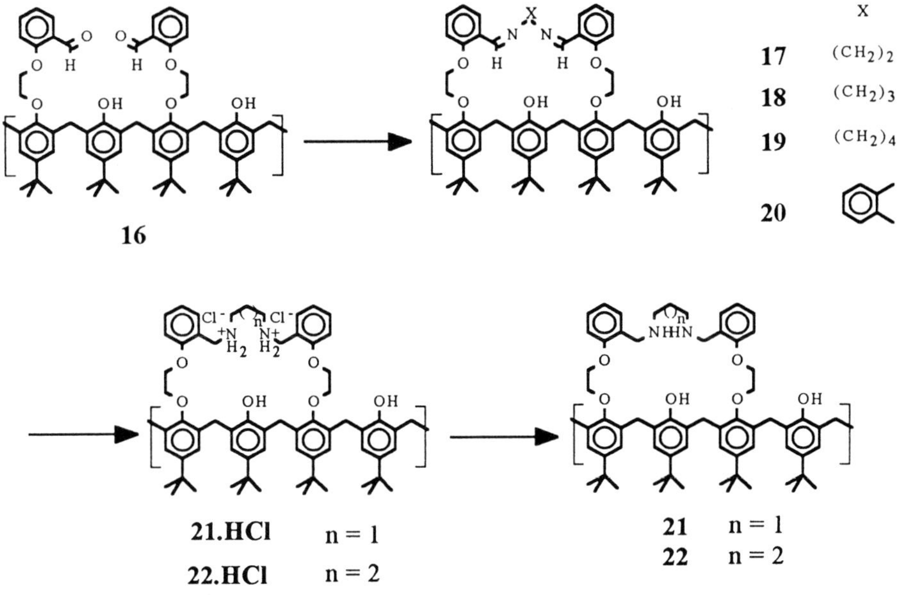

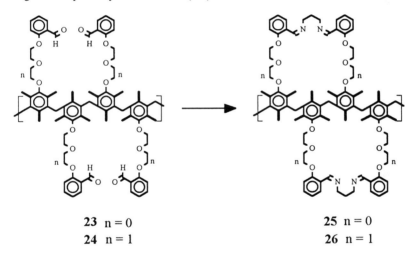

Fig. 6. Stepwise synthesis of calix(aza)crowns.

23 n = 0
24 n = 1

25 n = 0
26 n = 1

Fig. 7. *bis*-Schiff base-calix[4]mesitylenes.

drate **30**·HCl. Deprotonation with NaOH lead to the di-aza-di-benzo-tetraethylene glycol **30** which was transformed into tetratosylate derivative **31**. The reaction of calix[4]arene with 2 equivalents of **31** using conditions previously described leads to the tetratosylate derivative **32**. Ditosylation of compound **32** with H_2SO_4 produced the expected 1,3-calix[4]-*bis*-(di-aza-di-benzo-crown-6) **27**.

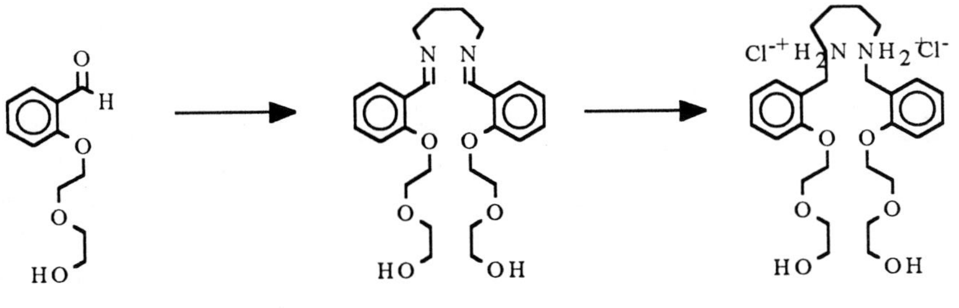

28 **29** **30.HCl**

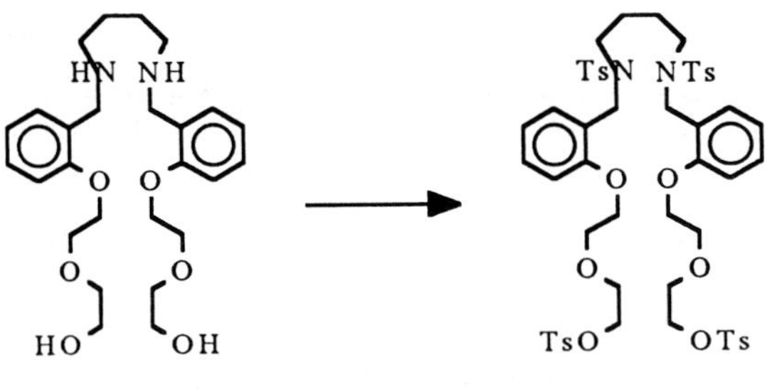

30 **31**

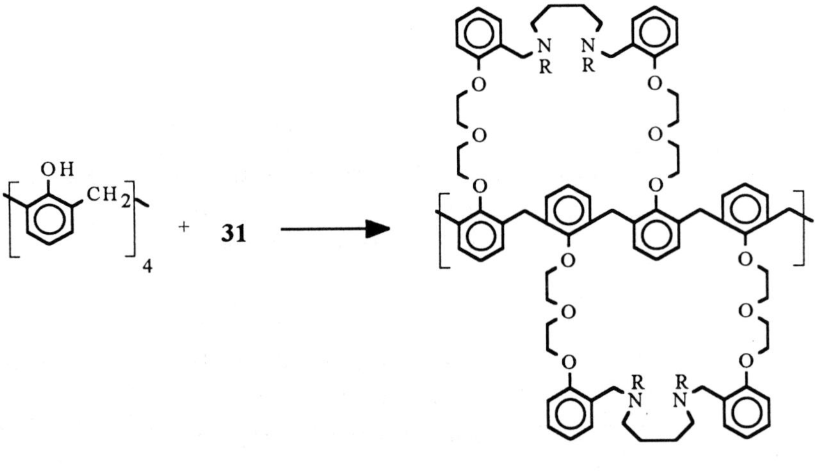

32 R = Ts
27 R = H

Fig. 8. Stepwise synthesis of hydrogenated *bis*-Schiff base-calix[4]arene **27**.

6. Metal Complexation

Preliminary binding properties of Schiff base-*p*-*tert*-butylcalix[4]arenes **17–20** were tested by solvent extraction of alkaline, alkaline earth metal, Mn, Fe, Co, Ni, Cu, Cd, Pb, Y, Pr, Nd, Eu, Gd, Yb picrates into dichloromethane under neutral conditions [29]. In a general manner **17** was observed to extract less than its 3- and 4-carbon-containing homologues [29]. Alkali and alkaline earth cations were poorly extracted [29]. Heavy metal cations are better extracted, with higher preference for Pb^{2+} (15% and 18%) with **18** and **19**, respectively [29]. In the lanthanide series, there is an extraction selectivity for Nd^{3+} (12%) and Eu^{3+} (13%) ions with **18** and for Eu^{3+} (14%) with **19**. The best extracted metal ions were tested with the aromatic ligand **20** which showed a good extracting ability for Cu^{2+} (24%) [29]. The more efficient systems in extraction were studied in complexation by UV-visible spectrometry. All the data were interpreted by the presence of 1 : 1 complexes [29]. As was foreseen from the extraction results, the binding ability depends on the length of the Schiff base bridge. The optical cavity size for Eu^{3+} and for the larger Pb^{2+} is the bridge containing three carbons [29].

Zinc cation complexation by di-aza-benzo-crown-ether-*p-tert*-butylcalix[4]-arene **21** in $CDCl_3/CH_3OH$ was studied by monitoring the ^{1}H-NMR upon addition of Zn^{2+}. The plot of data obtained after mixing indicated a well-defined titration curve with mole ratio 1 : 1 and 2 : 1 until complete addition of 6 equivalents of Zn^{2+}. The reaction solution reached equilibrium after 26 days, fitting with a total formation of a 2 : 1 complex. The data were analyzed by employing two equations:

$$M + L \rightleftharpoons ML$$

$$ML + L \rightleftharpoons ML_2$$

with $\log K_1 = 3.6$ and $\log K_2 = 3.0$. All the results were rationalized by a rapid chelation of a first Zn^{2+} by the two nitrogens followed by a slower entry of a second Zn^{2+} into the remaining six-oxygen array [31].

7. Conclusions

To summarize, in this paper we have presented the synthesis of calixcrowns and calix(aza)crowns. Calixcrowns were prepared by a one-pot procedure. Depending on the nature of the reactants and on the experimental conditions one can induce the reaction towards the formation of calix-*bis*-crowns, double calixarenes or double calixcrowns. Alkali metal complexation ability of the calixcrowns was studied and the selectivity of complexation was observed to depend on the number of oxygens in the polyether chain. This fundamental study found an application in the transport of Cs^+ through SLMs with selectivity $Na^+/Cs^+ \sim 1/45\,000$.

Calix(aza)crowns and 1,3-calix[4]-*bis*-(aza)crowns were constructed via the Schiff base intermediates. The Schiff base-calixarenes were observed to preferentially complex soft cations, probably due to the presence of nitrogen atoms. The best systems were for Pb^{2+} and Eu^{3+} with stability constants ~ 5 log units. Zn^{2+} complexation by a calix(aza)crown was studied, showing that the receptor occludes two metals by two distinguishable steps.

Acknowledgements

Research support from the Commissariat à l'Energie Atomique and the Centre National de la Recherche Scientifique is gratefully acknowledged.

References

1. C. J. Pedersen: *J. Am. Chem. Soc.* **89**, 7017 (1967).
2. S. R. Cooper: in *Crown Compounds, Toward Future Applications*, VCH Publishers Inc., New York (1992).
3. H. An, J. S. Bradshaw, and R. M. Izatt: *Chem. Rev.* **92**, 543 (1992).
4. R. C. Helgeson, T. L. Tarnowski, and D. J. Cram: *J. Org. Chem.* **44**, 2538 (1979).
5. E. Weber: *Angew. Chem. Int. Ed. Engl.* **22**, 616 (1983).
6. R. P. Sijbesma and R. J. M. Nolte: *J. Am. Chem. Soc.* **113**, 6695 (1991).
7. A. Nakano, Q. Xie, J. Mallen, L. Etchegoyen, and G. W. Gokel: *J. Am. Chem. Soc.* **112**, 1287 (1990).
8. S. Rumney IV and E. T. Kool: *Angew. Chem. Int. Ed. Engl.* **31**, 1617 (1992).
9. Y. Kokube, Y. Sumida, M. Hayashi, and H. Ogoshi: *Angew. Chem. Int. Ed. Engl.* **30**, 1496 (1991).
10. Z. Asfari, J. Harrowfield, M. I. Ogden, J. Vicens, and A. H. White: *Angew. Chem. Int. Ed. Engl.* **30**, 854 (1991).
11. C. Alfieri, E. Dradi, A. Pochini, R. Ungaro, and G. D. Andreetti: *J. Chem. Soc., Chem. Commun.* 1075 (1983).
12. C. Bressot, Z. Asfari, and J. Vicens: *J. Incl. Phenom.* (to appear).
13. P. J. Dijkstra, J. A. Brunink, K.-E. Bugle, and D. N. Reinhoudt: *J. Am. Chem. Soc.* **112**, 1597 (1990).
14. C. Bressot, J. P. Astier, Z. Asfari, J. Estienne, G. Pepe, and J. Vicens: *J. Incl. Phenom.* **19**, 291 (1994).
15. J. A. J. Brunink, W. Verboom, J. F. J. Engbersen, S. Harkema, and D. N. Reinhoudt: *Rec. Trav. Chim. Pays-Bas* **111/112**, 511 (1992).
16. D. Kraft, R. Arnecke, V. Böhmer, and W. Vogt: *Tetrahedron* **49**, 6019 (1993).
17. Z. Asfari, J. Harrowfield, and J. Vicens: *Aust. J. Chem.* **47**, 757 (1994).
18. Z. Asfari, J. Weiss, S. Pappalardo, and J. Vicens: *Pure Appl. Chem.* **68**, 585 (1993).
19. F. Ohseto and S. Shinkai: *Chem. Lett.* 2045 (1993).
20. Z. Asfari, J. Weiss, and J. Vicens: *Polish J. Chem.* **66**, 709 (1992).
21. Z. Asfari, R. Abidi, F. Arnaud, and J. Vicens: *J. Incl. Phenom.* **13**, 163 (1992).
22. S. Pappalardo, G. Ferguson, and J. F. Gallagher: *J. Org. Chem.* **57**, 7102 (1992).
23. Z. Asfari, S. Pappalardo, and J. Vicens: *J. Incl. Phenom* **14**, 189 (1992).
24. Z. Asfari, J. Weiss, and J. Vicens: *Synlett* 719 (1993).
25. I. H. Gerow, J. E. Smith Jr., and M. W. Davis, Jr.: *Sep. Sci. Techn.* **16**, 519 (1981).
26. E. Graf and J.-M. Lehn: *J. Am. Chem. Soc.* **98**, 6403 (1976).
27. B. Souley, Z. Asfari, F. Arnaud, and J. Vicens: unpublished results.
28. J. F. Dozol, Z. Asfari, C. Hill, and J. Vicens: *French Patent No. 92 14245*, 26 November (1992).
29. R. Seangparsertkij, Z. Asfari, F. Arnaud, and J. Vicens: *J. Org. Chem.* **59**, 1741 (1994).
30. R. Seangparsertkij, Z. Asfari, F. Arnaud, and J. Vicens: *J. Incl. Phenom.* **14**, 141 (1992).
31. R. Seangparsertkij, Z. Asfari, and J. Vicens: *J. Incl. Phenom.* **17**, 111 (1994).

Calixarene-Based Sensing Agents*

DERMOT DIAMOND
School of Chemical Sciences, Dublin City University, Ireland

(Received: 8 June 1994; in final form: 8 August 1994)

Abstract. The well-known selective receptor properties and ease of structural modification makes calixarene derivatives attractive materials for use in chemical sensors. This review looks at the history of sensor-related calixarene research prior to 1994 and identifies current trends in sensor research which are influencing the types of derivatives being synthesised and methods of evaluation, such as the increasing popularity of optical modes of transducing host-guest interaction. Future possibilities are briefly discussed and the need for more fundamental studies highlighted.

Key words: Calixarenes, ion-selective electrodes, sensors, optrodes.

1. Introduction

The development of new and more efficient means of performing real-time monitoring of species by means of sensors is one of the most difficult challenges facing modern science. The problems involved are multifaceted, requiring a broad understanding of many areas of expertise (ranging from organic synthetic chemistry, to thin layer deposition and surface analysis technologies, to computer-based data acquisition and signal processing) and their solution therefore demands a multi-disciplinary research effort. Central to determining the overall performance of any chemical sensor is the nature of the sensing component used to generate the analytical signal, as this will largely (but not absolutely) define the critical characteristics of the resulting device, namely the selectivity, lifetime and response time. However, despite a huge effort over the past 30 years or so, the number of really efficient individual sensors remains disappointingly small, mainly due to the *ad-hoc* nature of the design and synthesis of possible agents.

To be fair, the difference between a really efficient sensing agent and a hopeless one is very difficult to predict as, on a molecular basis, the processes which together define the overall preference of a host for a target guest in preference to all other interferents interact in subtle ways. However, recent improvements in the power of computer systems, and refinements in the algorithms used to minimise molecular energies have enabled more accurate predictions of 3-D structures to be made, and to probe in advance how a ligand might interact in a dynamic sense with certain guest species in different solvents. In addition, the large amount of information now

---

* This paper is dedicated to the commemorative issue on the 50th anniversary of calixarenes.

available should perhaps enable statistical tools and pattern recognition techniques to give us more insight into the structural factors which influence selectivity.

2. Transduction Modes

The role of the sensing agent in a chemical sensor is to provide a transduction mechanism which enables an analytical signal to be obtained. The vast majority of calixarenes investigated as chemical sensors have employed an electrochemical transduction mechanism, either potentiometric or voltammetric/amperometric, although more recently there has been a move towards optical transduction. As the vast majority of working devices described have been potentiometric, this paper will focus mainly on these devices.

3. Electrochemical Transduction

POTENTIOMETRIC

In this case, spontaneous processes occur in the electrochemical cell leading to the generation of a cell potential which reaches a steady state when the net current flowing in the cell and measurement circuitry is zero, i.e. the processes occurring at the electrodes are at equilibrium.

AMPEROMETRIC

In this case, the electrochemical reactions are forced to happen at the working electrode under the influence of an externally poised potential controlled by a potentiostat.

Of these, calixarenes have been most extensively investigated as sensing agents for potentiometric sensors, that is, ion-selective electrodes (ISEs). This was perhaps inevitable, given the requirements of ideal sensing agents (ionophores, ligands) for these devices, which many of the early calix[4]arene derivatives such as the esters and ketones clearly possessed (see below). Pioneering work by Simon and coworkers [1] over a period of almost 30 years had demonstrated that neutral carriers were suitable ligands for use in potentiometric sensors aimed at cation analysis. Although most ISEs these days utilise PVC or some other support material to give a 'pseudo-solid' sensing membrane, the signal generation process still involves partitioning the primary ion into a nonpolar sensing liquid-membrane phase. Hence the role of the ionophore can be summarised as:

1. selectively complex the target or primary cation;
2. reject other cations and anions;
3. retain the complex within the membrane phase;
4. allow the complex to diffuse freely in the direction of the potential gradient; and finally,

5. the stability constant for the ion–ligand complex should not be too large.

Requirements 1 and 2 above are vital in determining the selectivity of the sensor, whereas 3 ensures an adequate lifetime. Requirement 4 (permselectivity) is necessary to provide a mechanism for charge transport through the membrane while 5 is a requirement for reversibility of the analytical signal (i.e. the signal should be able to return to baseline after being exposed to a more concentrated solution of the primary ion). In addition to these, the kinetics of complexation–decomplexation has to be sufficiently fast to guarantee an adequate response time to fluctuations in the primary ion concentration during continuous monitoring applications.

The above requirements define in turn the structural features needed by ligands which would render them possible contenders for use in ISE membranes. These can be summarised as:

(i) Polar ligating groups arranged spatially in such a manner that they would interact strongly with the target ion. The ions in question are usually alkali metals (or alkaline earth), with the commercial driving force over the past 30 years being the development of sensing technology for use in blood analysis. Suitable groups would be carbonyl oxygen atoms (as part of amide, ester or ketone functionalities) arranged so as to define a 3-D polar cavity which would give a best-fit with the target ion. The polar cavity would have to be sufficiently rigid to maximise selectivity and the ion–dipole interaction strong enough to ensure that the process represented by Equation (1) below favoured the right-hand side to the degree defined in Equation (2).

$$L_{(m)} + M^+_{(aq)} \rightleftharpoons LM^+_{(m)} \qquad (1)$$

$$\beta = \frac{[LM^+_{(m)}]}{[L_{(m)}][M^+_{(aq)}]} \approx 10^5 \qquad (2)$$

where L = ligand, M^+ is the metal ion, LM^+ is the complex, β is the stability constant and (m) and (aq) denote the sensor membrane and sample aqueous phases, respectively. From thermodynamics, we can predict from Equation (2) that;

$$\Delta G = -RT \ln \beta \approx -30 \text{ kJ mol}^{-1}. \qquad (3)$$

In other words, the ligating groups have to provide 30 kJ mol^{-1} surplus energy above that required to replace the water molecules of the hydrated ions.

(ii) However, notwithstanding the above discussion, in order to comply with requirement 3, these ligands, and the resulting positively charged complex, would have to be preferentially retained in the nonpolar membrane phase of the sensor. This could only be achieved by combining large, nonpolar groups

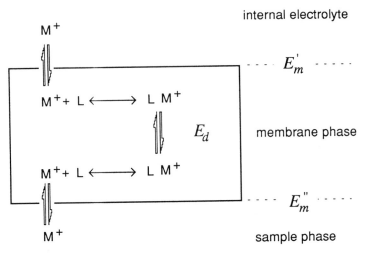

Fig. 1. Processes occurring in an idealised ion-selective electrode membrane.

around the polar cavity, so as to shield the effect of the polar groups and charged complex from the nonpolar membrane environment. However, the molecule could not be made too bulky, or diffusion of the complex through the membrane would become a problem (requirement 4).

The processes occurring in an idealised ion-selective electrode (ISE) membrane are summarised in Figure 1. The aqueous analyte ions ($M^+_{(aq)}$) are in equilibrium with the electrode membrane phase under the control of the ion–ligand complexation reaction, generating a boundary potential (E''_m). The complexes generate a diffusion-migration potential (E_d) across the membrane phase which is normally constant, and an internal boundary potential (E'_m) which is fixed, assuming that the concentration of the analyte ions on both sides of the internal boundary remains invariant. Hence fluctuations in the overall electrode potential arise from changes in the sample/membrane boundary potential (E''_m) alone, which are related to the concentration (more correctly, activity) of the analyte ions in the sample phase.

4. Synthetic Ligands for ISEs

As mentioned above, Simon and coworkers led the way in the design of synthetic receptors for metal ions which would be suitable for use in ion-selective electrodes. Besides these systems (primarily based on amides), other workers were looking at alternatives in the various groups of neutral ligating families which were being developed during the 1970s, such as crowns and spherands. Given this activity, it was inevitable that calixarenes would eventually be examined, as their structural features met many of the requirements outlined above. This was recognised by McKervey and Svehla, who initiated the first investigation into the use of calixarenes as potentiometric sensors in 1985. Luckily as it turns out, the compounds

TABLE I.

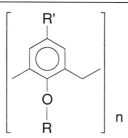

Index	R	R'	n
1	CH$_2$CO$_2$Et	t-But	4
2	CH$_2$CO$_2$Me	t-But	4
3	CH$_2$CO$_2$Et	t-But	6
4	CH$_2$CO$_2$Et	H	6
5	CH$_2$CO$_2$C$_{10}$H$_{21}$	t-But	4
6	CH$_2$CON(nC$_4$H$_9$)$_2$	t-But	4
7	CH$_2$CO$_2$-C$_6$H$_5$ (benzyl ester)	t-But	4
8	C$_2$H$_5$CO-But	t-But	4
9	C$_2$H$_5$CO-C$_6$H$_5$ (phenyl ester)	t-But	4
10	CH$_2$CO$_2$CH$_2$CH$_2$-N(pyrrolidinone)	t-But	4
11	CH$_2$C(S)N(C$_2$H$_5$)$_2$	t-But	4
12	CH$_2$CO$_2$CH$_2$CH$_2$SMe	t-But	4
13	CH$_2$CO$_2$CH$_2$CH$_2$OMe	t-But	4
14	CH$_2$CO$_2$(CH$_2$)$_3$(SiO$_2$)SiCH$_3$ with CH$_3$ groups	t-But	4
15	CH$_2$CH$_2$SCH$_3$	t-But	4
16	CH$_2$CH$_2$SC(S)N(C$_2$H$_5$)$_2$	t-But	4
17	CH$_2$CH$_2$OC(S)N(CH$_3$)$_2$	t-But	4

chosen for the study included two tetrameric esters (Table I; **1** tetraethyl-*p*-*t*-butylcalix[4]arene tetraactetate and **2** tetramethyl-*p*-*t*-butylcalix[4]arene tetraactetate), which turned out to be excellent ionophores for sodium ions. Initial screening experiments carried out with liquid membrane sensors confirmed that excellent sensors for sodium could be produced with these ligands (Figures 2a and 2b, respectively), and that devices exhibiting caesium selectivity (Figures 2c and 2d) could be made with the hexamer esters (Table I; **3** tetraethyl-*p*-*t*-butylcalix[6]arene tetraactetate and **4** tetraethylcalix[6]arene tetraactetate). The results were presented at an international conference held in Dublin, Ireland, in June 1986, and were published as part of the proceedings of this meeting later the same year [2]. To this author's knowledge, this was the first reported use of calixarenes as sensing agents to appear in the literature. A second publication early in 1987 by Diamond and Svehla [3] again highlighted the excellent selectivity of ligand **2** for sodium against potassium and a range of other interferents which can affect the measurement of sodium in blood (the most important commercial application for sodium measurements). Detailed studies on the properties of PVC membrane ISEs based on the same compounds confirmed their usefulness as sodium and caesium sensors [4].

Kimura and coworkers published results on the related ligands **5–7** (Table I) as sodium sensors [5], and the effect of varying the substituent groups on the esters was examined for a series of compounds by Cadogan *et al.* in two papers for the tetramers [6] and the hexamers [7], respectively. The former were confirmed to be markedly sodium selective, while the latter group again showed peak selectivity for caesium. More recently, Cunningham *et al.* examined 12 tetramers and one hexamer as sensing agents in PVC membrane electrodes and reported the tetra-*p*-*t*-butylcalix[4] methoxymethyl ester (Table I, **13**) as producing the most selective sodium electrodes so far [8].

The obvious application of these sodium sensors was in the clinical analysis of sodium in body fluids. Although sodium is present in blood at elevated levels (typically 120–150 mM), the range of the sodium concentration is relatively limited, compared to potassium (1–4 mM). It follows that the signal obtained will have a limited range of a few mV over which the entire normal sodium distribution will occur. Hence, careful experimental design, and attention to sampling and signal processing is required in order to obtain acceptable accuracy and precision in the analytical results.

Initial studies on the performance of mini-PVC membrane electrodes for blood analysis were encouraging [9]. In this study, 44 plasma samples were analysed for sodium with the calixarene methyl tetraester-based PVC electrodes, and the results compared with those obtained with a SMAC-Technion Analyser. Good correlation was found ($r = 0.95$), but a systematic bias was apparent due to the calibration regime used in the study. A more detailed report of these investigations published the following year confirmed the utility of applying the sensors based on **2** to the analysis of sodium in blood [10].

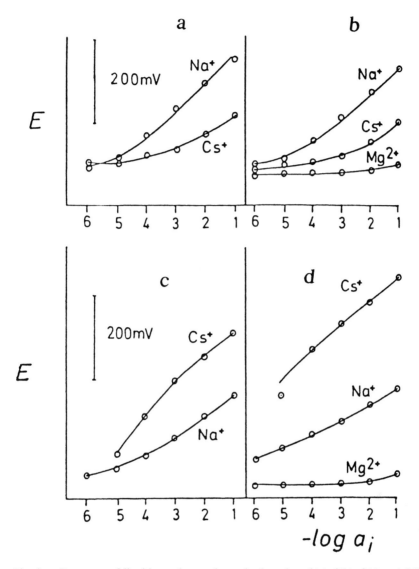

Fig. 2. Response of liquid-membrane electrodes based on **1**(a), **2**(b), **3**(c) and **4**(d). These were the first results demonstrating the selectivity of ISEs based on calix[4]arene tetraester and calix[6]arene hexaester derivatives. The tetramers are clearly sodium selective and the hexamers caesium selective. Taken from reference [2] with permission.

In parallel with these studies, other ligands were assessed for use in sodium-selective electrodes, including the methyl, butyl, and adamantyl ketone tetramers of *p-t*-butylcalix[4]arene. However, only the methyl ketone derivative produced satisfactory PVC membrane electrodes [6]. These were subsequently applied to the analysis of sodium in plasma samples [11]. Excellent correlations ($r = 0.979$, $r = 0.987$ and $r = 0.951$, $n = 10$) were found in comparative tests with three reference instruments (Hitachi 704 Analyser, Flame Photometer, SMAC Technicon Analyser, respectively). However, as before, a bias in the results was apparent in each case. Interestingly, in a paper by Kimura and coworkers, PVC electrodes based on **5** were also applied to the determination of blood sodium. Although only five samples were processed, a positive bias of around 2–3 mM was evident in all but one sample [12].

Obviously, when trying to establish a new analytical device in the face of existing technology, any bias in the results is unacceptable. Bias in analytical determinations commonly arises from systematic errors in calibration. In the above investigations, drift during the calibration and analytical measurements is a well-known problem. Its effect is further magnified by the very restricted range found in blood sodium samples, which leads to a narrow voltage range over which the measurements must be made, and, perhaps more importantly, significant 'bunching' of the concentration distribution in the samples. Hence most concentrations are focused in a very narrow range (135–140 mM) with a few outliers on either side which extend the range to perhaps 120–150 mM sodium. These outliers have a significant influence on the slope of the regression line, and must therefore be determined with particular care.

One way to reduce the effect of drift and to give very reproducible sample handling is to use flow-injection analysis (FIA). PVC membranes incorporating the methylketone and methyl ester derivatives were assessed as detectors in a FIA system for blood sodium analysis and the results demonstrated that the bias described above could be greatly reduced while still maintaining excellent correlation [13].

However, the best results were presented in a series of papers where the methylester tetramer was used as an element in an ISE array both in conventional dip-type measurements [14] and in a flow-injection analysis system [15, 16]. Using sophisticated calibration and sensor modelling techniques, these papers rigorously demonstrated that the methylester tetramer (Table I, **2**) could be applied to blood sodium analysis with excellent results (Figure 3). Furthermore, the same ISE was shown to be suitable for the analysis of sodium in mineral water samples.

More recently, sodium selective PVC membrane electrodes incorporating the methylester tetramer have been assessed using batch injection analysis (BIA). This technique differs from FIA in that the sample is injected directly onto the sensor surface, and a dilution/mixing effect sweeps the sample quickly away, resulting in high-speed transient signals which can be used for analytical measurements. Initially, a single sodium electrode was investigated and shown to have excellent characteristics for this technique [17]. Subsequently, the electrode was used in a $3 \times$ ISE array (Na, K, Ca) and successfully applied to the analysis of these ions in mineral water samples [18]. In the array study, the excellent selectivity of the calixarene–PVC membrane was apparent in carryover studies performed during the evaluation of the array, as virtually no response to the interfering ions was indicated.

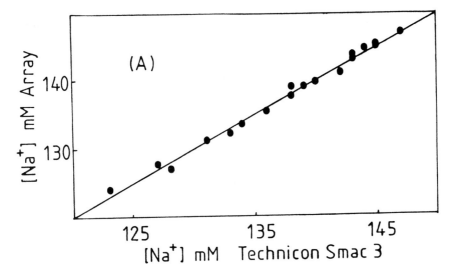

Fig. 3. Plasma sodium analysis results obtained with a PVC membrane electrode based on **1** compared with a SMAC analyser. The electrode was part of an array of ISEs which were carefully calibrated before use. Bias in the results was eliminated through the use of a flow-injection analysis (FIA) approach which enabled the entire sample handling and analysis to be automated. These results demonstrate clearly the successful application of calix[4]arene-based ISEs to this important assay (from reference [15] with permission).

From the above, it is clear that the tetraesters and close derivatives can form the basis of excellent sodium ISEs. Studies on device lifetime showed that the sensors could be expected to be used for months at a time [19], and could analyse several thousand blood samples before the signal would become unacceptably affected by membrane coating or leaching of membrane components [20].

5. Solid-State Sodium-Selective Sensors

In addition to the traditional ISE configuration discussed above, researchers are interested in solid-state designs of these sensors, such as ISFETs (ion-selective field-effect transistors) or coated wire electrodes (CWEs), as these are expected to be easier to mass produce and will be more compatible with the planar fabrication technologies used in the semiconductor and related industries. It is not surprising, therefore, that studies on the performance of ISFETs incorporating calix[4]arene derivatives have recently appeared in the literature [21, 22]. One paper [23] describes the characteristics of ISFETs based on the ketones **8** and **9** (Table I). These gave Nernstian slopes and good selectivity against other Group I and Group II cations. A well-known problem with these devices is the lack of a well-defined internal boundary potential (i.e. E'_m in Figure 1 above) due to the lack of an internal filling solution or compensating mechanism by which charge can be exchanged across the internal boundary.

The same problem occurs with coated wire electrodes (CWEs), which differ from ISEs in that the sensing membrane is deposited directly onto a metallic conductor. This leads to a blocked internal interface between the membrane and the metal, as the former conducts only by means of ion movement, while the latter is an electronic conductor. Hence CWEs, while simpler in make up than equivalent ISEs, are generally much less stable, and exhibit greatly reduced effective lifetimes. The ISFET design proposed by Brunink *et al.* involved using a poly(2-hydroxyethyl methacrylate) (polyHEMA) hydrogel layer to help anchor the PVC membrane on the gate region of the device and simultaneously reduce the effect of interferents such as CO_2 which can diffuse through the PVC layer and affect the internal boundary potential. An alternative proposed by Tsujimura *et al.* [24] was to use calix[4]arenes bearing oligosiloxane moieties (e.g. **14**) in silicone rubber membrane ISFETs. These groups promoted the dispersibility of the ligands within the rubber membrane leading to more stable responses compared to similar devices based on the ethyl ester tetramer (**1**). However, no data are given on the performance of the device in real samples such as plasma.

One strategy which might overcome this limitation is to substitute a conductor of mixed character which is capable of transferring charge by means of either ion or electron movement. With this in mind, PVC membranes incorporating ligand **1** have been deposited on polypyrrole which was electrochemically formed on platinum substrates [25]. The resulting sodium-selective solid-state sensors were shown to be much more stable than CWE equivalents, and were unaffected by the presence/absence of redox active species in the sample solution which react on polypyrrole surfaces. Impedance studies confirmed a dramatic reduction in the charge transfer resistance through the device compared to CWE devices which had no polypyrrole layer between the Pt layer and the PVC.

6. Potentiometric Sensors for Other Ions

One of the main reasons for the great interest in calixarenes is the variety of possibilities for structural modification and elaboration. In particular, one can vary the dimensions of the polar cavity defined by the pendant ester, ketone and amide groups described for the sodium selective sensors discussed above. As the selectivity of these devices is determined by a best-fit mechanism which ideally gives a large difference in the Gibbs free energy of the primary ion complex compared to interfering ions, this ability to vary the cavity size raises the prospect of developing ligands suitable for use in sensors aimed at other ions.

An example of these are the caesium electrodes mentioned above based on the hexamers **3** and **4** (Table I). X-ray crystal structures [26] confirm that these ligands are much more open, and define larger polar cavities than the sodium selective tetramers. PVC membrane electrodes based on these ligands were found to be caesium selective against a wide range of possible interfering ions, with ligand **4** being the better of the two.

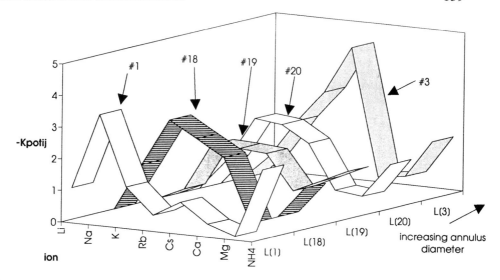

Fig. 4. Variation in selectivity of PVC membrane ISEs as calixarene annulus size increases. The tetramer **1** is clearly sodium selective, while the three oxocalixarenes **18**, **19** and **20** have intermediate and rather poor overall selectivity. On the other hand, the hexamer **3** is obviously caesium selective.

An alternate strategy is to increase the annulus size by extending the methylene linking groups to dimethylene ether groups, rather than increasing the number of repeating calixarene groups. These so-called 'oxacalixarenes' have also been investigated for use in ion-selective electrodes for potassium [27]. While functioning electrodes have been made, the potassium selectivity is not comparable to that available with the well-established valinomycin electrode. It was found that introducing these spacer groups into the annulus, while it did lead to changes in the selectivity, also led to a drastic reduction in the ability of the ligand to discriminate [28]. The basis of the excellent selectivity of the tetramers discussed above lies in their rigid cone conformation, which does not allow ions larger or smaller than the optimum size (i.e. sodium ions) to interact strongly with the polar carbonyl ester, ketone or amide atoms. In contrast, although we see a shift in selectivity from sodium to potassium when one compares electrodes based on the tetraesters and tetraketones with their mono and, more clearly, their dioxa equivalents, the ability to discriminate between ions of different sizes is greatly reduced. This is clearly demonstrated in Figure 4, where the peak selectivity obtained for sodium and caesium with the tetraester **1** and hexaester **3** (Table I) contrasts strikingly with the relative lack of selectivity of the oxacalixarenes **18**, **19** and **20** (Table II). This is indicative of a much more flexible ligating process which is able to interact relatively strongly with cations of differing sizes. This in turn is a consequence of introducing the longer spacer groups between the calixarene units which enables them to swivel and presents a larger volume through which the ligating groups can move.

TABLE II.

Index	R	R'	n
18	$CH_2CO_2C_2H_5$	t-but	1
19	$CH_2CO_2C_2H_5$	t-but	2
20	$CH_2CO_2C_4H_9$	t-but	2

7. Silver and Other 'Soft' Ions

O'Connor et al. have investigated the performance of ISEs based on calix[4]arenes containing silver and nitrogen atoms such as the amide, thioamide and thioester **10**, **11**, and **12** respectively (Table I). Of these, the thioester **12** gave the best results in terms of selectivity as this was the only one which produced electrodes selective for silver against sodium ions ($\log K^{Pot}_{Ag,Na} = -1.16$) [29, 30]. A glassy carbon electrode coated with PVC containing **12** was subsequently successfully used to follow potentiometric titrations of mixtures of I^-, Br^-, and Cl^- [29].

Cobben et al. [31] have described calixarene-based ISFETs targeted at Ag^+, Cu^{2+}, Cd^{2+} and Pb^{2+}. Twenty-nine calix[4]arene derivatives were investigated in this study. The silver-selective ISFETs were based on PVC membranes doped with calix[4]arene derivatives with S-containing groups on the lower rim, such as **15**. On the other hand, ligand **16** resulted in devices more selective for Cu^{2+} (although limited data are given) and significantly none for the selectivity against Ag^+ or Hg^{2+}, which would be probable interferents for this type of membrane sensor. Preliminary data are also reported in this paper for Cd^{2+} selective ISFETs based on several ligands (e.g. **17**) and Pb^{2+} selective ISFETs based on similar oxamide derivatives. The behaviour of the oxamide based sensor is curious, giving a slope of nearly 60 mV/decade change in Pb^{2+} concentration (double the theoretical value) and the calibration curve shows a response linearly decreasing below 10^{-6} M, in contrast to that obtained with most conventional PVC membranes ISEs (and the other ISFETs reported in the paper), which generally level off around 10^{-5} M. Unfortunately, no data regarding the lifetime of the devices is given.

8. Voltammetric and Amperometric Sensors

Many ligands have been used as analyte accumulation agents for stripping analysis. The only publication in this area to date involving calixarenes has been the use of a linear polymeric form of a calix[4]arene tetraester for the preconcentration of Pb^{2+}, Cu^{2+} and Hg^{2+} [32]. In this study, the polymeric calixarene was dispersed in a carbon paste which was then packed into an electrode body. The polymer form of the ligand was used as this was not likely to diffuse into the aqueous sample phase, even when complexed to positively charged ions. The electrode was left at open circuit in the presence of the aqueous sample containing the analyte (accumulation cycle) for various times, and the accumulated metal subsequently stripped off by applying a voltammetric stripping waveform to the carbon paste electrode. Good sensitivity was obtained for Pb^{2+} in particular, with good analytical signals being obtained for 10^{-6} M samples after 3 min accumulation.

9. Sensors for Organic Guests

As is clear from the above discussion, calixarene derivatives have been thoroughly investigated as sensing agents for simple inorganic ions. However, calixarene-based sensors for organic amines have begun to appear in the literature. Both studies have involved the use of calix[6]arene esters as hosts for organic amines. The protonated amine is able to form a stable inclusion complex by formation of tripodal hydrogen bonds between the NH_3^+ group of the guest and the C=O groups of the host, resulting in good potentiometric signals from PVC membrane electrodes. Furthermore, the electrodes are selective for primary amines against secondary and tertiary amines [33]. Another study has revealed strongest potentiometric responses to primary amines bearing no substituent adjacent to the amino group such as 1-octylamine and dopamine [34].

10. Optical Transduction

One of the most striking trends in sensor research over the past five years or so has been the tremendous interest in optical methods of transduction. Optical-based sensing is attractive for many reasons, including inherent safety, less noise pick up in signal transmission over long distances, and the possibility of obtaining much more comprehensive information from a single probe (full spectrum vs. one channel of electrochemical information). While there have been many recent publications describing new calixarene derivatives capable of optically signalling the presence of metal ions, there are no functioning 'optodes' or optical equivalents of the electrochemical sensors described above. However, we can expect to see some of these systems being developed over the next few years.

The optically responsive ligands described in the literature contain either a chromophore or fluorophore which is either joined to or part of the calixarene backbone. On complexation, the environment of the chromophore or fluorophore is sufficiently perturbed to produce a significant change in the UV-vis or fluorescence emission spectrum, respectively. In the case of the former, this is usually achieved by using an acidic chromophore in the presence of a base which is not strong enough to deprotonate the free ligand (i.e. the equilibrium in Equation (4) below favours the left-hand side). However, on complexation with a cation (Equation (5)), the additional positive charge residing in the calixarene cavity makes the chromophore a much stronger acid (i.e. $K_{a(C)} > K_{a(L)}$), and in the presence of the base, is deprotonated (i.e. the equilibrium in Equation (6) favours the right-hand side), shifting the UV-vis absorption to longer wavelengths (Figure 5). This shift in wavelength is easily monitored, and provided the selectivity of the parent calixarene is retained (not usually the case as the observed selectivity depends on three factors, the complexation of the metal ion, the effect of the metal ion on the acidity of the chromophore and the stability of the resulting deprotonated complex). These processes can be summarised as:

$$L-COH + B \overset{K_{a(L)}}{\rightleftharpoons} L-CO^- + BH^+ \quad (4)$$

$$L-COH + M^+ \rightleftharpoons LM^+-COH \quad (5)$$

$$LM^+-COH + B \overset{K_{a(C)}}{\rightleftharpoons} LM^+-CO^- + BH^+ \quad (6)$$

where L is the calixarene ligand, —COH is the acidic chromophore, B is the base, and M^+ is the metal ion. A variety of chromophores have been investigated, including nitrophenol [35, 36] and azophenol [37, 38] derivatives, and in most cases selective colour transduction on metal complexation has been achieved, although the selectivity is generally not as good as with the electrochemical equivalents (an exception is the chromogenic calixarene selective for potassium described by King et al. [39]). Further papers on similar systems are continuing to appear regularly in the literature [40].

Fluorescent derivatives have also been investigated for selective optical transduction of metal complexation [41–44]. Perturbation of the fluorophore is achieved through variation in the rigidity of the molecule accompanying complexation, or through the judicious positioning of fluorophores and quenching groups in such a way that the degree of quenching changes markedly due to conformational adjustments accompanying complexation. Once the effect of metal ion complexation on the emission spectrum is known, monitoring a responsive emission wavelength as a function of time enables transient changes in the concentration of metal ions to be detected (Figure 6).

11. Patents

A search of the patent literature has revealed 83 patents filed which include the use of the keyword 'calixarene'. These patents cover methods of preparation/synthesis

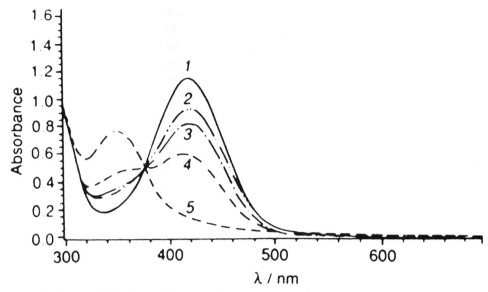

Fig. 5. Typical shifts in UV-vis absorbance accompanying complexation of a metal ion (in this case Li^+) by calixarene chromoionophore bearing acidic (ionisable) groups (in this case nitrophenyl-type moieties) in the presence of a base (0.1 M morpholine). The results show shifts brought about by additions of $LiClO_4$ to a 5×10^{-5} M solution of the ligand in THF containing 20 mm³ morpholine to give final Li^+ concentrations of 0.1 M (1), 10^{-2} M (2), 4×10^{-3} M (3), 8×10^{-4} M (4), 2×10^{-5} M (5). Addition of the metal ions changes the solution from colourless to yellow. Data from references [35] with permission. Ligand: Tetramer as in Table I with R = $OCH_2-\text{C}_6H_3(OH)-NO_2$

of certain calixarene derivatives, or their use in a wide variety of applications including corrosion inhibitors, fuel additives, hair dyes, charge controlling agents for developing electrostatic images, additives in epoxy resins and adhesives, developer solution for photographic negatives, extraction of uranium ions, deodorant additive, and stabilisers in rubbers. Very recently, the first chemical sensor related patents have begun to appear, and this trend will probably continue, given the obvious commercial importance of these devices. Of the four sensor patents, three relate to electrochemical sensors [45–47] for alkali metal ions, and one to the use of cation complexing dyes (chromoionophores) in optical sensors [48].

12. The Future

Sensing applications of calixarene derivatives are only beginning to develop. The potentiometric and other electrochemical sensors for metal ions can be regarded as the first generation of these sensors. Calixarenes capable of optically signalling complexation with metal ions, while valuable in their own right, could be the precursors of much more interesting sensing materials (see for example the fluorescence signalling of encapsulation of a flavin, pteridine, by calix[4]arene host

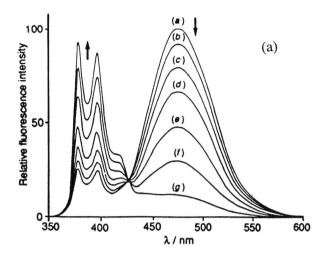

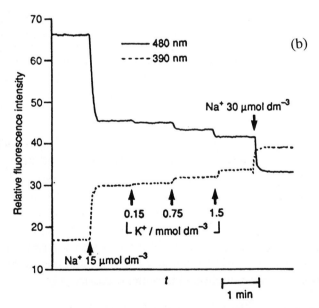

Fig. 6. (a) Changes in fluorescence emission spectra obtained with fluorescent labelled calix[4]arene derivatives on addition of NaSCN and (b) selectivity of fluorescent response measured at 480 nm and 390 nm on addition of Na^+ and K^+ ions. Data from reference [43] with permission.

capable of changing from a 'closed' to an 'open' form [49]). It is now recognized that calixarenes can be used as building blocks of much more substantial structures which could be used to sense a huge number of potential hosts, ranging from neutral gaseous molecules (e.g. toxic solvent vapours) to amino acids or more complex biological molecules. The key to achieving these new and exciting possibilities lies with a deeper understanding of the fundamental behaviour of these systems at the molecular level through detailed NMR and molecular modelling/molecular dynamics studies [50], and statistical analysis of the molecular basis of macroscopic properties [51]. Given the rapidly growing amount of basic information on calixarene host–guest behaviour, this will hopefully become a reality sooner rather than later.

References

1. For reviews of Professor Simon's contribution to the development of chemical sensors see H. M. Widner: *Analytical Methods and Instrumentation* **1**, 3 (1993), and M. E. Meyerhoff, Guest Editorial, *Electroanalysis* **5**, No 9/10 (1993).
2. D. Diamond: *Anal. Chem. Symp. Ser.* **25**, 155 (1986).
3. D. Diamond and G. Svehla: *Trends Anal. Chem.* **6**, 46 (1987).
4. D. Diamond, G. Svehla, E. Seward, and A. M. McKervey: *Anal. Chim. Acta* **204**, 223 (1988).
5. K. Kimura, M. Matsuo, and T. Shono: *Chem. Lett.* 615 (1988).
6. A. Cadogan, D. Diamond, M. R. Smyth, M. Deasy, M. A. McKervey, and S. J. Harris: *Analyst* **114**, 1551 (1989).
7. A. Cadogan, D. Diamond, M. Smyth, G. Svehla, M. A. McKervey, E. Seward, and S. J. Harris: *Analyst* **115**, 1207 (1990).
8. K. Cunningham, G. Svehla, S. J. Harris, and M. A. McKervey: *Analyst* **111**, 341 (1993).
9. M. Telting Diaz, D. Diamond, M. R. Smyth, E. Seward, M. A. McKervey, and G. Svehla: *Anal. Proc.* **26**, 29 (1989).
10. M. Telting Diaz, F. Regan, D. Diamond, and M. R. Smyth: *J. Pharm. Biomed. Anal.* **8**, 695 (1990).
11. M. Telting Diaz, D. Diamond, M. R. Smyth, E. Seward, and A. M. McKervey: *Electroanalysis* **3**, 371 (1991).
12. K. Kimura, T. Miura, M. Matsuo, and T. Shono: *Anal. Chem.* **62**, 1510 (1990).
13. M. Telting Diaz, F. Regan, D. Diamond, and M. R. Smyth: *Anal. Chim. Acta* **251**, 149 (1991).
14. R. J. Forster, F. Regan, and D. Diamond: *Anal. Chem.* **62**, 876 (1991).
15. R. J. Forster and D. Diamond: *Anal. Chem.* **64**, 1721 (1992).
16. D. Diamond and R. J. Forster: *Anal. Chim. Acta* **276**, 75 (1993).
17. J. Lu, Q. Chen, D. Diamond, and J. Wang: *Analyst* **118**, 1131 (1993).
18. D. Diamond, J. Lu, Q. Chen, and J. Wang: *Anal. Chim. Acta* **281**, 629 (1993).
19. D. Diamond and F. Regan: *Electroanalysis* **2**, 113 (1990).
20. A. Lewenstam: personal communication.
21. D. N. Reinhoudt: *Sens. Actuators* **B6**, 179 (1992).
22. K. Kimura, T. Miura, M. Matsuo, and T. Shono: *Anal. Chem.* **62**, 1510 (1990).
23. J. A. J. Brunink, J. R. Haak, J. G. Bomer, D. N. Reinhoudt, M. A. McKervey, and S. J. Harris: *Anal. Chim. Acta* **254**, 75 (1991).
24. Y. Tsujimura, M. Yokoyama, and K. Kimura: *Electroanalysis* **5**, 803 (1993).
25. A. Cadogan, Z. Gao, A. Lewenstam, A. Ivaska, and D. Diamond: *Anal. Chem.* **64**, 2496 (1992).
26. M. A. McKervey, E. M. Seward, B. Ruhl, and S. J. Harris: *J. Chem. Soc., Chem. Commun.* 388 (1985).
27. A. Cadogan, D. Diamond, S. Cremin, M. A. McKervey, and S. J. Harris: *Anal. Proc.* **28**, 13 (1991).
28. R. J. Forster, F. Regan, and D. Diamond: *Sensors and Actuators* **B4**, 325 (1991).
29. K. O'Connor, G. Svehla, S. J. Harris, and M. A. McKervey: *Talanta* **39**, 1549 (1992).
30. K. O'Connor, G. Svehla, S. J. Harris, and M. A. McKervey: *Anal. Proc.* **30**, 137 (1993).
31. P. L. H. M. Cobben, R. J. M. Egberick, J. G. Bomer, P. Bergveld, W. Verboom, and D. N. Reinhoudt: *J. Am. Chem. Soc.* **114**, 10573 (1992).

32. D. Arrigan, G. Svehla, S. J. Harris, and M. A. McKervey: *Anal. Proc.* **29**, 27 (1992).
33. W. H. Chan, K. K. Shiu, and X. H. Gu: *Analyst* **118**, 863 (1993).
34. K. Odasima, K. Yagi, K. Tohda, and Y. Umezawa: *Anal. Chem.* **65**, 1074 (1993).
35. M. McCarrick, B. Wu, S. J. Harris, D. Diamond, G. Barrett, and M. A. McKervey: *J. Chem. Soc., Chem. Commun.* 1287 (1992).
36. M. McCarrick, S. J. Harris, D. Diamond, G. Barrett, and M. A. McKervey: *J. Chem. Soc., Perkin Trans. 2* 1963 (1993).
37. H. Shimizu, K. Iwamoto, K. Fujimoto, and S. Shinkai: *Chem. Lett.* 2147 (1991).
38. M. McCarrick, S. J. Harris, D. Diamond, G. Barrett, and M. A. McKervey: *Analyst* **118** 1127 (1993).
39. M. A. King, C. P. Moore, K. R. A. Samankumara Sandayake, and I. O. Sutherland: *J. Chem. Soc., Chem. Commun.* 582 (1992).
40. Y. Kubo, S. Hamaguchi, A. Niimi, K. Yoshida, and S. Tokita: *J. Chem. Soc., Chem. Commun.* 305 (1993).
41. I. Aoki, T. Sakaki, and S. Shinkai: *J. Chem. Soc., Chem. Commun.* 730 (1992).
42. I. Aoki, H. Kawabati, K. Nakashima, and S. Shinkai: *J. Chem. Soc., Chem. Commun.* 1771 (1991).
43. T. Jin, K. Ichikawa, and T. Koyama: *J. Chem. Soc., Chem. Commun.* 499 (1992).
44. C. Perez-Jimenez, S. J. Harris, and D. Diamond: *J. Chem. Soc., Chem. Commun.* 480 (1993).
45. S. Watanabe and H. Yanagi, Tokuyama Soda Co.: *Japanese Patent* No. 04342733, 30 Nov. 1992 (*CA*119(17):177143a).
46. H. Yamamoto and T. Ogata, Tokuyama Soda Co.: *Japanese Patent* No. 04339251, 26 Nov. 1992 (*CA*118(26):265838r).
47. S. J. Harris, M. A. McKervey, G. Svehla, and D. Diamond: *European Patent* No. 490631, *USA Patent* No. 07/624, 623 and *European Patent* No. 91 311 462.5, 17 Jun. 1992 (*CA*118(8):72762c).
48. C. P. Moore, I. O. Sutherland, and A. King: *MARPAT* 119:205465, CS Kodak Ltd.; Eastman Kodak Co., No. 9312428, 24 Jun. 1993 (*CA*119(20):205465c).
49. H. Murakami and S. Shinkai: *J. Chem. Soc., Chem. Commun.* 1533 (1993).
50. P. Guilbaud, A. Varnek, and G. Wipff: *J. Am. Chem. Soc.* **115**, 8298 (1993).
51. M. Careri, A. Casnati, A. Guarinoni, A. Mangia, G. Mori, A. Pochini, and R. Ungaro: *Anal. Chem.* **65**, 3156 (1993).

Proximally Functionalized Cavitands and Synthesis of a Flexible Hemicarcerand*

PETER TIMMERMAN[a], HAROLD BOERRIGTER[a], WILLEM VERBOOM[a], GERRIT J. VAN HUMMEL[b], SYBOLT HARKEMA[b] and DAVID N. REINHOUDT[a,**]
Laboratories of Organic Chemistry[a] and Chemical Physics[b], University of Twente, P.O. Box 217, 7500 AE Enschede, the Netherlands

(Received: 3 March 1994; in final form: 30 May 1994)

Abstract. A general study on the synthesis of partly bridged octols **3a–d** and **4c–d** is described. Tri-bridged diol **3c** can be prepared in 54% yield in DMSO at 70°C with excess CH_2BrCl or in 52% yield in DMF at 70°C with only 4 equiv. of CH_2BrCl. 1,3-Di-bridged tetrol **4a**, one of the two possible di-bridged isomers formed in preference to the other, was obtained in 30% yield. Tri-bridged diols **3c** and **d** can be selectively debrominated in one step by treatment with 5 equiv. of *n*-BuLi in THF to afford the corresponding dibromo derivatives **8a** and **b** in 77% and 76% yields, respectively. After incorporation of the fourth bridge, the remaining two bromines can be replaced by C(O)OMe to give **9c** (60%), by OH to give **9d** (62%) or by CN to give **9f** (> 95%). When the lithiated derivatives of **3c** and **d** are quenched with electrophiles other than H^+, a selectively functionalized tri-bridged diol with hydroxyl (**8c**, 47%) and selectively functionalized cavitands with thiomethyl (**9g**, 25%) or iodo (**9h**, 20%) groups can be synthesized. Two molecules of **9d** were coupled with CH_2BrCl in DMSO/THF under high dilution conditions to give the flexible hemicarcerand **10** in 71% yield.

Key words: Partly bridged octols, selective functionalization, hemicarcerand.

Supplementary Data. A list of observed and calculated structure factors have been deposited with the British Document Supply Centre as Supplementary Publication No. SUP 82170 (50 pages).

1. Introduction

After calixarenes were rediscovered and their structures were elucidated by Gutsche [1, 2], the molecules have been studied in great detail thanks to their vase-like shape and their ability to complex guest molecules. At first, most efforts dealt with the synthesis of tetrafunctionalized calix[4]arenes [3–7], but more recently attention has increasingly been focused on the introduction of functional groups at *specific* positions in the molecule. This has resulted in a number of convenient syntheses for selectively functionalized calix[4]arenes [8], viz. mono- [9, 10], 1,2-di- [10–12], 1,3-di [13–15] and trifunctionalized [10, 16, 17] calix[4]arenes, making these molecules valuable building blocks in supramolecular chemistry.

---

* This paper is dedicated to the commemorative issue on the 50th anniversary of calixarenes.
** Author for correspondence.

In contrast to the parent calix[4]arenes, selective functionalization of the resorcinol-based calix[4]arenes [18] has hardly been studied. Partly functionalized cavitands have only been isolated as side products in the synthesis of tetrafunctionalized cavitands [19, 20]. Although such dissymmetric cavitands have been used in the synthesis of chiral cavitands [21] and hemicarcerands [20] with promising applications [22], their synthesis was never studied in detail. While our work was in progress [23], Sorrell and Richards showed that cavitands can be selectively functionalized at the upper rim by means of a Claisen rearrangement [24]. In this paper we describe the synthesis of partly bridged octols [25] together with a novel route to functionalize these cavitands selectively in a proximal way.

2. Experimental

GENERAL

Melting points were determined with a Reichert melting point apparatus and are uncorrected. ^{1}H-NMR and ^{13}C-NMR spectra were recorded with a Bruker AC 250 spectrometer with Me$_4$Si as internal standard. Mass spectra were recorded with a Finnigan MAT 90 spectrometer using m-NBA as a matrix. IR spectra were obtained using a Nicolet 5SXC FT-IR spectrophotometer. All reagents were distilled before use. THF was distilled from Na/benzophenone, PE 60-80, CH$_2$Cl$_2$ and ethyl acetate from K$_2$CO$_3$, DMSO from CaH$_2$. DMF and acetonitrile were dried over molecular sieves for at least 3 days. N-Bromosuccinimide was recrystallized from H$_2$O and dried over CaCl$_2$. n-BuLi was used as a 1.5 M solution in hexane. Prior to lithiation, the starting compound was first dissolved in THF and evaporated to dryness. This process was repeated until no other guest molecule than THF was present. Flash column chromatography was performed using silica 60 (0.040–0.063 mm, 230–400 mesh) or when necessary silica 60H (0.005–0.040 mm). Compounds **1a** [19], **1b** [26], **1c** [27] and **1d** [28] were prepared according to literature procedures.

5,11,17,23-Tetrabromo-2,8,14,20-tetraundecylpentacyclo[19.3.1.1^{3,7}.1^{9,13}.1^{15,19}]-octacosa-1(25),3,5,7(28),9,11,13(27),15,17,19(26),21,23-dodecaen-4,6,10,12,16,18,22,24-octol (**1e**). To a solution of octol **1d** (32.6 g, 29.5 mmol) in DMF (500 mL) was added N-bromosuccinimide (21.3 g, 120 mmol) and the mixture was stirred for 24 h at room temperature with protection from light. The reaction mixture was poured out into water (1.15 L), the precipitate was filtered off and washed several times with water. After drying the solid for 20 h at 15 mm Hg, it was dissolved in acetone (2.3 L), the hot solution was filtered and the volume was reduced to 115 mL. The precipitate formed after standing for 18 h at -20°C was filtered off, washed with acetone and dried. Yield 82%, mp 288–290°C (acetone). Mass spectrum (FAB, NBA): m/z 1265.5 [(M—C$_{11}$H$_{23}$)$^+$, *calc.* 1265.3]. ^{1}H-NMR (acetone-d_6): δ 8.33 (s, 8H, OH), 7.63 (s, 4H, ArH), 4.45 [t, 4H, J = 7.7 Hz, CH(CH$_2$)$_{10}$CH$_3$], 2.4–2.2 [m, 8H, CHCH$_2$(CH$_2$)$_9$CH$_3$], 1.5–1.2 [m, 72H, CHCH$_2$(CH$_2$)$_9$CH$_3$], 0.88 [t, 12H, J = 6.8 Hz, CH(CH$_2$)$_{10}$CH$_3$]. ^{13}C-NMR (DMSO-d_6): δ 149.7 (s, ArC—OH), 126.0

(s, ArCCH), 124.1 (d, ArCH), 102.0 (s, ArCBr). *Anal. Found*: C, 60.78; H, 7.94. *Calc.* for $C_{72}H_{108}Br_4O_8$: C, 60.84; H, 7.66.

General Procedures for the Preparation of Compounds **2d**, **3a–d** *and* **4c–d**. A solution of octol **1** (5.0 mmol), K_2CO_3 (1.73 g, 12.5 mmol) and CH_2BrCl in *degassed* DMF or DMSO (250 mL) was stirred; experimental details are given in Table IV.

Work-up Procedure A. The solvent was removed *in vacuo*, the residue was poured into demineralized water and the precipitate was filtered off through Celite. After drying the filtered solid at 100°C under vacuo for 6 h, a Soxhlet extraction was performed for several days using $CHCl_3$ as the solvent, having 5 g of silica per gram **1** used in the collecting flask. The $CHCl_3$ solution was evaporated to dryness and the product, absorbed on silica, was further purified by flash column chromatography.

Work-up Procedure B. The solvent was removed *in vacuo*, the residue was dissolved in $CHCl_3$ and washed with H_2O (3×), brine and subsequently dried over $MgSO_4$. After removal of the solvent, the product was further purified by flash column chromatography.

7,11,15,28-Tetrabromo-1,21,23,25-tetraundecyl-2,20:3,19-dimetheno-1H,21H,23 H,25H-bis[1,3]dioxocino[5,4-i:5',4'-i']benzo[1,2-d:5,4-d']bis[1,3]benzodioxocin (**2d**) was isolated after flash column chromatography (SiO_2, CH_2Cl_2) as a black viscous oil. It was further purified by dissolving it in a 1 : 1 (v/v) mixture of CH_2Cl_2/hexane and filtering it over silica, to give **2d** as a colorless oil. Mass spectrum (FAB, NBA, high resolution): m/z 1468.494 (M^+, *calc.* 1468.474). ^{1}H-NMR ($CDCl_3$): δ 7.02 (s, 4H, ArH), 5.95 (d, 4H, $J = 7.4$ Hz, outer OCH_2O), 4.84 [t, 4H, $J = 4.0$ Hz, CH(CH_2)$_{10}CH_3$], 4.38 (d, 4H, $J = 7.4$ Hz, inner OCH_2O), 2.3–2.2 [m, 8H, CHCH_2(CH_2)$_9CH_3$], 1.5–1.2 [m, 72H, CHCH_2(CH_2)$_9CH_3$], 0.88 (t, 12H, $J = 6.5$ Hz, CH_3). ^{13}C-NMR ($CDCl_3$): δ 152.1 (s, ArCOCH_2O), 139.3 (s, ArCCH), 119.1 (d, ArCH), 113.5 (s, ArCBr), 98.5 (t, OCH_2O), 37.6 (d, ArCHAr).

4,8,12,16-Tetrabromo-20,22,24,25-tetraphenyl-2,18-methano-20H,22H,24H-dibenzo[d,d'][1,3]dioxocino[5,4-i:7,8-i]bis[1,3]benzodioxocin-3,17-diol (**3b**) was isolated after flash column chromatography (SiO_2, EtOAc/CH_2Cl_2, 0 : 100–10 : 90) as a white solid. mp > 290°C (CH_2Cl_2/MeOH). Mass spectrum (FAB, NBA): m/z 1144.9 [$(M+H)^+$, *calc.* 1144.9]. ^{1}H-NMR ($CDCl_3$): δ 7.4–7.1 (m, 20H, C_6H_5), 6.91 (s, 2H, ArH meta to OCH_2O), 6.83 (s, 2H, ArH meta to OH), 6.52, 6.42, 6.10 [3s (1 : 2 : 1), 4H, CHPh], 6.07, 6.05 (2d, 3H, $J = 7.3$ Hz, outer OCH_2O), 4.57, 4.49 (2d, 3H, $J = 7.3$ Hz, inner OCH_2O). ^{13}C-NMR ($CDCl_3$): δ 152.9–152.3 (s, ArCOCH_2O), 150.9 (s, ArCOH), 126.3, 126.2 (d, ArCH), 113.1, 107.6 (s, ArCBr), 98.4 (t, OCH_2O), 42.5, 42.3, 41.0 (d, ArCHAr). *Anal. Found*: C, 57.40; H, 3.55. *Calc.* for $C_{55}H_{36}Br_4O_8$: C, 57.72; H, 3.17.

4,8,12,16–Tetrabromo-20,22,24,25-tetrakis(2-phenylethyl)-2,18-methano-20H,22 H,24H-dibenzo[d,d'][1,3]dioxocino[5,4-i:7,8-i]bis[1,3]benzodioxocin-3,17-diol (**3c**) was isolated after flash column chromatography (SiO_2, $EtOAc/CH_2Cl_2$, 0 : 100–5 : 95) as a white solid. mp > 290°C ($CH_2Cl_2/MeOH$). Mass spectrum (FAB, NBA): m/z 1256.3 (M^+, *calc.* 1256.0). ^{1}H-NMR ($CDCl_3$): δ 7.3–7.0 (m, 24H, C_6H_5 + ArH), 5.92, 5.89 (2d, 3H, J = 7.4 Hz, outer OCH_2O), 4.89, 4.84, 4.40 [3t, 4H (2 : 1 : 1), J = 8.0, 7.6 and 7.6 Hz, CH(CH_2)$_2C_6H_5$], 4.36, 4.28 (2d, 3H, J = 7.4 Hz, inner OCH_2O), 2.8–2.4 [m, 16H, CH(CH_2)$_2C_6H_5$]. ^{13}C-NMR ($CDCl_3$): δ 152.5–152.0 (s, ArCOCH_2O), 148.4 (s, ArCOH), 121.1, 118.5 (d, ArCH), 114.0, 107.2 (s, ArCBr), 98.6 (t, OCH_2O), 37.5, 35.6 (d, ArCHAr). *Anal. Found*: C, 60.35; H, 4.42. *Calc.* for $C_{63}H_{52}Br_4O_8$: C, 60.21; H, 4.17.

4,8,12,16-Tetrabromo-20,22,24,25-tetraundecyl-2,18-methano-20H,22H,24H-dibenzo[d,d'][1,3]dioxocino[5,4-i:7,8-i]bis[1,3]benzodioxocin-3,17-diol (**3d**) was isolated after flash column chromatography (SiO_2, CH_2Cl_2) as a light-brown solid which could not be recrystallized. mp 80–82°C. Mass spectrum (FAB, NBA): m/z 1456.4 (M^+, *calc.* 1456.5). ^{1}H-NMR ($CDCl_3$): δ 7.01 (s, 4H, ArH + OH), 6.95 (s, 2H, ArH), 5.90, 5.87 (2d, 3H, J = 7.4 Hz, outer OCH_2O), 4.78, 4.68, 4.32 [3t, 4H (2 : 1 : 1), J = 8.0 Hz, CH(CH_2)$_{10}CH_3$], 4.34, 4.25 (2d, 3H, J = 7.4 Hz, inner OCH_2O), 2.3–2.0 [m, 8H, CHCH_2(CH_2)$_9CH_3$], 1.5–1.2 [m, 72H, CHCH$_2$(CH_2)$_9CH_3$], 0.81 [t, 12H, J = 6.8Hz, CH(CH_2)$_{10}$CH_3]. ^{13}C-NMR ($CDCl_3$): δ 152.1–151.6 (s, ArCOCH_2O), 148.0 (s, ArCOH), 121.1, 118.5 (d, ArCH), 113.5, 106.7 (s, ArCBr), 98.4 (t, OCH_2O), 37.4–35.7 (d, ArCHAr). *Anal. Found*: C, 61.99; H, 7.44. *Calc.* for $C_{75}H_{108}Br_4O_8$: C, 61.82; H, 7.47.

5,11,17,23-Tetrabromo-2,14,27,32-tetrakis(2-phenylethyl)-7,9,19,21-tetraoxaheptacyclo[13.9.5.5^{3,13}.0^{6,33}.0^{10,31}.0^{18,28}.0^{22,26}]tetratriaconta-3,5,10,12,15,17,22,24, 25,28,30,33-dodecaene-4,12,16,24-tetrol (**4c**) was isolated after flash column chromatography (SiO_2, $EtOAc/CH_2Cl_2$, 5 : 95) as a white solid, mp 187–188°C ($CHCl_3$). Mass spectrum (FAB, NBA): m/z 1244.0 (M^+, *calc.* 1244.1). ^{1}H-NMR ($CDCl_3$): δ 7.3–7.1 [m, 24H, CH(CH_2)$_2C_6H_5$ + ArH], 6.95 (s, 4H, OH), 5.98 (d, 2H, J = 7.4 Hz, outer OCH_2O), 4.89, 4.47 [2t, 4H, J = 7.9 Hz, CH(CH_2)C_6H_5], 4.36 (d, 2H, J = 7.4 Hz, inner OCH_2O), 2.8–2.4 [m, 16H, CH(CH_2)$_2C_6H_5$]. ^{13}C-NMR ($CDCl_3$): δ 152.2 (s, ArCOCH_2O), 148.4 (s, ArCOH), 120.6 (d, ArCH), 107.1 (s, ArCBr), 98.7 (t, OCH_2O), 37.2–36.1 (d, ArCHAr). *Anal. Found*: C, 53.68; H, 3.77. *Calc.* for $C_{62}H_{52}Br_4O_8 \cdot 1.5CHCl_3$: C, 53.57; H, 3.79.

5,11,17,23-Tetrabromo-2,14,27,32-tetraundecyl-7,9,19,21-tetraoxaheptacyclo[13. 9.5.5^{3,13}.0^{6,33}.0^{10,31}.0^{18,28}.0^{22,26}]tetratriaconta-3,5,10,12,15,17,22,24,25,28,30,33-dodecaene-4,12,16,24-tetrol (**4d**) was isolated after flash column chromatography (SiO_2, $EtOAc/CH_2Cl_2$, 5 : 95) as an off-white solid which could not be recrystallized. mp 117–120°C. Mass spectrum (FAB, NBA, negative mode): m/z 1443.2 [$(M-H)^-$, *calc.* 1443.5]. ^{1}H-NMR ($CDCl_3$): δ 7.06 (s, 4H, ArH), 6.76 (s, 4H,

OH), 5.90 (d, 2H, $J = 7.4$ Hz, outer OCH$_2$O), 4.75, 4.32 [2t, 4H, $J = 7.9$ Hz, CH(CH$_2$)$_{10}$CH$_3$], 4.34 (d, 2H, $J = 7.4$ Hz, inner OCH$_2$O), 2.3–2.0 [m, 8H, CHCH_2(CH$_2$))$_9$CH$_3$], 1.4–1.1 [m, 72H, CHCH$_2$(CH_2)$_9$CH$_3$], 0.81 [t, 12H, $J = 6.4$ Hz, CHCH$_2$(CH$_2$)$_9$CH_3]. ^{13}C-NMR (CDCl$_3$): δ 151.9 (s, ArCOCH$_2$O), 148.1 (s, ArCOH), 120.6 (d, ArCH), 106.8 (s, ArCBr), 98.7 (t, OCH$_2$O), 37.2–36.1 (d, ArCHAr). *Anal. Found*: C, 61.85; H, 8.00. *Calc.* for C$_{74}$H$_{108}$Br$_4$O$_8$: C, 61.49; H, 7.53.

2,2',2'',2'''[[5,11,17,23-Tetrabromo-2,14,27,32-tetramethyl-7,9,19,21-tetraoxa-heptacyclo[13.9.5.5^{3,13}.0^{6,33}.0^{10,31}.0^{18,28}.0^{22,26}]tetratriaconta-3,5,10,12,15,17,22, 24,25,28,30,33-dodecaene-4,12,16,24-tetryl]tetrakis(oxy)]tetraacetic acid, tetramethyl ester (**6a**). The reaction mixture was worked up according to procedure A, with the exception that the Soxhlet extraction was carried out without silica in the collecting flask. The CHCl$_3$ solution was evaporated to dryness and the residue was suspended in CH$_3$CN (200 mL). To this suspension was added K$_2$CO$_3$ (8.04 g, 58.2 mmol) and methyl bromoacetate (3.11 mL, 31.7 mmol) and the mixture was refluxed for 18 h. The solvent was evaporated, the residue was dissolved in CH$_2$Cl$_2$ (250 mL), successively washed with sat. NH$_4$Cl (50 mL), with H$_2$O (2 × 50 mL) and brine (50 mL), and dried over MgSO$_4$. After removal of the solvent, the crude product was separated by flash column chromatography (SiO$_2$ 60H, CH$_2$Cl$_2$/hexane, 98 : 2) to give pure **6a** as a white solid. mp 291–293°C (CH$_2$Cl$_2$/pentane). Mass spectrum (FAB, NBA): m/z 1173.0 ([(M+H)$^+$, *calc.* 1173.4). ^{1}H-NMR (CDCl$_3$): δ 7.59 (s, 4H, ArH), 6.00 (d, 2H, $J = 7.4$ Hz, outer OCH$_2$O), 5.38, 5.07 (2q, 4H, $J = 7.4$ Hz, —CHCH$_3$), 4.68 and 4.39 [ABq, 8H, $J = 15.1$ Hz, OCH_2C(O)OCH$_3$], 4.36 (d, 2H, $J = 7.4$ Hz, inner OCH$_2$O), 3.81 [s, 12H, OCH$_2$C(O)OCH_3], 1.87, 1.59 (2d, 12H, $J = 7.4$ Hz, CHCH_3). ^{13}C-NMR (CDCl$_3$): δ 168.8 (s, C=O), 152.6, 151.2 [s, ArCOCH$_2$O + ArCOCH$_2$C(O)OCH$_3$], 122.2 (d, ArCH), 113.1 (s, ArCBr), 98.9 (t, OCH$_2$O), 70.1 [t, OCH$_2$C(O)OCH$_3$], 52.1 (q, OCH$_3$), 32.2, 28.4 (d, ArCHAr), 24.4, 16.6 (q, CHCH$_3$). *Anal. Found*: C, 47.58; H, 3.86. *Calc.* for C$_{46}$H$_{44}$Br$_4$O$_{16}$: C, 47.12; H, 3.78.

2,2',2'',2'''[[4,8,12,24-Tetrabromo-16,18,19,26-tetramethyl-2,14-(methano[1,3]-benzenomethano)-16H,18H-benzo[1,2-d:5,4-d']bis[1,3]benzodioxocin-3,13,23, 25-tetryl]tetrakis(oxy)tetraacetic acid, tetramethyl ester (**7a**) was isolated as a minor product in the synthesis of **6a**. mp 214–216°C (CH$_2$Cl$_2$/hexane). Mass spectrum (FAB, NBA): m/z 1173.3 [(M+H)$^+$, *calc.* 1173.4). ^{1}H-NMR (CDCl$_3$): δ 7.35, 7.24, 6.70 [3s, 4H (2 : 1 : 1), ArH], 5.96 (d, 2H, $J = 7.3$ Hz, outer OCH$_2$O), 5.3–5.1 (m, 4H, CHCH$_3$), 4.84, 4.67 and 4.44, 4.32 [2ABq, 8H, $J = 15.1$ Hz, OCH_2C(O)OCH$_3$], 4.44 (d, 2H, $J = 7.4$ Hz, inner OCH$_2$O), 3.81, 3.77 [2s, 12H, OCH$_2$(O)OCH_3], 1.79, 1.51 (2d, 12H, $J = 7.4$ Hz, CHCH_3). ^{13}C-NMR (CDCl$_3$): δ 168.8, 168.7 (s, C=O), 152.1–151.5 [s, ArCOCH$_2$O + ArCOCH$_2$C(O)OCH$_3$], 125.2, 120.9, 118.5 (d, ArCH), 112.8, 122.2 (s, ArCBr), 98.3 (t, OCH$_2$O), 70.1, 69.0 [t, OCH$_2$C(O)OCH$_3$], 52.1, 52.0 (q, OCH$_3$), 32.1, 30.3 (d, ArCHAr), 23.9,

15.6 (q, CHCH$_3$). *Anal. Found*: C, 47.34; H, 3.92. *Calc.* for C$_{46}$H$_{44}$Br$_4$O$_{16}$: C, 47.12; H, 3.78.

4,16-Dibromo-20,22,24,25-tetrakis(2-phenylethyl)-2,18-methano-20H,22H,24H-dibenzo[d,d'][1,3]dioxocino[5,4-i:7,8-i]bis[1,3]benzodioxocin-3,17-diol (**8a**). To a solution of diol **3c** (0.40 g, 0.32 mmol) in THF (15 mL) was quickly added *n*-BuLi (0.73 mL, 0.95 mmol) at -70°C. After 15 sec the reaction mixture was quenched with excess water. The solution was allowed to warm to room temperature and the solvent was removed *in vacuo*. The residue was taken up in CH$_2$Cl$_2$ (25 mL) and successively washed with 1 N HCl (10 mL), H$_2$O (3 × 10 mL), brine (10 mL) and dried over MgSO$_4$. The crude product was purified by flash column chromatography (SiO$_2$, CH$_2$Cl$_2$) to give **8a** as a white solid in 77% yield. mp > 290°C (CHCl$_3$/CH$_3$CN). Mass spectrum (FAB, NBA): m/z 1098.9 [(M+H)$^+$, *calc.* 1099.2]. ^{1}H-NMR (CDCl$_3$): δ 7.3–7.1 (m, 24H, CH(CH$_2$)$_2$C$_6$H$_5$ + ArH), 6.57 (s, 2H, ArH), 5.89, 5.74 (2d, 3H, J = 7.3 Hz, outer OCH$_2$O), 4.90, 4.79, 4.46 (3t, 4H, (2 : 1 : 1), J = 7.7, 7.8 and 7.8 Hz, C*H*(CH$_2$)$_2$C$_6$H$_5$], 4.44, 4.39 (2d, 3H, J = 7.3 Hz, inner OCH$_2$O), 2.8–2.4 (m, 16H, CH(C*H*$_2$)$_2$C$_6$H$_5$]. ^{13}C-NMR (CDCl$_3$): δ 155.4–152.6 (s, ArCOCH$_2$O), 148.2 (s, ArCOH), 121.1, 120.1, 117.1 (d, ArCH), 106.9 (s, ArCBr), 99.2 (t, OCH$_2$O), 37.0–36.3 (d, ArCHAr). *Anal. Found*: C, 68.87; H, 5.04. *Calc.* for C$_{63}$H$_{54}$Br$_2$O$_8$: C, 68.86; H, 4.95.

4,16-Dibromo-20,22,24,25-tetraundecyl-2,18-methano-20H,22H,24H-dibenzo[d, d'][1,3]dioxocino[5,4-i:7,8-i]bis[1,3]benzodioxocin-3,17-diol (**8b**). The reaction was carried out following the procedure for **8a**, using **3d** (1.0 g, 0.69 mmol), *n*-BuLi (2.1 mL, 3.15 mmol) and THF (150 mL) to give pure **8b** after column chromatography (SiO$_2$, CH$_2$Cl$_2$) in 76% yield. The compound could not be recrystallized. mp 154–156°C. Mass spectrum (FAB, NBA): m/z 1298.1 [(M—H)$^+$, *calc.* 1297.7]. ^{1}H-NMR (CDCl$_3$): δ 7.07, 6.98 (2s, 4H, ArH), 6.43 (s, 2H, ArH), 5.79, 5.64 (2d, 3H, J = 7.3 Hz, outer OCH$_2$O), 4.71, 4.60, 4.32 [3t, 4H, (2 : 1 : 1), J = 8.0 Hz, C*H*(CH$_2$)$_{10}$CH$_3$], 4.34, 4.29 (2d, 3H, J = 7.3 Hz, inner OCH$_2$O), 2.3–2.0 [m, 8H, CHC*H*$_2$(CH$_2$)$_9$CH$_3$], 1.5–1.1 [m, 72H, CHCH$_2$(C*H*$_2$)$_9$CH$_3$], 0.81 [t, 12H, J = 6.8 Hz, CH(CH$_2$)$_{10}$C*H*$_3$]. ^{13}C-NMR (CDCl$_3$): δ 154.9–152.2 (s, ArCOCH$_2$O), 147.6 (s, ArCOH), 121.1, 120.1, 116.5 (d, ArCH), 106.4 (s, ArCBr), 99.0 (t, OCH$_2$O), 36.5–36.1 (d, ArCHAr). *Anal. Found*: C, 69.13; H, 8.60. *Calc.* for C$_{75}$H$_{110}$Br$_2$O$_8$: C, 69.32; H, 8.53.

4,16-Dibromo-20,22,24,25-tetraundecyl-2,18-methano-20H,22H,24H-dibenzo[d, d'][1,3]dioxocino[5,4-i:7,8-i]bis[1,3]benzodioxocin-3,8,12,17-tetrol (**8c**). To a solution of diol **3d** (2.05 g, 1.41 mmol) in THF (150 mL) was added excess NaH (0.17 g, 5.6 mmol) and the solution was stirred at room temperature until the evolution of hydrogen stopped. The reaction mixture was cooled down to -70°C and *n*-BuLi (4.7 mL, 7.0 mmol) was quickly added. After 15 sec, the reaction mixture was quenched with B(OMe)$_3$ (1.6 mL, 14 mmol). The solution was warmed to room temperature and stirred for 1 h. After cooling the reaction mixture again to -70°C, a 15% solution of H_2O_2 in 1.5 M NaOH (14 mL, 70 mmol) was added and the mixture was stirred overnight. Excess H_2O_2 was destroyed by adding $Na_2S_2O_5$ (13 g, 70 mmol) to the solution. The solvent was removed *in vacuo*, H_2O (100 mL) was added and the precipitated solid was filtered off and washed with H_2O (3 × 50 mL). After drying the solid at 80°C under vacuum for 3 h, it was dissolved in THF (100 mL), silica (4 g) was added and the solution was evaporated to dryness. The crude product, absorbed on silica, was purified by flash column chromatography (SiO_2, $EtOAc/CH_2Cl_2$, 20 : 80) to give pure **8c** in 47% yield. The compound could not be recrystallized. mp 72–75°C. Mass spectrum (FAB, NBA): m/z 1330.8 (M^+, calc. 1330.7). ^{1}H-NMR ($CDCl_3$): δ 7.04 (s, 2H, ArH para to Br), 6.67 (s, 2H, ArH para to OH), 5.95, 5.92 (2d, 3H, J = 7.3 Hz, outer OCH_2O), 5.6 (bs, 2H, OH), 4.77, 4.63, 4.40 [3t, 4H (2 : 1 : 1), J = 8.0 Hz, CH(CH_2)$_{10}CH_3$], 4.42, 4.38 (2d, 3H, J = 7.2 and 6.8 Hz, inner OCH_2O), 2.3–2.1 [m, 8H, CHCH_2(CH_2)$_9CH_3$], 1.5–1.2 [m, 72H, CHCH_2(CH_2)$_9CH_3$], 0.87 [t, 12H, J = 6.5 Hz, CH(CH_2)$_{10}CH_3$]. ^{13}C-NMR ($CDCl_3$): δ 152.3 (s, ArCOCH_2O), 148.0 (s, ArCOH ortho to Br), 142.3 (s, ArCOH), 121.5 (d, ArCH para to Br), 109.5 (d, ArCH para to OH), 106.6 (s, ArCBr), 99.2 (t, OCH_2O), 37.0–35.9 (t, ArCHAr). *Anal. Found*: C, 67.60; H, 8.73. *Calc.* for $C_{75}H_{110}Br_2O_{10}$: C, 67.65; H, 8.33.

7,11-Dibromo-1,21,23,25-tetrakis(2-phenylethyl)-2,20:3,19-dimetheno-1H,21H, 23H,25H-bis[1,3]dioxocino[5,4-i:5',4'-i']benzo[1,2-d:5,4-d']bis[1,3]benzodioxocin (**9a**). To a suspension of **8a** (0.92 g, 0.84 mmol) and K_2CO_3 (1.2 g, 8.4 mmol) in CH_3CN (100 mL) was added CH_2BrCl (2.2 mL, 33 mmol) and the mixture was refluxed for 24 h. The solvent was removed and the residue was taken up in CH_2Cl_2 (50 mL). The solution was successively washed with 1 N HCl (20 mL), H_2O (3 × 10 mL), brine (10 mL), dried over $MgSO_4$, whereupon the solvent was removed under vacuum. The crude product was dissolved in a 1/9 (v/v) hexane/CH_2Cl_2 mixture and filtered over silica. After removal of the solvent, pure **9a** was obtained as a white solid in 92% yield. mp 270–272°C (CH_2Cl_2/CH_3CN). Mass spectrum (FAB, NBA): m/z 1110.2 (M^+, calc. 1110.3). ^{1}H-NMR ($CDCl_3$): δ 7.3–7.1 [m, 24H, CH(CH_2)$_2C_6H_5$ + ArH], 6.57 (s, 2H, ArH), 5.98, 5.88, 5.77 [3d, 4H, (1 : 2 : 1), J = 7.3 Hz, outer OCH_2O], 5.0–4.8 [m, 4H, CH(CH_2)$_2C_6H_5$], 4.48, 4.43, 4.40 [3d, 4H (1 : 2 : 1), J = 7.3 Hz, inner OCH_2O], 2.8–2.4 [m, 16H, CH(CH_2)$_2C_6H_5$]. ^{13}C-NMR ($CDCl_3$): δ 155.2–152.1 (s, ArCOCH_2O), 120.5, 118.9, 117.1 (d, ArCH), 113.6 (s, ArCBr), 99.5–98.6 (t, OCH_2O), 37.8–36.5 (d, ArCHAr). *Anal. Found*: C, 68.84; H, 4.89. *Calc.* for $C_{64}H_{54}Br_2O_8$: C, 69.19; H, 4.90.

7,11-Dibromo-1,21,23,25-tetraundecyl-2,20:3,19-dimetheno-1H,21H,23H,25H-bis[1,3]dioxocino[5,4-i:5',4'-i']benzo[1,2-d:5,4-d']bis[1,3]benzodioxocin (**9b**). The reaction was carried out following the procedure for **9a**, using **8b** (0.25 g, 0.19 mmol), K_2CO_3 (0.31 g, 2.3 mmol), CH_2BrCl (0.65 mL, 10 mmol) and CH_3CN (55 mL). Pure **9b** was obtained as a glass in 99% yield. mp 65–67°C. Mass spectrum (FAB, NBA): m/z 1310.9 (M$^+$, *calc.* 1310.7). ^{1}H-NMR (CDCl$_3$): δ 7.01, 6.99 (2s, 4H, ArH), 6.44 (s, 2H, ArH), 5.88, 5.78, 5.68 [3d, 4H (1 : 2 : 1), J = 7.3 Hz, outer OCH$_2$O], 4.78, 4.71, 4.65 [3t, 4H (1 : 2 : 1), J = 8.0 Hz, C*H*(CH$_2$)$_{10}$CH$_3$], 4.38, 4.33, 4.30 [3d, 4H (1 : 2 : 1), J = 7.3 Hz, inner OCH$_2$O], 2.3–2.0 [m, 8H, CHC*H*$_2$(CH$_2$)$_9$CH$_3$], 1.5–1.1 [m, 72H, CHCH$_2$(C*H*$_2$)$_9$CH$_3$], 0.81 [t, 12H, J = 6.8 Hz, CH(CH$_2$)$_{10}$C*H*$_3$]. ^{13}C-NMR (CDCl$_3$): δ 155.0–151.8 (s, ArCOCH$_2$O), 120.6, 119.1, 116.7 (d, ArCH), 113.2 (s, ArCBr), 99.0, 98.6 (t, OCH$_2$O), 37.7–36.3 (d, ArCHAr). *Anal. Found*: C, 68.97; H, 8.63. *Calc.* for C$_{76}$H$_{110}$Br$_2$O$_8$·0.25H$_2$O: C, 69.36; H, 8.46. Karl-Fischer found: 0.42; calc. for 0.25H$_2$O: 0.34.

1,21,23,25-Tetraundecyl-2,20:3,19-dimetheno-1H,21H,23H,25H-bis[1,3]dioxo-cino[5,4-i:5′,4′-i′]benzo[1,2-d:5,4-d′]bis[1,3]benzodioxocin-7,11-dicarboxylic acid, dimethyl ester (**9c**). To a solution of **9b** (0.21 g, 0.16 mmol) in THF (50 mL) was added *n*-BuLi (1.2 mL, 1.8 mmol) at -100°C (using EtOH/N$_2$). After stirring for 15 min at this temperature, the reaction was quenched with ClC(O)OMe (0.45 mL, 5.8 mmol). The mixture was allowed to warm to room temperature, the solvent was removed *in vacuo* and the residue was dissolved in CH$_2$Cl$_2$ (25 mL). The solution was successively washed with 1 N HCl (10 mL), H$_2$O (3 × 10 mL), brine (10 mL), dried over MgSO$_4$ and evaporated to dryness. The crude product was purified with flash column chromatography (SiO$_2$, CH$_2$Cl$_2$) to give pure **9c** in 65% yield. The compound could not be recrystallized. mp 175°C. Mass spectrum (FAB, NBA): m/z 1269.7 [(M+H)$^+$, *calc.* 1269.9]. ^{1}H-NMR (CDCl$_3$): δ 7.10, 7.00 (2s, 4H, ArH), 6.46 (s, 2H, ArH), 5.66, 5.61, 5.58 [3d, 4H (1 : 2 : 1), J = 7.4 Hz, outer OCH$_2$O], 4.7–4.6 [m, 4H, C*H*(CH$_2$)$_{10}$CH$_3$], 4.52, 4.41, 4.33 [3d, 4H (1 : 2 : 1), J = 7.4 Hz, inner OCH$_2$O], 3.78 [s, 6H, C(O)OCH$_3$], 2.2–2.0 [m, 8H, CHC*H*$_2$(CH$_2$)$_9$CH$_3$], 1.4–1.1 [m, 72H, CHCH$_2$(C*H*$_2$)$_9$CH$_3$], 0.81 [t, 12H, J = 6.4 Hz, CH(CH$_2$)$_{10}$C*H*$_3$]. ^{13}C–NMR (CDCl$_3$): δ 166.2 [s, *C*(O)OCH$_3$], 155.1–150.9 (s, ArCOCH$_2$O), 123.4 [s, Ar*C*C(O)OCH$_3$], 121.7, 120.3, 116.9 (d, ArCH), 99.5 (t, OCH$_2$O), 52.7 [q, C(O)O*C*H$_3$], 36.3 (d, ArCHAr). *Anal. Found*: C, 75.53; H, 9.61. *Calc.* for C$_{80}$H$_{116}$O$_{12}$: C, 75.67; H, 9.21.

1,21,23,25-Tetraundecyl-2,20:3,19-dimetheno-1H,21H,23H,25H-bis[1,3]dioxo-cino[5,4-i:5′,4′-i′]benzo[1,2-d:5,4-d′]bis[1,3]benzodioxocin-7,11-diol (**9d**). To a solution of **9b** (0.21 g, 0.16 mmol) in THF (50 mL) was quickly added *n*-BuLi (0.50 mL, 0.75 mmol) at -70°C. After 30 sec, the reaction mixture was quenched with B(OMe)$_3$ (0.27 mL, 2.4 mmol). The solution was warmed to room temperature and stirred for 1 h. After cooling the reaction mixture again to -70°C, a 15% solution of H$_2$O$_2$ in 1.5 M NaOH (4.2 mL, 20 mmol) was added and the mixture allowed to warm to room temperature and was stirred overnight. Excess H$_2$O$_2$ was destroyed by adding Na$_2$S$_2$O$_5$ (3.8 g, 20 mmol) to the solution. The THF was removed *in vacuo*, the residue was dissolved in CH$_2$Cl$_2$ (50 mL) and the solution was washed with H$_2$O (3 × 10 mL), with brine (10 mL), dried over MgSO$_4$ and evaporated to dryness. The crude product was purified with preparative TLC

(SiO$_2$, EtOAc/CH$_2$Cl$_2$, 30 : 70) to give pure **9d** in 62% yield. mp 150–152°C (CH$_2$Cl$_2$/EtOH). Mass spectrum (FAB, NBA): m/z 1184.6 (M$^+$, *calc.* 1184.8). ^{1}H-NMR (CDCl$_3$): δ 7.10 (s, 2H, ArH para to H), 6.64 (s, 2H, ArH para to OH), 6.50 (s, 2H, ArH ortho to OCH$_2$O), 5.94, 5.84, 5.75 [3d, 4H (1 : 2 : 1), J = 6.9, 7.0 and 7.2 Hz, outer OCH$_2$O], 5.65 (s, 2H, OH), 4.8–4.6 [m, 4H, C*H*(CH$_2$)$_{10}$CH$_3$], 4.5–4.4 (m, 4H, inner OCH$_2$O), 2.3–2.1 [m, 8H, CHC*H*$_2$(CH$_2$)$_9$CH$_3$], 1.5–1.2 [m, 72H, CHCH$_2$(C*H*$_2$)$_9$CH$_3$], 0.88 [t, 12H, J = 6.5 Hz, CH(CH$_2$)$_{10}$C*H*$_3$]. ^{13}C-NMR (CDCl$_3$): δ 154.9–154.8 (s, ArCOCH$_2$O), 140.9 (s, ArCOH), 120.9, 116.5, 109.8 (d, ArCH), 99.7 (t, OCH$_2$O), 36.8–36.3 (d, ArCHAr). *Anal. Found*: C, 76.32; H, 9.50. *Calc.* for C$_{76}$H$_{112}$O$_{10}$·0.5H$_2$O: C, 76.41; H, 9.53. Karl-Fischer found: 0.71; calc. for 0.5H$_2$O: 0.75.

1,21,23,25-Tetraundecyl-2,20:3,19-dimetheno-1H,21H23H,25H-bis[1,3]dioxocino[5,4-i:5',4'-i']benzo[1,2-d:5,4-d']bis[1,3]benzodioxocin-7,11-dicarbonitrile (**9f**). A solution of **9b** (0.82 g, 0.63 mmol), CuCN (0.44 g, 4.9 mmol) in *N*-methylpyrrolidone (5 mL) was refluxed for 21 h. After cooling the reaction mixture to room temperature, FeCl$_3$·6H$_2$O (1.0 g) and 1 N HCl (2 mL) were added and the mixture was stirred at 75°C for 30 min. After cooling the reaction mixture to room temperature, H$_2$O (25 mL) was added and the mixture was extracted twice with CHCl$_3$ (2 × 50 mL). The combined organic layers were washed with sat. NH$_4$Cl (4 × 10 mL), with H$_2$O (3 × 10 mL), with brine (10 mL) and dried over MgSO$_4$. After removal of the solvent the crude product was recrystallized from CH$_2$Cl$_2$/MeOH to give pure **9f** in 99% yield. mp 112–114°C (CH$_2$Cl$_2$/MeOH). Mass spectrum (FAB, NBA): m/z 1203.8 [(M+H)$^+$, *calc.* 1203.9]. ^{1}H-NMR (CDCl$_3$): δ 7.23, 7.00 (2s, 4H, ArH), 6.47 (s, 2H, ArH), 5.99, 5.84, 5.69 [3d, 4H (1 : 2 : 1), J = 7.2, 7.3 and 7.4 Hz, outer OCH$_2$O], 4.8–4.6 [m, 4H, C*H*(CH$_2$)$_{10}$CH$_3$], 4.55, 4.44, 4.30 [3d, 4H (1 : 2 : 1), J = 7.4, 7.3 and 7.2 Hz, inner OCH$_2$O], 2.3–2.0 [m, 8H, CHC*H*$_2$(CH$_2$)$_9$CH$_3$], 1.5–1.1 [m, 72H, CHCH$_2$(C*H*$_2$)$_9$CH$_3$], 0.80 [t, 12H, J = 6.5 Hz, CH(CH$_2$)$_{10}$C*H*$_3$]. ^{13}C-NMR (CDCl$_3$): δ 156.7–154.6 (s, ArCOCH$_2$O), 125.0, 120.3, 116.8 (d, ArCH), 112.7 (s, ArCCN), 103.9 (s, ArCCN), 99.6–98.7 (t, OCH$_2$O), 36.3 (d, ArCHAr). *Anal. Found*: C, 77.65; N, 1.96; H, 9.57. *Calc.* for C$_{78}$H$_{110}$N$_2$O$_8$: C, 77.83; N, 2.33; H, 9.21.

7,11-Dibromo-1,21,23,25-tetrakis(2-phenylethyl)-15,28-bis(thiomethyl)-2,20:3,19-dimetheno-1H,21H,23H,25H-bis[1,3]dioxocino[5,4-i:5',4'-i']benzo[1,2-d:5,4-d']-bis[1,3]benzodioxocin (**9g**). To a solution of diol **3c** (0.50g, 0.40 mmol) in THF (25 mL) was added excess NaH (1.6 mmol) and the solution was stirred at room temperature until the evolution of hydrogen stopped. The reaction mixture was cooled to -70°C and *n*-BuLi (1.3 mL, 20 mmol) was quickly added. After 15 sec the reaction mixture was quenched with excess CH$_3$SSCH$_3$ (0.36 mL, 4.0 mmol). The mixture was allowed to warm to room temperature and the solvent was removed *in vacuo*. The residue was taken up in CH$_2$Cl$_2$ (25 mL) and the solution was successively washed with 1 N HCl (10 mL), H$_2$O (3 × 10 mL), brine (10 mL), dried over $MgSO_4$ and evaporated to dryness. To a suspension of the crude product in CH_3CN (25 mL) were added K_2CO_3 (0.55 g, 4.0 mmol) and CH_2BrCl (1.0 mL, 16 mmol) whereupon the mixture was refluxed for 24 h. The solvent was removed and the residue was taken up in CH_2Cl_2 (25 mL). The solution was successively washed with 1 N HCl (10 mL), H_2O (3 × 10 mL), and brine (10 mL), dried over $MgSO_4$, and the solvent was removed under vacuum. The crude product was purified with flash column chromatography (SiO_2 60H, CH_2Cl_2/hexane, 55 : 45) to give pure **9g** as a white solid in 25% yield: mp 178–180°C (CH_2Cl_2/hexane). Mass spectrum (FAB, NBA): m/z 1203.5 [(M+H)$^+$, calc. 1203.2]. ^{1}H-NMR (CDCl$_3$): δ 7.3–7.0 [m, 24H, CH(CH$_2$)$_2$C$_6$H$_5$ + ArH], 5.97 (d, 4H, J = 7.3 Hz, outer OCH$_2$O), 5.0–4.8 [m, 4H, CH(CH$_2$)$_2$C$_6$H$_5$], 4.41, 4.38, 4.35 [3d, 4H (1 : 2 : 1), J = 7.4 Hz, inner OCH$_2$O], 2.8–2.4 [m, 16H, CH(CH_2)$_2$C$_6$H$_5$], 2.43 (s, 6H, ArSCH$_3$). ^{13}C-NMR (CDCl$_3$): δ 155.9–152.2 (s, ArCOCH$_2$O), 124.9 (s, ArCSCH$_3$), 119.7, 118.9 (d, ArCH), 113.9 (s, ArCBr), 98.9 (t, OCH$_2$O), 37.8–37.6 (d, ArCHAr), 17.9 (q, SCH$_3$). *Anal. Found*: C, 66.22; H, 4.78. *Calc.* for C$_{66}$H$_{58}$Br$_2$O$_8$S$_2$: C, 65.89; H, 4.86.

7,11-Dibromo-15,28-diiodo-1,21,23,25-tetrakis(2-phenylethyl)-2,20:3,19-dimetheno-1H,21H,23H,25H-bis[1,3]dioxocino[5,4-i:5',4'-i']benzo[1,2-d:5,4-d']bis[1, 3]benzodioxocin (**9h**) was synthesized according to the procedure for **9g**, with the exception that the lithiation reaction was quenched with I_2 instead of CH_3SSCH_3. The reaction was carried out using **3c** (0.56 g, 0.45 mmol), *n*-BuLi (1.5 mL, 2.2 mmol), THF (25 mL), I_2 (1.15 g, 4.5 mmol), K_2CO_3 (1.23 g, 8.9 mmol), CH_2BrCl (0.58 mL, 8.9 mmol) and CH_3CN (25 mL) to give pure **9h** after flash column chromatography (SiO_2 60H, CH_2Cl_2/hexane, 40 : 60) in 20% yield. mp 198–201°C (CH_2Cl_2). Mass spectrum (FAB, NBA): m/z 1361.9 (M$^+$, calc. 1362.1). ^{1}H-NMR (CDCl$_3$): δ 7.2–7.0 [m, 24H, CH(CH$_2$)$_2$C$_6$H$_5$ + ArH], 5.92, 5.91, 5.90 [3d, 4H (1 : 2 : 1), J = 7.4, 7.4 and 7.3 Hz, outer OCH$_2$O], 4.89 [t, 4H, J = 7.7 Hz, CH(CH$_2$)$_2$C$_6$H$_5$], 4.33, 4.31, 4.27 [3d, 4H (1 : 2 : 1), J = 7.4, 7.4 and 7.5 Hz, inner OCH$_2$O], 2.7–2.3 [m, 16H, CH(CH_2)$_2$C$_6$H$_5$]. ^{13}C-NMR (CDCl$_3$): δ 155.2, 152.3 (s, ArCOCH$_2$O), 120.5, 118.9 (d, ArCH), 113.9 (s, ArCBr), 98.7–98.5 (t, OCH$_2$O), 93.5 (s, ArCI), 38.1–37.8 (d, ArCHAr). *Anal. Found*: C, 53.47; H, 3.74. *Calc.* for C$_{64}$H$_{52}$Br$_2$I$_2$O$_8$·CH$_2$Cl$_2$: C, 53.10; H, 3.76.

1,19,21,29,47,49,57,62-Octaundecyl-23,27:51,55-dimethano-2,46:3,45:17,31:18, 30-tetrametheno-1H,19H,21H,29H,47H,49H-bis[1,3]benzodioxocino[9,8-d:9',8'-d''][1,3,6,8,11,13,16,18]octaoxacycloeicosino[4,5-j:10,9-j':14,15-j'':20,19-j''']tetrakis[1,3]benzodioxocin (C-isomer) (**10**). A solution of diol **9d** (75 mg, 0.063 mmol) and CH_2BrCl (10 μL, 0.14 mmol) in a mixture of DMSO (5 mL) and THF (1 mL) was added to a solution of Cs_2CO_3 (0.21 g, 0.63 mmol) in DMSO (25 mL) at 60°C over a 4 h period. After 17 h, another portion of CH_2BrCl (15 μL, 0.23 mmol) was added and the temperature was raised to 100°C. The addition of CH_2BrCl (15 μL, 0.23 mmol) was repeated every 24 h. After a total reaction

TABLE I. Crystallographic data.

Experimental	
Crystal data	
$C_{46}H_{44}Br_2O_{16} \cdot 2C_5H_{10}$	$D_c = 1.61$ Mg m^{-3}
$M_r = 1172.4$	MoK_α radiation
triclinic	$P\bar{1}$
$a = 9.777(3)$ Å	$\alpha = 105.46(2)°$
$b = 13.601(4)$ Å	$\beta = 106.65(2)°$
$c = 22.271(8)$ Å	$\gamma = 105.47(2)°$
$V = 2686.1(7)$ Å^3	$T = 173(1)$ K
$Z = 2$	
Refinement	
Final $R = 0.065$	$wR = 0.091$
3872 reflections	325 parameters

time of 72 h, the solvent was removed *in vacuo*. The residue was dissolved in CH_2Cl_2 (50 mL), washed with H_2O (2 × 10 mL), with brine (10 mL) and dried over $MgSO_4$. The solution was evaporated to dryness and the product was purified by flash column chromatography (SiO_2, CH_2Cl_2/hexane, 50 : 50) to give pure **10** as a white solid in 71% yield. mp 178–182°C (CH_2Cl_2/MeOH). Mass spectrum (FAB, NBA): m/z 2395.0 (M$^+$, calc. 2394.7). ^{1}H-NMR (CDCl$_2$CDCl$_2$, 90°C): δ 7.12 (s, 4H, ArH para to H), 6.86 (s, 4H, ArH para to OCH$_2$O), 6.52 (s, 4H, ArH ortho to OCH$_2$O), 5.95 (d, 2H, $J = 6.4$ Hz, outer OCH$_2$O), 5.89 (d, 2H, $J = 7.4$ Hz, outer OCH$_2$O), 5.4, 5.2 [2bs, 3H (2 : 1), OCH$_2$O connecting the two parts], 5.2–4.9 [m, 9H, outer OCH$_2$O (4H) + OCH$_2$O connecting the two parts (1H) + inner OCH$_2$O (2H) + CH(CH$_2$)$_{10}$CH$_3$ (2H)], 4.92 [t, 2H, $J = 7.9$ Hz, CH(CH$_2$)$_{10}$CH$_3$], 4.73 (d, 2H, $J = 7.4$ Hz, inner OCH$_2$O), 4.46 [t, 4H, $J = 7.9$ Hz, CH(CH$_2$)$_{10}$CH$_3$], 3.51 (d, 4H, $J = 7.4$ Hz, inner OCH$_2$O), 2.4–2.2 [m, 8H, CHCH_2(CH$_2$)$_9$CH$_3$], 2.2–2.0 [m, 8H, CHCH_2(CH$_2$)$_9$CH$_3$], 1.6–1.4 [m, 16H, CHCH$_2$CH_2(CH$_2$)$_8$CH$_3$], 1.4–1.2 [m, 128H, CH(CH$_2$)$_2$(CH_2)$_8$CH$_3$], 0.88 [t, 24H, $J = 6.5$ Hz, CH(CH$_2$)$_{10}$CH_3]. Anal. Found: C, 77.37; H, 10.29. Calc. for $C_{154}H_{224}O_{20}$: C, 77.21; H, 9.43.

X-Ray Structure Determination

A small sample of **6a** was recrystallized from CH_2Cl_2/pentane and its crystal structure was determined by X-ray diffraction. The most important crystallographic data are collected in Table I.

Data were collected in the $\omega/2\theta$ scan mode (scan width (ω): $1.10 + 0.34 \tan\theta$), using graphite monochromated MoK_α radiation. The intensity data were corrected for Lorentz and polarization effects and for long-time-scale variation. No absorption correction was applied. The structure was solved with MULTAN [29] and refined by

TABLE II. Fractional atomic coordinates and equivalent isotropic thermal parameters for non-H atoms, with e.s.ds in parentheses. $B_{eq} = (8\pi^2/3)\Sigma_i\Sigma_j U_{ij} a_i^* a_j^* a_i \cdot a_j$.

Atom	x	y	z	B_{eq}
Br(1)	0.7932(1)	0.30815(9)	0.02473(6)	2.20(3)
Br(2)	0.0286(1)	-0.18739(9)	0.40527(6)	2.34(3)
Br(3)	0.1860(1)	0.44472(9)	0.52321(6)	2.31(3)
Br(4)	0.0158(1)	-0.03727(9)	0.13681(6)	2.54(3)
O(7)	0.9078(7)	0.5433(5)	0.1112(3)	1.5(1)
O(10)	0.6583(8)	0.5993(6)	0.0379(4)	2.4(2)
O(12)	0.8203(8)	0.7362(6)	0.1142(4)	2.9(2)
O(27)	1.0138(7)	0.7533(5)	0.2611(3)	1.5(1)
O(30)	0.8128(8)	0.8384(6)	0.2583(4)	2.1(2)
O(32)	0.6337(8)	0.6811(6)	0.2164(4)	2.2(2)
O(33)	1.2974(7)	0.7735(5)	0.4582(4)	1.9(2)
O(35)	1.3559(8)	0.6478(5)	0.5009(4)	2.0(2)
O(47)	1.2523(8)	0.2668(6)	0.4282(4)	2.1(2)
O(50)	0.8930(8)	0.0973(6)	0.4018(4)	2.8(2)
O(52)	1.0928(8)	0.1144(6)	0.4712(4)	2.7(2)
O(67)	1.1824(8)	0.0757(6)	0.2735(4)	2.1(2)
O(70)	1.450(1)	-0.0232(8)	0.3246(5)	5.1(2)
O(71)	1.2961(8)	0.0338(6)	0.3777(4)	2.7(2)
O(73)	1.1376(8)	0.1152(5)	0.0663(4)	1.9(2)
O(75)	1.0608(7)	0.2378(5)	0.0252(4)	1.8(2)
C(1)	0.985(1)	0.3877(8)	0.0724(5)	1.3(2)
C(2)	1.019(1)	0.4939(8)	0.1104(5)	1.5(2)
C(3)	1.158(1)	0.5498(8)	0.1474(2)	1.6(2)
C(4)	1.260(1)	0.4952(8)	0.1472(5)	1.3(2)
C(5)	1.228(1)	0.3904(8)	0.1088(5)	1.3(2)
C(6)	1.092(1)	0.3389(8)	0.0714(5)	1.5(2)
C(8)	0.895(1)	0.5864(8)	0.0592(6)	1.8(2)
C(9)	0.788(1)	0.6500(8)	0.0741(6)	1.8(2)
C(11)	0.545(1)	0.652(1)	0.0494(7)	3.3(3)
C(13)	1.195(1)	0.6700(8)	0.1868(5)	1.5(2)
C(14)	1.329(1)	0.7372(8)	0.1689(6)	1.9(2)
C(21)	1.152(1)	0.7555(8)	0.3591(5)	1.4(2)
C(22)	1.126(1)	0.7293(8)	0.2934(5)	1.5(2)
C(23)	1.220(1)	0.6880(8)	0.2583(5)	1.3(2)
C(24)	1.341(1)	0.6732(8)	0.2942(5)	1.3(2)
C(25)	1.367(1)	0.6971(8)	0.3599(5)	1.2(2)
C(26)	1.272(1)	0.7395(8)	0.3924(5)	1.4(2)
C(28)	0.869(1)	0.6775(8)	0.2502(5)	1.4(2)
C(29)	0.761(1)	0.7327(8)	0.2386(6)	1.9(2)
C(31)	0.704(1)	0.896(1)	0.2455(7)	3.1(3)
C(34)	1.246(1)	0.6943(8)	0.4869(6)	1.9(2)

TABLE II. *(continued)*

Atom	x	y	z	B_{eq}
C(36)	1.492(1)	0.6799(8)	0.3988(6)	1.8(2)
C(37)	1.630(1)	0.6929(8)	0.3700(6)	1.9(2)
C(41)	1.297(1)	0.4559(8)	0.4609(5)	1.4(2)
C(42)	1.360(1)	0.5577(8)	0.4550(5)	1.5(2)
C(43)	1.436(1)	0.5679(8)	0.4071(5)	1.3(2)
C(44)	1.451(1)	0.4750(8)	0.3671(5)	1.3(2)
C(45)	1.390(1)	0.3727(8)	0.3720(5)	1.5(2)
C(46)	1.311(1)	0.3648(8)	0.4190(5)	1.2(2)
C(48)	1.104(1)	0.2164(9)	0.3967(6)	2.1(3)
C(49)	1.036(1)	0.1380(8)	0.4296(6)	1.8(2)
C(51)	0.805(1)	0.014(1)	0.4249(7)	3.0(3)
C(53)	1.415(1)	0.2713(8)	0.3300(5)	1.5(2)
C(54)	1.579(1)	0.2907(9)	0.3336(6)	2.2(3)
C(61)	1.160(1)	0.0981(8)	0.1703(5)	1.5(2)
C(62)	1.227(1)	0.1368(8)	0.2346(5)	1.5(2)
C(63)	1.334(1)	0.2366(8)	0.2616(5)	1.4(2)
C(64)	1.370(1)	0.2981(8)	0.2203(5)	1.5(2)
C(65)	1.305(1)	0.2617(8)	0.1550(5)	1.5(2)
C(66)	1.200(1)	0.1589(8)	0.1311(5)	1.3(2)
C(68)	1.230(1)	-0.020(1)	0.2643(6)	3.0(3)
C(69)	1.339(1)	-0.0035(9)	0.3248(6)	2.4(3)
C(72)	1.394(1)	0.054(1)	0.4382(7)	3.6(3)
C(74)	1.013(1)	0.1435(9)	0.0432(6)	2.2(3)
C(76)	1.340(1)	0.3281(8)	0.1099(6)	1.9(2)
C(77)	1.499(1)	0.4015(8)	0.1277(6)	1.9(2)
C(80)	0.403(2)	-0.036(1)	0.0919(8)	4.5(4)
C(81)	0.516(2)	0.051(1)	0.1447(9)	5.9(4)
C(82)	0.652(2)	0.062(1)	0.1126(8)	5.6(4)
C(83)	0.625(2)	-0.056(1)	0.0683(8)	5.2(4)
C(84)	0.476(1)	-0.1063(8)	0.0589(5)	1.5(2)
C(90)	0.167(1)	0.5597(8)	0.7260(5)	1.1(2)
C(91)	0.055(2)	0.544(1)	0.6781(8)	5.0(4)
C(92)	0.051(2)	0.412(1)	0.292(1)	7.1(5)
C(93)	0.961(2)	0.321(1)	0.2341(9)	6.0(4)
C(94)	0.172(2)	0.641(1)	0.7788(8)	5.5(4)

full-matrix least-squares methods. Weights for each reflection in the refinement (on F) were $w = 4F_0^2/\sigma(F_0^2)$, $\sigma(F_0^2) = \sigma^2(I) + (pF_0^2)^2$; the value of the instability factor p was determined to be 0.04. All calculations were done with SDP [30]. Atomic scattering factors were taken from the *International Tables for X-Ray Crystallography* [31]. Atomic parameters are given in Table II. Bond distances and angles are given in Table III. The atom numbering and structure determined

TABLE III. Bond distances (Å) and angles (°) for the heavy atoms with e.s.ds in parentheses

Atom 1	Atom 2	Distance	Atom 1	Atom 2	Distance
Br(1)	C(1)	1.900(9)	C(13)	C(14)	1.55(2)
Br(2)	C(21)	1.88(1)	C(13)	C(23)	1.52(2)
Br(3)	C(41)	1.88(1)	C(21)	C(22)	1.38(2)
Br(4)	C(61)	1.884(9)	C(21)	C(26)	1.40(2)
O(7)	C(2)	1.43(1)	C(22)	C(23)	1.42(2)
O(7)	C(8)	1.44(2)	C(23)	C(24)	1.43(2)
O(10)	C(9)	1.31(1)	C(24)	C(25)	1.38(2)
O(10)	C(11)	1.49(2)	C(25)	C(26)	1.41(2)
O(12)	C(9)	1.20(1)	C(25)	C(36)	1.52(2)
O(27)	C(22)	1.40(1)	C(28)	C(29)	1.49(2)
O(27)	C(28)	1.46(1)	C(36)	C(37)	1.55(2)
O(30)	C(29)	1.32(1)	C(36)	C(43)	1.55(2)
O(30)	C(31)	1.51(2)	C(41)	C(42)	1.41(2)
O(32)	C(29)	1.22(1)	C(41)	C(46)	1.39(1)
O(33)	C(26)	1.38(1)	C(42)	C(43)	1.39(2)
O(33)	C(34)	1.41(2)	C(43)	C(44)	1.39(1)
O(35)	C(34)	1.43(2)	C(44)	C(45)	1.40(2)
O(35)	C(42)	1.38(1)	C(45)	C(46)	1.39(2)
O(47)	C(46)	1.39(1)	C(45)	C(53)	1.54(2)
O(47)	C(48)	1.42(1)	C(48)	C(49)	1.51(2)
O(50)	C(49)	1.35(1)	C(53)	C(54)	1.54(2)
O(50)	C(51)	1.48(2)	C(53)	C(63)	1.51(2)
O(52)	C(49)	1.18(2)	C(61)	C(62)	1.39(2)
O(67)	C(62)	1.39(2)	C(61)	C(66)	1.38(2)
O(67)	C(68)	1.47(2)	C(62)	C(63)	1.39(1)
O(70)	C(69)	1.18(2)	C(63)	C(64)	1.42(2)
O(71)	C(69)	1.32(2)	C(64)	C(65)	1.40(2)
O(71)	C(72)	1.46(2)	C(65)	C(66)	1.42(1)
O(73)	C(66)	1.39(1)	C(65)	C(76)	1.53(2)
O(73)	C(74)	1.44(2)	C(68)	C(69)	1.53(2)
O(75)	C(6)	1.41(1)	C(76)	C(77)	1.54(1)
O(75)	C(74)	1.42(1)	C(80)	C(81)	1.50(2)
C(1)	C(2)	1.39(1)	C(80)	C(84)	1.45(3)
C(1)	C(6)	1.38(2)	C(81)	C(82)	1.57(3)
C(2)	C(3)	1.39(1)	C(82)	C(83)	1.58(3)
C(3)	C(4)	1.40(2)	C(83)	C(84)	1.40(2)
C(3)	C(13)	1.55(1)	C(90)	C(91)	1.35(2)
C(4)	C(5)	1.38(1)	C(90)	C(94)	1.37(2)
C(5)	C(6)	1.36(1)	C(91)	C(92)	1.48(3)
C(5)	C(76)	1.55(2)	C(92)	C(93)	1.50(3)
C(8)	C(9)	1.54(2)	C(93)	C(94)	1.54(3)

TABLE III. *(continued)*

Atom 1	Atom 2	Atom 3	Angle	Atom 1	Atom 2	Atom 3	Angle
C(2)	O(7)	C(8)	111.5(9)	O(33)	C(26)	C(25)	121(1)
C(9)	O(10)	C(11)	116.8(8)	C(21)	C(26)	C(25)	121(2)
C(22)	O(27)	C(28)	115.4(8)	O(27)	C(28)	C(29)	108.2(9)
C(29)	O(30)	C(31)	115.3(8)	O(30)	C(29)	C(32)	125(1)
C(26)	O(33)	C(34)	116.4(7)	O(30)	C(29)	C(28)	114.6(8)
C(34)	O(35)	C(42)	117.8(8)	O(32)	C(29)	C(28)	120(1)
C(46)	O(47)	C(48)	113(1)	O(33)	C(34)	O(35)	110.3(9)
C(49)	O(50)	C(51)	117(1)	C(25)	C(36)	C(37)	114(2)
C(62)	O(67)	C(68)	114.4(9)	C(25)	C(36)	C(43)	107.3(8)
C(69)	O(71)	C(72)	117(2)	C(37)	C(36)	C(43)	112(1)
C(66)	O(73)	C(74)	117.3(9)	Br(3)	C(41)	C(42)	118.9(8)
C(6)	O(75)	C(74)	119.0(9)	Br(3)	C(41)	C(46)	121.0(8)
Br(1)	C(1)	C(2)	120.9(9)	C(42)	C(41)	C(46)	120(2)
Br(1)	C(1)	C(6)	120.0(7)	O(35)	C(42)	C(41)	119(2)
C(2)	C(1)	C(6)	119.1(9)	O(35)	C(42)	C(43)	120.6(9)
O(7)	C(2)	C(1)	118.7(9)	C(41)	C(42)	C(43)	120(1)
O(7)	C(2)	C(3)	120.8(9)	C(36)	C(43)	C(42)	118.7(9)
C(1)	C(2)	C(3)	120(2)	C(36)	C(43)	C(44)	123(2)
C(2)	C(3)	C(4)	118.2(9)	C(42)	C(43)	C(44)	119(2)
C(2)	C(3)	C(13)	120(2)	C(43)	C(44)	C(45)	123(1)
C(4)	C(3)	C(13)	122.3(8)	C(44)	C(45)	C(46)	118(1)
C(3)	C(4)	C(5)	121.6(9)	C(44)	C(45)	C(53)	123(2)
C(4)	C(5)	C(6)	119(2)	C(46)	C(45)	C(53)	120(1)
C(4)	C(5)	C(76)	121.8(8)	O(47)	C(46)	C(41)	118(2)
C(6)	C(5)	C(76)	119.4(9)	O(47)	C(46)	C(45)	121.0(9)
O(75)	C(6)	C(1)	117.4(9)	C(41)	C(46)	C(45)	121(1)
O(75)	C(6)	C(5)	120(2)	O(47)	C(48)	C(49)	108(1)
C(1)	C(6)	C(5)	122(1)	O(50)	C(49)	O(52)	125(2)
O(7)	C(8)	C(9)	103(1)	O(50)	C(49)	C(48)	107(2)
O(10)	C(9)	O(12)	124(1)	O(52)	C(49)	C(48)	128.3(9)
O(10)	C(9)	C(8)	112.1(8)	C(45)	C(53)	C(54)	109.8(7)
O(12)	C(9)	C(8)	124.2(9)	C(45)	C(53)	C(63)	113(1)
C(3)	C(13)	C(14)	109.2(9)	C(54)	C(53)	C(63)	110(1)
C(3)	C(13)	C(23)	112(1)	Br(4)	C(61)	C(62)	120.0(9)
O(14)	C(13)	C(23)	111.3(8)	Br(4)	C(61)	C(66)	120.2(7)
Br(2)	C(21)	C(22)	120.8(9)	C(62)	C(61)	C(66)	119.8(8)
Br(2)	C(21)	C(26)	118.8(9)	O(67)	C(62)	C(61)	118.8(8)
C(22)	C(21)	C(26)	120(2)	O(67)	C(62)	C(63)	120(1)
O(27)	C(22)	C(21)	120(2)	C(61)	C(62)	C(63)	122(1)
O(27)	C(22)	C(23)	119(1)	C(53)	C(63)	C(62)	121(2)
C(21)	C(22)	C(23)	121(2)	C(53)	C(63)	C(64)	122.5(8)
C(13)	C(23)	C(22)	119(2)	C(62)	C(63)	C(64)	117(1)
C(13)	C(23)	C(24)	124(2)	C(63)	C(64)	C(65)	122.8(8)

TABLE III. *(continued)*

Atom 1	Atom 2	Atom 3	Angle	Atom 1	Atom 2	Atom 3	Angle
C(22)	C(23)	C(24)	117(2)	C(64)	C(65)	C(66)	118(2)
C(23)	C(24)	C(25)	123(2)	C(64)	C(65)	C(76)	123.7(8)
C(24)	C(25)	C(26)	118(2)	C(66)	C(65)	C(76)	119.4(9)
C(24)	C(25)	C(36)	124(2)	O(73)	C(66)	C(61)	118.5(8)
C(26)	C(25)	C(36)	118(1)	O(73)	C(66)	C(65)	120(2)
O(33)	C(26)	C(21)	118(2)	C(61)	C(66)	C(65)	122(1)
O(67)	C(68)	C(69)	108.1(9)	C(80)	C(81)	C(82)	100(1)
O(70)	C(69)	O(71)	123(1)	C(81)	C(82)	C(83)	103(1)
O(70)	C(69)	C(68)	124(1)	C(82)	C(83)	C(84)	105(1)
O(71)	C(69)	C(68)	113(2)	C(80)	C(84)	C(83)	113(1)
O(73)	C(74)	O(75)	109.7(8)	C(91)	C(90)	C(94)	111(1)
C(5)	C(76)	C(65)	107(1)	C(90)	C(91)	C(92)	107(1)
C(5)	C(76)	C(77)	113.4(9)	C(91)	C(92)	C(93)	105(1)
C(65)	C(76)	C(77)	113(1)	C(92)	C(93)	C(94)	100(1)
C(81)	C(80)	C(84)	106(1)	C(90)	C(94)	C(93)	109(1)

are shown in Figure 1. The bromine atoms were refined anisotropically. Hydrogen atoms were not resolved.

3. Results and Discussion

Cavitands containing an enforced cavity [32] are generally prepared via a four-fold bridging reaction of an octol (**1**) with a dihalide to give compounds of general structure **2** (Scheme 1) [19, 20, 33, 34]. Several years ago, tri-bridged diol **3a** was isolated in a yield of 27% as a side product in the synthesis of cavitand **2a** [19]. Such tri-bridged diols, in which three out of four pairs of hydroxyl groups are connected via a methylene spacer, are interesting molecules with respect to selective functionalization. They have a lower degree of symmetry than tetra-bridged cavitands, which means that the reactivity of the aromatic rings bearing the hydroxyl groups will be different from the others.

First, we optimized the synthesis of tri-bridged diol **3** by variation of the reaction conditions, temperature (and the reaction time), solvent, and amount of CH_2BrCl. These reactions were carried out with octols with different side chains (R_1). The results are summarized in Table IV.

One of the first things that becomes clear from the data in Table IV is that reactions run at 70°C (entries 2, 6, 7, 10, 12 and 14–16) generally give higher overall yields than the corresponding reactions at room temperature (entries 1, 5, 8, 9, 11 and 13). With respect to the formation of tri-bridged diol **3** there is a marked difference between the solvents DMF (entries 4, 11, 12, 15 and 16) and DMSO (entries 1–3, 5–10, 13 and 14).

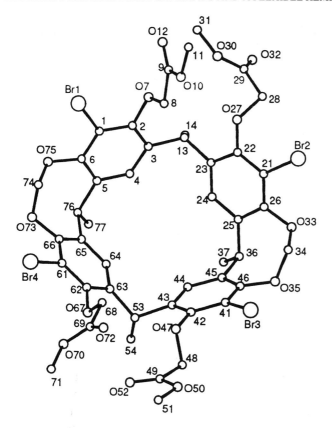

Fig. 1. View of the X-ray crystal structure of tetramethyl ester **6a** (the cyclopentane molecules are not located in the cavity of **6a** and have therefore been omitted for clarity) made using PLUTO [39].

In DMF the reaction at room temperature (entry 11) gives mainly diol **3** and only a small amount of cavitand **2** is formed. At this temperature the introduction of the fourth methylene bridge is considerably slower than the formation of the other three bridges. The reason for this decreased reactivity has been attributed to the fact that cavitands become progressively more rigid when more bridges are introduced and they are therefore less adaptable to the linear transition state required for bridging [35]. If the temperature is raised to 70°C (entry 15), the fourth bridge can be easily introduced when excess CH_2BrCl is present, leading to high yields of **2**. In order to facilitate the isolation of reasonable amounts of **3**, the amount of CH_2BrCl was lowered to 4 equiv. (entries 12 and 16).

In DMSO the reaction temperature has hardly any influence. Both at room temperature and at 70°C the tri-bridged diol **3** is formed as the main product, regardless of the amount of excess CH_2BrCl used. The introduction of the fourth bridge took place, to some extent, only in the case of octol **1a** (entry 2). This result is quite surprising, because DMSO is one of the best solvents for S_N2 reactions [36].

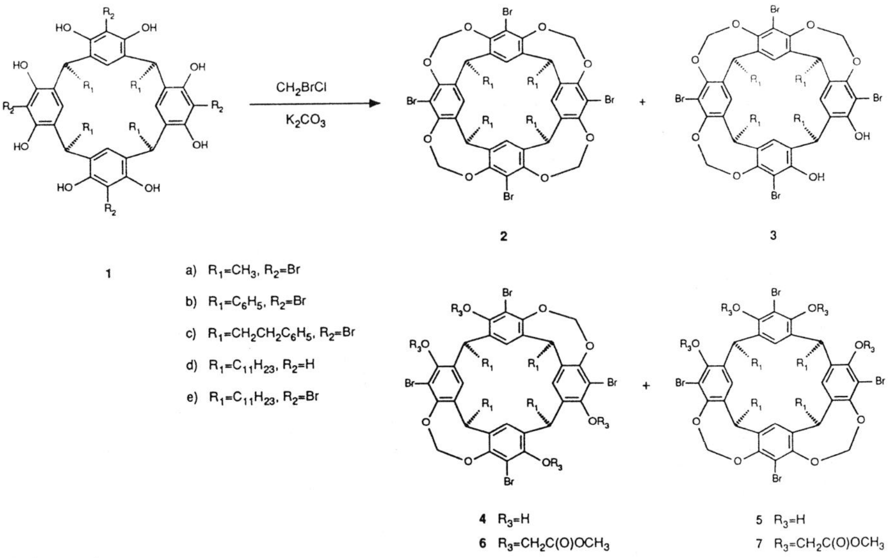

Scheme 1.

1 a) $R_1=CH_3$, $R_2=Br$
   b) $R_1=C_6H_5$, $R_2=Br$
   c) $R_1=CH_2CH_2C_6H_5$, $R_2=Br$
   d) $R_1=C_{11}H_{23}$, $R_2=H$
   e) $R_1=C_{11}H_{23}$, $R_2=Br$

4 $R_3=H$
6 $R_3=CH_2C(O)OCH_3$

5 $R_3=H$
7 $R_3=CH_2C(O)OCH_3$

However, sulfoxides are also good nucleophiles that can be alkylated by methyl halides to form the corresponding sulfoxonium salts [37]. This reaction normally takes place at room temperature with methyl iodide, but apparently requires a higher temperature when a less reactive halide like CH_2BrCl is involved.

These results show that the optimum reaction conditions for the formation of tri-bridged diols **3** are: (a) reaction at 70°C for 3 days in DMSO with excess CH_2BrCl; or (b) in DMF with 4 equiv. of CH_2BrCl; comparable results can be obtained by reaction at room temperature in DMF of DMSO with excess CH_2BrCl for 8 days.

From the reactions of **1c** in DMSO at room temperature (entries 8 and 9), in addition to the compounds **2** and **3** we isolated considerable amounts of a di-bridged tetrol. According to the ^{1}H-NMR spectrum, this product appears to be the more symmetrical A,C-bridged isomer **4c** (*vide infra*). No A,B-bridged isomer could be isolated in these reactions. In an attempt to optimize the reaction conditions for the formation of **4a** (entries 3 and 4), using only 2 equiv. of CH_2BrCl, we observed that, in addition to the A,C-di-bridged tetrol **4a**, a substantial amount of A,B-di-bridged tetrol **5a** was formed. Since these products could not be isolated in a pure state, due to their low solubility, they were first converted to the corresponding tetramethyl esters **6a** and **7a**, respectively, by reaction with 4 equiv. of methyl bromoacetate [38]. The structure of **6a**·2C_5H_{10} was unambiguously proven by solving the X-ray crystal structure (Figure 1). The structure of this molecule is rather rigid because all four aromatic rings are fixed to a neighbouring ring via a methylene bridge. The molecule therefore possesses a crown-like conformation. Compound **7a** is much more flexible, because only three aromatic rings are bridged and one ring is free in

TABLE IV. Synthesis of cavitands **2–5** under different reaction conditions.

Entry	R₁	Solvent	Temperature (°C)	Reaction time (days)	Equiv. CH₂BrCl	Yield (%)	Product distribution (%)			
							2	**3**	**4**	**5**
1	CH₃	DMSO	25	8	3.3 + 7(5d)	47[b]	<2	45	-[c]	-[c]
2	"	"	70	3	8	60[a]	40	20	0	0
3	"	"	70	1	2	48[a]	0	0	30[d]	18[d]
4	"	DMF	70	2	2	45[a]	0	9[d]	24[d,e]	9[d]
5	C₆H₅	DMSO	25	8	4 + 8(5d)	13[a]	<1	12	-[c]	-[c]
6	"	"	70 / 70–85 / 95	2.5 / 1 / 0.5	8 / 4(3.5d) / –	65[a]	22	43	0	0
7	"	"	70	3.5	8	49[b]	7	42	0	0
8	(CH₂)₂Ph	"	25	8	4 + 4(5d)	59[a]	1	41	17	0
9	"	"	25	8	4 + 4(5d)	80[b]	2	49	29	0
10	"	"	70	3	8	65[b]	11	54	0	0
11	"	DMF	25	18	8 + 8(6d)	53[b]	11	42	0	0
12	"	"	70	4	4	62[b]	16	46	0	0
13	C₁₁H₂₃	DMSO	25	8	8 + 8(5d)	43[a]	6	34	4	0
14	"	"	70	3	8	61[a]	22	39	-[c]	-[c]
15	"	DMF	70	3	8	80[a]	73	7	0	0
16	"	"	70	3	4	82[a]	30	52	-[c]	-[c]

[a] Work-up according to procedure A.
[b] Work-up according to procedure B.
[c] Compounds **4** and **5** were not isolated.
[d] Isolated as the tetramethyl esters **6** and **7**.
[e] An additional 3% of the corresponding tribromo derivative was also isolated.

this molecule. This ring is considerably tilted with respect to the others according to the marked upfield shift of the only Ar—H proton in this aryl ring (from 7.3 to 6.7 ppm) in the ^{1}H-NMR spectrum, compared to the same protons in the other aryl rings.

The formation of the tetrols **4** and **5** is generally believed to occur via the introduction of a methylene bridge at the distal position C (formation of A,C-di-bridged tetrol **4**) or at the proximal position B (formation of A,B-di-bridged tetrol **5**) of a mono-bridged hexol. The formation of an A,B-di-bridged tetrol without bromo atoms has been reported by Cram *et al.* [35]. In this case, no A,C-di-bridged tetrol could be detected. This selectivity can be attributed to the different acidities of the different phenolic moieties in the mono-bridged hexol. Due to the presence of the methylenedioxy bridge, the acidity of the phenolic hydroxyl group meta to this bridge increases, because the electron-releasing effect of a methyleneoxy substituent is generally smaller than that of a hydroxyl group [40].

When bromo substituents are present at the ortho positions of the phenolic hydroxyl groups, the selectivity is reversed. In this case the nucleophilic attack at the distal phenolic groups seems to be more favorable and the A,C-di-bridged tetrol **4a** is preferably formed. We believe that this change in selectivity can be attributed both to an inductive and a steric effect of the bulky bromo substituents. It is known that the pK_a of 2-bromophenol is 1.5 pK_a units lower than that of phenol [41]. Therefore the much smaller increase of the acidity due to the introduction of a methyleneoxy substituent is simply overruled. The bulky bromo substituents will also shield the hydroxyl groups and therefore hinder the attack of the incoming nucleophile. The flexibility of the different aromatic rings will now play an important role in determining which phenolic group will be alkylated first. Considering the mono-bridged hexol, which is the precursor of the tetrols, we believe that the *decreased* flexibility of the two aromatic rings connected via the methylenedioxy spacer mainly governs the selectivity of the reaction, i.e. that nucleophilic attack takes place preferentially at the distal phenolic groups.

TRANSFER OF FUNCTIONALITY

As was stated earlier (*vide supra*), **3** is an interesting compound in regard to selective functionalization of the upper rim, because of its lower degree of symmetry. Several strategies can be used to exploit the different chemical reactivity of the aromatic rings involved, one of which was recently used by Sorrell and Richard [24]. We have developed a very simple and straightforward procedure for the one-step synthesis of proximally functionalized cavitands **8** and **9** (Charts 1 and 2), in which a variety of functional groups can be easily introduced.

In an attempt to fully debrominate **3a** by treatment with excess *n*-BuLi in THF at -78°C, we observed that bromo-lithium exchange is not equally fast at all four positions. The same reaction with diol **3c** using only 5 equiv. of *n*-BuLi and quenching the reaction mixture with excess H_2O after a reaction time of only

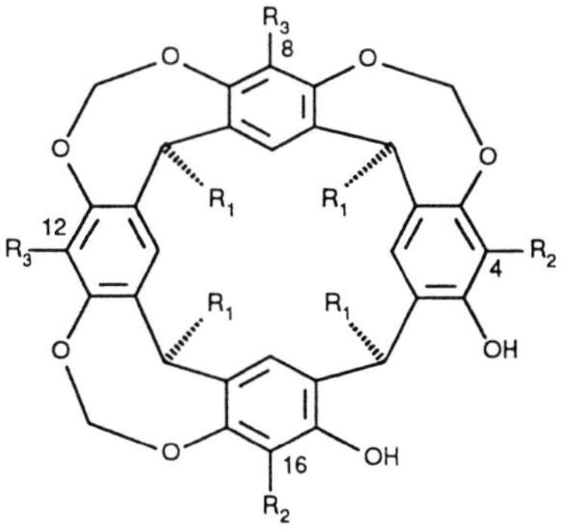

Chart 1.

8a $R_1 = CH_2CH_2C_6H_5$, $R_2 = Br$, $R_3 = H$

8b $R_1 = CH_2(CH_2)_9CH_3$, $R_2 = Br$, $R_3 = H$

8c $R_1 = CH_2(CH_2)_9CH_3$, $R_2 = Br$, $R_3 = OH$

8d $R_1 = CH_2CH_2C_6H_5$, $R_2 = Br$, $R_3 = SCH_3$

8e $R_1 = CH_2CH_2C_6H_5$, $R_2 = Br$, $R_3 = I$

Chart 2.

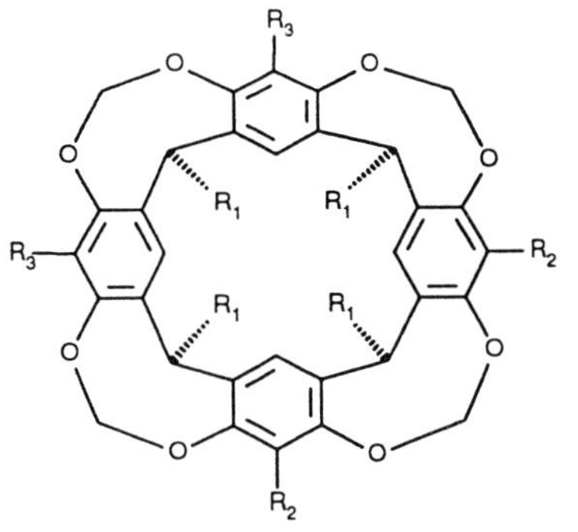

9a $R_1 = CH_2CH_2C_6H_5$, $R_2 = Br$, $R_3 = H$

9b $R_1 = CH_2(CH_2)_9CH_3$, $R_2 = Br$, $R_3 = H$

9c $R_1 = CH_2(CH_2)_9CH_3$, $R_2 = C(O)OMe$, $R_3 = H$

9d $R_1 = CH_2(CH_2)_9CH_3$, $R_2 = OH$, $R_3 = H$

9e $R_1 = CH_2CH_2C_6H_5$, $R_2 = OH$, $R_3 = H$

9f $R_1 = CH_2(CH_2)_9CH_3$, $R_2 = CN$, $R_3 = H$

9g $R_1 = CH_2CH_2C_6H_5$, $R_2 = Br$, $R_3 = SCH_3$

9h $R_1 = CH_2CH_2C_6H_5$, $R_2 = Br$, $R_3 = I$

15 sec yielded the selectivity debrominated diol **8a** in 77% yield. In a similar way **8b** could be obtained in 76% yield from **3b**.

The ^{1}H-NMR spectra of **8a** and **8b** exhibit a characteristic singlet at about 6.5 ppm for the protons that are introduced. Definite proof for the regioselectivity was given by a single-crystal X-ray analysis of **8a** [23], which clearly showed that the remaining bromo substituents are adjacent to the free hydroxyl groups. The inhibition of bromo-lithium exchange at these two aromatic rings is due to the presence of the hydroxyl groups. Upon addition of *n*-BuLi, the hydroxyl groups are deprotonated first and this increases the electron density on the aromatic ring to such an extent, that bromo-lithium exchange is strongly disfavored. A similar lack of reactivity in bromo-lithium exchange reactions has previously been observed in the case of bromoanilines and cyanomethylphenyl bromides [42].

The tri-bridged diols **8a** and **8b** were converted quantitatively into the cavitands **9a** and **9b** with excess of CH_2BrCl and K_2CO_3 in refluxing CH_3CN for 24 h. Subsequently, the two remaining bromines could be easily substituted for a variety of functional groups, leaving the 8- and 12-positions unaffected. Treatment of **9b** with *n*-BuLi in THF at -100°C for 15 min followed by quenching with ClC(O)OMe gave diester **9c** in 60% yield. Quenching the lithiated product at -78°C with $B(OMe)_3$

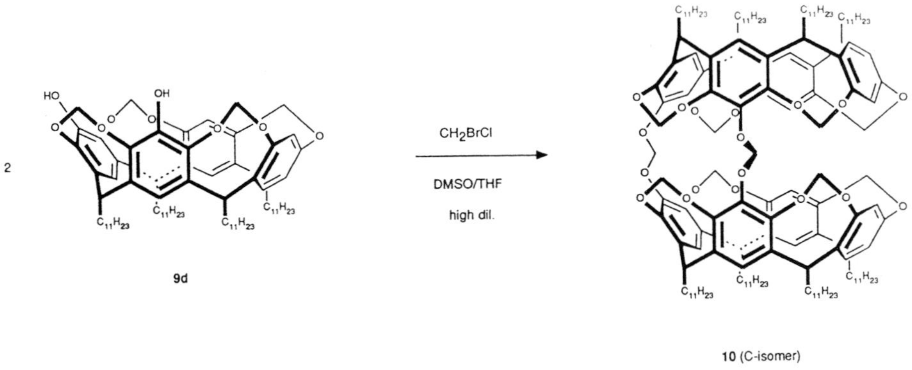

Scheme 2.

followed by oxidation with basic H$_2$O$_2$ afforded diol **9d** in 62% yield. In this way diol **9e**, which was reported previously by Cram *et al.* [20] in an overall yield of 0.7% starting from the corresponding tetrabromooctol **1c**, could be prepared in an overall yield of 19%. Heating **9b** with CuCN in refluxing *N*-methylpyrrolidone for several hours gave the dicyano cavitand **9f** in almost quantitative (> 95%) yield.

Attempts to substitute **9a** at the 15- and 28-positions by direct iodination [43] or nitration failed and only starting material was recovered. This clearly demonstrates that the debrominated aromatic rings are not able to undergo electrophilic aromatic substitutions, although they are expected to be activated for this type of reaction. It is possible that the non-planar intermediate formed in electrophilic aromatic substitution reactions is energetically too unfavorable in the case of a fully bridged cavitand. Even bromination with NBS, a reaction which is generally believed to occur via a radical mechanism, did not give any reaction on these rings. Apparently, these positions have lost their reactivity upon bridging the *o*-hydroxyl groups with methylene bridges.

Another way to introduce functional groups at the debrominated rings is to quench the lithiated products, obtained by reaction of **3** with *n*-BuLi (*vide supra*), with electrophiles other than H$^+$. In order to carry out these reactions successfully, it turned out to be necessary to treat diol **3** with NaH, prior to lithiation, since considerable amounts of hydrogen were otherwise incorporated at the 8- and 12-positions. According to ^{1}H-NMR spectroscopy of the crude reaction mixtures, pretreatment of diol **3** with NaH decreased the amount of hydrogen incorporated to roughly half the original amount. This strongly suggests that the hydroxyl groups protonate the lithiated positions before reaction with another electrophile can take place. Such *in situ* protonation of the lithiated species formed has been described by Beak *et al.* [44].

When the lithiated product, obtained by reaction of **3d** with *n*-BuLi, was quenched with B(OMe)$_3$, we were able to isolate tri-bridged diol **8c** functionalized at the 8- and 12-positions with hydroxyl groups in 47% yield. In the same way, starting from **3c**, two thiomethyl (**8d**) or iodo (**8e**) groups could be intro-

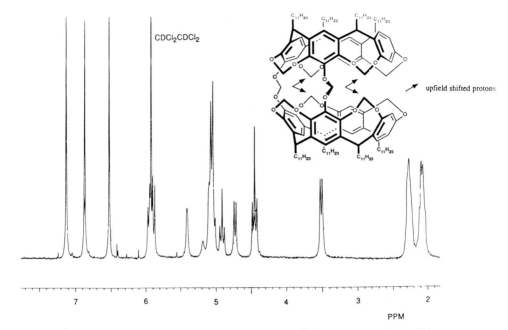

Fig. 2. ^{1}H-NMR spectrum of hemicarcerand **10** (isomer C) in CDCl$_2$CDCl$_2$ at 90°C.

duced. Because of serious problems in the purification of these compounds they were converted to the corresponding selectively functionalized cavitands **9g** and **9h** in overall yields of 25% and 20%, respectively. In these cases the yields were considerably lower than in the reaction starting from **3d** (47%). We cannot explain this decrease in yield, but we cannot rule out that, in addition to bromo-lithium exchange, reaction takes place at the CH$_2$CH$_2$Ph chains.

To illustrate the utility of selectively functionalized cavitands, we subjected diol **9d** to a coupling reaction with CH$_2$BrCl in order to synthesize for the first time a hemicarcerand which is only coupled by two proximal bridges (Scheme 2). The reaction was performed under Cram's high dilution conditions [20] in DMSO/THF. After a reaction time of 4 days, a colorless compound was obtained in 71% yield. This was identified by MS-FAB as the expected coupled product (M$^+$ = 2395). Although two diastereomers (denoted C and Z, analogously to recently published related hemicarcerands [35]) can be formed, ^{1}H-NMR spectroscopy strongly suggests that only one isomer is actually present. The spectrum is broad in CDCl$_3$ at room temperature, but sharpens up when measured in CDCl$_2$CDCl$_2$ at 90°C (Figure 2). This indicates that the molecule is flexible to a certain extent, something which was not seen for the hemicarcerands with three spacers prepared by Cram [20]. The most significant changes in the spectrum are the upfield shift for half of the inner and outer protons of the methylene bridges, shifting from 4.4 to 3.5 ppm and from 5.8 to 5.1 ppm, respectively. We believe that these marked upfield shifts can only be present in **10** (isomer C), because only in this isomer are the methylene bridges able to enter into the molecular cavity of the opposite octol fragment, causing the upfield shift. This also implies that the hemicarcerand does not contain an enforced molecular cavity in which a guest molecule is complexed. This is in accordance with the ^{1}H-NMR and MS FAB spectra, in which no guest molecule could be detected.

The present work shows that in the bridging reaction of octols of type **1** with CH_2BrCl, the introduction of the fourth bridge is generally slower than the other three, and that the rate of the reaction is solvent dependent. This provides the synthetic methodology to prepare tri-bridged diols **3**. Reduction of the amount of CH_2BrCl to 2 equiv. yields considerable amounts of doubly bridged cavitands, preferentially the 1,3-bridged product **4**.

The different aromatic rings in the tri-bridged diols (**3**) exhibit different reactivities. Bromo-lithium exchange of two bromo substituents with n-BuLi provides an easy way to selectively functionalized cavitands of type **8**. After incorporation of the fourth bridge, the remaining two bromines can be substituted for a variety of functional groups, leading to selectively functionalized cavitands **9c–h**.

References

1. C. D. Gutsche: *Calixarenes*. Monographs in Supramolecular Chemistry, Vol. 1, J. F. Stoddart (Ed.), The Royal Society of Chemistry, Cambridge (1989).
2. J. Vicens and V. Böhmer (Eds.): *Calixarenes: A Versatile Class of Macrocyclic Compounds*, Kluwer Academic Publishers, Dordrecht (1991).
3. C. D. Gutsche and P. F. Pagoria: *J. Org. Chem.* **50**, 5795 (1985). C. D. Gutsche, J. A. Levine, and P. K. Sujeeth: *J. Org. Chem.* **50**, 5802 (1985).
4. K. No and Y. Noh: *Bull. Korean Chem. Soc.* **7**, 314 (1986).
5. K. No, Y. Noh, and Y. Kim: *Bull. Korean Chem. Soc.* **7**, 442 (1986).
6. S. Shinkai, K. Araki, T. Tsubaki, T. Arimura, and O. Manabe: *J. Chem. Soc., Perkin Trans. 1* 2297 (1987).
7. M. Almi, A. Arduini, A. Casnati, A. Pochini, and R. Ungaro: *Tetrahedron* **45**, 2177 (1989).
8. J.-D. van Loon, W. Verboom, and D. N. Reinhoudt: *Org. Prep. Proc. Int.* **24**, 437 (1992).
9. L. C. Groenen, B. H. M. Rüel, A. Casnati, W. Verboom, A. Pochini, R. Ungaro, and D. N. Reinhoudt: *Tetrahedron* **39**, 8379 (1991).
10. E. Kelderman, L. Derhaeg, G. J. T. Heesink, W. Verboom, J. F. J. Engbersen, N. F. van Hulst, A. Persoons, and D. N. Reinhoudt: *Angew. Chem. Int. Ed. Engl.* **31**, 1075 (1992).
11. L. C. Groenen, B. H. M. Rüel, A. Casnati, P. Timmerman, W. Verboom, S. Harkema, A. Pochini, R. Ungaro, and D. N. Reinhoudt: *Tetrahedron Lett.* **32**, 2675 (1991).
12. S. Pappalardo, L. Giunta, M. Foti, G. Ferguson, J. F. Gallagher, and B. Kaitner: *J. Org. Chem.* **57**, 2611 (1992).
13. J.-D. van Loon, A. Arduini, L. Coppi, W. Verboom, A. Pochini, R. Ungaro, S. Harkema, and D. N. Reinhoudt: *J. Org. Chem.* **55**, 5639 (1990).
14. Y. Ting, W. Verboom, L. C. Groenen, J.-D. van Loon, and D. N. Reinhoudt: *J. Chem. Soc., Chem. Commun.* 1432 (1990).
15. Z. Goren and S. Biali: *J. Chem. Soc., Perkin Trans. 1* 1484 (1990).
16. C. D. Gutsche, B. Dhawan, J. A. Levine, K. H. No, and L. J. Bauer: *Tetrahedron* **39**, 409 (1983).
17. K. Iwamoto, K. Araki, and S. Shinkai: *Tetrahedron* **47**, 4325 (1991).
18. H. Erdtman, S. Högberg, S. Abrahamsson, and B. Nilsson: *Tetrahedron Lett.* 1679 (1968).
19. D. J. Cram, S. Karbach, H.-S. Kim, C. B. Knobler, E. F. Maverick, J. L. Ericson, and R. C. Helgeson: *J. Am. Chem. Soc.* **110**, 2229 (1988).
20. D. J. Cram, M. E. Tanner, and C. B. Knobler: *J. Am. Chem. Soc.* **113**, 7717 (1991).

21. P. Soncini, S. Bonsignore, E. Dalcanale, and F. Ugozzoli: *J. Org. Chem.* **57**, 4608 (1992).
22. D. J. Cram, M. E. Tanner, and R. Thomas: *Angew. Chem. Int. Ed. Engl.* **30**, 1024 (1991).
23. Part of this work was published as a preliminary communication: P. Timmerman, M. G. A. van Mook, W. Verboom, G. J. van Hummel, S. Harkema, and D. N. Reinhoudt: *Tetrahedron Lett.* **33**, 3377 (1992).
24. T. N. Sorrell and J. L. Richards: *Synlett* 155 (1992).
25. Throughout this paper we use the nomenclature proposed by Cram [35]: starting from a resorcinol-aldehyde tetramer, named an *octol*, the introduction of one bridge gives a *mono-bridged hexol*, two bridges give an *A,B* or *A,C-di-bridged tetrol*, and three bridges give a *tri-bridged diol*; the name *cavitand* refers to an octol rigidified by four bridges.
26. J. A. Tucker, C. B. Knobler, K. N. Trueblood, and D. J. Cram: *J. Am. Chem. Soc.* **111**, 3688 (1989).
27. J. C. Sherman, C. B. Knobler, and D. J. Cram: *J. Am. Chem. Soc.* **113**, 2194 (1991).
28. L. M. Tunstad, J. A. Tucker, E. Dalcanale, J. Weiser, J. A. Bryant, J. C. Sherman, J. C. Helgeson, C. B. Knobler, and D. J. Cram: *J. Org. Chem.* **54**, 1305 (1989).
29. G. Germain, P. Main, and M. M. Woofson: *Acta Crystallogr.* **A27**, 368 (1971).
30. *Structure Determination Package*: B. A. Frenz and Associates, Inc., College Station, TX and Enraf-Nonius, Delft (1983).
31. *International Tables for X-Ray Crystallography*, Vol. IV, Kynoch Press, Birmingham, England (1974). Distr. Kluwer Academic Publishers, Dordrecht, the Netherlands.
32. D. J. Cram: *Science* **219**, 1177 (1983).
33. J. A. Bryant, M. T. Blanda, M. Vincenti, and D. J. Cram: *J. Am. Chem. Soc.* **113**, 2167 (1991).
34. D. J. Cram, H.-J. Choi, J. A. Bryant, and C. B. Knobler: *J. Am. Chem. Soc.* **114**, 7748 (1992).
35. D. J. Cram, L. M. Tunstad, and C. B. Knobler: *J. Org. Chem.* **57**, 528 (1992).
36. F. A. Carey and R. J. Sundberg: *Advanced Organic Chemistry* Part B, 3rd edition, Ch. 3, Plenum Press, New York (1990).
37. R. Kuhn and H. Trischmann: *Liebigs Ann. Chem.* **611**, 117 (1958).
38. These tetraesters are used in the reaction with upper rim functionalized calix[4]arene derivatives, which will be published elsewhere.
39. W. D. S. Motherwell and W. Clegg: *PLUTO*, Program for plotting molecular and crystal structures, University of Cambridge, England (1978).
40. F. A. Carey and R. J. Sundberg: *Advanced Organic Chemistry*, Part A, 3rd edition, Ch. 10, Plenum Press, New York (1990).
41. A. I. Biggs and R. A. Robinson: *J. Chem. Soc.* 388 (1960).
42. W. E. Parham and R. M. Piccirilli: *J. Org. Chem.* **42**, 257 (1977), and references therein.
43. The procedure for iodination is described elsewhere: P. Timmerman, W. Verboom, D. N. Reinhoudt, A. Arduini, S. Grandi, A. R. Sicuri, A. Pochini, and R. Ungaro: *Synthesis* 185 (1994).
44. D. J. Gallagher and P. Beak: *J. Am. Chem. Soc.* **113**, 7984 (1991), and references therein.

Synthesis and Properties of N-Substituted Azacalix[n]arenes[*]

HIROYUKI TAKEMURA[**]
Department of Chemistry, Faculty of Science, Kyushu University Ropponmatsu, Ropponmatsu 4-2-1, Chuo-ku, Fukuoka, 810 Japan

TERUO SHINMYOZU
Institute for Fundamental Research of Organic Chemistry, Kyushu University, Hakozaki 6-10-1, Higashi-ku, Fukuoka, 812 Japan

and

HIROKAZU MIURA, ISLAM ULLAH KHAN and TAKAHIKO INAZU
Department of Chemistry, Faculty of Science, Kyushu University, Hakozaki 6-10-1, Higashi-ku, Fukuoka, 812 Japan

(Received: 3 March 1994; in final form: 19 July 1994)

Abstract. Side arm modifications of hexahomotriazacalix[3]arene (**1**) were achieved by simple synthetic methods. Compound **5** has picolyl side arms and liquid–liquid extraction experiments showed that the alkali cation affinity of **5** is much stronger than that of **1**. A chiral group was also introduced into the azacalixarene structure. Calix[4]arene was converted into dihomoazacalix[4]arene (**2**) in 8% yield. Clathrate formation of **2** with various solvents is described. MM3 calculations were carried out on p-substituted analogs of **2**. The 'self-filled' structure, in which the benzyl side arm is placed in its cavity, is the most stable structure when the p-positions of the aromatic rings carry small substituents. Strong hydrogen bonds between nitrogen and phenolic hydroxyl groups in dihomoazacalix[4]arene (**2**) were observed at low temperatures. The ^{1}H-NMR signals of phenolic hydroxyl groups appeared as six singlets in the range of 9.8 $\sim$ 17.1 ppm at -70°C.

Key words: Azacalixarene, azacyclophane, chiral molecule, MM3 calculation, strong hydrogen bond.

1. Introduction

Azacalixarenes are a recently developed branch of the calixarene family [1, 2]. They have phenolic oxygen and nitrogen as donor atoms. Some of the compounds in this series are fixed in rigid cone conformations by strong hydrogen bonds between OH and N atoms. One-pot reactions between bis(hydroxymethyl)phenol derivatives and benzylamine produced the desired azacalixarenes in satisfactory yields, and the macrocyclic compounds were isolated in a simple manner. By this procedure, the functionalized side arms can be easily introduced into the macrocyclic structures using the appropriate amines as starting materials. In this

---

[*] This paper is dedicated to the commemorative issue on the 50th anniversary of calixarenes.
[**] Author for correspondence.

report, we describe the one-step synthesis of hexahomotriazacalix[3]arenes with chiral benzyl arms and picolyl side arms. Previously, we described the synthesis of **2** by a one-pot procedure [2]. Here, a newly developed conversion reaction from calix[4]arene into dihomoazacalix[4]arene (**2**) is reported. Basic studies on compound **2** concerning structural analyses by MM3 calculations, the hydrogen bonds between N and OH groups, and the clathrate formation are described.

2. Experimental

2.1. APPARATUS

Melting points were measured with a Yanaco MP-500D apparatus and are uncorrected. NMR spectra were recorded on a JEOL GSX-270 (270 MHz for ^{1}H) spectrometer. UV spectra were recorded on a Shimadzu UV-2200 spectrometer.

2.2. SYNTHESIS OF CHIRAL AZACALIXARENE (**4**)

A mixture of 2,6-bis(hydroxymethyl)-4-methylphenol (4.02 g, 23.9 mmol) and (S)-(−)-α-methylbenzylamine (Tokyo Kasei, 2.98 g, 24.6 mmol) in 100 mL of toluene was refluxed for 24 h, and the water generated was removed during the course of the reaction. The toluene was evaporated, and the residual oily material was heated at 120°C for an additional 20 h. After the reaction mixture cooled, the oil was dispersed in 100 mL of methanol by vigorous stirring. The yellow powder thus obtained was collected by filtration, washed with methanol, and then chromatographed on silica gel with CH_2Cl_2/MeOH (98/2) as an eluent. Recrystallization of the resulting yellow powder from a mixture of C_6H_6–CH_3OH–CH_2Cl_2 afforded **4** as pale yellow powder (3.23 g, 53%), m.p. 138–140.9°C; ^{1}H-NMR (270 MHz, $CDCl_3$) δ 11.0 (bs, 3H, OH), 7.32 (s, 15H, aromatic), 6.87–6.68 (m, 6H, aromatic), 3.80 (s, 3H, —(Me)—CH—Ph), 3.75 (s, 12H, CH_2N), 2.16 (s, 9H, CH_3—Ar), 1.48 (s, 9H, —CH_3); FAB-Mass m/z 760 (M+H)$^+$; $[\alpha]_D^{24} = 67.9\pm0.5°$ (optical purity unknown). *Anal. Calcd.* for $C_{51}H_{57}N_3O_3 \cdot C_6H_6 \cdot 1/2\ CH_2Cl_2$: C, 78.43; H, 7.33; N, 4.77. *Found*: C, 78.46; H, 7.24; N, 4.62. The contents of the solvents in the analytical sample were confirmed by the ^{1}H-NMR spectrum.

2.3. SYNTHESIS OF **5**

The preparation of **5** was achieved in a way similar to that of compound **1** [1]. The reaction products obtained from 3.37 g (20.0 mmol) of 2,6-bis(hydroxymethyl)-4-methylphenol and 2.31 g (21.4 mmol) of 2-picolylamine were chromatographed on silica gel with CH_2Cl_2/CH_3OH (98/2) as an eluent. Recrystallization of the resultant powder from cyclohexane gave 1.59 g (29%) of **5**, m.p. 129–135° (decomp.); ^{1}H-NMR (270 MHz, $CDCl_3$) δ 11.8 (bs, 3H, OH), 8.51 (bs, 3H, picolyl-H_6), 7.52 (bs, 3H, picolyl-H_4), 7.28 (s, 3H, picolyl-H_3), 7.09 (bs, 3H, picolyl-H_5), 6.84–6.73 (m, 6H, aromatic), 3.68 (s, 12H, —CH_2—), 3.75 (s, 6H, —CH_2—picolyl), 2.15 (s, 9H,

CH$_3$—Ar); FAB-Mass m/z 721 (M+H)$^+$. *Anal. Calcd.* for C$_{45}$H$_{48}$N$_6$O$_3$·H$_2$O: C, 73.15; H, 6.82; N, 11.37. *Found*: C, 73.35; H, 6.47; N, 10.95.

2.4. CONVERSION OF CALIX[4]ARENE INTO DIHOMOAZACALIX[4]-ARENE (2)

A mixture of *p-tert*-butylcalix[4]arene (171 mg, 0.27 mmol), benzylamine (0.55 g, 5.1 mmol), paraformaldehyde (0.30 g, 10 mmol), and aq. KOH (5 N, 0.01 mL) in 30 mL of xylene was heated under reflux for 150 h. The solution was washed with water and evaporated under reduced pressure. Column chromatography on silica gel (C$_6$H$_6$, followed by CH$_2$Cl$_2$) of the resultant mixture gave 16 mg (8%) of dihomoazacalix[4]arene (**2**) and 74 mg of recovered *p-tert*-butylcalix[4]arene. The product was confirmed by the ^{1}H-NMR spectrum and FAB mass spectrum, which were the same as those previously reported [2].

2.5. LIQUID–LIQUID EXTRACTION

Solvent extraction experiments were carried out as follows. A water-saturated chloroform solution of the ligand (1.00×10^{-4}, 5 mL) and an aqueous alkali metal chloride solution (0.10–0.11 M of Li$^+$–Cs$^+$ containing 2.1×10^{-4} M of potassium picrate, 5 mL) were introduced into Teflon-sealed tubes, which were shaken for 5 min for three times $25 \pm 0.1°$C. Prolonged times of shaking gave identical results. The resulting mixtures were allowed to stand for 30 min at $25 \pm 0.1°$C and then centrifuged. The aqueous phases were carefully withdrawn and the concentrations of the picrates were determined spectrophotometrically at 353 nm ($\varepsilon = 1.54 \times 10^4$). The extraction ratio, E, was calculated using the following equation:

$$E = ([\text{Pic}^-]_i - [\text{Pic}^-])/[\text{Pic}^-]_i.$$

Here [Pic$^-$]$_i$ is the initial concentration of aqueous picrate and [Pic$^-$] is the observed concentration after the extraction.

2.6. FORMATION OF CLATHRATE COMPOUNDS

Compound **2** was recrystallized from the solvents listed in Table I. The methanol or ethanol solutions were prepared using CH$_2$Cl$_2$ as a cosolvent. Alternatively, compound **2** was dissolved in a mixture of DMF and DMSO (1/2, v/v) by heating, and allowed to stand at room temperature. The crystals thus formed were collected and dried in a vacuum for 12 h at room temperature. The host : solvent ratios were confirmed by NMR spectra.

TABLE I. Clathrate formation of **2** with various solvents.

Solvents	Crystal form	Molar ratios (host : solvents)
Ethyl acetate	Colorless granules	– (no clathrate formation)
Dioxane	Colorless granules	1 : 1
Methanol	Colorless needles	–
Ethanol	Colorless needles	–
Acetone	Colorless resin	–
Acetonitrile	Colorless needles	3 : 2
Cyclohexane	Colorless granules	1 : 1
Hexane	Colorless granules	–
Chloroform	Colorless needles	–
Benzene	Colorless needles	2 : 3
Toluene	Colorless granules	1 : 1
p-Xylene	Colorless needles	1 : 1
m-Xylene	Colorless needles	1 : 1
o-Xylene	Colorless granules	2 : 3
Tetrahydrofuran	Colorless granules	1 : 1
Nitromethane	Yellow granules	1 : 1
DMF/DMSO	Yellow plates	1 : 1 (DMF only)

3. Results and Discussion

3.1. SYNTHESIS OF AZACALIX[n]ARENES

In our research on the azacyclophane series, cyclization methods using toluenesulfonamide and trifluoroacetamide were developed [3, 4]. However, to apply these methods to the synthesis of the azacalixarenes, the phenolic group had to be protected.

Starting from 4-*tert*-butyl-2,6-bis(bromomethyl)anisole, 2,11-diaza[3.3]- and 2,11,20,29-tetraaza[3.3.3.3]metacyclophanes were obtained in 10 and 13% yields, respectively [5]. However, the desired triaza[3.3.3]cyclophane was not obtained (Scheme 1). Subsequent deprotections of two functional groups, the methoxy and Ts groups, were not only troublesome but also led to poor results. Thus, the newly developed direct coupling method using 2,6-bis(hydroxymethyl)-4-alkylphenols and amines has many advantages for the construction of azacalixarenes. Compounds **1–3** were synthesized by direct coupling between bis(hydroxymethyl)-phenols and benzylamine (Figure 1) [1, 2]. The cyclization reaction proceeds in nonpolar solvents like toluene or xylene. In a polar solvent, however, only a mixture of oligomeric acyclic compounds was obtained, and generation of cyclic compounds was not observed [1]. The template effects of the OH—OH and OH—N hydrogen bonds play important roles in the formation of cyclic compounds.

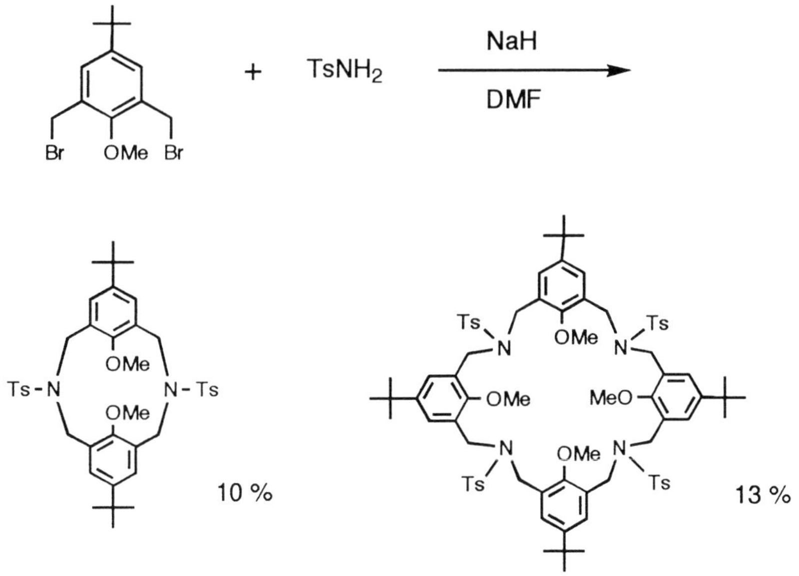

Scheme 1. Synthetic scheme for azacyclophanes: cyclization with toluenesulfonamide.

3.2. SIDE ARM FUNCTIONALIZATION

3.2.1. *Introduction of Picolyl Side Arms into Hexahomotriazacalix[3]arene*

Various kinds of functionalization methods of the phenol OH groups of calix[n]-arenes have been reported to yield ionophores [6]. Also, armed crown ethers with functional groups on nitrogen have been synthesized and effective recognitions of ions have been reported [7].

In the case of hexahomotriazacalix[3]arene (**1**), O-alkylation by picolyl chloride was easily achieved with sodium hydride as a base in DMF (78%). However, even without O-alkylation, functional groups can be introduced into the azacalixarene structures by the one-step procedure described here. Attachment of various side arms to bifunctional azacalixarene systems should allow the construction of new macrocyclic host molecules with multi-binding sites.

3.2.2. *Alkali Metal Ion Complex Formation*

As previously reported, the hexahomotriazacalix[3]arene (**1**) has cation binding ability toward alkali and alkaline-earth metal ions [1]. Moreover, compound **1** showed pH-dependent and effective extraction of uranyl ion in the presence of a high concentration of NaCl. However, dihomoazacalix[4]arene (**2**) and tetrahomodiazacalix[4]arene (**3**) have poor cation affinity: the extraction of alkali and alkaline-earth metal ions from a neutral aqueous solution into the organic phase was not observed. In order to increase the cation affinity, we introduced picolyl side arms on the bridging nitrogen atoms of the hexahomotriazacalix[3]arene structure.

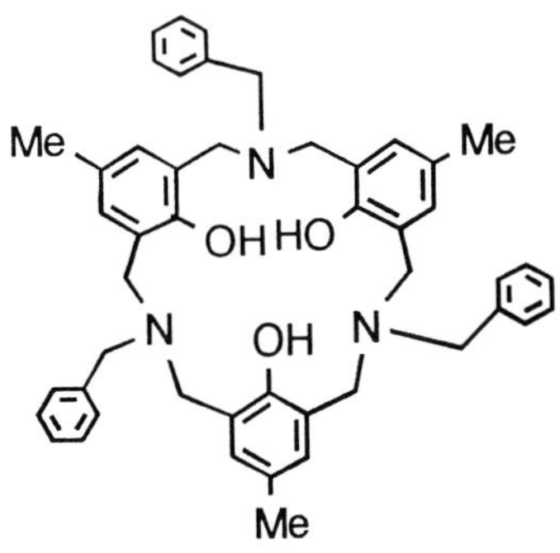

Fig. 1. Structures of azacalixarenes.

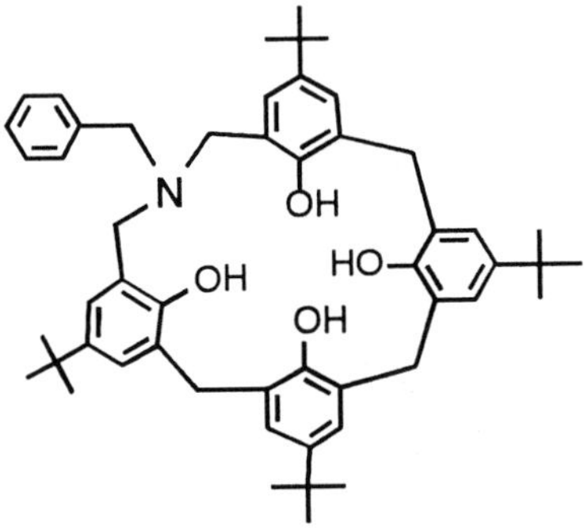

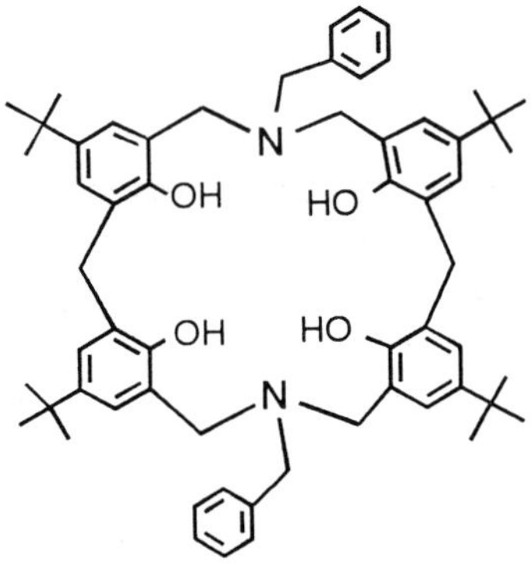

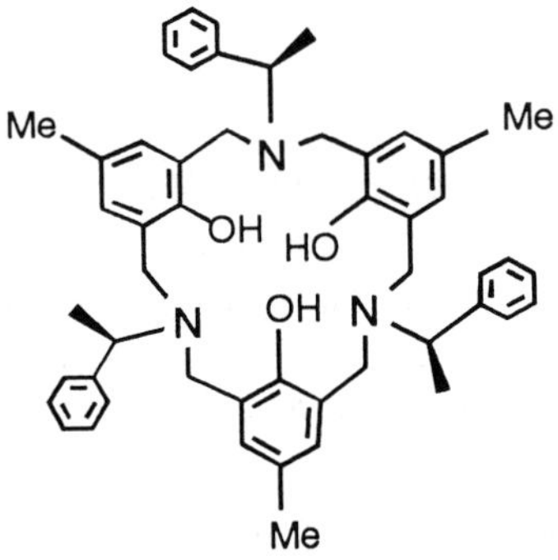

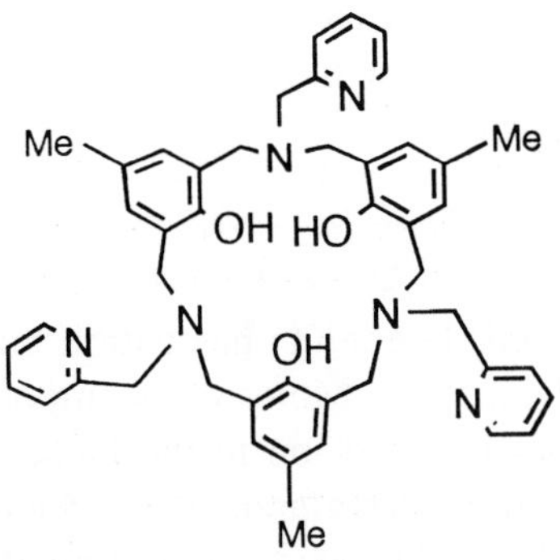

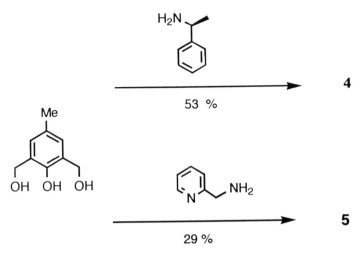

Scheme 2. Syntheses of compounds **4** and **5**.

Synthesis of the ligand, **5**, was easily achieved by the direct coupling reaction between picolylamine and 2,6-bis(hydroxymethyl)-4-methylphenol. The structure and the synthetic scheme of **5** are shown in Figure 1 and Scheme 2, respectively. Liquid–liquid alkali metal picrates extraction experiments were carried out using **5** as a ligand. Figure 2 shows the plots of extraction ratios, E (%), vs. ionic radii of alkali metal ions. Compared to compound **1**, the extraction ratios became 50–60 times greater. However, the ion selectivity decreased remarkably, showing the flexibility of the picolyl side arms. Similar to **1**, compound **5** showed affinity to K^+ ion, but Cs^+ was extracted in greater amounts than Rb^+. This result may be due to the change in the ligand to metal ratio of the complex.

3.2.3. A Chiral Molecule

Although syntheses of calixarenes which have no plane of symmetry have been reported, optical resolutions were unsuccessful because of the rapid inversion of aromatic rings at room temperature [8]. The first chiral resolution was shown to have an asymmetrically substituted calixarene fixed in a cone conformation by O-alkylation [9].

On the other hand, chiral side arms can easily be introduced into the hexa-homotriazacalix[3]arene structure starting from a chiral benzylamine derivative: S-(−)-α-methylbenzylamine ($[\alpha]_D^{20}$ = -39°). Optical rotation of the isolated chiral azacalixarene, **4**, was $[\alpha]_D^{24}$ = -67.9 ± 0.5° (c = 0.25, CHCl$_3$). The synthetic scheme and structure are shown in Scheme 2 and Figure 1, respectively. Starting from various kinds of chiral primary amines, this synthetic method can be applied to other chiral azacalix[n]arene systems.

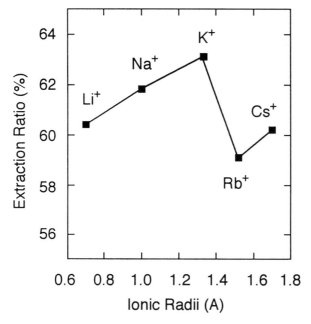

Fig. 2. Plots of extraction ratios (%) of alkali metal picrates from aqueous to organic phase.

3.3. CONVERSION OF CALIX[4]ARENE INTO DIHOMOAZACALIX[4]-ARENE

Considering the reaction pathway of calix[n]arene formations [11a] and the transformations of calix[8]arene or calix[6]arene into calix[4]arene [11b, c], we expected the generation of dihomoazacalix[4]arene (**2**) by nitrogen atom insertion into calix[4]arene. Indeed, when a mixture of *p-tert*-butylcalix[4]arene, paraformaldehyde, and benzylamine was heated in refluxing xylene in the presence of catalytic amounts of base (KOH) for 150 h, dihomoazacalix[4]arene (**2**) was obtained in 8% yield (Scheme 3), and 42% of the starting *p-tert*-butylcalix[4]arene was recovered. Even in the presence of a larger excess of benzylamine, the isolated azacalixarene was only compound **2**, while other possible (aza)$_n$calixarenes were not isolated. Calix[4]arene is stable enough in refluxing xylene, even in the presence of a base, and the ring opening–closing reaction hardly occurs. This might be one of the reasons for the low rate of nitrogen insertion into calix[4]arene. Under similar conditions, the reaction starting from *p-tert*-butylcalix[8]arene gave a complex mixture, and azacalixarenes were not isolated.

3.4. CLATHRATE FORMATION

All azacalixarenes are readily soluble in organic solvents except methanol and ethanol. Dihomoazacalix[4]arene (**2**) formed clathrate compounds with various solvent molecules (Table I). Among these clathrate compounds, **2**·CH_3NO_2 and **2**·DMF were yellow crystals, while the others were colorless. Interestingly, from

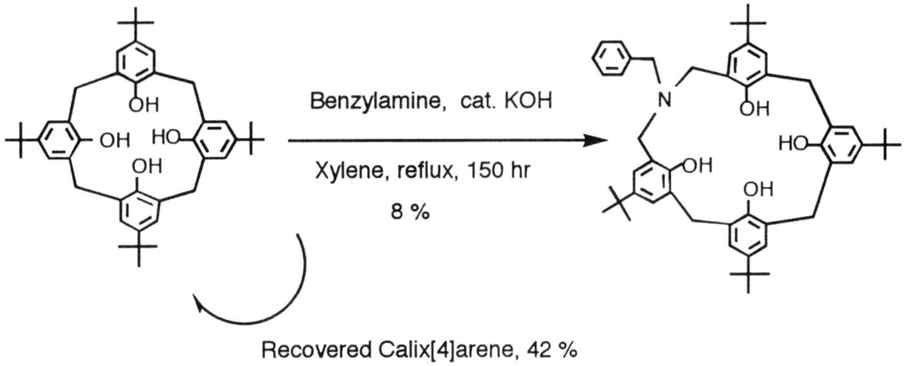

Scheme 3. Conversion of *p-tert*-butylcalix[4]arene into dihomoazacalix[4]arene (**2**).

a solution of **2** in a DMF/DMSO mixture, the **2**·DMF complex was selectively obtained as yellow plates. Hydrogen bonding between the host and alcohols seems to be possible, but ethanol and methanol did not form clathrates with **2**, probably because of the insolubility of the host in alcohols. At this stage, the driving forces of clathrate formation are unclear, but they seem to be unrelated to several factors (size-fit, hydrogen bonding, n–π, π–π interactions).

3.5. Hydrogen Bonds in the Cavity

3.5.1. *MM3 Calculations*

Previously, we showed that the conformations of **2** and **3** are fixed in the 'cone' conformation in solution: in spite of the elongation of one methylene bridge of calix[4]arene into an azamethylene bridge, the cone conformer of **2** is more stable than that of calix[4]arene [2]. Unfortunately, the X-ray crystallographic analysis of the **2**·DMF complex was unfruitful. Only the cone conformation and inclusion of a DMF molecule in the crystal lattice were confirmed. Further attempts at X-ray crystallographic analysis are being made. To support the results of the experiments and predict the optimal structure of derivatives of **2**, MM3 calculations were carried out [12]. The calculations started from their conformational isomers, i.e., cone, partial cone, 1,2-, and 1,3-alternate.

The stability of these conformations, expressed in terms of their steric energy is as follows: cone < 1,2-alternate ≅ 1,3-alternate ≅ 2-partial cone < 2,3-alternate < 1-partial cone. Among all optimal structures, the 'cone' conformation was the most stable. The phenolic OH groups formed a cyclic array of hydrogen bonds. When bulky isopropyl groups are introduced at the *p*-position of aromatic rings adjacent to the —CH_2—N—CH_2-bridge, the 'cone' conformation is the most stable, and the benzyl group is located outside the molecule (Figure 3) [12]. The NMR spectrum of **2** showed that the position of the benzyl group is on the outer side of the molecule; thus, the results of the calculations coincide with the NMR spec-

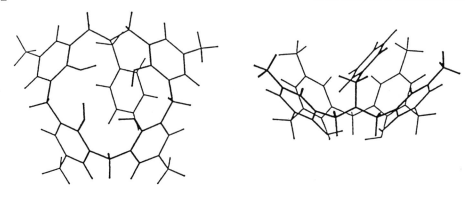

Conformer A

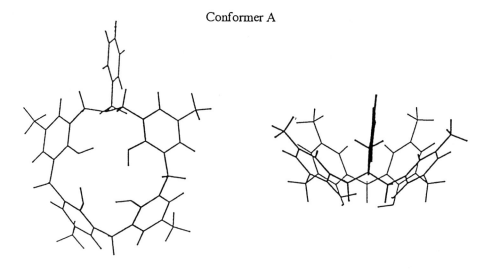

Conformer B

Fig. 3. Optimal structures of *p*-tetramethyldihomoazacalix[4]arene by MM3 calculations. Left: bottom view from OH side. Right: side view from benzyl side arm.

tra. Interestingly, the side arm benzyl groups of *p*-tetrabromo, *p*-tetramethyl and unsubstituted compounds are located inside the cavity in their optimal structures. Inside benzyl conformers of *p*-tetrabromo, *p*-tetramethyl and unsubstituted compounds are more stable than their outside conformers by 8.3, 8.9 and 6.2 kcal mol^{-1} ($\Delta\Delta G^0$), respectively (Figure 3, conformer A and B). If these conformers actually exist, functionalized benzyl groups can interact with a guest molecule in the cavity. To confirm the existence of such a 'self-filled' conformer, the syntheses and con-

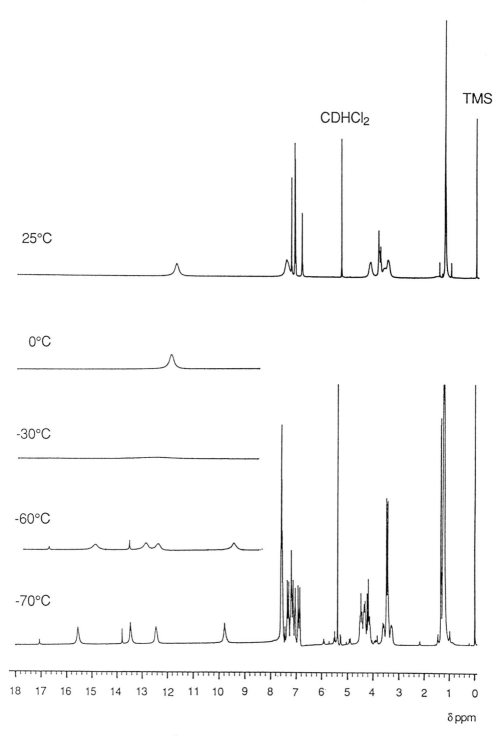

Fig. 4. Temperature-dependent ^{1}H-NMR (270 MHz, CD_2Cl_2) spectra of **2**.

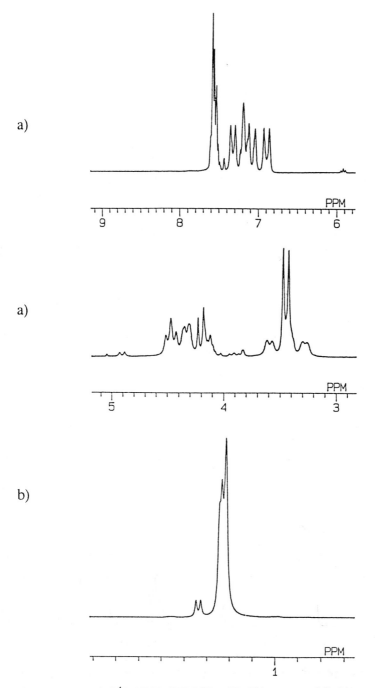

Fig. 5. Expanded ^{1}H-NMR (270 MHz, CD$_2$Cl$_2$) spectra of **2**: (a) aromatic and benzylic protons at -70°C, (b) *tert*-butyl protons at -80°C.

formational analyses of *p*-tetrabromo- and *p*-tetramethyldihomoazacalix[4]arenes are currently in progress.

3.5.2. Variable Temperature NMR Measurements

In contrast to the sharp OH signals of calix[*n*]arenes in ^{1}H-NMR spectra [10a], the OH signals of azacalixarenes appear as relatively broad singlets at lower fields than those of calix[*n*]arenes. The nitrogen lone pairs form favorable hydrogen bonds with the phenolic OH protons and thus weaken the phenolic O—H bonds. This intramolecular acid–base system provides the strong hydrogen bonds [1, 2]. The phenolic OH signal of **2** appears at 11.6 ppm as a broad singlet in CDCl$_3$ at 25°C. At low temperatures, the signal coalesced at circa -30° and finally appeared as six singlets in ^{1}H-NMR spectra below -60°C. The sharp singlets at 17.1 and 13.8 ppm are small, while the singlets at 15.6, 13.5, 12.5 and 9.8 ppm have equal intensities and are relatively broad (Figure 4). The splittings of the OH signal is concomitant with the splitting of aromatic, benzyl and *tert*-butyl proton signals. A small doublet of *tert*-butyl protons shows the existence of a minor conformational isomer at -70°C and the two small sharp singlets (17.1 and 13.8 ppm) should correspond to that minor conformational isomer. The four OH signals (15.6, 13.5, 12.5 and 9.8 ppm) originate from the four OH groups of the major conformer. Probably, the four OH groups are in different spacial arrangements. This speculation agrees with the splitting pattern of aromatic protons: they split into eight singlets, showing that the circumstances surrounding each aromatic proton are different. Moreover, all methylene protons are nonequivalent and the major *tert*-butyl protons split into three singlets at -80° (Figure 5). At the present stage, the interpretation of the splitting pattern of benzyl protons is difficult, but these results show that the major conformer of **2** is the 1,2-alternate, and the minor signals correspond to those of a cone conformation. These results are different from those of the MM3 calculations, in which the most stable conformation is the 'cone' of C_s symmetry (without considering the cyclic array of the OH groups). The MM3 calculations showed that the difference in free energy (ΔG^0) between the cone and the 1,2-alternate conformations is 10.0 kcal mol^{-1}. Further conformational analyses of **2** by calculations and dynamic NMR studies are in progress. Similar splittings of OH groups were also observed in **1** and **3**, but were not as remarkable as those in **2**.

References

1. H. Takemura, K. Yoshimura, I. U. Khan, T. Shinmyozu, and T. Inazu: *Tetrahedron Lett.* **33**, 5775 (1992).
2. I. U. Khan, H. Takemura, M. Suenaga, T. Shinmyozu, and T. Inazu: *J. Org. Chem.* **58**, 3158 (1993).
3. H. Takemura, M. Suenaga, K. Sakai, H. Kawachi, T. Shinmyozu, Y. Miyahara, and T. Inazu: *J. Incl. Phenom.* **2**, 207 (1984).
4. T. Shinmyozu, N. Shibakawa, K. Sugimoto, H. Sakane, H. Takemura, K. Sako, and T. Inazu: *Synthesis* 1257 (1993).

5. Islam Ullah Khan: *Azacalixarenes: Novel Macrocyclic Host Molecules*, Thesis, Kyushu University, Japan (1993).
6. (a) P. Neri and S. Pappalardo: *J. Org. Chem.* **58**, 1048 (1993). (b) S. Kanamathareddy and C. D. Gutsche: *J. Org. Chem.* **57**, 3160 (1992). (c) J. S. Rogers and C. D. Gutsche: *J. Org. Chem.* **57**, 3152 (1992). (d) A. Casnati, P. Minari, A. Pochini, and R. Ungaro: *J. Chem. Soc., Chem. Commun.* 1413 (1991).
7. (a) V. Balzani and J.-M. Lehn: *Angew. Chem. Int. Ed. Engl.* **30**, 190 (1991). (b) H. Tsukube, K. Takagi, T. Higashiyama, T. Iwachido, and N. Hayama: *Tetrahedron Lett.* **26**, 881 (1985). (c) H. Tsukube, H. Minatogawa, M. Munakata, M. Toda, and K. Matsumoto: *J. Org. Chem.* **57**, 542 (1992).
8. (a) V. Böhmer, F. Marschollek, and L. Zetta: *J. Org. Chem.* **57**, 3200 (1987). (b) H. Casablanca, J. Royer, A. Satrallah, A. Taty-C and J. Vicens: *Tetrahedron Lett.* **28**, 6595 (1987).
9. (a) S. Shinkai, T. Arimura, H. Satoh, and O. Manabe: *J. Chem. Soc., Chem. Commun.* 1495 (1987). (b) S. Shinkai, T. Arimura, H. Kawabata, K. Araki, K. Iwamoto, and T. Matsuda: *J. Chem. Soc., Chem. Commun.* 1734 (1990). (c) T. Arimura, H. Kawabata, T. Matsuda, T. Muramatsu, H. Satoh, K. Fujio, O. Manabe, and S. Shinkai: *J. Org. Chem.* **56**, 301 (1991). (d) S. Shinkai, T. Arimura, H. Kawabata, H. Murakami, and K. Iwamoto: *J. Chem. Soc., Perkin Trans. 1* 2429 (1991).
10. (a) G. J.-C. Bünzli and J. M. Harrowfield: in *Calixarenes*, J. Vicens and V. Böhmer (Eds.), Kluwer Academic Publishers, the Netherlands (1990), pp. 211–231. (b) N. Sabbatini, M. Guardini, A. Mecati, V. Balzani, R. Ungaro, E. Ghidini, A. Casnati, and A. Pochini: *J. Chem. Soc., Chem. Commun.* 878 (1990).
11. (a) C. D. Gutsche: in *Calixarenes*, J. F. Stoddart (Ed.), Monographs in Supramolecular Chemistry, Royal Society of Chemistry, Great Britain (1989), pp. 50–59. (b) C. D. Gutsche, M. Iqbal, and D. Stewart: *J. Org. Chem.* **51**, 742 (1986). (c) B. Dhawan, S.-I. Chen, and C. D. Gutsche: *Macromol. Chem.* **188**, 921 (1987).
12. MM3 calculations were carried out with a MM3 ('92) program in its ANCHOR II implementation. ANCHOR II is a molecular design support system for SUN computers from Fujitsu Limited. Trade Mark of Fujitsu Limited and Kureha Chemical Company, 1992. In these calculations, the memory size of the hardware was limited, thus, molecules larger than *p*-diisopropyldihomoazacalix[4]arene were not considered for the model.

Journal of Inclusion Phenomena and Molecular Recognition in Chemistry **19:** 207–225, 1994.

Arranging Coordination Sites around Cyclotriveratrylene*

JENNIFER A. WYTKO and JEAN WEISS**
Laboratoire d'Electrochimie et de Chimie Physique du Corps Solide, URA No. 405 au CNRS, Université Louis Pasteur, 4 rue Blaise Pascal, 67000 Strasbourg, France

(Received: 3 March 1994; in final form: 15 June 1994)

Abstract. This article describes the attachment of coordination sites around a rigid matrix: cyclotriveratrylene (CTV). The synthetic approaches leading to these new ligands possessing pyridines and bipyridines as coordinating sites are discussed and full synthetic details are given. One expanded CTV derivative bearing three 3-pyridyl groups has been characterized by X-ray crystallography and the structure shows that the conformation adopted by the CTV matrix is appropriate for the coordination of transition metals, and inclusion of a range of molecules in the hydrophobic pocket.

Key words: Cyclotriveratrylene, cavities, pyridines, 2,2′-bipyridine.

1. Introduction

For more than twenty-five years, researchers in the constantly expanding field of molecular recognition have been trying to adjust the shape, size, and nature of synthetic receptors in order to bind, extract and activate various substrates [1a–e]. Ultimately, host–guest chemistry and molecular recognition will create simple models for complicated, naturally occurring active sites of enzymes and proteins. Regarding this particular goal, families of compounds possessing hydrophobic cavities such as calixarenes, cyclodextrins [2–9] and cyclophanes [10, 11] are of great interest. These compounds are able to select guests depending on the type of interactions developed with a given substrate, depending on its size and/or its shape. In addition, these motion-restricted cavities offer the tremendous advantage of providing architectural control of the location and directionality of functional groups attached on an edge or in a cleft. Many different binding sites for alkaline and alkaline-earth metals have been preorganized with the help of rigid matrices such as calixarenes [12–14] or cyclotriveratrylene [15], but only a few researchers have taken advantage of rigid cavities to preorganize binding sites for transition metals.

* This paper is dedicated to the commemorative issue on the 50th anniversary of calixarenes.
** Author for correspondence.

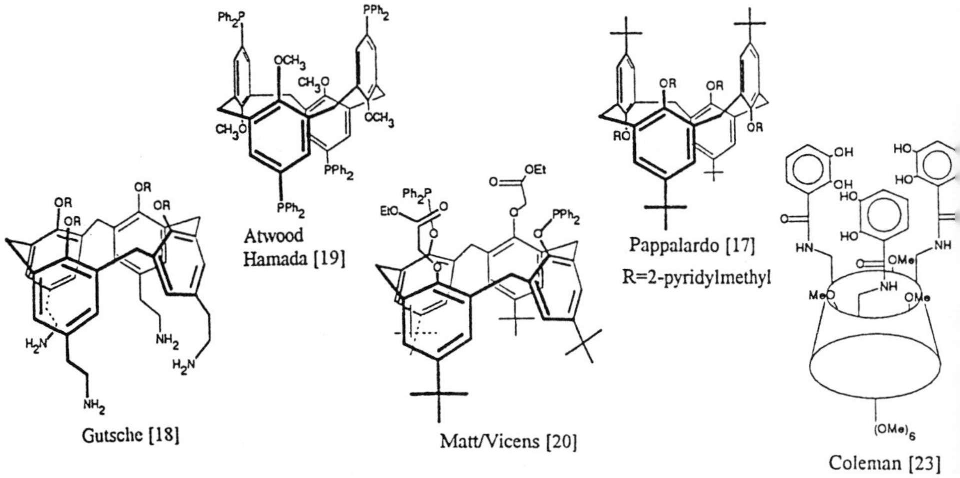

Fig. 1. Arranging binding groups around cavities: selected examples.

1.1. COMBINATIONS OF CAVITIES AND TRANSITION METALS

The first consequence of organizing phosphorus or nitrogen containing groups around a cavity is the generation of a tridimensional chelate effect, similar to the widely used chelation observed when binding alkaline and alkaline-earth metal cations with functionalized calixarenes. Some examples of cavities preorganizing pyridines, amines and phosphines are represented in Figure 1.

Pappalardo [16] has used the fixed partial cone conformation of a tetra-substituted calix[4]arene to generate chiral ligands bearing two pyridine and two quinoline rings. The synthesis of calix[4]arenes bearing pendant pyridine groups at the lower rim as potential ligands for transition metals has been described in an earlier work [17]. Gutsche [18] has used primary, secondary and tertiary amino groups, arranged in a C_4 symmetry around the upper rim of calix[4]arenes, to coordinate several transition metals in a tetradentate ligand. The possibility of an interaction between an auxiliary ligand and the hydrophobic cavity of the calixarene matrix has been mentioned but no experimental evidence for such a phenomenon has been obtained. The chelate effect with four diphenyl phosphine groups attached on a 1,3 alternate calix[4]arene has been used by Atwood and Hamada [19] for the binding and extraction of alkali and transition metal picrates in organic solvents. More recently and directly oriented towards homogeneous catalysis, calixarenes bearing phosphinites have attracted several groups [20–22]. For example, Matt and Vicens [20] have oriented their synthesis towards sterically hindered phosphine ligands to modify the properties of a metallic center, probably Rh(I) or Ir(I). Finally, Coleman [23] has designed a siderophore analogue built around a cyclodextrin, exhibiting a strong chelate effect in the complexation of iron(III). In their discussion, the authors mentioned the advantage offered by the

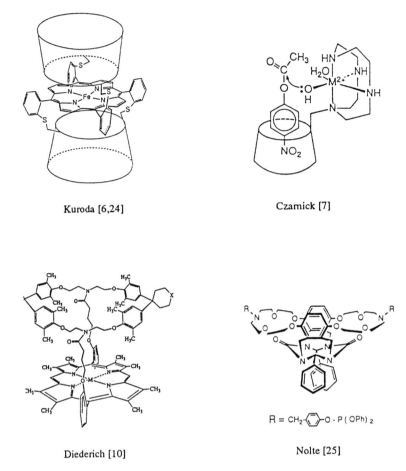

Fig. 2. Catalysis with combination of a cavity and binding sites for transitions metals.

cyclodextrin cavity, which is able to carry an organic substrate together with the iron(III) center.

Building ligands by preorganization of binding sites for transition metals around a rigid cavity used as a matrix has a second interest. It provides the opportunity to combine the chemical and physical properties of the hydrophobic cavity with the reactivity of the metallic center. The ability of the cavity to develop specific interactions with substrates can be used to select, bind and properly orient an organic fragment. Thus, the metal complexed within the binding sites at the edge or in a cleft of the cavity can interact with the substrate in the cavity. Two examples emphasizing this useful cooperation, using cyclodextrin derivatives are represented in Figure 2. Besides the classical hydrophobic cavities like cyclodextrins, calix[4]arenes and Högberg bowl derivatives, new functionalized molecular clefts or tweezers will offer chemists a wide range of architectural possibilities as illustrated by the bis-phosphine ligand represented in Figure 2.

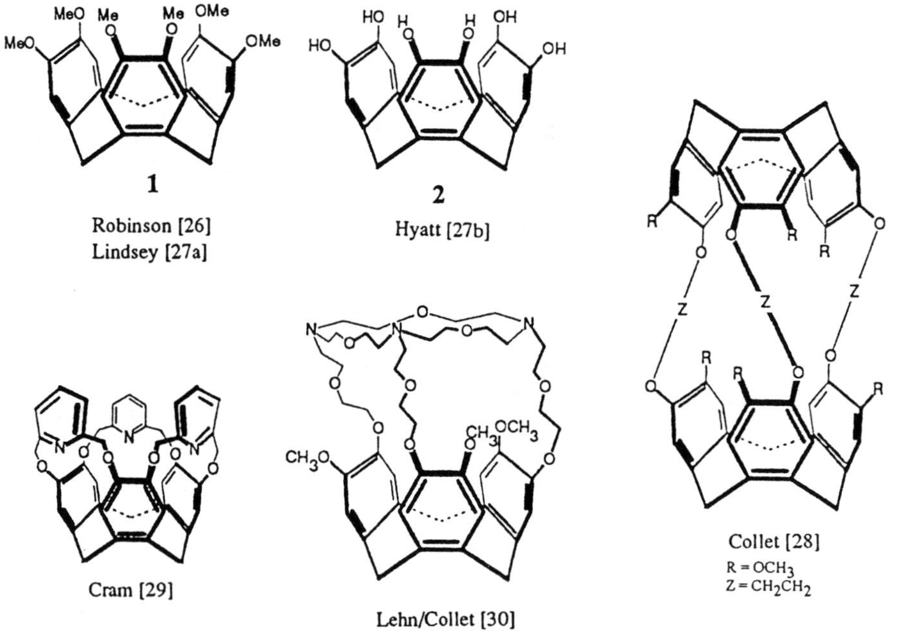

Fig. 3. Cyclotriveratrylene: a hydrophobic cavity for molecular recognition.

When an iron-porphyrin complex was located in a cyclodextrin sandwich, the cyclodextrin cavity could be used for the selection of terminal olefins versus internal olefins [6]. Thus, Kuroda was able to selectively perform the oxidation of olefins with iron-porphyrin complexes, in analogy with cytochrome P-450. When a photosensitive free base porphyrin was sandwiched between two cyclodextrins [24], the cavity was used to select, on a size criterion, an electron acceptor for quenching the excited state of the photosensitive unit. Still using cyclodextrins, Czarnick [7] has described a ligand in which transition metal complexes were attached at the edge of the hydrophobic cavity (see also [9]). These 'cavity-complex' combinations display a significant rate enhancement of the transacylation of both activated and inactivated aromatic esters. The substrate was properly oriented by the cyclodextrin in order to favor the interaction of the ester function with the metallic center. Diederich [10] has described a cytochrome P-450 model in which a cyclophane was used to bind and properly orient some polycyclic aromatics towards a hemic iron(IV)oxo. Finally, the use of hydrophobic cavities to promote interactions between a substrate and a metallic center is not restricted to calixarenes and cyclodextrins derivatives, as demonstrated by Nolte [25] who designed a system combining a cage structure with a phosphine chelate. The corresponding rhodium(I) complex selectively catalyses the hydrogenation of allyl catechol versus allyl veratrole, the former being strongly H-bonded to the receptor and the latter showing no H-bond interaction with the receptor.

As summarized above, it is quite surprising that, among the compounds that qualify as cavitands according to the generic name introduced by Cram at the end of the seventies ([1a] and references therein), only a few of these compounds have been used for organizing coordination sites for transition metal cations. The chemistry of cyclotriveratrylene (CTV) [26, 27] has afforded numerous chiral hosts, the cryptophanes [28], Figure 3, as well as hosts for alkali cations [15]. The chemistry of chiral derivatives of CTV has been reviewed by Collet [28], and detailed data concerning the restricted mobility of this bowl shaped molecule are available in the literature. Although CTV derivatives bearing three phenanthroline units and three pyridine units have been reported by Cram, Figure 3, the heteroatoms were not properly located to allow the formation of complexes, it has never been combined with transition metal complexation. In our approach to CTV functionalization, only achiral derivatives are obtained, thus allowing quick and easy characterization of the target compounds. The attachment of aromatic imines or polyimines to the cyclotriveratrylene and its derivatives are described hereafter.

The first example of a cyclization reaction involving non-adjacent oxygen atoms of the hexademethylated CTV derivative trivially named hexaphenol, Figure 3, was reported by Cram in 1988 [29]. Aromatic spacers have been used to expand the hydrophobic cavity of cyclotriveratrylene by a triple cyclization, under 'pseudo' high dilution conditions, bridging oxygen atoms located on neighbouring catechol subunits. We have extended the macrocycle formation to the synthesis of various functionalized expanded CTVs and, after a brief discussion of the strategies available for attaching binding sites on cyclotriveratrylene, the synthesis of the bridges and the cyclization reaction will be described in detail.

2. Experimental

GENERAL

THF was distilled from LiAlH$_4$ under an argon atmosphere. DMSO was purified according to the literature [37]. Dioxane was distilled from NaBH$_4$ under argon. DMF was dried over molecular sieves before use. All other chemicals were reagent grade and used without further purification. When anhydrous conditions were required the glassware was flame-dried under a dry argon stream. ^{1}H-NMR spectra were recorded on Bruker WP-200 (200 MHz) and Bruker AM-400 (400 MHz) spectrometers. Chemical shifts were determined by taking the solvent as a reference: CHCl$_3$ (7.26 ppm), CHDCl$_2$ (5.32 ppm), DMSO-d_5 (2.49 ppm). Melting points were determined on a Kofler Heating Plate type WME and are uncorrected. Elemental analyses were performed by the Service d'Analyse Elementaire de Composés Solides à l'Institut Universitaire de Technologie, Strasbourg Sud and the Service de Microanalyse de l'Institut de Chimie de Strasbourg. Mass spectra were obtained on a FAB: ZAB-HF mass spectrometer. Thin-layer chromatography (TLC) was performed using Macherey-Nagel Polygram Sil G/UV$_{254}$ (0.25 mm) and Polygram Alox N/UV$_{254}$ analytical polyethylene coated plates. E. Merck silica gel 60 (70–230 mesh) and aluminium oxide 90 (70–230 mesh) were used for column chromatography.

5-BROMOISOPHTHALIC ACID (3)

Isophthalic acid (18.3 g, 110 mmol) was dissolved in 250 mL concentrated H_2SO_4. Ag_2SO_4 (20 g, 64 mmol) and bromine (7.5 mL, 140 mmol) were added and the suspension was heated at 110°C for 24 h. Excess bromine was removed by bubbling an argon stream through the solution. The cooled solution was poured into 500 mL of iced water. The resulting white precipitate was isolated and then dissolved in a sodium hydroxide solution. The green AgOH precipitate was removed and the filtrate was acidified with concentrated HCl. The resulting white solid was filtered and dried by azeotrope distillation with ethanol and toluene to afford (24.5 g, 100 mmol, 91%) a white solid. White crystals could be obtained by recrystallization from an acetone/water mixture. This product was characterized by comparison with literature data [31].

1-BROMO-3,5-BIS(HYDROXYMETHYL)BENZENE (4)

To a degassed solution of 5-bromo-isophthalic acid 3 (12.5 g, 51 mmol) in 190 mL of THF, a 1 M solution of BH_3 in THF (200 mmol, 200 mL) was added via cannula under argon. The clear solution was refluxed for 3.5 h, then allowed to cool to room temperature before carefully quenching with 50 mL of MeOH. The solvents were removed under vacuum and two additional portions of MeOH were added and evaporated. The resulting oil was dissolved in 200 mL of Et_2O and extracted with 200 mL of 5% aqueous Na_2CO_3 and then with H_2O. All aqueous phases were washed with Et_2O. The combined organic layers were dried over $MgSO_4$, filtered, and evaporated to dryness. Filtration over silica gel (CH_2Cl_2), 10% MeOH) yielded a white solid (10.8 g, 51 mmol, 100%). White crystals could be obtained from recrystallization in hot benzene. ^{1}H-NMR (DMSO-d_6) δ ppm: 7.34 (s, 2H, $H_{2,6}$), 7.23 (s, 1H, H_4), 5.31 (t, J = 6 Hz, 2H, OH), 4.66 (d, J = 6 Hz, 4H, H_{Bz}). *Anal. Calcd.* for $C_8H_9BrO_2$ (217.1): C, 44.26; H, 419. *Found*: C, 44.08; H, 4.07. Melting point: 93–95°C.

PROTECTION OF 1-BROMO-3,5-BIS(HYDROXYMETHYL)BENZENE (5)

A degassed solution of 4 (9.8 g, 46 mmol), dihydropyran (13 mL, 140 mmol), and pyridinium *p*-toluene sulfonate (PPTS) (514 mg, 2 mmol) in 400 mL of CH_2Cl_2 was stirred under argon for 40 h. The solution was extracted with water. The organic layer yielded, after drying ($MgSO_4$) and removal of solvents under vacuum, a yellow oil (17.6 g, 45 mmol, 98%). If necessary the oil could be purified by filtration over aluminia (CH_2Cl_2). ^{1}H-NMR (CD_2Cl_2) δ ppm: 7.46 (s, 2H, $H_{4,6}$), 7.27 (s, 1H, H_2), 4.77 (d, J = 10 Hz, 2H, H_{Bz}), 4.70 (m, 2H, $H_{1',1''}$), 4.46 (d, J = 10 Hz, 2H, H_{Bz}), 3.90 and 3.60 (two m, 4H$_{total}$, $H_{5',5''}$), 1.96–1.43 (m, 12H,

$H_{2',2',4',2'',3'',4''}$). *Anal. Calcd.* for $C_{18}H_{25}BrO_4$ (385.3): C, 56.11; H, 6.54. *Found*: C, 56.09; H, 6.58.

BORONIC ACID (7)

To a degassed solution of **5** (7.6 g, 20 mmol) in 120 mL of THF at 78°C two equivalents of 1.5 M *t*-butyl lithium (27 mL, 40 mmol) were added dropwise under argon, maintaining the temperature below -70°C. At the end of the addition, the red solution was rapidly transferred via cannula to a degassed solution of B(OMe)$_3$ (4.5 mL, 40 mmol) in 120 mL of THF at -78°C. The resulting solution was stirred under argon for 30 min. After warming the solution to room temperature the solvent and excess B(OMe)$_3$ were removed under reduced pressure. The crude product, dissolved in 150 mL of Et$_2$O was washed with 100 mL of 5% aqueous HCl solution, then with water. The organic phase was dried over MgSO$_4$, filtered, and evaporated to dryness. The resulting yellow oil (6.7 g, 20 mmol, 100%) was used without further purification.

BRIDGE SYNTHESES VIA SUZUKI COUPLING REACTIONS

*Synthesis of Protected 2-Pyridine Bridge (**8**).* To a degassed solution of 2-bromopyridine (2.9 g, 18.4 mmol) and Pd(PPh$_3$)$_4$ (0.751 g, 0.65 mmol) in 300 mL of toluene, a degassed 2 M aqueous solution of Na$_2$CO$_3$ (150 mL), and a degassed solution of **7** (6.7 g, 20 mmol) in 75 mL of methanol was added. The vigorously stirred suspension was refluxed under argon for 12 h. After cooling to room temperature, the reaction mixture was extracted twice with 250 mL of 2 M aqueous Na$_2$CO$_3$ containing 50 mL of concentrated NH$_4$OH. The organic phase was dried over MgSO$_4$, filtered, and the solvent was removed under reduced pressure. Purification by column chromatography over alumina (CH$_2$Cl$_2$) afforded **8** as a colorless oil (4.87 g, 12.9 mmol, 70%). ^{1}H-NMR (DMSO-d_6) δ ppm: 8.69 (d, J = 5 Hz, 1H, H$_\alpha$ py), 7.89 (s, 2H, H$_{2,6}$ xyl), 7.74 (t, J_1 = 5 Hz, J_2 = 4 Hz, J_3 = 1 Hz, 2H, H$_{\beta,\delta}$ py), 7.46 (s, 2H, H$_4$ xyl), 7.24 (m, 1H, H$_\gamma$ py), 4.87 (d, J = 12 Hz, 2H, H$_{Bz}$), 4.74 (m, 4H, H$_{1',1''}$), 4.58 (d, J = 12 Hz, 2H, H$_{Bz}$), 3.91–3.56 (two m, 4H, H$_{5',5''}$), 1.85–1.60 (m, 12H, H$_{2',2'',3',3'',4',4''}$). *Anal. Calcd.* for $C_{23}H_{29}NO_4$ (383.5): C, 72.03; H, 7.62. *Found*: C, 71.82; H, 7.74.

*Synthesis of Protected 3-Pyridine Bridge (**9**).* The procedure described for the preparation of **8** was followed using the following quantities: 3-bromopyridine (3.6 g, 22.7 mmol) and Pd(PPh$_3$)$_4$ (0.939 g, 0.81 mmol) in 375 mL of toluene, 2 M aqueous Na$_2$CO$_3$ (190 mL, degassed), and **7** (8.5 g, 25 mmol) in 100 mL of degassed methanol. Reflux time: 11 h. Purification by column chromatography over alumina (CH$_2$Cl$_2$) afforded **9** (8.4 g, 22 mmol, 89%) as a colorless oil. ^{1}H-NMR (DMSO-d_6) δ ppm: 8.86 (d, J = 2 Hz, 1H, H$_{\alpha'}$ py), 8.57 (dd, J_1 = 4.5 Hz, J_2 = 2 Hz, 1H, H$_\alpha$ py), 8.04 (dd, J_1 = 5 Hz, J_2 = 2 Hz, 1H, H$_\gamma$ py), 7.57 (s, 2H, H$_{2,6}$ xyl), 7.48 (dd, J_1 = 4.5 Hz, J_2 = 5 Hz, 1H, H$_\beta$ py), 7.39 (s, 1H, H$_4$ xyl),

4.74 (d, J = 10 Hz, 2H, H_{Bz}), 4.72 (m, 2H, $H_{1',1''}$), 4.52 (d, J = 10 Hz, 2H, H_{Bz}), 3.87–3.40 (two m, 4H, $H_{5',5''}$), 1.85–1.40 (m, 12H, $H_{2',2'',3',3'',4',4''}$). *Anal. Calcd.* for $C_{23}H_{29}NO_4$ (383.5): C, 72.03; H, 7.62. *Found*: C, 72.23; H, 7.48.

2(3',5'-bis(Hydroxymethyl)phenyl)pyridine) Bridge (**10**). The protected 2-pyridine bridge **8** (1.55 g, 4 mmol) was dissolved in 50 mL of absolute ethanol, degassed, and the PPTS catalyst (82 mg, 0.32 mmol) was added. The stirred solution was refluxed under argon for 11.5 h. After cooling to ambient temperature, the solvents were removed *in vacuo*. The yellow oil was taken in 50 mL of CH_2Cl_2 and washed with water and 50 mL of a saturated aqueous NaCl solution. The aqueous phases were repeatedly washed with 50 mL of CH_2Cl_2 as the product was sparingly soluble in dichloromethane. The combined organic layers were dried over $MgSO_4$, filtered, and evaporated to dryness. A colorless oil of **10** (600 mg, 2.8 mmol, 70%) was isolated after filtration over silica gel (CH_2Cl_2, 5% MeOH). The product was used without further purification for the next step. ^{1}H-NMR (DMSO-d_6) δ ppm: 8.65 (dd, J_1 = 1 Hz, J_2 = 3 Hz, 1H, H_α py), 7.88 (m, 4H, $H_{\beta,\delta}$ py, $H_{2,6}$ xyl) 7.35 (m, 2H, H_γ py, H_4 xyl), 5.27 (t, J_1 = 12 Hz, J_2 = 6 Hz, 2H, OH), 4.56 (d, J = 6 Hz, 4H, H_{Bz}).

3-(3',5'-bis(Hydroxymethyl)phenyl)pyridine Bridge (**11**). The procedure described for the preparation of **10** via the deprotection method was followed using the following quantities: **9** (8.4 g, 22 mmol), and PPTS (452 mg, 1.76 mmol) in 250 mL of ethanol. Reflux time: 13 h. Filtration over silica gel (CH_2Cl_2, 5% MeOH) afforded **11** as a yellow oil (3.5 g, 16 mmol, 74%) which was used without further purification for the next step. ^{1}H-NMR (DMSO-d_6) δ ppm: 8.87 (s, 1H, $H_{\alpha'}$ py), 8.57 (d, J = 4 Hz, 1H, H_α py), 8.06 (d, J = 8 Hz, H_γ py), 7.51 (m, 1H, H_β py), 7.51 (s, 2H, $H_{2,6}$ xyl), 7.35 (s, 1H, H_4 xyl), 6.54 (br. s, 2H, OH), 4.57 (s, 4H, H_{Bz}).

BRIDGE SYNTHESIS VIA ORGANOTIN COUPLING REACTIONS

2-(3',5'-bis(Hydroxymethyl)phenyl)pyridine Bridge (**10**). A degassed mixture of 2-trimethylstannylpyridine **14** (630 mg, 2.6 mmol), bromo-3,5-bis(hydroxymethyl)-benzene **4** (500 mg, 2.3 mmol), and $PdCl_2(PPh_3)_2$ (81 mg, 0.12 mmol) in 10 mL of distilled dioxane was refluxed under argon for 25 h. The volume of the crude mixture was reduced to 5 mL, then filtered over a column of alumina (CH_2Cl_2, 0–5% MeOH) to afford **10** (380 mg, 1.77 mmol, 76%) as a colorless oil. This compound was characterized by comparison with data from the product synthesized via the Suzuki cross coupling method described above.

3-(3',5'-bis(Hydroxymethyl)phenyl)pyridine Bridge (**11**). The reaction procedure described for the preparation of **10** was followed using the following quantities: 3-trimethylstannylpyridine **15** (11.82 g, 48.8 mmol), bromo-3,5-bis(hydroxymethyl)-benzene **4** (9.29 g, 43.2 mmol), and $PdCl_2(PPh_3)_2$ (1.52 g, 2.16 mmol) in 85 mL of distilled dioxane. After 23 h under reflux, 1.52 g (48.8 mmol) of PdCl$_2$(PPh$_3$)$_2$ were added to the reaction mixture. After an addition 5 h under reflux, the volume of the black mixture was reduced to 10 mL under reduced pressure. A pale yellow oil of **11** (6.8 g, 31.9 mmol, 74%) was isolated after filtration over a column of alumina (CH$_2$Cl$_2$, 2% MeOH). This compound was characterized by comparison with data from the product synthesized via the Suzuki cross coupling method described above.

2(3′,5′-bis(Chloromethyl)phenyl)pyridine (**12**). To a degassed suspension of **10** (40 mg, 0.18 mmol) in 10 mL of CH$_2$Cl$_2$, thionyl chloride (0.5 mL, 5.4 mmol) and two drops of DMF were added. After stirring under argon for 3 h, the solvents were removed under reduced pressure. The crude product was taken in 75 mL of CH$_2$Cl$_2$ and extracted with 15 mL of H$_2$O containing several drops of triethylamime (pH 10). The organic layer was dried over MgSO$_4$, filtered, and evaporated to dryness. Filtration over silica gel (CH$_2$Cl$_2$/Hexane: 9/1) afforded a white solid (42 mg, 0.16 mmol, 90%). ^{1}H-NMR (CDCl$_3$) δ ppm: 8.70 (s, 1H, H$_\alpha$ py), 7.98 (d, J = 1.6 Hz, 2H, H$_{2,6}$ xyl), 7.78 (m, 2H, H$_{\beta,\delta}$ py), 7.49 (s, 1H, H$_4$ xyl), 7.28 (m, 1H, H$_\gamma$ py), 4.67 (s, 4H, CH$_2$ xyl). Melting point: 65–67°C. *Anal. Calcd.* for C$_{13}$H$_{11}$NCl$_2$·1/3C$_6$H$_{14}$ (280.88): C, 64.14; H, 5.62; N, 4.99. *Found*: C, 64.13; H, 5.45; N, 4.45.

3-(3′,5′-bis(Chloromethyl)phenyl)pyridine (**13**). The reaction procedure described for the preparation of **12** was followed using the following quantities: **11** (400 mg, 1.8 mmol) in 100 mL of CH$_2$Cl$_2$, thionyl chloride (2 mL, 21.6 mmol), one drop of DMF. Reaction time: 30 min. The crude crystalline product **13** (445 mg, 1.77 mmol, 98%) was pure by TLC and used without further purification. If necessary, the product could be purified by filtration over silica gel (CH$_2$Cl$_2$, 10% MeOH). ^{1}H-NMR (CDCl$_3$) δ ppm: 8.84 (d, J = 2 Hz, 1H, H$_{\alpha'}$ py), 8.63 (d, J = 4 Hz, 1H, H$_\alpha$ py), 7.88 (dd, J_1 = 4.5 Hz, J_2 = 2 Hz, 1H, H$_\gamma$ py), 7.56 (s, 2H, H$_{2,6}$ xyl) 7.47 (s, 1H, H$_4$ xyl), 7.37 (dd, J_1 = 4 Hz, J_2 = 4.5 Hz, 1H, H$_\beta$ py), 4.66 (s, 4H, ArCH$_2$Cl). *Anal. Calcd.* for C$_{13}$H$_{11}$NCl$_2$ + 1.5 H$_2$O (279.15): C, 61.93; H, 4.40; N, 5.55. *Found*: C, 61.48; H, 4.35; N, 5.32. Melting point: 124–126°C.

2-(3′,5′)-bis(Bromomethyl)phenyl)pyridine (**18**). A solution of **10** in 10 mL of 48% aqueous HBr and 5 mL of acetic acid was heated at 40°C for 20 h. The solution was cooled, neutralized with Na$_2$CO$_3$, and washed twice with 20 mL of CH$_2$Cl$_2$. The combined organic phases were dried over MgSO$_4$, filtered, and evaporated to dryness. The crude product was further dried by azeotrope distillation with benzene. Filtration over a short column of silica gel (CH$_2$Cl$_2$) afforded **18** (25 mg, 0.073 mmol, 31%) as a crystalline solid. ^{1}H-NMR (CDCl$_3$) δ ppm: 8.70 (d, J = 4 Hz, 1H, H$_{\alpha'}$ py), 7.96 (s, 2H, H$_{4,6}$ xyl), 7.76 (m, 2H, H$_{\beta,\delta}$ py), 7.48 (s, 1H, H$_2$ xyl), 7.28 (m, 1H, H$_\gamma$ py), 4.55 (s, 4H, ArCH$_2$Br). *Anal. Calcd.* for C$_{13}$H$_{11}$Br$_2$N +

1/6 C$_6$H$_6$ (354.06): C, 47.79; H, 3.42; N, 3.96. *Found*: C, 47.28; H, 3.11; N, 3.85. Melting point: 112–114°C.

BRIDGING REACTIONS

Synthesis of **19**. A solution of hexaphenol **2** (0.792 g, 2.16 mmol) and **16** (1.8 g, 7.15 mmol) were dissolved in 100 mL of degassed DMF and added, under argon, over 27 h to a vigorously stirred suspension of Cs$_2$CO$_3$ (11.9 g, 36 mmol) in 250 mL of degassed DMF at 70°C. The mixture was stirred for an additional 15 h at 70°C. The solvent was removed by distillation under reduced pressure. The crude product was taken in 400 mL of CH$_2$Cl$_2$ and extracted with 400 mL of water. The aqueous layer was washed twice with 300 mL of CH$_2$Cl$_2$. The combined organic phases were dried over MgSO$_4$, filtered, and evaporated to dryness. The purified product (column chromatography over silica gel/CH$_2$Cl$_2$) was dissolved in 75 mL of THF. The solvent was removed under reduced pressure. This process was repeated twice to afford beige crystalline **19** (500 mg, 0.55 mmol, 28%). This compound was characterized by comparison with literature data [29].

Synthesis of **20**. The reaction procedure for the preparation of **19** was followed using the following quantities: hexaphenol **2** (437 mg, 1.3 mmol) and **18** (1.77 g, 5.2 mmol) in 130 mL of degassed DMF, and Cs$_2$CO$_3$ (7.6 g, 23.4 mmol) in 350 mL of degassed DMF at 70°C. Addition time: 18.5 h followed by an additional 15 h of vigorous stirring under argon at 70°C. Purification by column chromatography over alumina (CH$_2$Cl$_2$) afforded the yellow crystalline product **20** (184 mg, 0.18 mmol, 14%). ^{1}H-NMR (CDCl$_3$) δ ppm: 8.55 (d, J = 4.7 Hz, 3H, H$_\alpha$ py), 8.01 (s, 6H, H$_{2,6}$ and 1H, DMF), 7.55 (m, 6H, H$_{\beta,\delta}$ py), 7.13 (m, 6H, H$_4$ xyl, H$_\gamma$ py), 6.53 (s, 6H, ArH CTV), 5.24 (d, J = 13 Hz, 6H, ArCH$_2$ xyl), 5.05 (d, J = 13 Hz, 6H, ArCH$_2$ xyl), 4.40 (d, J = 13.6 Hz, 3H, ArCH$_2$ CTV), 3.20 (d, J = 13.6 Hz, 3H, ArCH$_2$ CTV), 2.91 (d, J = 14.1 Hz, 6H, CH$_3$ DMF). *Anal. Calcd.* for C$_{63}$H$_{52}$N$_4$O$_7$ (977.14), i.e, product **20** + DMF: C, 77.45; H, 5.36; N, 5.73. *Found*: C, 77.36; H, 5.35; N, 5.46. Melting point: 207–208°C. Mass spectroscopy: *Calcd.* for C$_{60}$H$_{45}$N$_3$O$_6$: 904.0. *Found*: 904.2 (47%), 180.1 (100%) FAB positive I = 6.9 V.

Synthesis of **21**. The reaction procedure for the preparation of **19** was followed using the following quantities: hexaphenol **2** (0.290 g, 0.8 mmol) and **13** (0.8 g, 3.2 mmol) in 150 mL of degassed DMSO, and Cs$_2$CO$_3$ (3.9 g, 11.9 mmol) in 300 mL of degassed DMSO at 80°C. Addition time: 25 h followed by an additional 14 h of vigorous stirring under argon at 80°C. Column chromatography over silica gel (CH$_2$Cl$_2$, 1% MeOH) afforded crystalline **21** (289 mg, 0.32 mmol, 40%). ^{1}H-NMR (DMSO-d_6) δ ppm: 8.68 (d, J = 2 Hz, 3H, H$_{\alpha'}$ py), 8.46 (dd, J_1 = 5 Hz, J_2 = 2 Hz, 3H, H$_\alpha$ py), 8.01 (s, 6H, H$_4$ xyl), 7.78 (d, 3H, J = 8 Hz, H$_\gamma$ py), 7.56 (s, 3H, H$_{2,6}$ xyl), 7.20 (dd, J_1 = 5 Hz, J_2 = 8 Hz, 3H, H$_\beta$ py), 6.91 (s, 6H, ArH CTV), 5.75 (s, 12H, ArCH$_2$ xyl), 4.47 (d, J = 14 Hz, 3H, ArCH$_2$ CTV), 3.38 (d, J =

14 Hz, 3H, ArCH$_2$ CTV). Melting point: 265–267°C. Mass spectroscopy: *Calcd.* for C$_{60}$H$_{45}$N$_3$O$_6$: 904.0. *Found*: 904.1 (100%) FAB positive I = 2.4 V.

Synthesis of **23**. A suspension of hexaphenol **2** (79 mg, 0.22 mmol), 6-bromomethyl-2,2'-bipyridine **22** (6.43 mg, 2.58 mmol), and potassium carbonate (547 mg, 6.96 mmol) in 30 mL of DMF was stirred under argon at 60°C for 21 h. After removing the solvent *in vacuo*, the residue was taken in CH$_2$Cl$_2$ and washed twice with 100 mL of H$_2$O. The organic layer was dried over MgSO$_4$, filtered, and evaporated to dryness. Purification by column chromatography over alumina (CH$_2$Cl$_2$, 1% MeOH), followed by slow crystallization in CH$_2$Cl$_2$ and several drops of methanol afforded white needles of **23** (216 mg, 0.157 mmol, 71%). ^{1}H-NMR 400 MHz (CHDCl$_2$) δ ppm: 8.57 (ddd, $J_{3'-6'}$ = 1.0 Hz, $J_{4'-6'}$ = 1.8 Hz, $J_{5'-6'}$ = 4.7 Hz, 6H, H$_{6'}$), 8.35 (dt, $J_{3'-6}$ = 1.0 Hz, $J_{3'-5'}$ = 1.0 Hz, $J_{3'-4'}$ = 7.8 Hz, 6H, H$_{3'}$), 8.27 (d, J_{3-4} = 7.85 Hz, 6H, H$_3$), 7.76 (t, J_{3-4} = 7.85 Hz, J_{4-5} = 5.7 Hz, 6H, H$_4$), 7.62 (td, $J_{4'-t'}$ = 1.8 Hz, $J_{4'-5'}$ = 7.75 Hz, $J_{3'-4'}$ = 7.8 Hz, 6H, $H_{4'}$), 7.53 (d, J_{4-5} = 7.75 Hz, 6H, H$_5$), 7.21 (ddd, $J_{3'-5'}$ = 1.0 Hz, $J_{5'-6'}$ = 4.7 Hz, $J_{4'-5'}$ = 7.75 hz, 6H, H$_{5'}$), 6.99 (s, 6H, ArCH$_2$ CTV), 5.24 (s, 12H, ArCH$_2$ xyl), 4.71 (d, J = 13.75 Hz, 3H, ArCH$_2$ CTV), 3.49 (d, J = 13.75 Hz, 3H, ArCH$_2$ CTV). Anal. Calcd. for C$_{87}$H$_{66}$N$_{12}$O$_6$ (1375.6): C, 75.97; H, 4.84; N, 12.22). *Found*: C, 75.78; H, 4.97; N, 11.99). Melting point: 120–121°C. Mass spectroscopy: *Calcd.* for C$_{87}$H$_{66}$N$_{12}$O$_6$: 1375.6. *Found*: 1375.2 (90%); 1207.2 (32%); 1037.1 (23%); 869.1 (7%); 699.0 (8%); 529.0 (8%); 338.1 (100%) FAB positive I = 4.4 V.

3. Results and Discussion

3.1. STRATEGIES

Two different approaches could be envisioned to attach binding groups to the CTV derivatives. The usually 'low' yield of the triple cyclization reaction may be placed at the end of the synthesis or at the beginning of the synthesis, as represented in Figure 4.

If functionalized bridges are available on a large scale, ending the synthesis with the cyclization should afford the target compound on a larger scale (Strategy A). However, this approach will be limited to the controlled synthesis of symmetrically substituted ligands, as the use of different bridges in the same high dilution reaction would lead to random mixtures of compounds. Strategy B could be used for the synthesis of reactive CTV derivatives (e.g. Z = Br) and the scope of the further functionalization would be broadened as it will allow the introduction of functional groups which are not compatible with the cyclization conditions. The last approach available, not depicted, would be the direct connection of the binding groups to the hexaphenol which will afford six coordinating sites arranged around the cavity of the CTV. Syntheses of ligands, with pyridines as binding groups, involving the first strategy have been used, together with the direct attachment of ligands to the CTV.

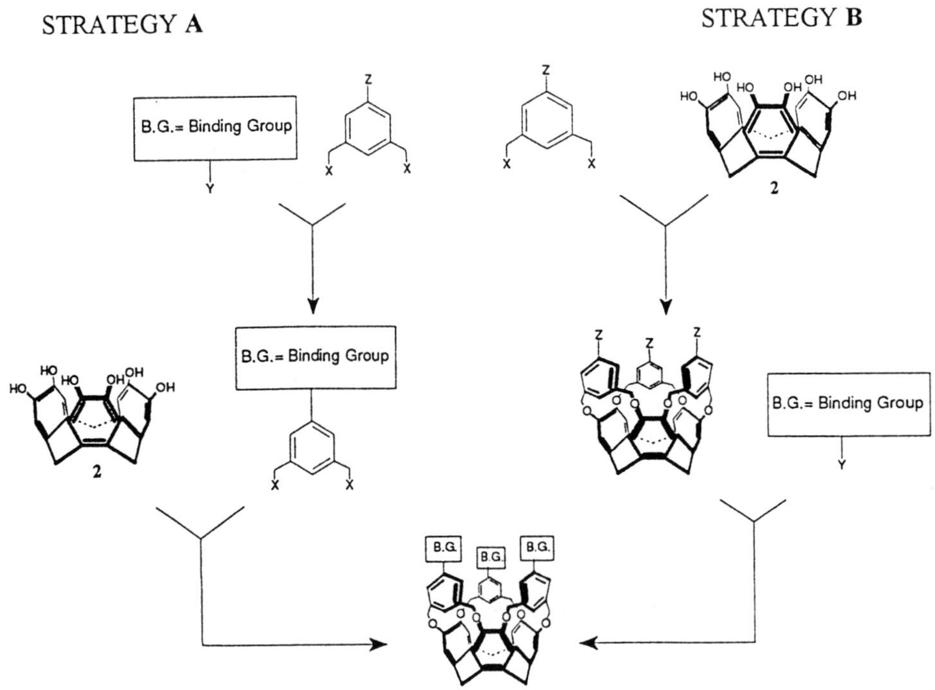

Fig. 4. Two strategies for arranging binding sites around cyclotriveratrylene.

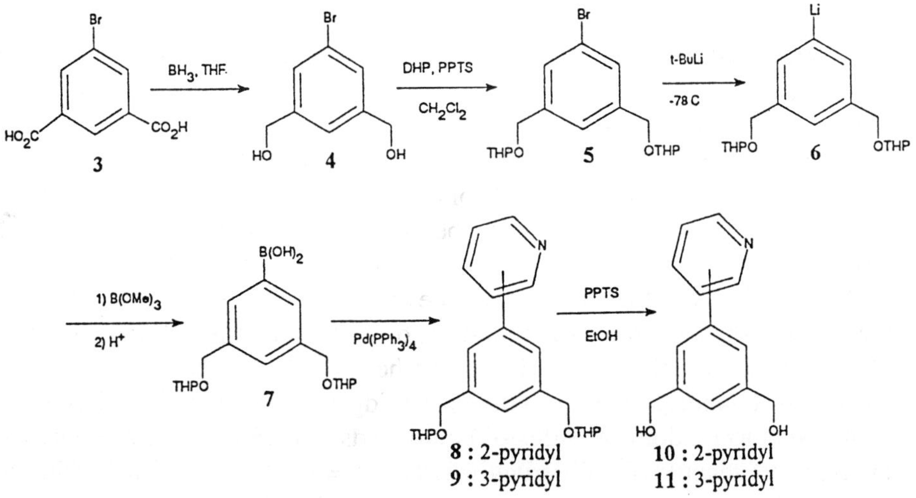

Fig. 5. Synthesis of functionalized bridges for expanded CTVs.

3.2. SYNTHESIS OF THE BRIDGING UNITS

The pyridyl *m*-xylene bridges were synthesized according to the reaction scheme in Figure 5. The 5-bromoisophthalic acid **3** [31] was reduced to the corresponding

10 : 2-pyridyl
11 : 3-pyridyl

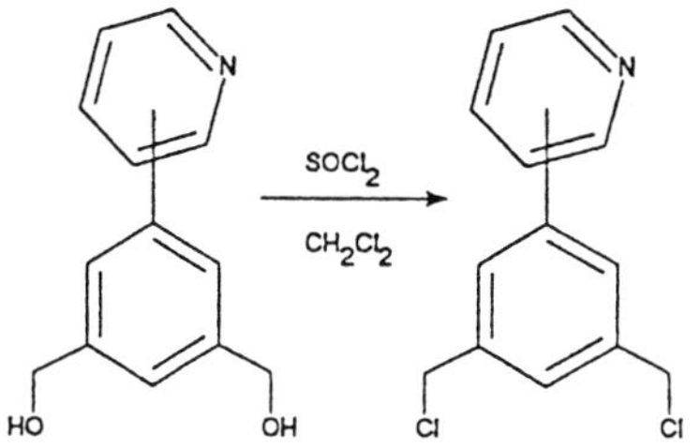

12 : 2-pyridyl
13 : 3-pyridyl

Fig. 6. Synthesis of substituted α, α'-dichloro *m*-xylenes.

14 : 2-pyridine
15 : 3-pyridine

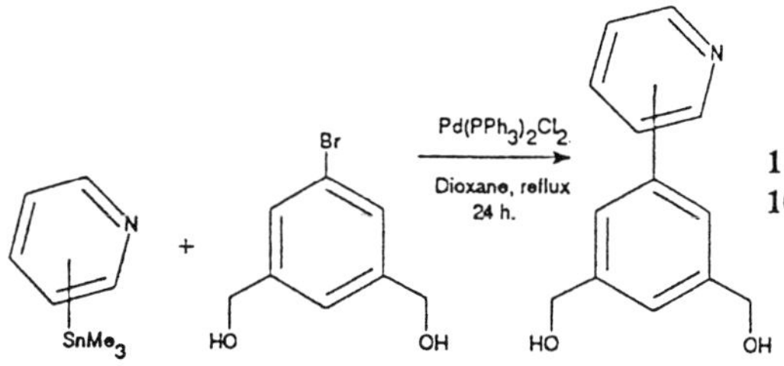

11 : 3-pyridyl
10 : 2-pyridyl

Fig. 7. Alternate route to functionalized bridges via organotin intermediates.

diol **4** by refluxing a solution of **3** in tetrahydrofuran (THF) for 4 h in the presence of an excess of $BH_3 \cdot THF$ complex. Before metallation, the diol **4** was protected with dihydropyran in the presence of pyridinium-*p*-toluene sulfonate (PPTS) as catalyst to afford **5** in 98% yield.

As the metallation of **5** using magnesium to form a Grignard reagent was unsuccessful due to lack of reaction of **5** with magnesium, we followed the previously reported [32] efficient preparation of aryl boronic acids via lithio derivatives. A solution of **5** in THF at -78°C was treated with two equivalents of *t*-butyllithium to generate a red solution of **6**. The immediate quenching of **6** with $B(OMe)_3$, afforded the desired boronic acid **7** in 100% yield after hydrolysis and rapid acid and base extraction. It should be noted that **7** was obtained quantitatively only when working with less than 20 mmol of **5**. The Suzuki coupling [33] of bromopyridines with crude **7** in the presence of $Pd(PPh_3)_4$ and 2 M aqueous Na_2CO_3 afforded **8** and **9** in yields of 70 and 89%, respectively.

The protecting groups were removed in refluxing ethanol (PPTS, 40 h). As shown in Figure 6, the corresponding dichlorides **12** and **13** were obtained by treatment of the diols in dichloromethane with an excess of $SOCl_2$ and triethylamine (TEA). For some reason, the dichloride **12** did not react as expected in the cyclization reactions described below, and the corresponding dibromide had to be prepared by treatment of the diol under the usual aqueous HBr conditions to afford the dibromide **18**.

Due to the limitation of the metal/halogen exchange reaction used for the preparation of the lithio derivatives **6** (much lower yields when working with more than 20 mmol), we have developed a quicker, general method for the preparation of the

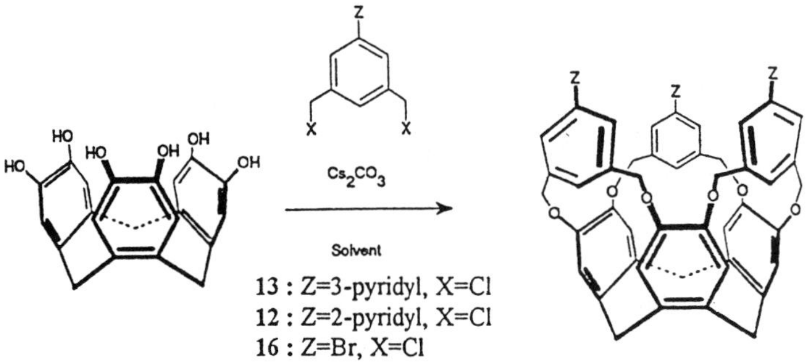

Fig. 8. General synthetic scheme of the formation of expanded CTVs.

13 : Z=3-pyridyl, X=Cl
12 : Z=2-pyridyl, X=Cl
16 : Z=Br, X=Cl
17 : Z=Br, X=Br
18 : Z=2-pyridyl, X=Br

19 : Z=Br
20 : Z=2-pyridyl
21 : Z=3-pyridyl

pyridyl-substituted xylenes. As depicted in Figure 7, the diols **10** and **11** are easily prepared by direct coupling of the arylbromide **4** and the subsequent trimethyl stannylpyridine derivative **14** or **15** [34], in yields of 74 and 76%, respectively.

Although the yields of the coupling reactions [35] are somewhat lower than those obtained under 'Suzuki' conditions, this method is much shorter than the original method. Furthermore, the overall yields of these syntheses are higher and no problems are encountered when running the reactions on larger scales.

3.3. TRIPLE CYCLIZATION REACTIONS

In order to minimize polymer formation, the bridging reactions, according to the general equation in Figure 8, were carried out in dilute conditions by the slow addition (over 25 h) of **2** and 3.3 equivalents of the subsequent bridge [36] to a suspension of an 18-fold excess of Cs_2CO_3 in DMSO [37] or DMF at 80°C. After purification by column chromatography the tribridged compounds **19–21** were isolated in yields listed in Table I; they strongly depend on the nature of the solvent and the type of leaving group used, Cl or Br.

Surprisingly, when using the bridging reaction with 5-bromo-1,3-bis(halomethyl)benzene as a test reaction, better yields were obtained with chlorine as a leaving group than with bromine. From literature data, the third cyclization seems to require a longer reaction time, probably because of the steric hindrance around the catecholate oxygen after the first two macrocycles have been closed. The dichloride bridge being more stable than the dibromide in the presence of a base like Cs_2CO_3, the stoichiometric bridge/hexaphenol ratio will be less time dependent in the case of the dichloride derivative. Concerning the pyridyl-xylene derivatives, the difference between the reactivity of **12**, **13** and **18** cannot yet be explained; however, in other experiments such as free radical bromination of 5-(3-pyridyl) *m*-xylene and 5-(2-pyridyl) *m*-xylene with *N*-bromosuccinimide, we have observed a similar difference

TABLE I.

Entry	Bridge	Solvent	T (°C)	Product (yield)
1	16	DMF	70	19 (28%)
2	16	DMSO	80	19 (42%)
3	17	DMF	70	19 (22%)
4	18	DMF	70	20 (14%)
5	12	DMSO	80	20 (0%)
6	13	DMSO	80	21 (40%)
7	13	DMF	70	21 (0%)

DMSO (dimethylsulfoxide): purified according to reference [37].
DMF (*N,N*-dimethylformamide): commercial grade dried over 4 Å molecular sieves.

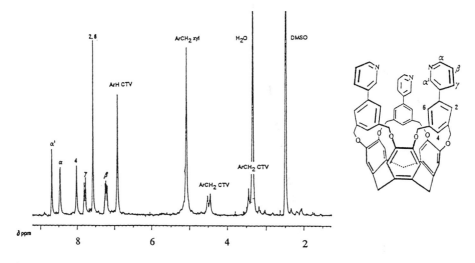

Fig. 9. ^{1}H-NMR (200 MHz) of the tripod **21**.

in reactivity depending on the position of the nitrogen atom within the pyridine ring.

The expanded CTV derivatives obtained have been characterized by the usual techniques, ^{1}H-NMR affording the best evidence for the C_{3v} symmetry of the tripods in solution. For example, in the case of **21**, the AB pattern (J_{AB} = 14 Hz) at 4.49 and 3.28 ppm and the integration ratios for the pyridine and CTV protons observed in the ^{1}H-NMR at 200 MHz, shown in Figure 9, confirms that the isolated product corresponds to a tribridged CTV ligand. The AB pattern indicates that the threefold symmetry axis existing in this type of CTV derivative [29] has been conserved.

The unambiguous structure of **21** has been proved by X-ray crystallographic measurements [38].

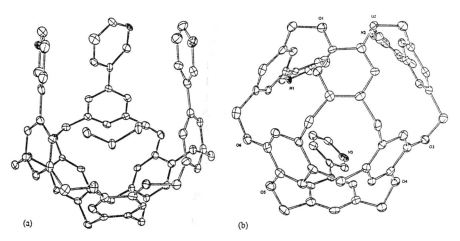

Fig. 10. Ortep drawings of **21**: (a) side view with methylene chloride guest; (b) bottom view of free ligand.

3.4. CRYSTAL STRUCTURE OF TRIPOD **21**

Colorless crystals of the tripod **21**[1] were obtained after several days in the absence of light by vapor diffusion of a hexane/methanol (one drop) layer through a dichloromethane/methanol (two drops) solution of **21**. In contrast to the previously reported [29] crystal structure of a xylene bridged CTV in which only two of the bridges point upward from the cavity whereas the third xylene bridge is bent outside the cavity, the crystal structure of **21** (Figure 10), shows that all three xylene bridges are oriented upward from the CTV cavity. From calculations [29] it appears that the energy difference between different bridge conformations is small (< 1 kcal). The pyridine rings are oriented so as to minimize the global dipole of **21** and the interactions between the three lone pair electrons of the nitrogens. The conformation adopted by ligand **21** should be primarily due to this type of weak interaction. The presence of the dichloromethane guest within the hydrophobic cavity may also favor this conformer in the solid state. A close contact is observed between the chlorine atoms and the uppermost xylene carbon of two of the bridges.

The hydrophobic cavity is quite large, on the order of 9.5 Å in diameter. A close contact between the hydrogen atoms of the CH_2Cl_2 guest and the π cloud of one xylene bridge may exist but the R factor of the structure does not allow for estimation of the H···π distance. As has been previously observed for an inclusion complex of methylene chloride in CTV [29], the CH_2Cl_2 guest is disordered in the crystal structure. The three bridges lean slightly towards one another which results in a smaller distance (~6.5 Å) between the pyridines. The tripod **21** is a terpyridine analogue with three electronically independent pyridine rings. Two molecules of this terpyridine analogue are able to coordinate octahedral transition metal cations such as cobalt(II), each subunit occupying opposite faces of the

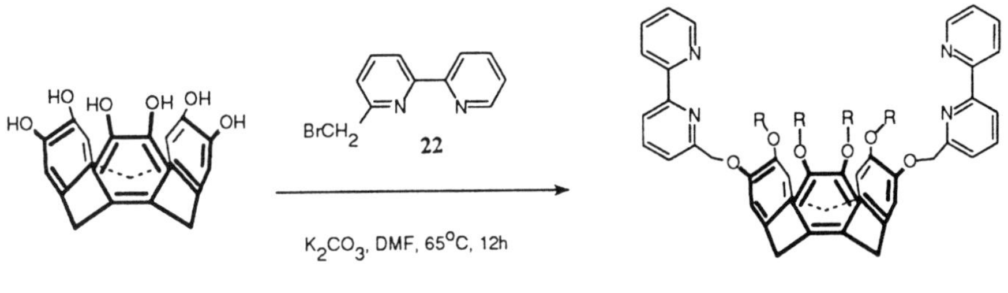

Fig. 11. Direct connection of binding sites to the cyclotriveratrylene matrix.

octahedron [39]. The corresponding tripod ligand with the 2-pyridine bridges is of interest because the nitrogen atoms would be oriented towards the CTV cavity. This orientation would increase the number of lone electron pairs, due to the oxygens around the CTV matrix, within the tripod's hydrophobic cavity. Furthermore, it should facilitate the coordination of metal centers or organic substrates capable of forming hydrogen bonds.

3.5. Direct connection of binding groups to the CTV matrix

As mentioned before, the CTV could also be used as a matrix for the direct preorganization of binding sites, without expansion of the hydrophobic cavity. Starting from the hexaphenol **2**, up to six oxygen atoms can undergo an alkylation with an appropriate functionalized binding group. Due to other research projects in progress in our group, a consequent amount of 6-bromomethyl-2,2′-bipyridine **22** [40] was available to carry out the reaction represented in Figure 11.

Using a stoichiometric amount of **22**, i.e. six equivalents, afforded a mixture of mono- to hexa-alkylated species from which the ligand **23** was isolated by crystallization. The hexa-alkylation process could be enhanced reacting a twofold excess of **22** with the hexaphenol. The ligand CTV(bipy)$_6$ was then isolated in 71% yield. Again, the ^{1}H-NMR provided evidence for the hexasubstitution and the C_{3v} symmetry of the isolated ligand. The compound obtained can accommodate either three square planar (copper(II)) or tetrahedral (copper(I)) transition metals, or, two octahedral transition metals (nickel(II)). The corresponding complexes are of interest in order to study, metal–metal interactions either by electrochemistry, EPR or photochemistry [39].

4. Conclusion

Cyclotriveratrylene, by analogy with calixarenes, cyclodextrins and a few other cyclophane-type compounds, can be used as a matrix to preorganize coordination sites for transition metals, either at the edge of the rigid structure, or around an expanded CTV's cavity. Achiral 2- and 3-pyridine tripod ligands have been synthesized together with a bipyridine multichelate. Studies of these new ligands, both by

UV-visible spectrophotometry and electrochemistry, regarding their coordination to transition metals and the properties of the corresponding complexes are under progress, as well as the extension of the synthetic method described above to the preorganization of other binding groups around the cyclotriveratrylene.

Acknowledgements

We warmly thank the Centre National de la Recherche Scientifique for financial support. Professor J. Fischer and Dr. A. Decian, URA CNRS No. 424 are gratefully acknowledged for the determination of the crystallographic structure of **21**.

Notes

[1] X-Ray Experimental Data for **21**·3CH$_2$Cl$_2$·H$_2$O: Molecular weight: 1352.3; Crystal system: monoclinic; a (Å): 25.735 (9); b (Å): 18.636 (6); c (Å): 12.994 (4); β (deg): 102.68 (2); Volume (Å^3): 6079.9; z: 4; D_{calc} (g cm^{-3}): 1.477; Wavelength (Å): 1.5418; μ (cm^{-1}): 43.892; Space group: $P2_1/n$; Diffractometer: Philips PW 1100/16; Crystal dim. (mm): 0.10 × 0.22 × 0.33; Temperature: -100°C; Radiation: CuK_α graphite monochromated; Mode: $\theta/2\theta$ flying step-scan; Scan speed (deg^{-1}): 0.020; Step width (deg): 0.05; Scan width (deg): 1.20 + 0.14 × tan(θ); Octants: $\pm h + k + l$; θ min/max (deg): 3.49; Number of data collected: 6536; Number of data with $I > 3\sigma$ (I): 4213; Abs min/max: 0.70/1.35; R (F): 0.107; R_w (F): 0.156; p: 0.08; GOF: 3.180.

References

1. For reviews see: (a) D. J. Cram: *Angew. Chem. Int. Ed. Engl.* **27**, 1009 (1988); (b) F. N. Diederich: *Angew. Chem. Int. Ed. Engl.* **27**, 362 (1988); (c) J. M. Lehn: *Angew. Chem. Int. Ed. Engl.* **27**, 89 (1988); (d) J. Rebek Jr.: *Angew. Chem. Int. Ed. Engl.* **29**, 245 (1990); (e) D. J. Cram: *Nature* **356**, 29 (1992).
2. E. Tsuchida and H. Nishide: *Top. Curr. Chem.* **132**, 63 (1986).
3. F. P. Schmidtchen: *Top. Curr. Chem.* **132**, 63 (1986) and references therein.
4. D. H. Busch and N. A. Stephenson: *J. Incl. Phenom.* **7**, 137 (1989).
5. R. Breslow, J. W. Canary, M. Varney, S. T. Waddle, and D. Young: *J. Am. Chem. Soc.* **112**, 5212 (1990).
6. Y. Kuroda, T. Hiroshige and H. Ogoshi: *J. Chem. Soc., Chem. Commun.* 1594 (1990).
7. M. I. Rosenthal and A. W. Czarnick: *J. Incl. Phenom.* **10**, 119 (1991).
8. H. J. Schneider and F. Xiao: *J. Chem. Soc., Perkin Trans.* 2, 387 (1992).
9. R. Fornasier, E. Scarpa, P. Scrimin, P. Tecilla, and U. Tonnelato: *J. Incl. Phenom.* **14**, 205 (1992).
10. D. R. Benson, R. Valentekovitch, S. W. Tam, and F. Diederich: *Helv. Chim. Acta* **76**, 2034 (1993) and references therein.
11. S. W. Tam-Chang, L. Jimenez, and F. Diederich: *Helv. Chim. Acta* **76**, 2616 (1993).
12. C. D. Gutsche: in *Calixarenes*, J. F. Stoddart (Ed.), Monographs in Supramolecular Chemistry, The Royal Society of Chemistry, Cambridge (1989).
13. J. Vicens and V. Böhmer: in *Calixarenes: A Versatile Class of Macrocyclic Compounds*, Kluwer Academic Publishers, Dordrecht, Holland (1991).
14. R. Ungaro and A. Pochini: in *Frontiers in Supramolecular Organic Chemistry and Photochemistry*, H. J. Schneider and H. Dürr (Eds.), VCH, Weinheim (1991).
15. K. Frensch and F. Vögtle: *J. Lieb. Ann. Chem.* 2121 (1979).
16. S. Pappalardo, L. Giunta, M. Foti, G. Ferguson, J. F. Gallagher, and B. Kaitner: *J. Org. Chem.* **57**, 2611 (1992).
17. F. Bottino, L. Giunta, and S. Pappalardo: *J. Org. Chem.* **54**, 5407 (1989).
18. C. D. Gutsche and K. C. Nam: *J. Am. Chem. Soc.* **110**, 6153 (1988).

19. F. Hamada, T. Fukugari, K. Murai, G. W. Orr, and J. L. Atwood: *J. Incl. Phenom.* **10**, 57 (1991).
20. D. Matt, C. Loeber, J. Vicens, and Z. Asfari: *J. Chem. Soc., Chem. Commun.* 604 (1993).
21. C. Floriani, D. Jacoby, A. Chiesi-Villa, and C. Guastini: *Angew. Chem. Int. Ed. Engl.* **10**, 1376 (1989).
22. J. K. Moran and M. Roundhill: *Inorg. Chem.* **31**, 4213 (1992).
23. A. W. Coleman, C. C. Ling, and M. Miocque: *Angew. Chem. Int. Ed. Engl.* **31**, 1381 (1992).
24. Y. Kuroda, M. Ito, T. Sera, and H. Ogoshi: *J. Am. Chem. Soc.* **115**, 7003 (1993).
25. H. K. A. C. Coolen, P. W. N. M. van Leeuwen, and R. J. M. Nolte: *Angew. Chem. Int. Ed. Engl.* **31**, 905 (1992).
26. G. M. Robinson: *J. Chem. Soc.* 267 (1915).
27. (a) A. S. Lindsey: *J. Chem. Soc.* 1685 (1965); (b) J. A. Hyatt: *J. Org. Chem.* **43**, 1808 (1978).
28. A. Collet: *Tetrahedron* **43**, 5725 (1987).
29. D. J. Cram, J. Weiss, R. Hegelson, C. B. Knobler, A. E. Dorigo, and K. N. Houk: *J. Chem. Soc., Chem. Commun.* 407 (1988).
30. J. Canceill, A. Collet, J. Gabard, F. Kotzyba-Hibert, and J. M. Lehn: *Helv. Chim. Acta* **65**, 1894 (1982).
31. E. W. Crandall and L. Harris: *Organic. Preparations and Procedures* **1**(3), 147 (1969).
32. (a) M. J. Sharp and V. Snieckus: *Tetrahedron Lett.* **26**, 5997 (198); (b) M. J. Sharp, W. Cheng, and V. Snieckus: *Tetrahedron Lett.* **28**, 5093 (1987); (c) W. Cheng and V. Snieckus: *Tetrahedron Lett.* **28**, 5097 (1987); (d) W. J. Gaudino and J. J. Thompson: *J. Org. Chem.* **49**, 5237 (1984).
33. N. Miyaura, T. Yanagi, and T. Suzuki: *Synth. Commun.* **11**, 513 (1981).
34. Y. Yamamoto and A. Yanagi: *Chem. Pharm. Bull.* **30**, 1731 (1982).
35. T. R. Bailey: *Tetrahedron Lett.* **27**, 4407 (1986).
36. A preliminary account of this work has already been published, see: J. A. Wytko and J. Weiss: *Tetrahedron Lett.* **49**, 7261 (1991).
37. D. D. Perrin, W. L. F. Armarego, and D. R. Perrin: in *Purification of Laboratory Chemicals*, Pergamon, 2nd Edition (1980).
38. J. Wytko, J. Fischer, A. Decian, and J. Weiss: *Unpublished Results*. Details about crystallographic data of the ligand will be published together with the structure of selected complexes.
39. C. Boudon, J. Wytko, M. Gross, and J. Weiss: Preliminary Communication at the *Journées d'Electrochimie*, June 7–10, Grenoble, France (1993).
40. G. R. Newkome, V. K. Gupta, and F. R. Fronczek: *Inorg. Chem.* **22**, 171 (1983).

X-Ray Crystallographic Studies of Tricarbonylchromium Complexes of Calix[4]arene Conformers: On an Unusual Conformation Which Appears in Cone Conformers *

HIDESHI IKI,[†] TAKETOSHI KIKUCHI,[†] HIROHISA TSUZUKI[§] and SEIJI SHINKAI[†]**

Department of Organic Synthesis, Faculty of Engineering, Kyushu University, Fukuoka 812, Japan, and Center of Advanced Instrumental Analysis, Kyushu University, Kasuga, Fukuoka 816, Japan

Abstract. The structures of three arene-tricarbonylchromium complexes prepared from cone and 1,3-alternate-25,26,27,28-tetrapropoxycalix[4]arene(**1**) and $Cr(CO)_6$ were determined by single crystal X-ray studies. Crystal data for 1,3-alternate-**1**•$Cr(CO)_3$ are space group $P2_1/a$, $a=19.496(3)$Å, $b=11.118(2)$Å, $c=19.121(2)$Å, $\beta=109.95°(1)$ and $V=3895$Å^3. The structure was refined to $Rw=0.068$. Crystal data for cone-**1**•$Cr(CO)_3$ are space group $P2_1/a$, $a=21.457(4)$Å, $b=12.184(1)$Å, $c=14.816(2)$Å, $\beta=91.61°(1)$ and $V=3872$Å^3. The structure was refined to $Rw=0.077$. Crystal data for cone-**1**•$2Cr(CO)_3$ are space group $P2_1/a$, $a=18.019(3)$Å, $b=41.347(4)$Å, $c=11.743(2)$Å, $\beta=97.39°(1)$ and $V=8676$Å^3. The single crystal included two similar but slightly different structures but the data were successfully refined to $Rw=0.092$. The structure of 1,3-alternate-**1**•$Cr(CO)_3$ differs only slightly from that of the regular 1,3-alternate calix[4]arene. In contrast, cone-**1**•$Cr(CO)_3$ and cone-**1**•$2Cr(CO)_3$ show an unusual conformation with a pair of faced gable-like roofs, which is considerably distorted from the regular cone calix[4]arene. The origin of this distortion is discussed in combination with the spectral studies.

Key Words: Calixarenes, arenetricarbonylchromium, crystal structure.

1. Introduction

Calix[n]arenes are a class of cavity-shaped macrocycles composed of n molecules of phenol and n molecules of formaldehyde. X-ray crystallographic studies of calix[n]arene derivatives have been continuously reported by Atwood et al. [1-4], Andreetti et al. [4-8], and others [9-11]. We have also carried out the structure determination of partial-cone conformers and a 1,3-alternate conformer of calix[4]arene derivatives [12-14]. More recently, we have synthesized arene-tricarbonylchromium complexes from cone, 1,2-alternate and 1,3-alternate conformers of 25,26,27,28-tetrapropoxycalix[4]arene (**1**) [15]. The purpose of the study was to selectively introduce the desired functional group into the desired benzene nucleus because tricarbonylchromium [$Cr(CO)_3$] forms stable η^6-arene complexes and the complexed benzene nucleus becomes extraordinarily "reactive" [16-20].

[†] Department of Organic Synthesis.
[§] Center of Advanced Instrumental Analysis.
* This paper is dedicated to the commemorative issue of the 50th anniversary of calixarenes.
** Author for correspondence.

This chemistry is also interesting from a stereochemical viewpoint: for example, (i) the molecular motion (*e.g.*, the rate of the oxygen-through-the-annulus rotation) of the benzene unit carrying a "heavy" $Cr(CO)_3$ on its back should be different from that of the "free" benzene unit, (ii) introduction of $Cr(CO)_3$ into the appropriate benzene unit in calix[4]arene conformers such as partial-cone and 1,2-alternate leads to the loss of molecular symmetry and therefore the products are optically-active [15] and (iii) the steric crowding increased by introduction of $Cr(CO)_3$ should cause some steric distortion from regular calix[4]arene structures. In order to obtain basic insights into such stereochemical characteristics of calix[4]arene-tricarbonylchromium complexes we have carried out X-ray crystallographic studies on three complexes: cone-**1**•$Cr(CO)_3$ (1:1 complex of **1** with a cone conformation), cone-**1**•2$Cr(CO)_3$ (distal 1:2 complex of **1** with a cone conformation) and 1,3-alternate-**1**•$Cr(CO)_3$ (1:1 complex of **1** with a 1,3-alternate conformation). We find that the basic skeletons of calix[4]arene are surprisingly distorted by complexation with $Cr(CO)_3$.

cone-**1**   1,3-alternate-**1**

2. Experimental

2.1. MATERIALS

Preparations of cone-**1**•$Cr(CO)_3$, cone-**1**•2$Cr(CO)_3$ and 1,3-alternate-**1**•$Cr(CO)_3$ have been described previously[15]. Single crystals of these complexes were prepared by recrystallization from methanol-dichloromethane for the cone-**1** complexes and from acetonitrile for the 1,3-alternate complex.

2.2. X-RAY CRYSTALLOGRAPHY

Integral intensities were collected by using Cu-K_α radiation by the ω-2θ scan technique up to $2\theta = 130°$. Absorption correction was made routinely. The intensities of the reflections in 1,3-alternate-**1**•$Cr(CO)_3$, cone-**1**•$Cr(CO)_3$ and cone-**1**•2$Cr(CO)_3$ decreased by 0.01%, 0.02% and 0.07% per hour, respectively. The structures were solved by direct methods (MULTAN 11/82 [21] for the 1,3-alternate-**1** complex and SIR88 [22] for the cone-**1** complexes) and then refined by the full-matrix least-squares procedure with anisotropic thermal parameters. Hydrogen atoms were included in refinement except for cone-**1**•2$Cr(CO)_3$. Crystal data thus obtained are summarized in Table I.

Table I. Summary of crystal data and intensity collection for complexes of **1**.

Compound	Cone-**1**•Cr(CO)$_3$	Cone-**1**•2Cr(CO)$_3$	1,3-Alternate-**1**•Cr(CO)$_3$
Formula	C$_{43}$H$_{48}$O$_7$Cr	C$_{46}$H$_{48}$O$_{10}$Cr$_2$	C$_{43}$H$_{48}$O$_7$Cr
Fw	728.85	864.88	728.85
Crystal system	monoclinic	monoclinic	monoclinic
Space group	$P2_1/a$	$P2_1/a$	$P2_1/a$
a, Å	21.457(4)	18.019(3)	19.496(3)
b, Å	12.184(1)	41.347(4)	11.118(2)
c, Å	14.816(2)	11.743(2)	19.121(2)
β, deg	91.61(1)	97.39(1)	109.95(1)
V, Å^3	3872	8676	3895
Dcalc, g cm^{-3}	1.250	1.324	1.243
Z	4	8	4
Crystal size, mm	0.5x0.3x0.3	0.4x0.4x0.4	0.4x0.2x0.15
μ, cm^{-1}	28.4	46.5	28.2
T, K	293	296	293
Rflns collected	7228	15773	7217
Independent rflns	6650	14207	6396
Std rflns	110	108	144
No.of data used in refinement	4018[$I_o>3\sigma(I_o)$]	7177[$I_o>3\sigma(I_o)$]	3059[$I_o>3\sigma(I_o)$]
Range of h, k, l	-25,0,0 to 25,14,17	-21,0,0 to 21,48,13	-22,0,0 to 22,13,22
R	0.058	0.067	0.056
Rw	0.077	0.092	0.068

3. Results and Discussion

3.1. STRUCTURE OF 1,3-ALTERNATE-**1**•Cr(CO)$_3$

X-Ray crystallographic studies on calix[4]arenes with a 1,3-alternate conformation have still been very limited. Bott et al. [4] found that conformationally-mobile 5,11,17,23-tetra-*tert*-butyl-25,26,27,28-tetramethoxycalix[4]arene forms complexes with alkyl aluminiums and adopts a 1,3-alternate conformation in that crystal. Fujimoto et al. [14] and Verboom et al. [23] have reported the X-ray structures of 5,11,17,23-tetra-*tert*-butyl-25,26-bis(ethoxycarbonylmethoxy)-26,28-bis(2-pyridylmethoxy)calix[4]arene and 25,26,27,28-tetraethoxycalix[4]arene, respectively. These studies consistently show that, as suggested by theoretical calculations [24,25], four phenyl units are more or less parallel to each other, the dihedral angles to the mean plane of the four methylene groups being close to 90°.

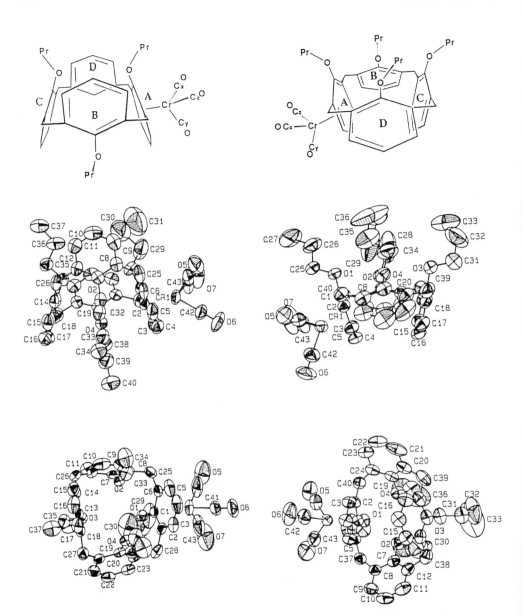

Fig. 1. Structure and ORTEP drawing of 1,3-alternate-**1**·Cr(CO)$_3$ with the thermal ellipsoids at the 50% probability level. Hydrogen atoms are deleted for clarity.

Fig. 2. Structure and ORTEP drawing of cone-**1**·Cr(CO)$_3$ with the thermal ellipsoids at the 50% probability level. Hydrogen atoms are deleted for clarity.

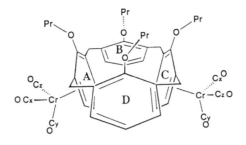

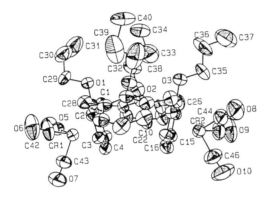

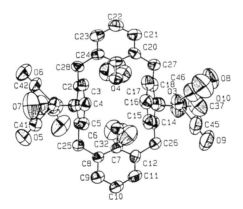

Fig. 3. Structure and ORTEP drawing of cone-**1**·2Cr(CO)$_3$ with the thermal ellipsoids at the 50% probability level. Hydrogen atoms are deleted for clarity.

The ORTEP drawing of 1,3-alternate-**1**·Cr(CO)$_3$ is shown in Fig. 1. Selected bond lengths and bond angles are summarized in Table II. In 1,3-alternate-**1**·Cr(CO)$_3$ the dihedral angles of benzenes A, B, C and D against the mean plane of the four methylene groups are 105.2°, -115.2°, 113.3° and -111.5°, respectively.[*] The absolute values of the dihedral angles obtained from previous X-ray studies are 93.5-106.1° [4] and 75.6-77.8° [23] and those obtained from theoretical calculations are 91.75-91.93° [25]. One can thus conclude that the phenyl units in 1,3-alternate-**1**•Cr(CO)$_3$ are "flattened" to some extent. Conceivably, the steric crowding increased by introduction of the Cr(CO)$_3$ group is relaxed by this conformational change. The bond angles for the ArCH$_2$Ar methylene groups were estimated to be 117.2-120.2°. These values are clearly greater than those for conventional 1,3-alternate calix[4]arenes (112.9-114.9° [4], 108.7-109.2° [23] and 115.1° [25]). This means that the flattening of the phenyl units inevitably causes the expansion of the ArCH$_2$Ar bond angles. It is also worthwhile mentioning that the O-CH$_2$-CH$_2$-CH$_3$ group in the Cr(CO)$_3$-complexed phenyl unit adopts a gauche conformation whereas those in the residual three phenyl units adopt an anti conformation. It seems that the gauche conformation acts to reduce the steric crowding.

In general, the structural change induced by Cr(CO)$_3$-complexation is relatively small, compared with those observed for cone-**1**•Cr(CO)$_3$ and cone-**1**•2Cr(CO)$_3$ (*vide post*).

3.2. STRUCTURE OF CONE-**1**•Cr(CO)$_3$ AND CONE-**1**•2Cr(CO)$_3$.

The ORTEP drawings of cone-**1**•Cr(CO)$_3$ and cone-**1**•2Cr(CO)$_3$ are shown in Figs. 2 and 3. The single crystal of cone-**1**•2Cr(CO)$_3$ included two similar but slightly different calix[4]arene structures in a 1:1 molar ratio. This explains $Z = 8$ for cone-**1**•2Cr(CO)$_3$ in contrast to $Z = 4$ for cone-**1**•Cr(CO)$_3$.

At the first glance it seems from Figs. 2 and 3 that the structures of the Cr(CO)$_3$ complexes differ appreciably from that of the regular cone conformation. In cone-**1**•Cr(CO)$_3$, a Cr(CO)$_3$-carrying phenyl unit (benzene A) and a phenyl unit (benzene C) distal to benzene A are turned towards the *exo*-annulus direction whereas the two remaining phenyl units (benzenes B and D) are flattened towards the *endo*-annulus direction (Fig. 2). The dihedral angles between the four phenyl rings and the mean plane of the four ArCH$_2$Ar methylene groups are 75.1°, 146.6°, 76.9°, and 144.3° (from A to D). In both structures of cone-**1**•2Cr(CO)$_3$, two Cr(CO)$_3$-carrying phenyl units (benzenes A and C) are turned towards the *exo*-annulus direction whereas the two remaining phenyl units (benzenes B and D) are flattened towards the *endo*-annulus direction. The dihedral angles are 74.6° (76.6°), 152.1° (143.5°), 76.2° (72.4°) and 142.1° (142.7°) (from A to D: the angles in parentheses are those for the second structure). Thus, 1:1 and 1:2 cone-**1**•Cr(CO)$_3$ complexes may be described as having two benzene roofs confronting each other (we call it a "bis-roof" structure for convenience). Cone-calix[4]arenes have a

---

[*] The dihedral angle is defined as follows.

Table II Selected structural data for 1·nCr(CO)₃.

Compound[a]		cone-1·Cr(CO)₃	cone-1·2Cr(CO)₃ [b]		1,3-alternate-1·Cr(CO)₃
Dihedral angle θ, deg.	A	75.1(1)	74.6(1)[76.7(1)]		105.2(1)
	B	146.6(1)	152.2(3)[143.5(3)]		–115.2(2)
	C	76.21(9)	76.2(2)[72.4(2)]		113.3(1)
	D	144.3(2)	142.1(2)[142.7(3)]		–111.5(1)
Bond angle for the ArCH₂Ar methylene groups, deg.	A-CH₂-B	110.2(3)	108.7(5)[106.1(6)]		117.9(5)
	B-CH₂-C	112.3(4)	109.3(6)[106.0(6)]		120.2(4)
	C-CH₂-D	111.4(4)	106.6(6)[106.2(6)]		119.0(5)
	D-CH₂-A	109.6(3)	107.2(5)[107.5(6)]		117.2(4)
Distance, Å	Cr-C(1)	2.246(4)	2.252(6)[2.245(6)]	2.251(6)[2.250(7)]	2.263(4)
	Cr-C(2)	2.248(4)	2.273(6)[2.260(7)]	2.279(6)[2.264(7)]	2.282(5)
	Cr-C(3)	2.200(4)	2.251(7)[2.211(8)]	2.238(6)[2.228(7)]	2.202(6)
	Cr-C(4)	2.209(4)	2.220(6)[2.203(7)]	2.193(7)[2.212(7)]	2.179(6)
	Cr-C(5)	2.237(4)	2.237(6)[2.249(7)]	2.236(7)[2.233(7)]	2.205(6)
	Cr-C(6)	2.271(4)	2.239(6)[2.254(7)]	2.253(7)[2.256(7)]	2.260(4)
Torsion angle, deg.[c]	C_x-Cr-C_g-C(2)	5.7	–0.2[–5.7]	–19.1[–13.7]	14.0
	C_y-Cr-C_g-C(4)	8.0	–4.0[–4.1]	–22.8[–12.8]	11.8
	C_z-Cr-C_g-C(6)	4.9	–2.0[–4.7]	–17.2[–10.5]	11.0
Distance between two p-carbons in benzenes A and C, Å		3.877(6)	3.81(1)[3.72(1)]		6.950(7)
Distance between two p-carbons in benzenes B and D, Å		10.136(7)	10.25(1)[10.11(1)]		7.237(9)

[a]Phenyl rings A-D and carbons x-z correspond to the lettering in shown in Figs. 1, 2 and 3.
[b]Numbers in [ ] are those for the second structure.
[c]C_g means an aromatic center of gravity.

skirt-like shape with an open upper side and a closed lower side. As a result, the four dihedral angles should be larger than 90°: for example, 135.48° and 96.68° for MM3-optimized 25,26,27,28-tetramethoxycalix[4]arene with C_{2v} symmetry [25] and 138°, 94°, 136° and 92° for X-ray-determined 5,11,17,23-tetra-*tert*-butyl-25,26,27,28-tetrakis(ethoxycarbonylmethoxy)calix[4]arene [9].

Why do cone-1·Cr(CO)₃ and cone-1·2Cr(CO)₃ adopt such an unusual "bis-roof" conformation? Careful examination of Figs. 2 and 3 reveals two structural characteristics of these complexes. The first characteristic is related to the position of Cr metal on the benzene ring. In conventional arene-tricarbonylchromium complexes Cr metal occupies the centro-position (C_g) on the benzene ring and the Cr-C_g line is perpendicular to the benzene plane [26-28]. In cone-1·Cr(CO)₃ and cone-1·Cr(CO)₃, on the other hand, Cr atom shifts to the *m*- or *p*-position side: in benzene A of cone-1·Cr(CO)₃ , for example, the distances from Cr metal to 3-, 4- and 5-carbons are 2.20, 2.21 and 2.24 Å, respectively whereas those from Cr metal to 1-, 2- and 6-carbons are 2.25, 2.25 and 2.27 Å, respectively. A similar characteristic is also observed for cone-1•2Cr(CO)$_3$: in benzenes A and C the distances from Cr to 3-carbon are 2.20 (2.19) and 2.20 (2.21) Å, respectively whereas those from Cr to 1- and 2-carbons are 2.25 (2.25) and 2.27 (2.28) Å in benzene A and 2.25 (2.28) and 2.28 (2.26) Å in benzene C. We consider that this shift is induced by steric repulsion between Cr(CO)$_3$ and the propyl group. In fact, van der Waals radii of 4- and 16-carbons are nearly in contact with each other (Figs. 2 and 3). The second characteristic is related to the conformation of the propyl group. In tetrapropoxycalix[4]arenes the propyl groups tend to adopt the most stable *anti-zigzag* conformation. In cone-1•Cr(CO)$_3$ and cone-1•2Cr(CO)$_3$, in contrast, some propyl groups adopt a less stable *gauche* conformation. We consider that this conformation is favorable to relax the steric crowding around the propyl group [29].

Foregoing results support the view that the "bis-roof" structure is brought forth by the Cr(CO)$_3$-enhanced steric crowding on the lower rim. One of the characteristics of this structure is that the *p*-carbon in benzene A is very close to that in benzene C (only 3.88 Å for cone-1•Cr(CO)$_3$ and 3.81 (3.72) Å for cone-1•2Cr(CO)$_3$) whereas two *p*-carbons in benzenes B and D are very far (10.14 Å for cone-1•Cr(CO)$_3$ and 10.25 (10.10) Å for cone-1•2Cr(CO)$_3$). The unusual intramolecular proximity between confronting benzene units may be reflected by some spectral properties of the complexes. We examined the IR spectra (CH$_2$Cl$_2$ solvent) of cone-1•Cr(CO)$_3$ and cone-1•2Cr(CO)$_3$. We found that the $v_{C=O}$ bands for cone-1•Cr(CO)$_3$ (1951 and 1868 cm^{-1}) shift to lower wavenumber than those for cone-1•2Cr(CO)$_3$ (1961 and 1880 cm^{-1}). The shift to lower wavenumber means that the π-basicity in the Cr(CO)$_3$-carrying benzene unit in cone-1•Cr(CO)$_3$ is higher than that in cone-1•2Cr(CO)$_3$[30]. We consider that the difference is induced by the difference in the transannular interaction: that is, in cone-1•Cr(CO)$_3$ benzene A and benzene C act as an electron-acceptor and an electron-donor, respectively and interact with each other transannularly.

The above-mentioned characteristic is well reflected by the ^{1}H NMR spectra in solution (30 °C, CDCl$_3$) [15]. In cone-1•Cr(CO)$_3$, for example, the δ_H values for *m*-H and *p*-H in benzene C shift to higher magnetic field by 0.44 and 0.23 ppm, respectively, from those in tetrapropoxycalix[4]arene. This indicates that these protons move into the shielding area of benzene A. On the other hand, the δ_H values for *m*-H and *p*-H in benzenes B and D shift to lower magnetic field by 0.58 and 0.38 ppm, respectively. This indicates that these protons leave from the shielding area of benzene nuclei. We thus believe that cone-1•Cr(CO)$_3$ and cone-1•2Cr(CO)$_3$ adopt a "bis-roof" structure also in solution.

4. Conclusion

The purpose of the present study was, when it was commenced, regioselective introduction of functional groups into calix[4]arene nuclei. Through the study, however, we noticed that introduction of Cr(CO)$_3$ causes large conformational changes in a calix[4]arene skeleton and without the information on the conformational changes one cannot predict the reactivities of nuclear substitution reactions. The X-ray crystallographic studies revealed that introduction of $Cr(CO)_3$ increases the steric crowding and the conformational changes take place to enable the skeleton to relax the steric crowding. In particular, the "bis-roof" conformation observed for cone-1-$Cr(CO)_3$ complexes is novel and can be classified into neither of the usual four conformations: cone, partial-cone, 1,2-alternate and 1,3-alternate. We believe that the structural information obtained here is useful for predicting the regioselective reaction course and the relative stability and steric crowding of calix[4]arene conformers.

References

1. S. G. Bott, A. Coleman and J. L. Atwood: *J. Chem. Soc., Chem. Commun.*, 610 (1986).
2. S. G. Bott, A. Coleman and J. L. Atwood: *J. Chem. Soc.*, **108**, 1709 (1986).
3. J. L. Atwood, G. W. Orr, F. Hamada, R. L. Vincent, S. G. Bott. and K. D. Robinson: *J. Am. Chem. Soc.*, **113**, 2760 (1991).
4. S. G. Bott, A. Coleman and J. L. Atwood: *J. Inclusion Phenom.*, **5**, 747 (1987).
5. G. D. Andreetti, A. Pochini and R. Ungaro: *J. Chem. Soc., Perkin Trans. 2*, 1773 (1983).
6. R. Ungaro, A. Pochini, G. D. Andreetti and P. Domiano: *J. Chem. Soc., Perkin Trans. 2*, 197 (1985).
7. A. Arduini, A. Pochini, S. Reverberi, R. Ungaro, G. D. Andreetti and F. Ugozzoli, *Tetrahedron*, **42**, 2089 (1986).
8. G. D. Andreetti, R. Ungaro and A. Pochini: *J. Chem. Soc., Chem. Commun.*, 4955 (1979).
9. F. Arnaud-Neu, E. M. Collins, M. Deasy, G. Ferguson, S. J. Harris, B. Kaitner, A. J. Lough, M. A. McKervey, E. Marques, B. L. Ruhl, M. J. Schwing-Weill and E. M. Seward: *J. Am. Chem. Soc.*, **111**, 8681 (1989).
10. P. D. J. Grootenhuis, .P. A. Kollman, L. C. Groenen, D. N. Reinhoudt, G. J. Van Hummel, F. Ugozzoli and G. D. Andreetti, *J. Am. Chem. Soc.*, **112**, 4165 (1990).
11. For a comprehensive review see J. L. Atwood and S. G. Bott, *Top Inclusion Sci.*, **3**, 199 (1991).
12. K. Iwamoto, K. Araki and S. Shinkai: *J. Org. Chem.*, **56**, 4955 (1991).
13. S. Shinkai, K. Fujimoto, T. Otsuka and H. L. Ammon, *J. Org. Chem.*, **57**, 1516 (1992).
14. K. Fujimoto, N. Nishiyama, H. Tsuzuki and S. Shinkai: *J. Chem. Soc., Perkin Trans. 2* 643 (1992).
15. H. Iki, T. Kikuchi and S. Shinkai: *J. Chem. Soc., Perkin Trans. 1*, 2109 (1992). Idem: *ibid.* 205 (1993).
16. For a comprehensive review see A. Solladie-Cavallo: *Polyhedron*, **4**, 901 (1985).
17. E. Langer, H. Lehner: *Tetrahedron*, **29**, 375 (1973).

18. M. Uemura: in *Advances in Metal-Organic Chemistry: Tricarbonyl(η^6-arene)-chromium Complexes in Organic Synthesis*, L. S. Liebeskind, (Eds.), JAI Press, Greenwich (1991) vol. 2, p. 195.
19. M. J. Aroney, M. K. Cooper, P. A. Englert and R. K. Pierens: *J. Mol. Str.*, **77**, 99 (1981).
20. T. Kikuchi, H. Iki, H. Tsuzuki and S. Shinkai: *Supramol. Chem.*, **1**, 103 (1993).
21. P. Main, S. J. Fiske, S. E. Hull, L. Lessinger, G. Germain, J. P. DeClerg and M. M. Woolfson: University of York England (1980).
22. M. C. Burla, M. Camalli, G. Cascarano, C. Giacovazzo, G. Poldori, R. Spagna and D. Viterbo: *J. Appl. Cryst.* **22**, 389 (1989).
23. W. Verboom, S. Datta, Z. Asfari and D. N. Reinhoudt: *J. Org. Chem.* **57**, 5384 (1992).
24. P. D. J. Grootenhuis, P. A. Koolman, D. N. Reinhoudt, G. J. van Hummel, F. Ugozzoli and G. D. Andreetti: *J. Am. Chem. Soc.* **112**, 4165 (1990).
25. T. Harada, J. M. Rudzinski and S. Shinkai: *J. Chem. Soc., Perkin Trans. 2* 2109 (1992).
26. B. Rees and P. Coppens: *Acta Cryst.* **B29**, 2515 (1973).
27. F. van Meurs and H. van Koningskveld: *J. Organomet. Chem.* **131**, 423 (1977).
28. O. L. Carter, A. T. McPhail and G. A. Sim: *J. Chem. Soc. (A)* 822 (1966).
29. The similar shift of Cr to the *p*-position side is also seen for 1,3-alternate-**1**•Cr(CO)$_3$: the distances from Cr to 3-, 4- and 5-carbons are 2.20, 2.18 and 2.21 Å, respectively whereas those to 1-, 2- and 6- carbons are 2.26, 2.28 and 2.26 Å, respectively.
30. H. Ohno, H. Horita, T. Otsubo, Y. Sakata and S. Misumi: *Tetrahedron Lett.* 265 (1977).

Preparation, Structure and Stereodynamics of Phosphorus-Bridged Calixarenes[*]

OLEG ALEKSIUK, FLAVIO GRYNSZPAN and SILVIO E. BIALI[**]
Department of Organic Chemistry, The Hebrew University of Jerusalem, Jerusalem 91904, Israel

(Received: 30 October 1993; in final form: 3 March 1994)

Abstract. The pyrolysis of several dialkylphosphate ester derivatives of calix[4]arenes yielded the same phosphorus bridged compound **7**. Under the pyrolytic conditions the phosphate groups may be cleaved or intermolecularly transferred. X-ray crystallography of the bridged calixarenes **7** and **8** shows that they exist in a chiral 'flattened cone' (fc) conformation. The bridged calixarenes undergo in solution a dynamic process with a barrier of about 10.1 kcal mol^{-1} for **7** and **8** and 13.1 kcal mol^{-1} for **10**, respectively. The dynamic processes result in enantiomerization of the systems. Pyrolysis of partially phosphorylated calix[6]arenes resulted in the formation of two products (**11** and **12**), each consisting of two subunits of three proximal rings bridged by a phosphate group. The rotational barriers for **11** and **12** are 14.4 and 8.8 kcal mol^{-1}, indicating that the bridged calix[6]arene system **12** is appreciably more flexible than **11**.

Key words: Calixarenes, preparation, conformation, dynamics.

Supplementary Data relating to this article are deposited with the British Library as Supplementary Publication No. SUP 82169 (51 pages).

1. Introduction

One attractive feature of the calixarenes is their ability to host small molecules in their molecular cavities [1]. The systems may in principle exist in a large number of conformations. The conformational preferences of a calix[4]arene are usually discussed in terms of four ideal conformations: 'cone', 'partial cone', '1,2-alternate' and '1,3-alternate'. These conformations are interconvertible by rotation of the phenolic rings [1]. The parent *p-tert*-butylcalix[4]arene (**1**) exists in solution in a 'cone' conformation which undergoes a cone to cone inversion process with a barrier (in CDCl$_3$) of 15.9 kcal mol^{-1} [2]. For larger calixarenes, the number of possible ideal conformations increases [1, 3].

One possible approach to improve the binding capabilities of the calixarenes is to freeze them in a conformation with a well defined cavity suitable for hosting small molecules [4]. For calix[4]arenes this can be achieved by replacement of the hydroxylic protons by a bulky group. Since for a calix[4]arene the interconversion

---

[*] This paper is dedicated to the commemorative issue on the 50th anniversary of calixarenes.
[**] Author for correspondence.

between the ideal conformations requires the passage of the OH groups through the ring cavity, the bulky group raises the barrier of ring inversion and freezes the calixarene in a given geometry. By this approach conformationally stable calix[4]arenes at the laboratory timescale were isolated [5]. However, for the larger calixarenes the replacement of the hydroxylic protons by a bulky group is not a sufficient condition to ensure rigidity, since the interconversion between the conformers may take place by the passage of the extraannular part of the phenyl rings (e.g., the *p-tert*-butyl group) through the ring annulus [6].

A second approach to achieving conformational rigidity is to introduce 'bridges' into the calixarene, i.e., to link covalently two or more points of the skeleton by a single atom or a chain of atoms. This severely restricts the possible motion of the calixarene skeleton and may freeze the calixarene in a given conformation. Calixarenes have been bridged at the extraannular positions (*para* to the OH) by an alkane chain [7], or at the intraannular positions, by connecting the oxygens by a chain of atoms [8] or by a single metal atom [9]. Several groups have recently reported the preparation of calixarenes bridged by nonmetallic atoms (phosphorus [10], silicon [11], and carbon [12]). In this paper we describe the preparation, conformations and dynamic processes of *p-tert*-butylcalix[4]arene and *p-tert*-butylcalix[6]arene multiply bridged by trivalent or pentavalent phosphorus atoms [13, 14].

2. Experimental

All NMR spectra were obtained with a Bruker AMX 400 spectrometer. Melting points were determined on a Mel-Temp II apparatus and are uncorrected. The pyrolysis was conducted in a Büchi GKR–51 oven. LDA was purchased from Aldrich. Dry tetrahydrofuran was freshly distilled from Na/benzophenone.

X-RAY CRYSTAL STRUCTURE ANALYSIS

Data were measured on an ENRAF-NONIUS CAD-4 and a PW1100/20 Philips Four-Circle computer controlled diffractometers. CuK_α ($\lambda = 1.54178$ Å) or MoK_α ($\lambda = 0.71069$ Å) radiation with a graphite crystal monochromator in the incident beam was used. Intensities were corrected for Lorentz, polarization and absorption effects. All nonhydrogen atoms were found by using the results of the SHELXS-86 direct method analysis. After several cycles of refinements the positions of the hydrogen atoms were calculated and added to the refinement process.

Crystal data for **7**. $C_{44}H_{52}O_7P_2 \cdot 1.5\ CH_3CN$, space group *Pbca*, a: 20.272(4) Å, b: 33.099(3) Å, c: 14.184(1) Å; V: 9517(1) Å^3, Z: 8, ρ_{calc}: 1.14 g cm^{-3}, $\mu(CuK_\alpha)$: 11.97 cm^{-1}, No. of unique reflections: 6599, No. of reflections with $I \geq 2\sigma_I$: 2666, R: 0.081, R_w: 0.099.

Crystal data for **8**. $C_{44}H_{52}P_2Cl_2$, space group $C2/c$. a: 9.182(2) Å, b: 26.625(3) Å,

c: 17.782(3) Å; β: 90.93(2)°, V: 4347(1) Å^3, Z: 4, ρ_{calc}: 1.19 g cm^{-3}, $\mu(\text{Mo}K_\alpha)$: 2.58 cm^{-1}, No. of unique reflections: 2834, No. of reflections with $I \geq 2\sigma_I$: 2325, R: 0.032, R_w: 0.045.

Calix[4]arene Monospirodienone Bis(diisopropyl phosphate ester)

A solution of 150 mg (0.23 mmol) **3** in 5 mL dry THF was cooled to −78°C, and under an argon atmosphere 0.5 mL of a 1.5 M solution (1 mmol) of LDA was slowly added. After 20 min 0.17 mL (1 mmol) of ClPO(Oi-Pr)$_2$ [26] were added and the mixture stirred for 1 h. The mixture was brought to room temperature and the solvent evaporated. The residue was dissolved in CH$_2$Cl$_2$ and washed with water. After evaporation of the CH$_2$Cl$_2$ the residue was chromatographed (silica, eluent 20 : 1 CHCl$_3$: MeOH) yielding 0.124 g (55%) of a yellow crystalline compound, mp 126–131°C (d).

^{1}H-NMR (400 MHz, CDCl$_3$) δ 0.64 (d, J = 6.1 Hz, 3H, Me), 0.78 (d, J = 6 Hz, 3H, Me), 0.86 (d, J = 6.1 Hz, 3H, Me), 1.04 (d, J = 6.1 Hz, 3H, Me), 1.05 (s, 9H, t-Bu), 1.13 (overlapping d, 12H, Me), 1.19 (s, 9H, t-Bu), 1.30 (s, 9H, t-Bu), 1.32 (s, 9H, t-Bu), 2.85 (d, J = 13.2 Hz, 1H, CH$_2$), 2.95 (d, J = 15.5 Hz, 1H, CH$_2$), 3.27 (d, J = 13.9 Hz, 1H, CH$_2$), 3.53 (d, J = 15.5 Hz, 1H, CH$_2$), 4.04 (d, J = 13.9 Hz, 1H, CH$_2$), 4.30 (s, 2H, CH$_2$), 4.48 (m, 2H, CHMe$_2$), 4.58 (m, 2H, CHMe$_2$), 4.63 (d, J = 12.9 Hz, 1H, CH$_2$), 5.75 (d, J = 2.3 Hz, 1H), 6.78 (d, J = 2 Hz, 1H), 6.96 (d, J = 1.4 Hz, 1H), 6.99 (d, J = 2.3 Hz, 1H), 7.09 (d, J = 2.3 Hz, 1H), 7.14 (d, J = 2.2 Hz, 1H), 7.17 (d, 1H), 7.36 (d, J = 2.3 Hz, 1H). CI MS (isobutane) m/z 975.3 (MH$^+$).

Pyrolysis of the Mono(diisopropyl phosphate ester) Derivative of 3

0.28 g of the phosphate ester were heated under vacuum to 240°C for 10 min. The product consisted of 35% **7** and 42% **1**.

5,11,17,23-Tetra-tert-butyl-25,26-dihydroxy-27,28-bis(diisopropoxy-phosphonoxy) Calix[4]arene (2d)

10 mg of the phosphate ester were refluxed for 15 min in 1.5 mL EtOH in the presence of 2 drops 48% HBr. After evaporation of the EtOH, the product was recrystallized from hexane, yielding 8.5 mg (85%) **2d**, mp 83–87°C.

^{1}H-NMR (400 MHz, CDCl$_3$) δ 0.75 (d, J = 6.2 Hz, 12H, CHMe_2), 0.99 (d, J = 6.2 Hz, 6H, CHMe_2), 1.04 (d, J = 6.1 Hz, 6H, CHMe_2), 1.20 (s, 18H, t-Bu), 1.28 (s, 18H, t-Bu), 3.58 (d, J = 14.9 Hz, 2H, CH$_2$), 3.64 (s, 2H, CH$_2$), 4.41 (m, 6H, CH$_2$ + CHMe$_2$), 4.48 (d, J = 15.0 Hz, 2H, CH$_2$), 6.98 (d, J = 2.3 Hz, 2H, Ar—H), 7.02 (d, J = 2.3 Hz, 2H, Ar—H), 7.15 (d, J = 2.5 Hz, 2H, Ar—H), 7.41 (d, J = 2.4 Hz, 2H, Ar—H). ^{13}C-NMR (100 MHz CDCl$_3$, RT) δ 22.77, 22.81, 22.87, 23.31, 23.36, 23.39, 23.45, 23.67, 29.68, 31.45, 31.54, 31.71, 33.87, 34.25, 35.40,

37.73, 73.11, 125.05, 125.32, 126.84, 127.12, 127.76, 128.93, 132.32 (d), 132.80 (d), 142.89, 144.71 (d), 147.51, 148.65.

5,11,17,23,29-Penta-tert-butyl-31,32,33,34,35-penta(diisopropoxyphosphonoxy) Calix[5]arene (**5**)

2.0 g of **4** (2.47 mmol), 0.3 g tetrabutyl ammonium bromide, and 10 mL diisopropylchlorophosphate were dissolved in 100 mL CH_2Cl_2 and 113 g of a solution of aqueous NaOH (50%) was slowly added. The mixture was refluxed and stirred for 6 h. After carefully adding 400 mL water, the phases were separated and the organic phase was washed with water and dried ($CaCl_2$). After evaporation of the organic solvent the residue was triturated with cold petroleum ether, yielding, after filtration, 2.94 g (73%) **5**, mp 225–230°C.

^{1}H-NMR (400 MHz, $CDCl_3$) δ 0.98 (s, 45H, t-Bu), 1.21 (d, 15H, J = 6.1 Hz, CH*Me*$_2$), 1.33 (d, 15H, J = 6.1Hz, CH*Me*$_2$), 3.47 (d, 5H, J = 15.3 Hz, CH_2), 4.71 (m, 20H, CH_2 and OC*H*Me$_2$), 6.94 (s, 10H, Ar—H). ^{13}C-NMR (100 MHz, $CDCl_3$) δ 23.63 (d, $^3J_{P-C}$ = 5.6 Hz, OCHMe*Me*), 23.82 (d, $^3J_{P-C}$ = 4.4 Hz, CH*Me*Me), 30.41, 31.35 (C*Me*$_3$), 34.00, 73.22 (d, $^2J_{P-C}$ = 5.9 Hz, OCHMeMe), 125.79, 133.26 (d, J_{P-C} = 3 Hz), 143.52, 146.06. CI MS (isobutane) 1632.3 (MH^+).

Preparation of **7** *by Pyrolysis. General Procedure*

0.39 g of phosphorylated calixarene were heated to 230°C under vacuum for 35 min. During the heating period the compound first melted and then solidified and a liquid distilled. The compound was chromatographed (eluent: $CHCl_3$) yielding the product, mp 375–376.5°C. The yields obtained from the pyrolysis of the phosphorus-containing calixarenes were 32% (**2b**), 38% (**2c**), 72% (**2d**).

^{1}H-NMR, (400 MHz, RT, $CDCl_3$) δ 1.18 (s, 36H, t-Bu), 3.42 (d, 2H, J = 13.7 Hz, CH_2), 3.61 (d, 2H, J = 15.6 Hz, CH_2), 4.48 (d, 2H, J = 13.6 Hz, CH_2), 4.55 (d, 2H, J = 15.5 Hz, CH_2), 6.99 (d, 2H, J = 2.1 Hz, Ar—H), 7.05 (d, J = 2.1 H_3, 2H, Ar—H). ^{1}H-NMR (400 MHz, 203 K, CD_2Cl_2) δ 1.03 (s, 18H, t-Bu), 1.20 (s, 18H, t-Bu), 3.48 (d, J = 13.8 Hz, 2H, CH_2), 3.68 (d, 2H, J = 15.6 Hz, CH_2), 4.37 (d, 2H, J = 13.5 Hz, CH_2), 4.43 (d, 2H, J = 15.5 Hz, CH_2), 7.00 (s, 2H, Ar—H), 7.04 (s, 2H, Ar—H), 7.14 (s, 2H, Ar—H), 7.32 (s, 2H, Ar—H). ^{13}C-NMR (100 MHz, RT, $CDCl_3$) δ 31.19, 31.67, 34.27, 34.71, 125.91, 126.53, 128.44, 130.89 (t), 146.39 (t), 148.67. ^{31}P-NMR δ -24.11. FAB MS: (sample dissolved in MeOH) m/z 785.9 $[M-H+MeOH]^-$, (sample dissolved in EtOH) m/z 799.8 $[M-H+EtOH]^-$.

5,11,17,23-Tetra-tert-butyl-µ-25,26-chlorophosphite-µ-27,28-chlorophosphite Calix[4]arene (**8**)

A mixture of 0.25 g **1**, 1 mL Et_3N and 8 mL PCl_3 were refluxed under a nitrogen atmosphere for 20 min. The mixture was poured into ice, and the white precipitate that resulted was filtered and washed with water yielding 0.28 g **8** (93%), mp 164–166°C (decompose).

^{1}H-NMR (400 MHz, 200 K, CD$_2$Cl$_2$) δ 0.83 (s, 18H, t-Bu), 1.09 (s, 18H, t-Bu), 3.31 (d, 2H, J = 13.6 Hz, CH$_2$), 3.38 (d, 2H, J = 15.1 Hz, CH$_2$), 4.22 (d, 2H, J = 13.5 Hz, CH$_2$), 4.66 (d, 2H, J = 15.0 Hz, CH$_2$), 6.72 (s, 2H, Ar—H), 6.78 (s, 2H, Ar—H), 6.90 (s, 2H, Ar—H), 7.20 (s, 2H, Ar—H), ^{1}H-NMR (400 MHz, RT, CDCl$_3$) δ 1.09 (s, 36H, t-Bu), 3.37 (d, 2H, J = 15.0 Hz, CH$_2$), 3.40 (d, 2H, J = 14.9 Hz, CH$_2$), 4.43 (d, 2H, J = 14.9 Hz, CH$_2$), 4.80 (d, 2H, J = 14.8 Hz, CH$_2$), 6.81 (s, 4H, Ar—H), 6.86 (s, 4H, Ar—H). ^{13}C-NMR (100 MHz, CDCl$_3$) δ 31.09, 31.26, 34.04, 38.62, 124.99, 126.72, 130.71, 132.24, 146.92 (t, J = 7 Hz) 147.07. ^{31}P-NMR (161.99 MHz, CDCl$_3$) δ 149.2 ppm. CI MS (isobutane): m/z 723.2.

5,11,17,23-Tetra-tert-butyl-25-hydroxy-26-dichlorophosphate-μ-27,28-chlorophosphate Calix[4]arene **(9)**

A solution of 2 mL POCl$_3$ in 20 mL CHCl$_3$ was added under a nitrogen atmosphere to a refluxing solution of 2 g **1** and 5 mL Et$_3$N dissolved in 150 mL CHCl$_3$. After reflux for 3 h the solution was washed with brine, dried (CaCl$_2$) and evaporated. The residue was triturated with 50 mL hot hexane and filtered, yielding 0.79 g (30%) of **9** as a white powder, mp 297–298°C.

^{1}H-NMR (400 MHz, CDCl$_3$) δ 0.90 (s, 9H), 1.18 (s, 9H), 1.34 (s, 9H), 1.36 (s, 9H), 3.49 (d, J = 12.8 Hz, 1H), 3.56 (d, J = 15.8 Hz, 1H), 3.76 (d, J = 15.6 Hz, 1H), 3.78 (s, 1H, OH), 3.93 (d, J = 15.6 Hz, 1H), 4.07 (d, J = 15.5 Hz, 1H), 4.18 (d, J = 14.2 Hz, 1H), 4.31 (d, J = 15.1 Hz, 1H), 4.64 (d, J = 15.7 Hz, 1H), 6.43 (br, 1H), 6.96 (d, J = 2.5 Hz, 1H), 7.06 (br, 1H), 7.14 (d, J = 2.2 Hz, 1H), 7.17 (d, J = 2.2 Hz, 1H), 7.26 (m, 1H), 7.30 (m, 2H). ^{13}C-NMR (100 MHz, CDCl$_3$) δ 30.84, 31.17, 31.42, 31.67, 32.17, 33.91, 34.02, 34.17, 34.40, 35.73, 37.27, 37.50, 125.44, 125.90 (d), 126.64 (d), 126.96 (d), 127.19, 127.44, 127.76, 128.21, 128.23, 128.33 (d), 130.01 (d), 130.17 (d), 132.10 (d), 132.52 (d), 142.40, 144.02 (d), 146.22 (d), 148.25 (d), 148.58 (d), 148.80 (d), 149.10, 151.77. ^{31}P-NMR δ 0.6, 3.2 ppm. MS (CI) m/z 845.2 (MH$^+$) *Microanalysis: Calcd.* for C$_{44}$H$_{53}$O$_6$Cl$_3$P$_2$ C: 62.45, H: 6.31, Cl: 12.57; *Found*: C: 62.74, H: 6.51, Cl: 12.96.

5,11,17,23-Tetra-tert-butyl-μ-25,26-chlorophosphate-μ-27,28-chlorophosphate Calix[4]arene **(10)**

A mixture of 1 g **1**, 5 mL Et$_3$N and 1 mL POCl$_3$ in 80 mL xylene was heated to 85°C for 30 min. The sample was concentrated by distilling about 30 mL of solvent at atmospheric pressure. The solution was washed with water and evaporated. The resulting solid was treated with 50 mL hot MeCN and filtered yielding 0.12 g **10** as a white powder (10%), decomposing without melting at 365–380°C.

^{1}H-NMR (400 MHz, RT, CDCl$_3$) δ 1.09 (broad, 36H, t-Bu), 3.51 (d, J = 14.8 Hz, 2H, CH$_2$), 3.54 (d, J = 16.2 Hz, 2H, CH$_2$), 4.51 (d, J = 14.8 Hz, 2H, CH$_2$), 5.06 (d, J = 16.1 Hz, 2H, CH$_2$), 6.79 (br, 8H, Ar—H). ^{1}H-NMR (400 MHz, 245 K, CDCl$_3$)

δ 0.77 (s, 18H, t-Bu), 1.36 (s, 18H, t-Bu), 3.51 (d, J = 14.8 Hz, 2H, CH_2), 3.56 (d, J = 16.4 Hz, 2H, CH_2), 4.44 (d, J = 14.8 Hz, 2H, CH_2), 4.93 (d, J = 16.2 Hz, 2H, CH_2), 6.11 (s, 2H, Ar—H), 6.45 (s, 2H, Ar—H), 7.10 (s, 2H, Ar—H), 7.27 (s, 2H, Ar—H). ^{13}C-NMR ($CDCl_3$, RT) δ 29.85, 31.26, 34.08, 37.85, 125.5 (br), 128.12, 130.27, 147.1 (d), 147.65. ^{31}P-NMR δ 1.0 ppm. MS (CI) m/z 809.3 (MH^+).

5,11,17,23,29,35-Hexa-tert-butyl-μ-37,38,39-phosphate-μ-40,41,42-phosphate Calix[4]arene (**11** and **12**)

1.25 g of the product obtained by the partial phosphorylation of **6** [20] was heated at 330°C under vacuum. The residue was recrystallized from MeCN. The first batch of crystals collected (6.4%) was **12**, mp > 450°C, the second fraction corresponded to **11** (47.4%) mp > 440°C.

11. ^{1}H-NMR (400 MHz, RT, $CDCl_3$) δ 1.13 (br, 36H, t-Bu), 1.27 (s, 18H, t-Bu), 3.66 (d, J = 14.1 Hz, 4H, CH_2), 3.93 (d, J = 17.8 Hz, 2H, CH_2), 4.62 (d, J = 17.8 Hz, 2H, CH_2), 4.72 (d, J = 14.2 Hz, 4H, CH_2), 6.77 (s, 4H, Ar—H), 7.13 (br, 4H, Ar—H), 7.24 (br, 4H, Ar—H). ^{31}P-NMR δ -22.35 ppm. MS *Calcd.* for $C_{66}H_{78}P_2O_8$ 1060.5; *Found*: 1060.

12 ^{1}H-NMR (400 MHz, RT, $CDCl_3$) δ 1.21 (s, 36H, t-Bu), 1.28 (s, 18H, t-Bu), 3.71 (d, J = 14 Hz, 4H, CH_2), 4.29 (s, 4H, CH_2), 4.69 (d, J = 14.1 Hz, 4H, CH_2), 6.82 (s, 4H, Ar—H), 7.22 (d, J = 2.2 Hz, 4H, Ar—H), 7.28 (s, 4H, Ar—H). ^{31}P-NMR δ -22.27 ppm. MS (CI) m/z 1061 (MH^+).

3. Results and Discussion

3.1. Preparation of Calixarene Phosphate Ester Derivatives

For the preparation of multiply phosphorus bridged systems, we studied the pyrolytic behavior of calixarene dialkyl phosphate ester derivatives. These compounds are useful intermediates for the OH-depletion of the calixarenes [15, 16] and for the preparation of aminocalixarenes [17]. These esters can be prepared by treatment of the calixarenes with a dialkyl chlorophosphate in the presence of base [18]. In the case of **1**, under mild conditions, the distal (i.e, 1,3-) bis(diethyl phosphate ester) derivative is obtained (**2a**) while under more drastic conditions (CH_2Cl_2/aq NaOH, phase transfer catalysis) the tetraphosphate **2b** is formed [15a]. Calixarene mono diisopropyl phosphate ester and 1,2-bis(diisopropyl phosphate ester) derivatives (**2c** and **2d**) were prepared by mono- or dideprotonation by LDA of the monospirodienone derivative **3** followed by treatment with the corresponding dialkyl chlorophosphate [15e, 17b]. Aromatization of the spirodienone products was achieved by heating or by their treatment with HBr yielding the dialkylphosphate esters derivatives **2c** and **2d** [15e, 17b].

p-tert-Butylcalix[5]arene [19] (**4**) was phosphorylated by its treatment with $ClPO(Oi-Pr)_2$ in the presence of base, yielding the pentaphosphate **5**. The compound displays a single *tert*-butyl signal, one pair of doublets for the methylene groups, and a single aromatic signal, indicating that at the NMR timescale the molecule exists in a symmetric cone-like conformation. The two methyl groups within a given isopropyl groups are symmetry nonequivalent and should be diastereotopic. Indeed, two doublets and two signals are observed for the isopropyl methyls of **5** in the ^{1}H- and ^{13}C-NMR spectrum, respectively.

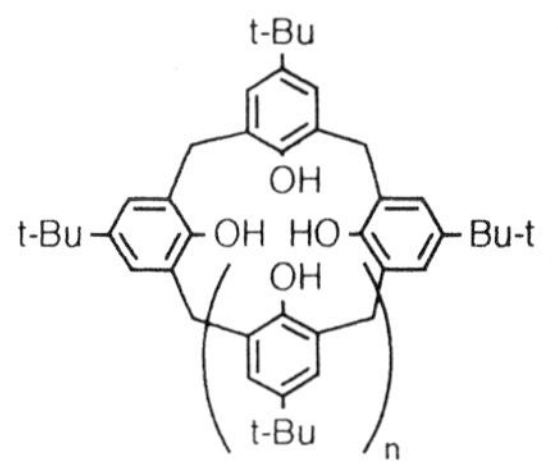

1 n=1
4 n=2
6 n=3

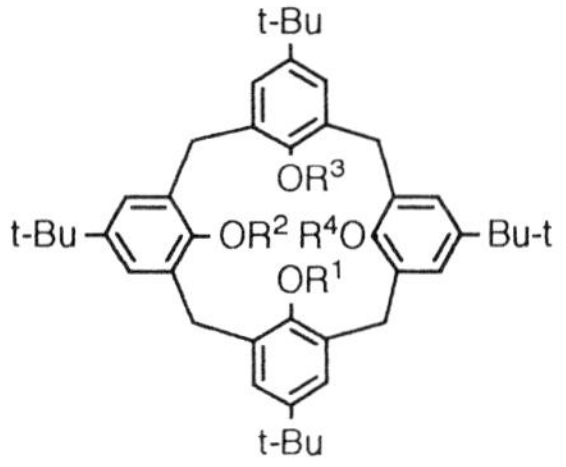

2a R^1=R^3=H, R^2=R^4=PO(OEt)$_2$
2b R^1=R^2=R^3=R^4= PO(OEt)$_2$
2c R^1=R^2=R^3=H, R^4=PO(OPr-i)$_2$
2d R^1=R^2=H, R^3=R^4=PO(OPr-i)$_2$

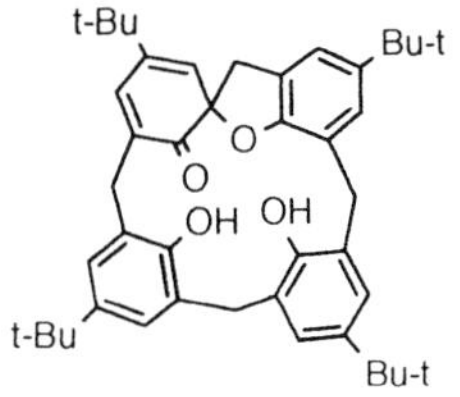

3

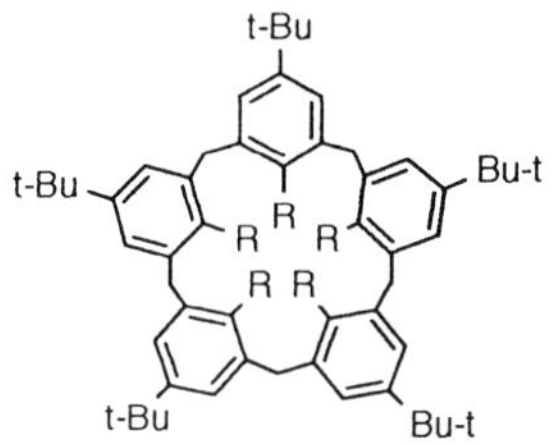

5 R = OPO(OPr-i)$_2$

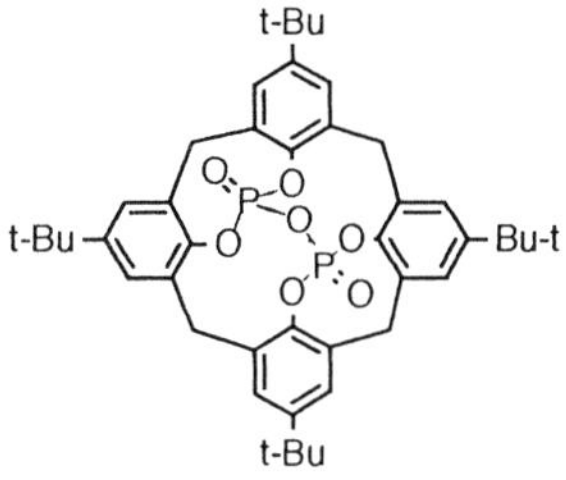

7

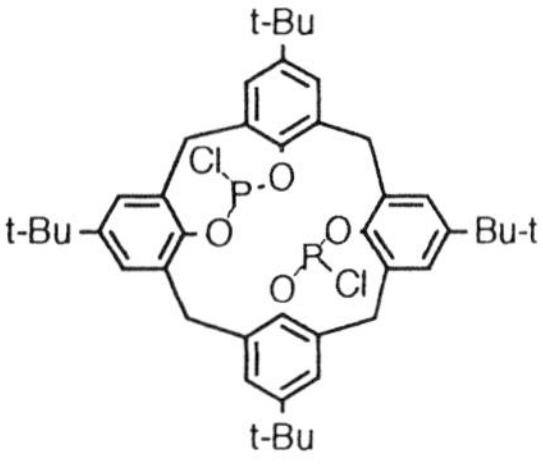

8

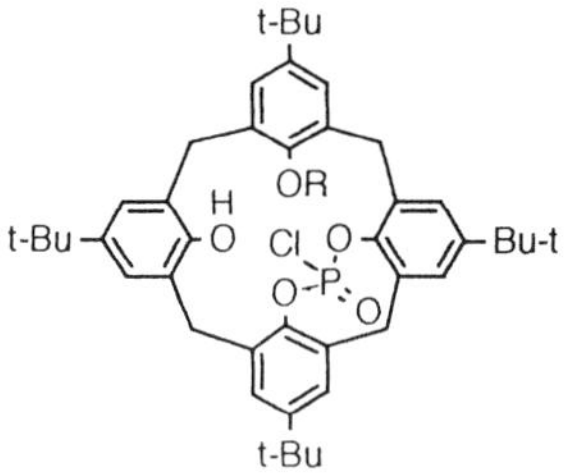

9 R = POCl$_2$

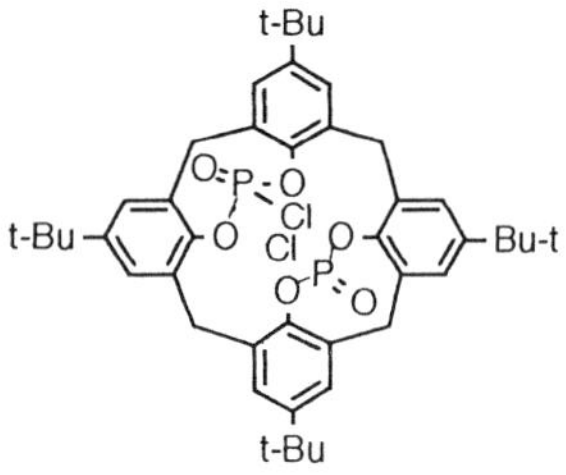

10

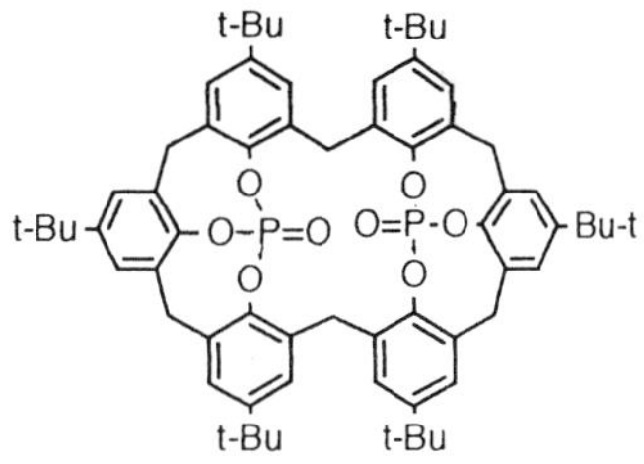

11 syn
12 anti

The partial phosphorylation of *p-tert*-butylcalix[6]arene (**6**) was first reported by Markovsky *et al.* [20] and recently studied by three groups [21]. We carried out the partial phosphorylation of **6** according to the literature procedure [20] and used the isolated product for the pyrolytic studies.

3.2. PREPARATION OF PHOSPHORUS BRIDGED CALIX[4]ARENES BY PYROLYSIS OR DERIVATIZATION

Vacuum pyrolysis of the phosphate ester derivatives was carried out by heating them above their melting point. Usually, the product solidified at the high temperature. Pyrolysis of the bis(diethyl phosphate ester) derivative **2a** (230°C) for 35 min yielded a main product of mp 375.5–376.5°C in 35% yield (Scheme 1) together with the formation of **1**. The product was characterized by its spectroscopic data and by X-ray crystallography as the pyrophosphate **7** (see below). The system contains two phosphorus bridges, each spanning over the oxygens of two proximal rings with an additional oxygen bridge connecting the two phosphorus atoms.

Since pyrolysis of the distal bis(dialkyl phosphate ester) derivative of **1** yielded the pyrophosphate **7**, it was of interest to examine also the pyrolytic behavior of other dialkyl phosphate ester derivatives of **1**. We examined first the pyrolysis of the tetraphosphate **2b**. The product obtained (Scheme 1) was pyrophosphate **7**, indicating that under the reaction conditions two phosphate groups were cleaved. Pyrolysis of the proximal diphosphate ester **2d** also gave **7**. We then examined the pyrolytic behavior of the mono(dialkyl phosphate ester) derivative **2c**. We expected that its pyrolysis will result in the formation of a phosphorus monobridged system, however, the product of the reaction (Scheme 1) was again pyrophosphate **7**, accompanied by **1**. Since the starting material has only a single phosphorus atom, while the product has two, obviously an intermolecular transfer of a phosphate ester group took place under the reaction conditions. Pyrolysis of the mono(diisopropyl phosphate ester) derivative of the spirodienone **3** resulted also in the formation of **7**. The spirodienone derivative has an interesting thermal behavior: at a lower temperature (170°C, 10 min), the spirodienone moiety is reduced to two phenol groups and **2c** is obtained, while at a higher temperature (240°C, 10 min) **7** is the product, most likely through the intermediacy of **2c**. Heating a mixture of **5** and **1** *in vacuo* for 15 min to 240°C resulted also in the exclusive formation of **7**, providing an additional evidence that under the reaction conditions intermolecular phosphate ester transfers take place. The pyrolytic studies are summarized in Scheme 1. It can be concluded that irrespectively of the starting calix[4]arene phosphate ester derivative, **7** is the bridged product formed. Its formation may involve fragmentation or intermolecular migrations of the phosphate groups. The multiple bridging present may account for the high stability of the compound.

The relatively facile formation of the phosphorus-bridged system led us to examine also the preparation by simple derivatization of **1** of systems with trivalent and pentavalent phosphorus bridges. Reaction of **1** with PCl_3 in the presence of base gave the bis P(III) bridged system **8**. In contrast with the kinetic stability of pyrophosphate **7**, this compound readily undergoes hydrolysis. For example, attempted recrystallization of **8** from MeOH regenerated **1**.

Reflux of a $CHCl_3$ solution of **1** and $POCl_3$ in the presence of Et_3N yielded the monobridged system **9**. Higher reaction temperatures (xylene, 85°C) yielded the

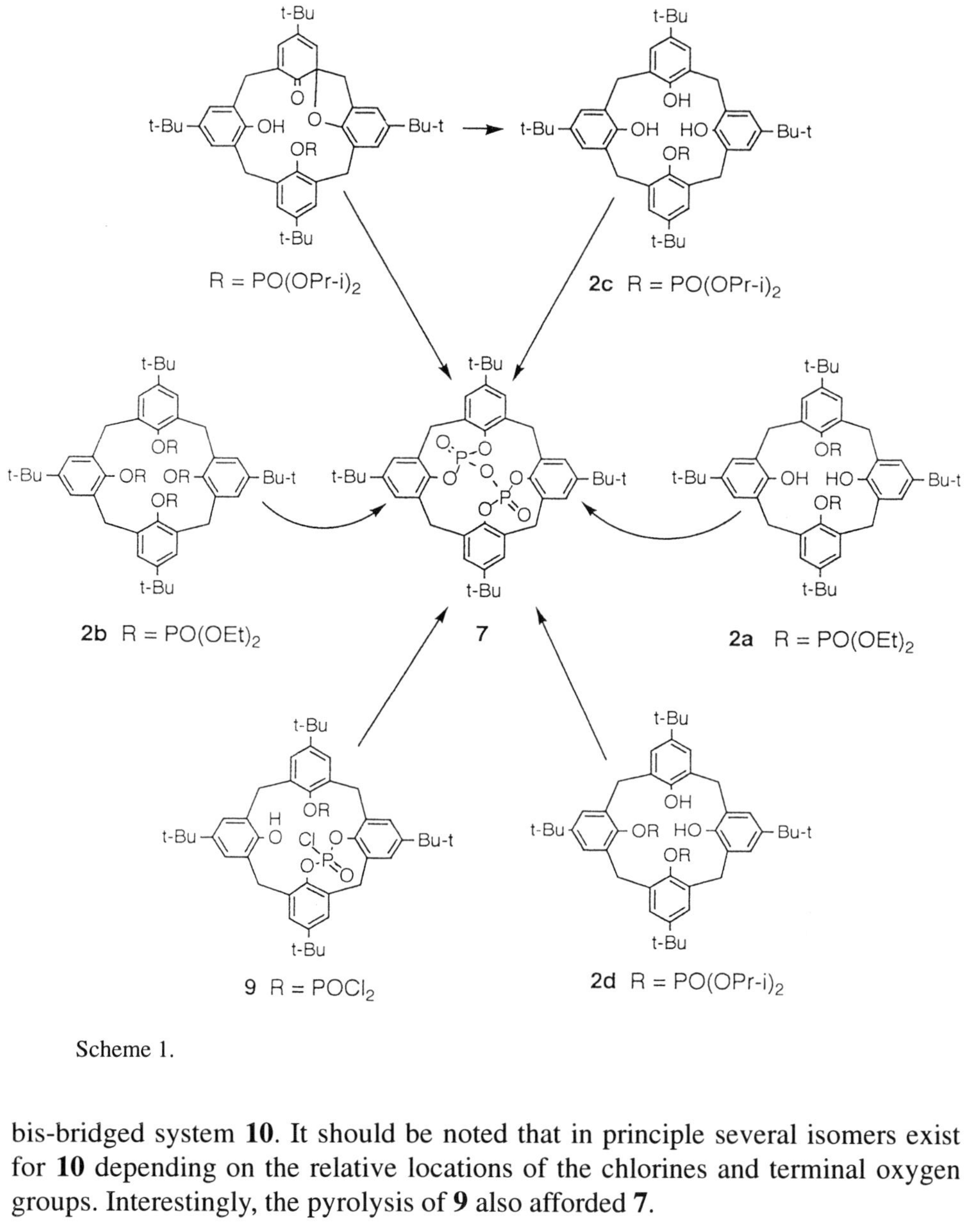

Scheme 1.

bis-bridged system **10**. It should be noted that in principle several isomers exist for **10** depending on the relative locations of the chlorines and terminal oxygen groups. Interestingly, the pyrolysis of **9** also afforded **7**.

3.3. SOLID STATE CONFORMATION OF **7** AND **8**

Bridging two proximal rings by a single atom severely limits the possible conformations of the moiety. It was therefore of interest to determine the solid state conformation of the systems by X-ray crystallography. A single crystal of **7** suitable for X-ray crystallography was grown from a MeCN solution. The molecule crys-

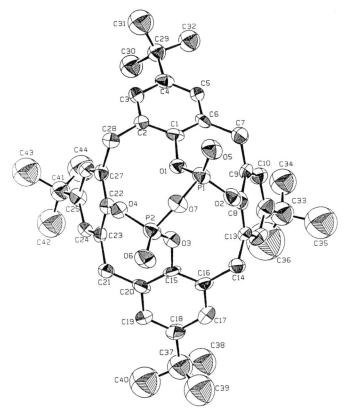

Fig. 1. Numbering scheme of the molecular structure of **7**. MeCN molecules are omitted.

tallizes with 1.5 molecules of MeCN. The numbering scheme and a stereoscopic view of the molecular structure are displayed in Figures 1 and 2. As shown in the figures, the two terminal P=O and the bridging P—O—P oxygen atoms (O(6), O(5) and O(7)), respectively) are pointing outside the molecular cavity. The dihedral angles between the plane of the rings and the mean plane defined by the four methylene carbons are 40° for the ring defined by the carbons C(1)—C(6), 66° for C(8)—C(13), 29° for C(15)—C(20), and 69° for C(22)—C(27). The conformation of the rings clearly departs from an ideal cone, with two nonvicinal rings being more twisted from the macrocyclic plane than the other two. This conformation sometimes called 'pinched' [22a] or 'boat' [22b] will be designated in the present paper as a 'flattened cone' (fc). For identification purposes the two rings with the larger torsional angle (C(8)—C(13) and C(22)—C(27)) will be dubbed 'twisted rings', whereas the rings with the smaller torsional angle will be dubbed 'coplanar rings'. Although the crystallographic conformation has strictly C_1 symmetry, for the bridged system this conformation should have ideally C_2 symmetry.

A single crystal of the bridged calixarene **8** was grown from hexane. The compound crystallizes in the $C2/c$ space group, with the center of the molecule located

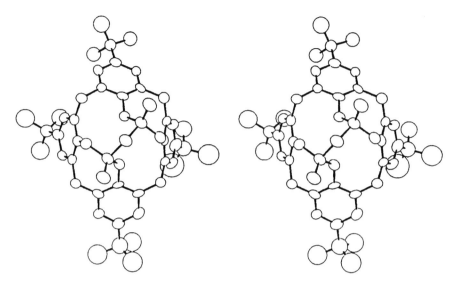

Fig. 2. Stereoscopic view of the crystal structure of **7**.

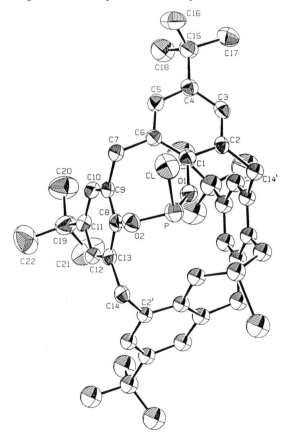

Fig. 3. Numbering scheme of the molecular structure of **8**.

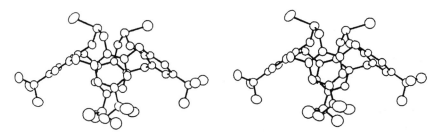

Fig. 4. Stereoscopic view of the crystal structure of **8**.

TABLE I. Selected bond distances and angles[a] for **8**.

Cl—P	2.095 (1)	P—O(1)	1.606(2)
P—O(2)	1.617(2)	O(1)C(1)	1.407(2)
O(2)C(8)	1.415(3)	Cl—P—O(1)	102.2(1)
Cl—P—O(2)	101.7(1)	O(1)—P—O(2)	102.8(1)
P—O(1)C(1)	135.2(2)	P—O(2)C(8)	124.2(1)
O(1)C(1)C(2)	115.5(2)	O(1)C(1)C(6)	121.8(2)
C(6)C(7)C(9)	112.4(2)	C(2′)C(14)C(13)	108.9(2)

[a] Bond length distances in Å, angles in degrees.

in the crystallographic C_2 axis. The compound also exists in a "fc" conformation very similar to that found for **7** but with exact C_2 symmetry. An ORTEP picture and a stereoscopic view of the structure are displayed in Figures 3 and 4. Selected bond lengths and angles are collected in Table I. The dihedral angles of the rings with the mean plane defined by the four methylenic carbons are 36° for C(1)—C(6) and 87° for C(8)—C(13). As shown in the figures, the two chlorine atoms are pointing away from the center of the molecule, while the two lone pairs at the phosphorus atoms are pointing to the C_2 axis. This feature makes the bridged calixarene **8** a potential bidentate ligand.

3.4. Room Temperature NMR Spectra of the Double Bridged Calix[4]arenes

Since the main difference between compounds **7**, **8** and **10** is in the nature of the phosphorus bridges it was interesting to examine the ^{31}P-NMR of the compounds. The calixarenes **7**, **8** and **10** display ^{31}P–NMR signals at δ -22.3, 149.2 and 1.0 ppm, respectively. The chemical shift of the ^{31}P-signal of **8** is in agreement with the shift observed for other trivalent phosphorus compounds [23]. The monobridged calixarene **9** displays two ^{31}P-signals of δ 0.6 and 3.2 ppm. In the nonbridged diphosphate ester **2a** the phosphorus atoms resonate at -6.4 ppm.

The ^{1}H-NMR spectra of compounds **7**, **8** and **10** show a similar pattern of signals. Compounds **7** and **8** display (400 MHz, CDCl$_3$, RT) one *tert*-butyl signal, four doublets for the methylene protons and two doublets for the aromatic protons. The ^{13}C-NMR (100 MHz, CDCl$_3$, RT) is relatively simple with four aliphatic and six aromatic signals for each molecule. The ^{1}H- and ^{13}C-NMR spectra display fewer signals than those expected for a frozen "fc" conformation, and indicate that at room temperature the molecules are conformationally flexible on the NMR timescale. Notably, although the bridges limit the motions of the phenyl rings, they do not freeze completely their internal rotation.

Calixarene **10** displays in the ^{1}H-NMR (400 MHz, CDCl$_3$, RT) two extensive broad signals for the *tert*-butyl and aromatic protons, and two pairs of doublets for the methylene signals, indicating that a dynamic process is taking place on the NMR timescale. Upon raising the temperature the t-Bu and aromatic signals became sharper. As in the case of **7** and **8**, the fast exchange NMR spectrum is in agreement with a species of dynamic C_{2v} symmetry. Lowering the temperature resulted in separation of the broad signals into two signals each. Based on the slow exchange NMR it is reasonable to assume that the system exists also in a fc conformation of C_2 symmetry. However, we cannot conclude whether the two symmetry-related chlorine atoms are both facing each other or not. The monobridged system **9** displays four t-Bu signals, eight doublets for the methylene signals, an OH signal at δ 3.77 ppm, and eight signals for the aromatic protons, in agreement with a frozen conformation of C_1 symmetry.

3.5. DYNAMIC NMR SPECTRA OF DOUBLE BRIDGED CALIX[4]ARENES

Upon lowering the temperature of a CD$_2$Cl$_2$ solution of **7** and **8** the ^{1}H-NMR spectrum changed. The single *tert*-butyl signal separated into two signals and the two aromatic signals into two pairs of signals while the methylene doublets remained unchanged. The ^{1}H-NMR spectrum of **7** at different temperatures is shown in Figure 5. The slow exchange NMR spectra for both compounds are in agreement with a fc conformation of C_2 symmetry. Coalescence processes were observed for the two pairs of aromatic protons and for the *tert*-butyl groups. Exchange rates at the coalescence temperatures were calculated using the Gutowsky–Holm equation [24]. The coalescence data for the compounds are collected in Table II. As shown in the table, the barriers measured from the coalescence of the different groups (t-Bu or aromatic protons) are identical within experimental error, indicating that the same process is being monitored at different parts of the molecule. Interestingly, the dynamic processes of **7** and **8** have barriers (10.1 kcal mol^{-1}) which are identical within experimental error. This indicates that the presence of an additional oxygen atom in **7** which connects the two phosphorus bridges does not result in an increase in the rigidity of the molecule. The rotational barrier of **10** is larger (13.1 kcal mol^{-1}), the higher barrier being probably the result of the steric inter-

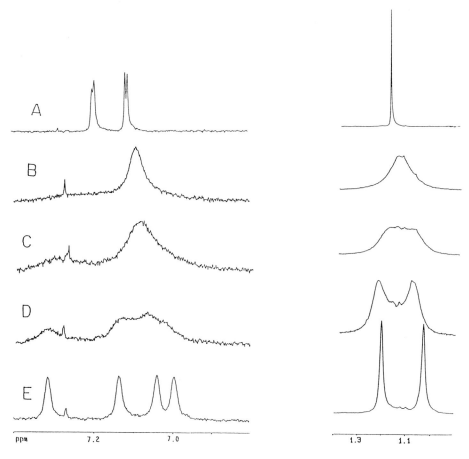

Fig. 5. 400 MHz ¹H-NMR spectrum of **7** at different temperatures. Left: aromatic region; right: *tert*-butyl region. A: 295 K, B: 217 K, C: 212 K, D: 209 K, E: 203 K.

ference in the transition state of the two groups in the bridges (Cl or O) which are facing each other.

The dynamic process which the systems **7, 8** and **10** undergo *is not* a ring inversion process since, even under fast exchange conditions, the two methylene protons within a given methylene group remain diastereotopic. The process observed must involve the rotation of the rings and should exchange the coplanar and twisted rings but without involving their passage through the macrocyclic plane. A possible dynamic process which is in agreement with the observed data is depicted in Figure 6, according to which the process results in the enantiomerization of the system, and therefore the measured barriers represent their enantiomerization barriers. Inspection of Dreiding models shows that in the case of **7**, the rotation of the rings in a given Ar—O—P—O—Ar bridge is accompanied by ca. 120° rotations about the two O(P)—O bonds. As viewed from the phosphorus, both

TABLE II. Coalescence data for the phosphorus bridged calixarenes.

Compound	Probe	$\Delta\nu$ (Hz)	T_c (K)	$\Delta G_c^{\neq}$ (kcal mol^{-1})
7[a]	t-Bu	68.5	212	10.1
	Ar—H	56.1	209	10.1
	Ar—H	243.7	217	10.2
8[a]	t-Bu	104.3	217	10.2
10[b]	t-Bu	233.4	285	13.1
11[c]	CH$_2$	108.7	303	14.4
	CH$_2$	111.0	303	14.4
12[d]	Ar—H	35.5	180	8.8
	Ar—H	85.8	174	8.4
	CH$_2$	108.7	180	8.7
	CH$_2$	68.5	180	8.8

[a] In CD$_2$Cl$_2$.
[b] In CDCl$_3$.
[c] In toluene-d_8.
[d] In CDCl$_2$F.

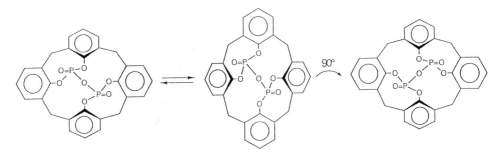

Fig. 6. Fc $\rightleftharpoons$ fc interconversion in **7**. The process results in an exchange of the twisted and coplanar rings. As is readily seen by 90° rigid body rotation of the central structure, the process results in enantiomerization.

rotations must occur in the same direction (clockwise or counterclockwise), the sense of rotation depending on the enantiomer undergoing the internal rotation. For example, for the enantiomer depicted in Figure 1, these rotations must be in a clockwise fashion. The experimental data do not allow us to decide whether the rotational process involves a single-step correlated rotation of the four rings or whether the rotation takes place through the intermediacy of a high energy form undetectable by NMR.

3.6. PYROLYSIS OF PARTIALLY PHOSPHORYLATED CALIX[6]ARENES

The multiple bridging of *p-tert*-butyl-calix[6]arene **6** is of special interest since, in contrast to **1**, the introduction of bulky groups in the intraannular positions is not sufficient for the elimination of the ring inversion process at the laboratory timescale [6]. Pyrolysis of the product obtained by the partial phosphorylation of **6** *in vacuo* at 330°C for 1.5 h resulted in the formation of a major product (**11**, 47%) and a minor product (**12**, 6.4%). Both compounds are high melting (mp > 440°C) and were characterized as multiple bridged calixarenes based on their mass spectral and spectroscopic properties. The phosphorus atoms of **11** and **12** resonate in the ^{31}P-NMR (CDCl$_3$, RT) at δ -20.35 and -20.38 ppm, respectively, indicating that both structures are closely related. Both compounds display in CI MS molecular peaks at m/z 1061 (MH$^+$), in agreement with structures of general formula $C_{66}H_{78}P_2O_8$. The observed molecular mass is in agreement with structures in which two P=O moieties span over three neighboring phenoxy groups.

3.7. STEREOCHEMISTRY OF THE MULTIPLY BRIDGED CALIX[6]ARENES

The attachment of three phenol groups to a phosphorus atom severely restricts the conformations possible for the system. In discussing the stereochemistry of the bridged derivatives of **6** two factors should be considered. Firstly, the bridging phosphate group may exist with the terminal P=O group oriented *endo* or *exo* to the molecular cavity. Secondly, the two subunits may exist with their cavities arranged in a *syn* or *anti* fashion.

The crystal structure of **11** was described in our preliminary communication [13]. The molecule adopts a conformation in which each of the two OP(OAr)$_3$ subunits exists in a 'flattened cone' conformation with one ring nearly coplanar to the reference plane and the two remaining rings twisted. The two P=O bonds are oriented in a parallel fashion and *exo* to the macrocyclic plane.

3.7.1. *NMR Spectra*

The ^{1}H-NMR spectrum of **11** (400 MHz, RT, CDCl$_3$) displays at room temperature both a sharp and a broad signal in the *tert*-butyl region, one pair of sharp doublets and one pair of somewhat broad doublets for the methylene groups, and one sharp singlet and two broad signals for the aromatic protons. The presence of some broad signals in the NMR suggests that the molecule is not rigid in the NMR timescale. The molecule displays under slow exchange conditions on the NMR timescale (C$_6$D$_5$CD$_3$, 250 K) three t-Bu signals and six methylenic doublets (Figure 7) in agreement with a frozen conformation of C_2 symmetry. The rotational barrier was calculated from the coalescence behavior of the signals in the methylenic region (Table II). Interestingly, whereas at a slow exchange six doublets are observed, four doublets are observed under fast exchange conditions.

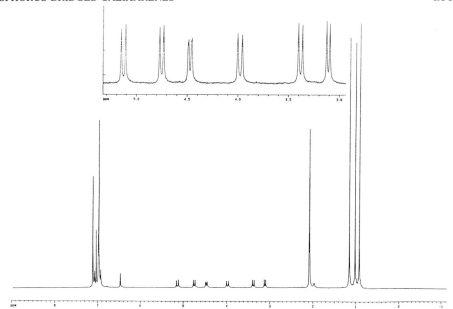

Fig. 7. Slow exchange ^{1}H-NMR spectrum (400 MHz, toluene-d_8, 250 K) of **11**. Top: expansion of the methylene region.

The ^{1}H-NMR spectrum of **12** (400 MHz, RT, CDCl$_3$) displays two *tert*-butyl signals in a 1 : 2 ratio, one pair of doublets and one singlet for the methylene protons and three aromatic signals. Lowering the temperature of a CDCl$_2$F [25] solution of **12** resulted in decoalescence of the largest t-Bu signal, and of the aromatic protons. One of the methylene doublets decoalesced into two close-spaced doublets while the central singlet showed some broadening but no detectable decoalescence process. At slow exchange conditions the exchanging t-Bu signals were too close to determine an accurate coalescence temperature. Rates of exchange at the coalescence temperatures were derived from the coalescence behavior of the methylene and aromatic protons and a barrier of 8.6±0.2 kcal mol^{-1} was calculated for the dynamic process followed by NMR. This barrier is significantly lower than the barrier of **11**.

3.7.2. Dynamic Processes of **11** and **12**

In the case of **11**, the protons within a methylene unit remain diastereotopic at the temperature range studied which rules out a ring inversion process. Assuming that the solution conformation is similar to the one present in the crystal, the dynamic process observed must exchange the coplanar ring with a nonvicinal twisted ring. This process, if present in both subunits, should result in enantiomerization of the system and should lead to mutual exchange of two pairs of methylene protons, as shown in Figure 8. The assignment of the dynamic process in **12** is difficult since we were unable to obtain crystals suitable for X-ray diffraction and therefore its solid

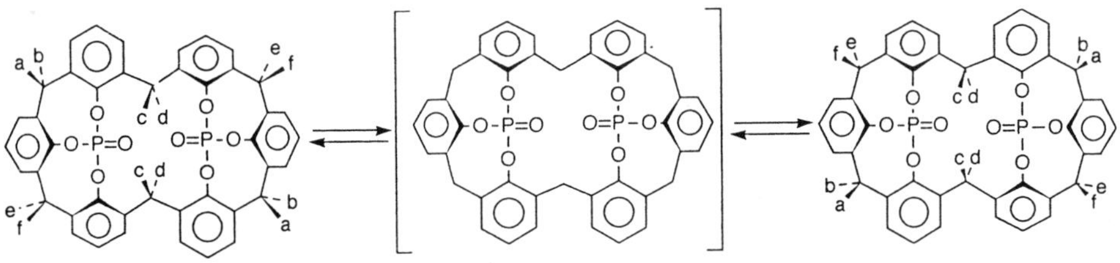

Fig. 8. The process of fc ⇌ fc interconversion **11**, leading to enantiomerization. The different methylene protons are denoted by letters (a–e). Homotopic sites are denoted by identical letters, enantiotopic sites are denoted by the same letter with a bar. As a result of the dynamic process the following exchanges take place: a ⇌ f̄, e ⇌ b̄, ā ⇌ f and ē ⇌ b̄. Note that the two protons within a given methylene group do not mutually exchange.

state conformation is not known. The appearance of four methylenic protons as a singlet indicates that a C_2 axis passes through two methylene groups. Assuming that **12** is a stereoisomer of **11**, these methylenes must be the ones which connect the two (ArO)$_3$PO subunits (the 'central' methylenes). The cavities of the two subunits must be oriented in an *anti* fashion. The appreciable lower barrier observed for **12** indicates than an analogous process to that observed for **11** is not taking place since in that case similar barriers should be observed. We therefore tentatively suggest that in **12** both P=O bonds are oriented *endo* to the cavities of the subunits. Inspection of space filling models indicates that in such arrangement, each subunit exists in a rigid cone-like conformation. The dynamic process observed does not involve changes in the torsional angles of the rings, but most likely involves partial rotations about the bonds linking the central methylenes to the two subunits. This is probably the reason why **12** has a lower rotational barrier than **11**.

4. Conclusions

Multiply bridged calixarenes can be easily obtained by pyrolysis of calixarene (dialkyl phosphate ester) derivatives. The multiple bridging of the calix[4]- and calix[6]arenes raises the barrier for ring inversion, but it does not completely freeze all possible dynamic processes.

Acknowledgements

We thank Dr. Shmuel Cohen for the X-ray structural determinations and the Mass Spectrometry Center at the Technion, Haifa for the mass spectra determinations. This work was supported by the Israel Science Foundation, administered by The Israeli Academy of Sciences and Humanities.

References

1. For comprehensive reviews on calixarenes see: C. D. Gutsche: *Calixarenes*, Royal Society of Chemistry, Cambridge (1989); *Calixarenes: A Versatile Class of Macrocyclic Compounds*, J. Vicens and V. Böhmer (Eds.), Kluwer, Dordrecht (1991).
2. C. D. Gutsche and L. J. Bauer: *J. Am. Chem. Soc.* **107**, 6052 (1985).
3. See for example H. Taniguchi, E. Nomura, and T. Hinomoto: *Chem. Express* **7**, 853 (1992).
4. For a recent example of a conformationally rigid 18-crown-6 derivative see: G. Li and W. C. Still: *J. Am. Chem. Soc.* **115**, 3804 (1993).
5. K. Iwamoto, K. Araki, and S. Shinkai: *J. Org. Chem.* **56**, 4955 (1991). S. Shinkai, K. Fujimoto, T. Otsuka, and H. L. Ammon: *J. Org. Chem.* **57**, 1526 (1992). H. Iki, T. Kikuchi, and S. Shinkai: *J. Chem. Soc. Perkin Trans. 1*, 205 (1993).
6. H. Otsuka, K. Araki, and S. Shinkai: *Chem. Express* **8**, 479 (1993).
7. For examples of calixarenes bridged by a chain of atoms see: H. Goldmann, W. Vogt, E. Paulus, and V. Böhmer: *J. Am. Chem. Soc.* **110**, 6811 (1988). V. Böhmer and W. Vogt: *Pure Appl. Chem.* **65**, 403 (1993).
8. A. Arduini, A. Casnati, L. Dodi, A. Pochini, and R. Ungaro: *J. Chem. Soc., Chem. Commun.* 1597 (1990). V. Böhmer, G. Ferguson, J. F. Gallagher, A. J. Lough, M. A. McKervey, E. Madigan, M. B. Moran, J. Phillips, and G. Williams: *J. Chem. Soc., Perkin Trans. 1* 1521 (1993). Z. Asfari, J. Weiss, S. Pappalardo, and J. Vicens: *Pure Appl. Chem.* **65**, 585 (1993).
9. See for example G. D. Andreetti, G. Calestani, F. Ugozzoli, A. Arduini, E. Ghidini, A. Pochini, and R. Ungaro: *J. Incl. Phenom.* **5**, 123 (1987). F. Corrazza, C. Floriani, A. Chiesi-Villa, and C. Guastini: *J. Chem. Soc., Chem. Commun.* 1083 (1990).
10. D. V. Khasnis, M. Lattman, and C. D. Gutsche: *J. Am. Chem. Soc.* **112**, 9422 (1990). D. V. Khasnis, J. M. Burton, M. Lattman, and H. Zhang: *J. Chem. Soc., Chem. Commun.* 562 (1991). D. V. Khasnis, J. M. Burton, J. D. McNeil, H. Zhang, and M. Lattman: *Phosphorus, Sulfur and Silicon* **75**, 253 (1993).
11. X. Delaigue, M. W. Hosseini, A. De Cian, J. Fischer, E. Leize, S. Kieffer, and A. Van Dorsselaer: *Tetrahedron Lett.* **34**, 3285 (1993).
12. P. Neri, J. F. Ferguson, G. Gallagher, and S. Pappalardo: *Tetrahedron Lett.* **33**, 7403 (1992).
13. For a preliminary communication of the present work see: F. Grynszpan, O. Aleksiuk, and S. E. Biali: *J. Chem. Soc., Chem. Commun.* 13 (1993).
14. A double bridged calix[6]arene has been reported recently. See: J. K. Moran and D. M. Roundhill: *Phosphorus, Sulfur and Silicon* **71**, 7 (1992).
15. (a) Z. Goren and S. E. Biali: *J. Chem. Soc., Perkin Trans. 1* 1484 (1990). (b) F. Grynszpan, Z. Goren, and S. E. Biali: *J. Org. Chem.* **56**, 532 (1991). (c) Y. Ting, W. Verboom, L. C. Groenen, J.-D. van Loon, and D. N. Reinhoudt: *J. Chem. Soc., Chem. Commun.* 1432 (1990). (d) J. E. McMurry and J. C. Phelan: *Tetrahedron Lett.* **41**, 5655 (1991). (e) O. Aleksiuk, F. Grynszpan, and S. E. Biali: *J. Chem. Soc., Chem. Commun.* 11 (1993).
16. For recent work on intraannular phosphorus containing calixarenes see for example: C. Floriani, D. Jacoby, A. Chiesi-Villa, and C. Guastini: *Angew. Chem. Int. Ed. Engl.* **28**, 1376 (1989). J. K. Moran and D. M. Roundhill: *Inorg. Chem.* **31**, 4213 (1992). D. Jacoby, C. Floriani, A. Chiesi-Villa, and C. Rizzoli: *J. Chem. Soc., Dalton Trans.* 813 (1993). D. Matt, C. Loeber, J. Vicens, and Z. Asfari: *J. Chem. Soc., Chem. Commun.* 604 (1993). L. N. Markovsky, V. I. Kalchenko, D. M. Rudkevich, and A. N. Shivanyuk: *Mendeleev Commun.* 106 (1992). For extraannular phosphorus substituted systems see: I. Kalchenko, L. I. Atamas, V. V. Pirozhenico, L. N. Markovsky, *Zh. Obschs. Khim.* **62**, 2623 (1992).
17. (a) F. Ohseto, H. Murakami, K. Araki, and S. Shinkai: *Tetrahedron Lett.* **33**, 1217 (1992). (b) F. Grynszpan, O. Aleksiuk, and S. E. Biali, *J. Org. Chem.* **59**, 2070 (1994).
18. G. W. Kenner and N. R. Williams: *J. Chem. Soc.* 523 (1955).
19. D. R. Stewart and C. D. Gutsche: *Org. Prep. Proceed. Int.* **25**, 137 (1993).
20. L. N. Markovsky, V. I. Kalchenko, and N.A. Parhomenko: *Zh. Obshch. Khim.* **60**, 2811 (1990).
21. R. G. Janssen, W. Verboom, S. Harkema, G. J. van Hummel, D. N. Reinhoudt, A. Pochini, R. Ungaro, P. Prados, and J. de Mendoza: *J. Chem. Soc., Chem. Commun.* 506 (1993).
22. (a) M. Conner, V. Janout, and S. Regen: *J. Am. Chem. Soc.* **113**, 9670 (1991). (b) E. Dahan and S. E. Biali: *J. Org. Chem.* **56**, 7269 (1991).

23. J. C. Tebby: in *Phosphorus-31 NMR Spectroscopy in Stereochemical Analysis*, J. G. Verkade and L.D. Quin (Eds.), Ch. 1, VCH, Deerfield Beach (1987).
24. H. S. Gutowsky and C. H. Holm: *J. Chem. Phys.* **25**, 1228 (1956).
25. J. S. Siegel and F. A. L. Anet: *J. Org. Chem.* **53**, 2629 (1988).
26. H. McCombrie, B. C. Saunders, and G. J. Stacey: *J. Chem. Soc.* 380 (1945).

Inclusion of Quaternary Ammonium Compounds by Calixarenes*

JACK M. HARROWFIELD**, WILLIAM R. RICHMOND and
ALEXANDER N. SOBOLEV‡
Research Centre for Advanced Minerals and Materials Processing, University of Western Australia, Nedlands, 6009, Western Australia

(Received: 3 March 1994; in final form: 10 June 1994)

Abstract. In weakly polar solvents, strong association occurs between calixarene anions and tetraalkylammonium cations, with the magnitude of the observed equilibrium constants depending upon the charge on the anion, the solvent, the ring size of the calixarene and the nature of the alkyl group of the cation. Large upfield shifts of the methyl resonances of the $[(CH_3)_4N]^+$ cation in solutions of $[(CH_3)_4N]_2[p\text{-}t\text{-butylcalix[6]arene} - 2H]$ indicate cation inclusion in a structure which is possibly identical with that found for the solid 'salt' by X-ray crystallography. This shows one of the cations to be included within a partial cone structure of a 'hinged 3-up, 3-down' conformation of the calixarene. The functionalised tetramethylammonium ions, choline and acetyl choline, are also strongly included by various calixarene anions but attempts to detect significant modification of the reactivity of acetyl choline resulting from inclusion have not been successful.

Key words: Calixarene anions, tetra-alkylammonium cations, association equilibria, inclusion, acetyl choline, catalysis.

1. Introduction

Thermodynamic [1], kinetic [2–4], spectroscopic [5, 6] and synthetic [7–9] studies have shown that calixarenes may be readily and sometimes extensively deprotonated by strong bases in nonaqueous solvents. Although solids containing six- and even seven-fold deprotonated calixarenes have been isolated as rare earth [7] or transition metal [8] ion complexes, solid salts of weakly or formally noncoordination cations are known only for degrees of deprotonation up to a maximum of three [9]. Obviously, these differences must in some way be associated with the nature of the cation–anion interactions occurring, and we have recently succeeded in characterising some aspects of these phenomena through structural studies of some alkali metal and tetraalkylammonium ion complexes [9–12]. It is apparent, as also suggested by theoretical calculations [13] and thermodynamic [1] measurements, that the negative charge on deprotonated calixarenes is delocalised and that cation

---

* This paper is dedicated to the commemorative issue on the 50th anniversary of calixarenes.
** Author for correspondence.
‡ On leave from the L. Karpov Institute of Physical Chemistry, ul. Obukha 10, Moscow 103064, Russia.

binding in both solution and solid phases can involve inclusion provided there is appropriate dimensional matching [2–4, 9–12]. The occurrence of inclusion raises the prospect of conducting reactions with the calixarene as catalyst, especially if charge delocalisation is such that the nucleophilicity of the calixarene itself may be ignored, so that inclusion of an electrophile does not simply result in attack on this negative charge. A relatively simple electrophile of interest because, in its natural role, it undergoes binding of its cationic centre into a hydrophobic cavity, is acetylcholine [14, 15]. We describe here measurements of equilibrium binding of some tetraalkylammonium ions and both choline and acetylcholine by calixarene anions, and our attempts to characterise the reactivity of included acetylcholine. The structure of the *bis*(tetramethylammonium) salt of p-t-butylcalix[6]arene is also reported as a model for the binding of acetylcholine by such a calixarene.

2. Experimental

2.1. SYNTHESIS

The salts [Pr$_4$N][calix[4]arene – H], [Et$_4$N][p-t-butyldihomooxacalix[4]arene – H], [(CH$_3$)$_4$N]$_2$[p-t-butylcalix[6]arene – 2H], [Et$_4$N]$_2$[p-t-butylcalix[6]arene – 2H], [Pr$_4$N][p-t-butylcalix[6]arene – H] and [Pr$_4$N]$_2$[calix[6]arene – 2H] (Et = CH$_3$CH$_2$, Pr = CH$_3$CH$_2$CH$_2$) were prepared as described previously [9]. Choline and acetylcholine perchlorates were precipitated by addition of NaClO$_4$ to aqueous solutions of the chlorides (Sigma Chemical Co.), and were recrystallised from methanol and from acetonitrile by the addition of ether, respectively. The previously unknown *bis*- and *tris*-(tetrapropylammonium) salts of p-t-butylcalix[6]arene, [Pr$_4$N]$_2$[p-t-butylcalix[6]arene – 2H] and [Pr$_4$N]$_3$[p-t-butylcalix[6]arene – 3H], were prepared by a slight variation of methods established for the other species, as follows.

[Pr$_4$N]$_2$[p-t-butylcalix[6]arene – 2H] Trihydrate

Approximately 1 mol L^{-1} methanolic Pr$_4$N$^+$OH$^-$ (4.0 mL) was added to a slurry of p-t-butylcalix[6]arene (1.93 g) in methanol (40 mL). The mixture was stirred vigorously for 60 min (without dissolution of the solid occurring) and the solvent was then evaporated under reduced pressure. The white residue was dissolved in CH$_2$Cl$_2$ (40 mL), the solution filtered and the product (1.75 g) precipitated as a white powder by the addition of hexane. Note that the attempts to recrystallise this material from hot methanol resulted in the deposition of [Pr$_4$N][p-t-butylcalix[6]arene – H]. *Analysis Calcd.* for C$_{90}$H$_{138}$N2O$_6$·3H$_2$O: C, 77.26; H, 10.37; N, 2.00. *Found*: C, 77.24; H, 9.88; N, 1.10%. Nitrogen analyses on these compounds were typically erratic and cation to anion ratios were more reliably established through integration of ^{1}H-NMR spectra.

[Pr₄N]₃[p-t-butylcalix[6]arene − 3H] Hexahydrate

Approximately 1 mol L^{-1} methanolic Pr$_4$N$^+$OH$^-$ (6.5 mL) was added to a slurry of *p-t*-butylcalix[6]arene (1.93 g) in methanol (40 mL). The mixture was stirred vigorously for 10 min and the nearly clear solution produced was then filtered and the solvent evaporated under reduced pressure. The white residue was dissolved in CH$_2$Cl$_2$ (40 mL) and the product was precipitated as clusters of fine, white needles by the addition of hexane. These were collected, washed with hexane and dried in air, efflorescing in this process to a white powder (2.45 g). *Analysis Calcd.* for C$_{102}$H$_{165}$N$_3$O$_6$·6H$_2$O: C, 74.81; H, 10.89; N, 2.57. *Found*: C, 74.73; H, 10.68; N, 2.02%.

Choline Trifluoromethanesulphonate (Choline Triflate)

Choline chloride (5 g) and trifluoromethane sulphonic acid (CF$_3$SO$_3$H, 4 mL) were mixed together and heated on a steam bath for 5 min, giving a clear, viscous solution. This was cooled to room temperature and added, with stirring, to ether (200 mL). A white, crystalline precipitate formed immediately. It was recrystallised from ethanol by the addition of ether to the point of permanent turbidity and storage at 4°C for 24 h. *Analysis Calcd.* for C$_6$H$_{14}$F$_3$NO$_4$S: C, 28.46; H, 5.57; N, 5.53. *Found*: C, 28.39; H, 5.79; N, 5.38%.

2.2. EQUILIBRIUM MEASUREMENTS

2.2.1. *Solution Spectrophotometry*

Acetonitrile (Ajax Chemicals HPLC Grade) solutions of constant calixarene anion (host) concentration and varying tetramethylammonium ion, choline or acetyl choline (guest) concentrations were prepared by appropriate dilutions of 10^{-2} mol L^{-1} stock solutions of the tetrapropylammonium salts of the calixarenes, tetramethylammonium triflate, choline triflate and acetylcholine perchlorate. Spectra were recorded in the range 200–350 nm on a Hewlett-Packard 8452 Diode Array spectrophotometer. Nonlinear least-squares fitting of absorbance vs. composition data to models of the reaction system was conducted using a program written locally by Dr. E. S. Kucharski.

2.2.2. *NMR Spectroscopy*

Solutions in acetone-d_6 of known concentration close to 3×10^{-2} mol L^{-1} in choline or acetylcholine perchlorates were prepared and used as solvents for calixarene salts so as to obtain series of solutions in which the host : guest (calixarene : quaternary ammonium group) ratio varied between 0 and 2.00. Proton chemical shifts were measured for the trimethylammonium resonances of all solutions using a Bruker WP80 instrument. Inclusion equilibrium constants were determined from nonlinear least-squares fitting of the chemical shift vs. host : guest ratio curves.

2.3. CRYSTALLOGRAPHY

Recrystallisation of [(CH$_3$)$_4$N]$_2$[p-t-butylcalix[6]arene – 2H] from acetonitrile provided a mixture of relatively large rhombs and some very fine needles. The structure determination was performed on a capillary-mounted specimen of the larger crystals.

2.3.1. Structure Determination

A unique diffractometer data set (Siemens P4; $2\theta_{max}$ 50°, $2\theta/\theta$ scan mode, filtered MoK_α radiation, $\lambda = 0.7107_3$ Å) was measured at 293(2) K, yielding 22347 independent reflections, 12637 of these with $I > 3\sigma(I)$ being considered 'observed' and used in the full-matrix least-squares refinement on F^2. The structure solution was obtained using SHELXS-86 [16], refinement being carried out with SHELXL-93 [17]. Anisotropic thermal parameters were refined for the ordered nonhydrogen atoms. The positions of the hydrogen atoms on carbon were calculated from geometrical considerations and were refined in constraint with the bonded carbon atoms. Phenolic hydrogen atoms were located by difference Fourier syntheses and were refined in constraint with the bonded oxygen atoms. Positional parameters for the nonhydrogen atoms are given in Table I and some selected torsional angles are given in Table II. Tables of bond lengths, bond angles, hydrogen positional parameters, least-squares planes, and structure factor amplitudes are available from the authors. Views of one calixarene dianion/tetramethylammonium cations unit, showing the inclusion of one of the cations by the calixarene (and the atom numbering scheme used), are given in Figure 1. There is extensive disorder within the structure; three of the six t-butyl groups on each independent calixarene anion are disordered, as are three of four cations associated with these calixarene moieties, and the lattice solvent molecules (water and acetonitrile). Thermal motion is high, especially for the solvent molecules, so that the crystal appears to be bordering on the liquid state. The formulation (below) based on the structure solution shows that preparation for elemental analysis [9] results in considerable loss of solvent. The two independent calixarene species are both identified as a dianion on the basis of the detection of two protons in association with each group of three commonly oriented phenyl rings. The final indices of fit were $R_1 = 0.139$, $wR_2 = 0.356$ with weight $= [\sigma^2 F_0^2 + (0.1551P)^2 + 28.1194P]^{-1}$.

2.3.2. Crystal Data

C$_{76}$H$_{115}$N$_3$O$_9$ ([(CH$_3$)$_4$N]$_2$[p-t-butylcalix[6]arene – 2H] · CH$_3$CN · 3H$_2$O), $M = 1214.7$, triclinic, space group $P\bar{1}$, $a = 17.424(3)$, $b = 18.447(4)$, $c = 27.229(6)$ Å, $\alpha = 82.10(2)$, $\beta = 98.27(3)$, $\gamma = 107.83(2)°$, $V = 8204(3)$ Å^3, D_c ($Z = 4$) $= 0.983$ g cm^{-3}, $F(000) = 2656$, $\mu_{Mo} = 0.63$ cm^{-1}, specimen $0.65 \times 0.60 \times 0.20$ mm^3.

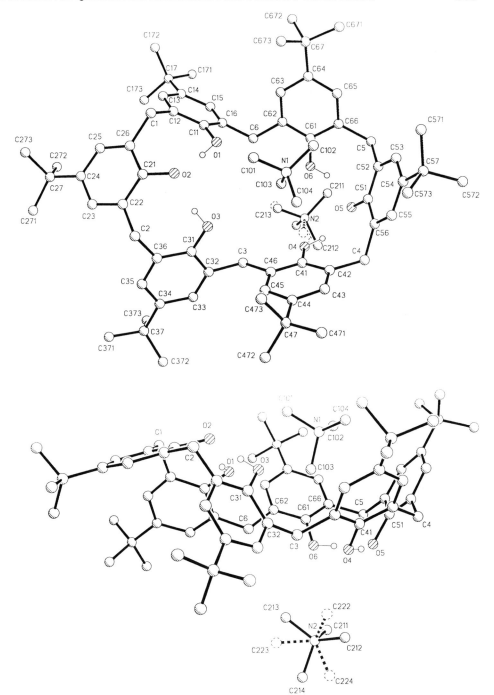

Fig. 1. Views of one of the two independent calixarene moieties in the unit cell of the salt [(CH$_3$)$_4$N]$_2$[p-t-butylcalix[6]arene − 2H], in which an included cation is ordered and another associated but nonincluded cation is disordered over two configurations. Only the phenolic hydrogen atoms are shown.

TABLE I. Atomic coordinates/Å, equivalent isotropic displacement parameters/Å^2 and site occupancies (population parameters) for [N(CH$_3$)$_4$]$_2$[p-t-butylcalix[6]arene − 2H]. $U_{(eq)}$ is defined as one-third of the trace of the orthogonalised U_{ij} tensor.

Atom	x	y	z	$U_{(eq)}$	Occupancy
C(1)A	0.5990(6)	0.5377(5)	0.2965(4)	0.060(3)	1.0
O(1)A	0.4800(5)	0.4101(4)	0.3440(3)	0.082(2)	1.0
C(11)A	0.4988(6)	0.4042(6)	0.2981(4)	0.061(3)	1.0
C(12)A	0.5557(5)	0.4646(5)	0.2738(4)	0.053(2)	1.0
C(13)A	0.5700(6)	0.4542(6)	0.2263(4)	0.058(3)	1.0
C(14)A	0.5329(6)	0.3882(6)	0.2035(4)	0.063(3)	1.0
C(15)A	0.4786(6)	0.3296(6)	0.2304(4)	0.061(3)	1.0
C(16)A	0.4595(6)	0.3367(6)	0.2762(4)	0.063(3)	1.0
C(17)A	0.5495(6)	0.3812(7)	0.1503(4)	0.071(3)	1.0
C(171)A	0.5144(8)	0.3009(8)	0.1351(5)	0.109(5)	1.0
C(172)A	0.6389(8)	0.404(1)	0.1460(6)	0.142(7)	1.0
C(173)A	0.508(1)	0.432(1)	0.1127(5)	0.131(6)	1.0
C(2)A	0.4006(6)	0.6547(5)	0.3515(4)	0.057(3)	1.0
O(2)A	0.4890(4)	0.5443(4)	0.3634(2)	0.067(2)	1.0
C(21)A	0.4976(6)	0.5961(5)	0.3223(4)	0.056(2)	1.0
C(22)A	0.4543	0.6506(5)	0.3139(4)	0.057(3)	1.0
C(23)A	0.4627(6)	0.7012(5)	0.2707(4)	0.058(3)	1.0
C(24)A	0.5136(6)	0.7010(5)	0.2356(4)	0.057(3)	1.0
C(25)A	0.5560(6)	0.6476(5)	0.2465(4)	0.061(3)	1.0
C(26)A	0.5500(5)	0.5958(5)	0.2873(4)	0.054(2)	1.0
C(27)A	0.5221(8)	0.7571(7)	0.1890(4)	0.078(3)	1.0
C(271)A	0.468(1)	0.810(1)	0.183(1)	0.15(1)	0.67
C(272)A	0.502(1)	0.713(1)	0.1427(8)	0.14(1)	0.67
C(273)A	0.611(1)	0.808(1)	0.1893(7)	0.109(7)	0.67
C(274)A	0.442(2)	0.759(2)	0.166(2)	0.15(1)	0.33
C(275)A	0.567(3)	0.743(3)	0.155(2)	0.14(1)	0.33
C(276)A	0.1899(7)	0.3981(6)	0.3691(3)	0.065(3)	1.0
O(3)A	0.3472(4)	0.4947(4)	0.3886(3)	0.076(2)	1.0
C(31)A	0.2931(6)	0.5278(6)	0.3631(3)	0.057(3)	1.0
C(32)A	0.2116(6)	0.4840(5)	0.3545(3)	0.055(3)	1.0
C(33)A	0.1542(7)	0.5160(6)	0.3284(4)	0.071(3)	1.0
C(34A)	0.1750(7)	0.5937(6)	0.3100(4)	0.070(3)	1.0
C(35)A	0.2554(7)	0.6356(6)	0.3184(4)	0.066(3)	1.0
C(36)A	0.3150(6)	0.6060(5)	0.3452(3)	0.054(2)	1.0
C(37)A	0.1076(8)	0.6254(7)	0.2795(6)	0.098(4)	1.0
C(371)A	0.139(1)	0.7096(9)	0.2665(9)	0.22(1)	1.0
C(372)A	0.035(1)	0.606(1)	0.3088(8)	0.167(8)	1.0
C(373)A	0.078(1)	0.586(1)	0.2318(7)	0.20(1)	1.0
C(4)A	0.1398(7)	0.1796(6)	0.5042(4)	0.072(3)	1.0
O(4)A	0.1569(5)	0.2479(4)	0.4016(2)	0.072(2)	1.0

TABLE I. (Continued)

Atom	x	y	z	$U_{(eq)}$	Occupancy
C(41)A	0.1651(6)	0.2932(5)	0.4389(4)	0.056(3)	1.0
C(42)A	0.1581(6)	0.2648(6)	0.4883(4)	0.058(3)	1.0
C(43)A	0.1657(3)	0.3161(6)	0.5226(4)	0.068(3)	1.0
C(44)A	0.1806(7)	0.3941(6)	0.5100(4)	0.072(3)	1.0
C(45)A	0.1900(7)	0.4209(6)	0.4598(4)	0.071(3)	1.0
C(46)A	0.1822(6)	0.3712(6)	0.4244(4)	0.058(3)	1.0
C(47)A	0.188(1)	0.4504(8)	0.5486(5)	0.107(5)	1.0
C(471)A	0.176(1)	0.410(1)	0.6028(7)	0.16(1)	0.67
C(472)A	0.130(2)	0.500(2)	0.533(1)	0.18(1)	0.67
C(473)A	0.280(1)	0.507(1)	0.5541(8)	0.135(9)	0.67
C(474)A	0.096(2)	0.411(2)	0.571(2)	0.16(1)	0.33
C(475)A	0.186(4)	0.517(3)	0.528(2)	0.18(1)	0.33
C(476)A	0.239(3)	0.433(2)	0.589(2)	0.135(9)	0.33
C(5)A	0.3295(7)	0.0616(5)	0.4407(4)	0.072(3)	1.0
O(5)A	0.1961(4)	0.1237(4)	0.4271(2)	0.064(2)	1.0
C(51)A	0.2369(6)	0.1213(5)	0.4735(4)	0.058(3)	1.0
C(52)A	0.3033(6)	0.0918(5)	0.4827(4)	0.058(3)	1.0
C(53)A	0.3436(6)	0.0906(5)	0.5305(4)	0.068(3)	1.0
C(54)A	0.3231(7)	0.1201(6)	0.5695(4)	0.065(3)	1.0
C(55)A	0.2560(7)	0.1483(5)	0.5586(4)	0.063(3)	1.0
C(56)A	0.2129(6)	0.1488(5)	0.5124(4)	0.056(2)	1.0
C(571)A	0.456(1)	0.113(1)	0.6176(9)	0.16(1)	0.67
C(572)A	0.320(1)	0.047(1)	0.6536(8)	0.16(1)	0.67
C(573)A	0.375(1)	0.194(1)	0.6438(7)	0.119(7)	0.67
C(574)A	0.385(2)	0.028(2)	0.637(2)	0.16(1)	0.33
C(575)A	0.333(3)	0.128(3)	0.662(2)	0.16(1)	0.33
C(576)A	0.461(3)	0.166(2)	0.621(2)	0.119(7)	0.33
C(6)A	0.3968(6)	0.2739(5)	0.3020(4)	0.068(3)	1.0
O(6)A	0.2901(5)	0.1654(4)	0.3575(3)	0.076(2)	1.0
C(61)A	0.3691(7)	0.1705(6)	0.3720(4)	0.062(3)	1.0
C(62)A	0.4279(7)	0.2236(5)	0.3457(4)	0.063(3)	1.0
C(63)A	0.5087(7)	0.2312(6)	0.3590(4)	0.068(3)	1.0
C(64)A	0.5350(7)	0.1834(7)	0.3938(4)	0.072(3)	1.0
C(65)A	0.4754(8)	0.1301(6)	0.4236(4)	0.068(3)	1.0
C(66)A	0.3931(8)	0.1218(6)	0.4108(4)	0.068(3)	1.0
C(67)A	0.6240(8)	0.1923(8)	0.4100(5)	0.089(4)	1.0
C(671)A	0.641(1)	0.135(1)	0.4525(6)	0.140(6)	1.0
C(672)A	0.658(1)	0.179(1)	0.3632(6)	0.140(6)	1.0
C(673)A	0.6694(9)	0.2725(9)	0.4241(7)	0.137(6)	1.0
C(1)B	0.0964(6)	0.8994(5)	0.1678(4)	0.061(3)	1.0
O(1)B	0.1303(4)	1.0577(4)	0.1221(3)	0.068(2)	1.0
C(11)B	0.1842(6)	1.0381(5)	0.1594(3)	0.049(2)	1.0
C(12)B	0.1719(6)	0.9637(5)	0.1818(3)	0.54(2)	1.0

TABLE I. (Continued)

Atom	x	y	z	$U_{(eq)}$	Occupancy
C(13)B	0.2306(6)	0.9488(6)	0.2182(3)	0.058(3)	1.0
C(14)B	0.3020(6)	1.0037(6)	0.2340(4)	0.060(3)	1.0
C(15)B	0.3091(6)	1.0782(6)	0.2104(4)	0.061(3)	1.0
C(16)B	0.2521(6)	1.0956(5)	0.1739(3)	0.054(2)	1.0
C(17)B	0.3635(7)	0.9833(7)	0.2734(4)	0.082(3)	1.0
C(171)B	0.4391(9)	1.0534(9)	0.2839(6)	0.137(6)	1.0
C(172)B	0.389(1)	0.921(1)	0.2588(9)	0.23(1)	1.0
C(173)B	0.327(1)	0.959(1)	0.3219(5)	0.160(8)	1.0
C(2)B	-0.1715(6)	0.9586(6)	0.1356(4)	0.063(2)	1.0
O(2)B	-0.0156(4)	0.9699(3)	0.1094(2)	0.053(2)	1.0
C(21)B	-0.0350(6)	0.9339(5)	0.1549(3)	0.048(2)	1.0
C(22)B	-0.1097(5)	0.9285(5)	0.1709(3)	0.050(2)	1.0
C(23)B	-0.1260(6)	0.8963(5)	0.2186(4)	0.059(3)	1.0
C(24)B	-0.0708(7)	0.8668(6)	0.2518(4)	0.064(3)	1.0
C(25)B	0.0002(6)	0.8708(5)	0.2339(4)	0.058(3)	1.0
C(26)B	0.0184(6)	0.9017(5)	0.1868(3)	0.053(2)	1.0
C(27)B	-0.0928(7)	0.8344(6)	0.3052(4)	0.075(3)	1.0
C(271)B	-0.173(1)	0.7725(8)	0.3029(5)	0.134(6)	1.0
C(272)B	-0.098(1)	0.8987(7)	0.3325(4)	0.110(5)	1.0
C(273)B	-0.025(1)	0.804(1)	0.3357(5)	0.148(7)	1.0
C(3)B	-0.0426(6)	1.2448(6)	0.1191(4)	0.065(3)	1.0
O(3)B	-0.0489(4)	1.0976(4)	0.1037(3)	0.072(2)	1.0
C(31)B	-0.1099(6)	1.1034(6)	0.1273(3)	0.055(3)	1.0
C(32)B	-0.1099(6)	1.760(5)	0.1384(4)	0.057(3)	1.0
C(33)B	-0.1727(8)	1.1834(6)	0.1574(4)	0.079(3)	1.0
C(34)B	-0.2365(8)	1.1217(8)	0.1719(6)	0.087(4)	1.0
C(35)B	-0.2331(7)	1.0500(7)	0.1650(4)	0.082(4)	1.0
C(36)B	-0.1712(6)	1.0386(6)	0.1424(3)	0.056(3)	1.0
C(37)B	-0.309(1)	1.1313(9)	-0.1938(7)	0.131(6)	1.0
C(371)B	-0.338(2)	1.069(1)	0.2358(9)	0.138(9)	0.67
C(372)B	-0.379(2)	1.128(2)	0.150(1)	0.19(1)	0.67
C(373)B	-0.287(3)	1.212(1)	0.2138(9)	0.147(9)	0.67
C(374)B	-0.385(2)	1.053(2)	0.189(2)	0.138(9)	0.33
C(375)B	-0.329(3)	1.197(4)	0.177(2)	0.19(1)	0.33
C(376)B	-0.277(3)	1.122(3)	0.252(3)	0.147(9)	0.33
C(4)B	0.0744(6)	1.4365(5)	-0.0217(4)	0.074(3)	1.0
O(4)B	0.786(4)	1.365(4)	0.0805(2)	0.070(2)	1.0
C(41)B	0.0120(6)	1.3381(5)	0.0475(4)	0.057(3)	1.0
C(42)B	0.0056(6)	1.3713(5)	-0.0022(4)	0.055(2)	1.0
C(43)B	-0.0628(6)	1.3398(6)	-0.0331(4)	0.061(3)	1.0
C(44)B	-0.1267(6)	1.2779(6)	-0.0184(4)	0.063(3)	1.0
C(45)B	-0.1188(6)	1.2479(6)	0.0317(4)	0.060(3)	1.0
C(46)B	-0.0513(6)	1.2773(5)	0.0648(4)	0.053(2)	1.0

TABLE I. (Continued)

Atom	x	y	z	$U_{(eq)}$	Occupancy
C(47)B	-0.2018(7)	1.2431(8)	-0.0524(5)	0.088(4)	1.0
C(471)B	-0.202(1)	1.285(1)	-0.1051(8)	0.21(2)	0.67
C(472)B	-0.280(1)	1.240(2)	-0.0310(9)	0.16(1)	0.67
C(473)B	-0.207(1)	1.158(1)	-0.0582(9)	0.14(1)	0.67
C(474)B	-0.231(2)	1.311(2)	-0.077(2)	0.21(2)	0.33
C(475)B	-0.262(3)	1.186(3)	-0.026(2)	0.15(1)	0.33
C(476)B	-0.174(3)	1.212(3)	-0.094(2)	0.14(1)	0.33
C(5)B	0.3418(6)	1.3721(6)	0.0040(4)	0.065(3)	1.0
O(5)B	0.2146(4)	1.4132(4)	0.0400(2)	0.063(2)	1.0
C(51)B	0.2070(6)	1.4027(5)	-0.0086(3)	0.054(2)	1.0
C(52)B	0.2666(6)	1.3816(5)	-0.0290(4)	0.055(2)	1.0
C(53)B	0.2556(6)	1.3699(5)	-0.0781(4)	0.055(2)	1.0
C(54)B	0.1884(5)	1.3771(5)	-0.1112(3)	0.050(2)	1.0
C(55)B	0.1319(6)	1.4003(5)	-0.0906(4)	0.057(3)	1.0
C(56)B	0.1397(6)	1.4133(5)	-0.0410(4)	0.053(2)	1.0
C(57)B	0.1752(6)	1.3608(6)	-0.1652(4)	0.064(3)	1.0
C(571)B	0.0964(7)	1.2998(6)	-0.1761(4)	0.084(3)	1.0
C(572)B	0.2424(8)	1.334(1)	-0.1800(5)	0.127(6)	1.0
C(573)B	0.1719(8)	1.4329(7)	-0.2001(4)	0.103(5)	1.0
C(6)B	0.2636(6)	1.1775(6)	0.1505(4)	0.062(3)	1.0
O(6)B	0.2728(4)	1.3157(4)	0.0979(2)	0.066(2)	1.0
C(61)B	0.3019(5)	1.2703(6)	0.755(4)	0.054(2)	1.0
C(62)B	0.2981(5)	1.1976(5)	0.1011(3)	0.051(2)	1.0
C(63)B	0.3292(6)	1.1495(6)	0.0804(4)	0.058(3)	1.0
C(64)B	0.3637(6)	1.1694(6)	0.0350(4)	0.061(3)	1.0
C(65)B	0.3660(6)	1.2426(6)	0.0114(4)	0.065(3)	1.0
C(66)B	0.3363(5)	1.2932(6)	0.0305(3)	0.053(3)	1.0
C(67)B	0.3979(8)	1.1157(7)	0.0138(5)	0.087(4)	1.0
C(671)B	0.487(1)	1.160(1)	0.0020(8)	0.115(7)	0.67
C(672)B	0.346(1)	1.095(1)	-0.0365(8)	0.129(8)	0.67
C(673)B	0.399(1)	1.042(1)	0.0477(8)	0.13(1)	0.67
C(674)B	0.410(2)	0.129(2)	-0.038(2)	0.115(7)	0.33
C(675)B	0.309(2)	1.028(2)	0.010(2)	0.129(8)	0.33
C(676)B	0.449(3)	1.087(2)	0.051(2)	0.13(1)	0.33
N(1)	0.4313(5)	0.3455(4)	0.4797(3)	0.062(2)	1.0
C(101)	0.4966(9)	0.4140(7)	0.4658(5)	0.116(5)	1.0
C(102)	0.465(1)	0.2832(8)	0.4987(6)	0.133(6)	1.0
C(103)	0.3723(7)	0.3240(6)	0.4352(4)	0.081(3)	1.0
C(104)	0.391(1)	0.361(1)	0.5186(5)	0.143(7)	1.0
N(2)	0.0858(4)	0.1195(3)	0.2696(2)	0.080(3)	1.0
C(211)	0.1236(6)	0.0545(5)	0.2856(4)	0.115(5)	1.0
C(212)	0.0348(9)	0.1239(8)	0.3094(5)	0.115(7)	0.5
C(213)	0.1525(7)	0.1947(6)	0.2628(7)	0.143(9)	0.5

TABLE I. (Continued)

Atom	x	y	z	$U_{(eq)}$	Occupancy
C(214)	0.0321(9)	0.1047(8)	0.2205(5)	0.126(8)	0.5
C(222)	0.097(1)	0.1671(8)	0.3130(5)	0.115(7)	0.5
C(223)	0.126(1)	0.1698(9)	0.2259(6)	0.143(9)	0.5
C(224)	-0.0045(7)	0.0864(7)	0.2539(7)	0.126(8)	0.5
N(3)	-0.0793(3)	0.8441(3)	0.0071(2)	0.048(2)	1.0
C(311)	-0.1026(4)	0.7972(4)	-0.0373(2)	0.066(3)	1.0
C(312)	-0.1513(5)	0.9693(8)	0.0164(5)	0.075(5)	0.3
C(313)	-0.0090(7)	0.9147(6)	-0.0037(4)	0.066(5)	0.3
C(314)	-0.054(1)	0.7962(6)	0.0529(3)	0.056(4)	0.3
C(322)	-0.1300(7)	0.8991(7)	0.0018(4)	0.075(5)	0.3
C(323)	0.0099(5)	0.8891(8)	0.0090(5)	0.066(5)	0.3
C(324)	-0.094(1)	0.7911(5)	0.0548(3)	0.056(4)	0.3
C(332)	-0.079(1)	0.9259(5)	-0.0103(3)	0.075(5)	0.3
C(333)	0.0048(6)	0.8436(9)	0.0309(5)	0.0665(5)	0.3
C(334)	-0.1399(8)	0.8098(7)	0.0449(4)	0.056(4)	0.3
N(4)	0.2952(3)	1.4730(3)	0.2043(2)	0.075(2)	1.0
C(411)	0.2927(6)	1.4289(6)	0.2567(3)	0.169(7)	1.0
C(412)	0.310(1)	1.5587(5)	0.2089(5)	0.17(1)	0.3
C(413)	0.2136(7)	1.4421(9)	0.1727(4)	0.144(9)	0.3
C(414)	0.3640(9)	1.462(1)	0.1789(5)	0.105(7)	0.3
C(422)	0.271(1)	1.5463(6)	0.2051(5)	0.17(1)	0.3
C(423)	0.2356(9)	1.4222(7)	0.1666(4)	0.144(9)	0.3
C(424)	0.3817(6)	1.495(1)	0.1889(5)	0.105(7)	0.3
C(432)	0.240(1)	1.525(1)	0.1986(6)	0.17(1)	0.3
C(433)	0.265(1)	1.4158(7)	0.1646(4)	0.144(9)	0.3
C(434)	0.3838(6)	1.522(1)	0.1974(5)	0.105(7)	0.3
N(5)	0.115(2)	-0.134(2)	0.636(1)	0.167(8)	0.5
C(511)	0.112(3)	-0.096(3)	0.605(2)	0.167(8)	0.5
C(512)	0.116(3)	-0.051(3)	0.571(2)	0.167(8)	0.5
N(6)	-0.146(2)	1.399(1)	0.135(1)	0.094(5)	0.5
C(611)	-0.168(2)	1.393(2)	0.105(1)	0.094(5)	0.5
C(612)	-0.209(2)	1.383(2)	0.067(1)	0.094(5)	0.5
N(7)	0.067(1)	0.342(1)	0.2416(8)	0.093(4)	0.5
C(711)	0.014(2)	0.333(2)	0.255(1)	0.093(4)	0.5
C(712)	-0.055(2)	0.327(2)	0.272(1)	0.093(4)	0.5
N(8)	0.148(2)	0.590(2)	0.658(1)	0.121(6)	0.5
C(811)	0.200(2)	0.597(2)	0.683(1)	0.121(6)	0.5
C(812)	0.263(2)	0.608(2)	0.713(1)	0.121(6)	0.5
O(1W)	0.058(2)	1.483(2)	0.133(1)	0.171(9)	0.5
O(2W)	0.065(1)	0.002(1)	0.4140(8)	0.133(7)	0.5
O(3W)	-0.041(2)	0.315(2)	0.444(1)	0.21(1)	0.5
O(4W)	0.673(2)	0.228(2)	0.573(1)	0.23(1)	0.5
O(5W)	0.688(3)	0.300(3)	0.555(2)	0.25(2)	0.5

TABLE I. (Continued)

Atom	x	y	z	$U_{(eq)}$	Occupancy
O(6W)	-0.107(2)	1.415(2)	0.144(1)	0.20(1)	0.5
O(7W)	-0.145(2)	1.397(2)	0.161(1)	0.18(1)	0.5
O(8W)	0.302(2)	1.557(1)	0.0584(9)	0.127(9)	0.5
O(9W)	0.338(2)	1.548(1)	0.0660(9)	0.125(9)	0.5
O(10W)	0.385(1)	1.585(1)	0.0302(8)	0.130(7)	0.5
O(11W)	-0.081(3)	0.258(3)	0.486(2)	0.22(2)	0.3
O(12W)	0.091(3)	-0.043(3)	0.426(2)	0.19(2)	0.3
O(13W)	0.772(3)	0.283(3)	0.573(2)	0.19(2)	0.3

3. Results and Discussion

3.1. CRYSTALLOGRAPHY

The structural information available for calix[6]arene and its derivatives is relatively limited compared to that for both calix[4]- and calix[8]arene compounds [18–20]. Nonetheless, conformers which may be described as 'pinched cone' and 'hinged 3-up, 3-down' species have been well characterised [18, 19]. In both of these, the molecule may be regarded as providing two cavities of similar dimensions to the cup of cone-form calix[4]arene with, however, one of the four walls of the cup missing in each. The present derivative, [(CH$_3$)$_4$N]$_2$[p-t-butylcalix[6]arene – 2H], provides another example of the hinged, 3-up, 3-down form, and inclusion of one of the two [(CH$_3$)$_4$N]$^+$ ions occurs in a manner which is similar to that observed in the simpler of the two inclusion moieties found in [(CH$_3$)$_4$N]$_2$[calix[4]arene – H]$_2$·calix[4]arene·H$_2$O [12] and is essentially as suggested by us previously on the basis of solution NMR measurements [9]. The two independent calixarene species in the unit cell differ principally in that the included cation is ordered in one (molecule A, Figure 1) and disordered in the other (molecule B); the nonincluded cations are disordered over two equally populated orientations. As in the structures determined for other tetra-alkylammonium ion complexes of calixarenes [9, 12], it is possible to ascribe the cation inclusion to CH$_3$–(phenyl)π electron interactions assisted by cation–dipole attractions. Thus, although the included cation can be seen as embedded within a cone-like structure formed by three phenyl rings, placing the hydrogen atoms of at least one of the methyl groups in close proximity to the π electrons, the 'open' side of the cone does provide access to the presumably negative array of three oxygen atoms of the other half of the calixarene. The importance of optimising both factors may be the reason why the calixarene moiety does not have C_2 symmetry (with respect to the C(3)–C(6) axis) and show inclusion of both cations. The nitrogen atom of the included cation is significantly closer to the oxygen atoms which are not part of its inclusion cone than to the three which are (in molecule A, for example, N(1)···O(1), N(1)···O(2), N(1)···O(3) are 3.86, 4.47,

TABLE II. Torsion angles/o defining calix[6]arene inner rings. Atom numbering follows the figure below; O(1) is attached to C(11), O(2) to C(21), etc.

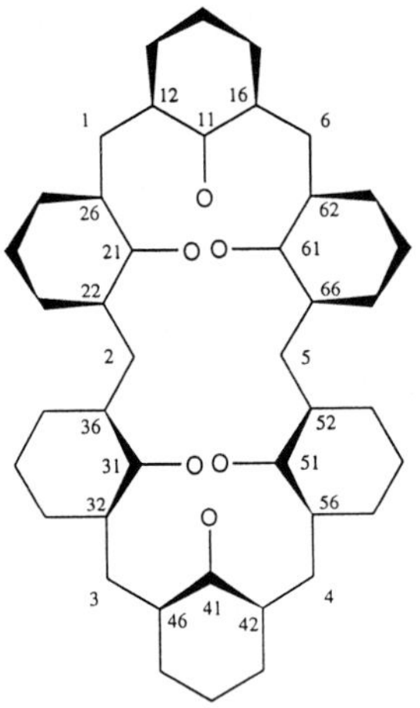

Torsion	Molecule A	Molecule B	Eu ligand [21]
C(26)-C(1)-C(12)-C(11)	-84.8(11)	-73.6(11)	-84.8
O(1)-C(11)-C(12)-C(1)	0.6(14)	-2.1(14)	-2.8
C(16)-C(11)-C(12)-C(1)	-179.9(9)	178.2(8)	-177.4
O(1)-C(11)-C(16)-C(6)	1.7(14)	2.7(13)	0.0
C(12)-C(11)-C(16)-C(6)	-177.8(9)	-177.7(8)	-177.4
C(36)-C(2)-C(22)-C(21)	-89.2(11)	-96.6(10)	-96.7
O(2)-C(21)-C(22)-C(2)	2.9(14)	3.9(12)	4.8
C(26)-C(21)-C(22)-C(2)	-177.4(8)	-176.2(8)	178.3
C(12)-C(1)-C(26)-C(21)	91.1(11)	97.3(10)	93.9
O(2)-C(21)-C(26)-C(1)	-1.5(13)	-6.7(12)	-3.3
C(22)-C(21)-C(26)-C(1)	178.8(8)	173.4(8)	-176.5
C(46)-C(3)-C(32)-C(31)	79.2(12)	84.5(12)	70.0
O(3)-C(31)-C(32)-C(3)	-7.3(13)	-2.7(14)	1.9
C(36)-C(31)-C(32)-C(3)	173.8(8)	178.8(8)	178.0
C(22)-C(2)-C(36)-C(31)	82.1(12)	73.4(12)	82.5
O(3)-C(31)-C(36)-C(2)	3.2(14)	2.8(14)	0.9
C(32)-C(31)-C(36)-C(2)	-177.8(8)	-178.8(9)	-175.0
C(56)-C(4)-C(42)-C(41)	82.8(12)	85.6(12)	86.6
O(4)-C(41)-C(42)-C(4)	-0.4(15)	3.0(14)	1.4
C(46)-C(41)-C(42)-C(4)	-179.7(9)	-179.1(8)	-179.3
C(32)-C(3)-C(46)-C(41)	-177.9(9)	-175.8(8)	-177.3
O(4)-C(41)-C(46)-C(3)	0.0(13)	-0.3(13)	0.5
C(42)-C(41)-C(46)-C(3)	179.3(9)	-178.3(8)	-178.9
C(66)-C(5)-C(52)-C(51)	90.4(12)	89.6(11)	94.1
O(5)-C(51)-C(52)-C(5)	-0.5(13)	-1.2(13)	5.9
C(56)-C(51)-C(52)-C(5)	179.8(8))	177.9(8)	-173.5
C(42)-C(4)-C(56)-C(51)	-90.4(12)	-89.0(11)	-96.3
O(5)-C(51)-C(56)-C(4)	1.8(13)	-0.3(13)	-4.4
C(52)-C(51)-C(56)-C(4)	-178.5(8)	-179.4(8)	175.1
C(11)-C(16)-C(6)-C(62)	-76.8(13)	-82.8(13)	-72.1
C(16)-C(6)-C(62)-C(61)	174.3(9)	173.4(8)	172.7
O(6)-C(61)-C(62)-C(6)	1.1(14)	-0.6(12)	4.9
C(66)-C(61)-C(62)-C(6)	178.0(9)	177.4(8)	179.6
C(52)-C(5)-C(66)–C(61)	-83.6(13)	-83.3(12)	-79.0
O(6)-C(61)-C(66)-C(5)	-2.8(15)	-0.6(13)	-3.0
C(62)-C(61)-C(66)-C(5)	-179.6(9)	-178.5(8)	-177.5

3.93 Å, respectively, whereas N(1)···O(4), N(1)···O(5), N(1)···O(6) are 4.85, 5.03, 4.88 Å, respectively) and, interestingly, the nitrogen atom of the nonincluded cation approaches the inclusion-cone array of oxygens in a very similar way to that in which the included cation nitrogen approaches the noninclusion-cone oxygens (in molecule A again, N(2)···O(5), N(2)···O(6), N(2)···O(6) are 4.40, 4.44, 3.92 Å, respectively).

Detailed comparison of the calixarene conformations observed in the present compound and in a 'true' metal complex of Eu(III) [21] where the cation appears to bind through a single coordinate bond can be made in terms of the macrocycle torsion angles given in Table II. It is apparent that the calixarene, which is formally dianionic in all three situations, also has very closely similar conformations in all. Flexibility and the presence of numerous energy minima associated with closely similar configurations are properties expected of the larger calixarenes in particular [13, 20, 22], so that it is somewhat remarkable that greater differences are not observed, although it is true that in the present structure phenoxide-oxygen separation indicate significantly higher symmetry than in the europium complex. Thus within the separate 'tripods' of the calixarene entities in the tetramethylammonium salt, the separation of adjacent oxygen atoms are O(2)···O(1) 2.57 Å (molecule A), 2.57 Å (molecule B), O(2)···O(3) 2.52 Å (molecule A), 2.57 Å (molecule B), and O(5)···O(4) 2.57 Å (molecule A), 2.61 Å (molecule B), O(5)···O(6) 2.57 Å (molecule A), 2.57 Å (molecule B), while the corresponding separations in the europium complex are 2.77, 2.76 Å (Eu end) and 2.51, 2.65 Å. Nonetheless, it has also been observed in calix[8]arene systems that the macrocycle conformation is seemingly more sensitive to the formal charge on the ligand than to the particular nature of the bound cation [8, 23, 24], and in this regard it is interesting to note that in $[Cl_3TiOTiCl_2]_2[p\text{-}t\text{-butylcalix[6]arenetetramethylether} - 2H]$ [25] the calixarene again has the hinged conformation, whereas in $[Ti_4O_2][p\text{-}t\text{-butylcalix[6]arene} - 6H]_2$ [26] both calixarene moieties are in the cone form. Further, it has recently been shown that the substituent conformation in a calix[4]arene derivative is remarkably sensitive to the ionisation of one phenolic group [27].

3.2. SOLUTION CHEMISTRY

A species depositing from a given solution may not necessarily be that which is most abundant in solution, and in the case of tetraalkylammonium salts of calixarenes it is apparent that large solubility increases (seemingly regardless of solvent over a large range of polarity) associated with increasing degrees of deprotonation can cause salts of the singly-charged anions to deposit even when those anions are minor components of the solution mixtures. Thus, in our first syntheses [9, 12], calixarene monoanion salts were frequently obtained from solutions containing quite large excesses of tetraalkylammonium hydroxide. Nonetheless, the nature of the isolated species is obviously determined by a variety of factors other than mere relative solubilities. We have, for example, been able to readily isolate salts presumed to contain the *p-t*-butylcalix[6]arene trianion even though equilibrium measurements [1] in benzonitrile indicate that it is difficult to remove more than two protons from this calixarene, and in the case of calix[4]arene [12], the isolated solid containing the tetramethylammonium salt of the monoanion also contains the unionised calixarene. The full complexity of this chemistry remains to be established but from the present and earlier work, a useful variety in stoichiometry is available.

The different solubility characteristics of the various calixarene salts presently studied made comparison of their properties in a common solvent difficult. A mixture of one volume $(CD_3)_2CO$ to ten volumes of $CDCl_3$, however, does provide a solvent suitable for at least qualitative study of the inclusion characteristics of most by ^{1}H-NMR spectroscopy. For the tetrapropylammonium salts of calix[4]-, *p-t*-butylcalix[4]-, *p-t*-butyldihomooxacalix[4]-, *p-t*-butylcalix[6]- and *p-t*-butylcalix[8]arene monoanions and of *p-t*-butylcalix[6]- and *p-t*-butylcalix[8]arene di- and trianions, the various propyl group resonances were little shifted, if at all (but see below), from those of [Pr$_4$N]Br in this solvent, indicating that the [Pr$_4$N]$^+$ cation is not included to a marked degree in any case. When choline triflate was added to these solutions, near-quantitative precipitation of the choline cation/calixarene anion salt occurred with calix[4]arene and *p-t*-butylcalix[4]arene, but, in all the other systems a solution was obtained for which the quaternary ammonium methyl resonance was shifted ~1.5–2 ppm upfield from the position in the absence of the calixarene anion. Thus, choline inclusion was presumed to be strongly favoured in all these cases. (Similar observations were made when tetramethylammonium triflate was used in place of the choline salt, but greater problems were experienced with product precipitation. Interestingly, close comparison of the *N*-propyl group resonances before and after addition of choline triflate revealed some subtle differences between the various systems, including those which formed precipitates. Small shifts were in fact detectable for both the NCH$_2$ and terminal methyl resonances, and these varied in both magnitude and direction. For all the anions derived from *p-t*-butylcalix[6]arene and *p-t*-butylcalix[8]arene, small (~0.1–0.2 ppm) downfield shifts appeared to result from choline inclusion, with the shift being greater for the methylene than the methyl protons except for the trianion of *p-t*-butylcalix[6]arene, where the methylene shift was barely detectable. In any case, these shifts are possibly indicative of a rather unfavourable inclusion equilibrium involving the simple [Pr$_4$N]$^+$ salts, an equilibrium which may involve two distinct forms, one where the N$^+$ centre can be regarded as inserted (so placing NCH$_2$ protons close to aromatic π electrons) and one where the terminal methyl group is simply inserted. For the tetrapropylammonium salts of *p-t*-butyldihomooxacalix[4]arene and calix[4]arene, addition of choline triflate resulted in very small (~0.05 ppm) downfield shifts of the methyl proton resonances only, perhaps indicating a very unfavourable inclusion equilibrium involving these methyl groups in the original solution. For the tetrapropylammonium (and the tetraethylammonium) salt of *p-t*-butylcalix[4]arene monoanion, a unique situation appears to prevail in that addition of choline triflate resulted in small (~0.1 ppm) *up*field shifts of all the propyl resonances. Since there is evidence (see below) that the *t*-butyl groups inhibit inclusion by this calixarene, this observation may indicate that ion association in the original solution placed the cation in close proximity to the electronegative phenol/phenolate oxygen atoms.

For the salts of the smaller calixarenes, acetone alone was a far superior solvent to acetone/chloroform mixtures, though it was generally apparent that cation inclusion was much weaker in acetone, perhaps as a result of competition for inclusion by the solvent itself [19, 28]. The solubilities of the monoanions of both *p-t*-butylcalix[4]arene and calix[4]arene in the presence of choline and acetylcholine cations were sufficiently high for NMR measurements to be readily made without interference from precipitation and so enabled demonstration of very marked selectivity in the interactions of these two calixarenes with either cation. Thus there was in fact no evidence at all of inclusion of either cation by *p-t*-butylcalix[4]arene, whereas the equilibrium constants (Table III) determined by fitting the marked upfield shifting of $N(^+)CH_3$ resonances in the presence of various amounts of calix[4]arene monoanion to a model of 1 : 1 complexation were so large as to indicate that under the given experimental conditions, effectively all cation that could be included was. The two calixarene anions appeared to be equally discriminatory towards the tetramethylammonium cation, though the solubilities of their salts with this cation were too low to allow extensive measurements, and it is worth noting that when the isolated salts were dissolved in $(CD_3)_2SO$, the $N(^+)CH_3$ resonances for both appeared slightly upfield (~0.2 ppm) of that for $[N(CH_3)_4][CF_3SO_3]$ in the same solvent. Nonetheless, unless acetone solvent inclusion is very much more favourable for *p-t*-butylcalix[4]arene than for calix[4]arene monoanion, it would appear that this is an instance of where inclusion by a calix[4]arene is inhibited rather than facilitated [19] by the addition of *t*-butyl substituents.

The binding of choline by calix[4]arene monoanion in acetone is appreciably stronger than that of acetylcholine, and the same relative affinity is seen for *p-t*-butyldihomooxacalix[4]arene monoanion (Table III). Note that the absolute values of the equilibrium constants for these two calixarene anions are not directly comparable due to the use of the tetraethylammonium salt of the latter species, in which some inclusion actually occurs initially. Since *p-t*-butyldihomooxacalix[4]arene is known to be sufficiently capacious to include two small molecules [28], it was considered that its monoanion might act as a catalyst for acetylation reactions involving acetylcholine, despite the fact that favourable binding of the choline product might be expected to ultimately inhibit reaction otherwise accelerated by the placement of reactants in close proximity [29]. No reaction in which such catalysis or even stoichiometric acetyl transfer can be observed has yet been found, however. The acetylation of benzylamine by acetylcholine, for example, is, if anything, slightly inhibited in the presence of *p-t*-butyldihomooxacalix[4]arene.

The apparently quite large values for inclusion equilibrium constants determined by NMR spectroscopy indicated that electronic spectroscopy might well be a more

TABLE III. Association equilibrium constants for calixarene anions and quaternary ammonium cations.

$$\text{Calixarene}^{n-} + \text{Cation}^+ \overset{K}{\rightleftharpoons} [\text{Calixarene} \cdot \text{Cation}]^{(n-1)-}$$

Cation	Anion	Solvent	Method	K/M^{-1}
Choline	[p-t-butyldihomooxacalix[4]arene − H]⁻	Acetone	¹H-NMR titration	$(2.50 \pm 0.37) \times 10^{3a,c}$
Choline	[calix[4]arene − H]⁻	Acetone	¹H-NMR titration	$(5.1 \pm 2.7) \times 10^{2a,c}$
Acetylcholine	[p-t-butyldihomooxacalix[4]arene − H]⁻	Acetone	¹H-NMR titration	$(8.5 \pm 2.2) \times 10^{2a,c}$
Acetylcholine	[calix[4]arene − H]⁻	Acetone	¹H-NMR titration	$(2.55 \pm 10^2)^{a,c}$
Acetylcholine	[p-t-butylcalix[6]arene − 3H]³⁻	Acetonitrile	Spectrophotometric titration	$(3.3 \pm 2.1) \times 10^{5b,c}$
[N(CH₃)₄]⁺	[p-t-butylcalix[6]arene − 3H]³⁻	Acetonitrile	Spectrophotometric titration	$(3.22 \pm 0.71) \times 10^{5b,c}$
[N(CH₃)₄]⁺	[p-t-butylcalix[6]arene − 2H]²⁻	Acetonitrile	Spectrophotometric titration	$(1.72 \pm 0.39) \times 10^{4b,c}$

[a] 303 ± 1 K
[b] 298.2 ± 0.1 K
[c] 1σ errors.

appropriate method for use in more extensive measurements involving more highly charged anions [30]. Again, however, problems arose in the choice of an appropriate solvent and ultimately in the sensitivity of the method. Acetonitrile is a relatively good solvent for all the compounds presently studied and is also convenient for measurements in the UV region but it is, of course, well known for its tendency to form inclusion complexes with calixarenes [19]. This may well be the reason that for the tetrapropylammonium salts of both *p-t*-butylcalix[4]arene and calix[4]arene monoanions, no effect of even a large molar excess of tetramethylammonium triflate on the UV spectra was discernible. For other anions, changes could be detected but were generally only just outside experimental errors unless a 50-fold or greater molar amount of $[(CH_3)_4N]^+$ was added. Indeed, only for the trianon of *p-t*-butylcalix[6]arene could reproducible effects of the addition of a similar quantity of tetramethylammonium ion (or acetylcholine) be determined and associated with an equilibrium constant for a 1 : 1 association (Table III), presumed to be an inclusion process. Surprisingly, the effects of choline addition, though giving rise to absorbance changes at least as great as those resulting from tetramethylammonium or acetylcholine ion additions, were not associated with well-defined isosbestic points (as the others were), and indicated that at least two new species were formed as the ratio of cation to anion varied between 0 and 1. Obviously, many complications may arise in these systems if, for example, inclusion results in the alteration of phenolic proton distributions or calixarene conformations, and more detailed measurements are in progress in an attempt to unravel these complexities. Only limited quantitative data on the binding of cations by uncharged calixarene derivatives are to be found in the literature [31], and any comparison with the present results is vitiated by the different solvent systems involved, though it is interesting to note the observation that cation binding by neutral calixarenes is also inhibited by the presence of *t*-butyl substituents.

Inclusion of acetylcholine by at least the di- and trianions of *p-t*-butylcalix[6]arene is clearly indicated by ^{1}H-NMR spectroscopy for acetone, dimethylsulphoxide and chloroform solvents and is presumed to be the cause of spectroscopic changes in acetonitrile. If the guest is bound in the same way as is the included cation in solid $[N(CH_3)_4]_2[p$-t-butylcalix[6]arene $-$ 2H], then the acetyl carbon electrophile would be brought into close proximity to a nucleophilic phenolate group. Further, in the incompletely deprotonated calixarene, adjacent acidic phenolic protons are available to facilitate alkoxide group departure. Figure 2 shows a possible structure for an acetylcholine : *p-t*-butylcalix[6]arene dianion inclusion complex obtained by using the crystal structure coordinates reported here and modifying the cation (to convert it to acetylcholine) in the molecular modelling program 'Chem3D$^+$' [32]. In this configuration, the separation between the acetyl carbon and the closest phenolic oxygen is 1.5 Å. Despite this further indication that inclusion of acetylcholine by a calixarene may provide a means of accelerating reaction, we have been unable to detect formation of the monoacetate of *p-t*-butylcalix[6]arene, even after extended periods of reaction in various

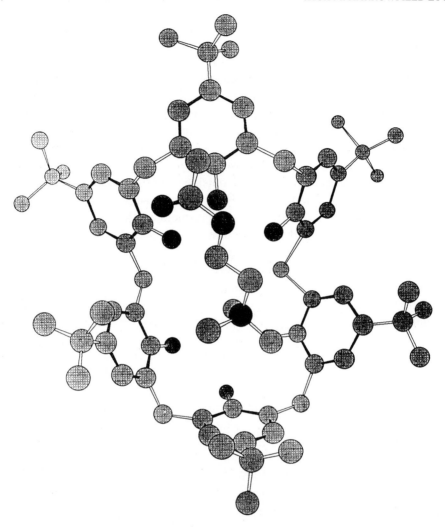

Fig. 2. A possible configuration for the inclusion complex formed between acetylcholine and the dianion of *p-t*-butylcalix[6]arene. (Viewed from 'above'.)

solvents. Thus, either the calixarene's oxygen nucleophilicity is too low for reaction to be favourable, or the inclusion complex adopts a conformation, such as the cone, which places the acetyl group remote from any nucleophilic centre. In regard to the latter point, it may be noted that although inclusion of acetylcholine is most easily monitored through the large chemical shift changes occurring for the $N(^+)CH_3$ resonances, the acetyl methyl also shows small upfield shifts, which could be indicative of simultaneous inclusion of this group within the half of a calix[6]arene cone not occupied by a N-methyl group. It is also worthy of note that it seems unlikely that proton transfer may be inhibitory, in that an incidental observation during the present work was that dissolution of the di- and trianionic species in $CDCl_3$ resulted in the growth of a signal due to $CHCl_3$ over 10–20 min at room temperature.

Overall, the present observations on acetylcholine support the notion that inclusion by CH_3/aromatic-π-electron interactions is an effective binding mechanism [14, 15] but they provide little information on how reaction subsequent to binding may be enhanced in an enzyme such as acetylcholine esterase. The calixarene framework does, however, provide a means to bring other nucleophiles into proximity with an included species and to attach activating units such as metal ions, and we are continuing to attempt to develop a more sophisticated model of an acylating enzyme. A recent report [33] of the development of an active acylation catalyst on the basis of principles similar to those described above vividly illustrates the complexities which may be encountered in this task.

Acknowledgement

Funding for this work was provided by the Australian Research Council. We thank Dr. E. Kucharski for assistance with data analysis and Professor M. Perrin for constructive criticism of the original manuscript.

References

1. A. F. Danil de Namor, M. T. Garrido Pardo, D. A. Pacheko Tanaka, F. J. Sueros Verlade, J. D. Cárdenas Garcia, M. C. Cabaleiro, and J. M. A. Al-Rawl: *J. Chem. Soc., Faraday Trans.* **89**, 2727 (1993) and references therein.
2. R. M. Izatt, J. D. Lamb, R. T. Hawkins, P. R. Brown, S. R. Izatt, and J. J. Christensen: *J. Am. Chem. Soc.* **105**, 1782 (1983).
3. S. R. Izatt, R. T. Hawkins, R. M. Izatt, and J. J. Christensen: *J. Am. Chem. Soc.* **107**, 63 (1985).
4. H. Goldmann, W. Vogt, E. Paulus, and V. Böhmer: *J. Am. Chem. Soc.* **110**, 6811 (1988).
5. C. D. Gutsche: *Progr. Macrocyclic Chem.* **3**, 93 (1987) (especially pp. 152–155) and references therein.
6. C. D. Gutsche, M. Iqbal, K. S. Nam, and I. Alam: *Pure Appl. Chem.* **60**, 483 (1988).
7. J.-C. G. Bünzli and J. M. Harrowfield: in *Calixarenes: A Versatile Class of Macrocyclic Compounds*, J. Vicens and V. Böhmer (Eds.), Kluwer Academic Publishers, Dordrecht, p. 211 (1991).
8. (a) G. E. Hofmeister, E. Alvarado, J. A. Leary, D. I. Yoon, and S. F. Pedersen: *J. Am. Chem. Soc.* **112**, 8843 (1990). (b) G. E. Hofmeister, F. E. Hahn, and S. F. Pedersen: *J. Am. Chem. Soc.* **111**, 2318 (1989).
9. J. M. Harrowfield, M. I. Ogden, W. R. Richmond, B. W. Skelton, and A. H. White: *J. Chem. Soc., Perkin Trans. 2* 2183 (1993) and references therein.
10. J. M. Harrowfield, M. I. Ogden, W. R. Richmond, and A. H. White: *J. Chem. Soc., Chem. Commun.* 1159 (1991).
11. R. Asmuss, V. Böhmer, J. M. Harrowfield, M. I. Ogden, W. R. Richmond, B. W. Skelton, and A. H. White: *J. Chem. Soc., Dalton Trans.* 2427 (1993).
12. J. M. Harrowfield, W. R. Richmond, A. N. Sobolev, and A. H. White: *J. Chem. Soc., Perkin Trans. 2* 5 (1994).
13. P. D. J. Grootenhuis, P. A. Kollman, L. C. Groenen, D. N. Reinhoudt, G. J. van Hummel, F. Ugozzoli, and G. D. Andreetti: *J. Am. Chem. Soc.* **112**, 4165 (1990).
14. J. L. Sussman, M. Harel, F. Frolow, C. Oefner, A. Goldman, L. Toker, and I. Silman: *Science* **253**, 872 (1991).
15. H. Soreq, A. Gnatt, Y. Loewenstein, and L. F. Neville: *TIBS* **17**, 353 (1992).

16. W. T. Robinson and G. M. Sheldrick: in *Crystallographic Computing 4: Techniques and New Technologies*, N. W. Isaacs and M. R. Taylor (Eds.), Oxford University Press (1988).
17. G. M. Sheldrick: in *Crystallographic Computing 5*, D. Moras, A. D. Podjarny, and J. C. Thierry (Eds.), Oxford University Press (1992).
18. M. Perrin and D. Oehler: in *Calixarenes: A Versatile Class of Macrocyclic Compounds*, Kluwer Academic Publishers, Dordrecht, p. 65 (1991).
19. G. D. Andreetti and F. Ugozzoli: in *Calixarenes: A Versatile Class of Macrocyclic Compounds*, Kluwer Academic Publishers, Dordrecht, p. 87 (1991).
20. P. Neri, M. Foti, G. Ferguson, J. F. Gallagher, B. Kaitner, M. Pons, M. A. Molins, L. Giunta, and S. Pappalardo: *J. Am. Chem. Soc.* **114**, 7814 (1992).
21. L. M. Engelhardt, B. M. Furphy, J. M. Harrowfield, D. L. Kepert, A. H. White, and F. R. Wilner: *Aust. J. Chem.* **44**, 1465 (1988).
22. C. D. Gutsche: in *Calixarenes*, Chemical Society Monographs in Supramolecular Chemistry, No. 1, J. F. Stoddart (Ed.), Royal Society of Chemistry, Ch. 4 (1989).
23. J. M. Harrowfield, M. I. Ogden, W. R. Richmond, and A. H. White: *J. Chem. Soc., Dalton Trans.* 2153 (1991).
24. J. M. Harrowfield, M. I. Ogden, and A. H. White: *J. Chem. Soc., Dalton Trans.* 2625 (1991).
25. S. G. Bott, A. W. Coleman, and J. L. Atwood: *J. Chem. Soc., Chem. Commun.* 610 (1986).
26. G. D. Andreetti, G. Calestani, F. Ugozzoli, A. Arduini, E. Ghidini, A. Pochini, and R. Ungaro: *J. Incl. Phenom.* **3**, 447 (1985).
27. J. L. Atwood, S. G. Bott, G. W. Orr, and K. D. Robinson: *Angew. Chem. Int. Ed. Engl.* **32**, 1093 (1993).
28. Z. Asfari, J. M. Harrowfield, M. I. Ogden, J. Vicens, and A. H. White: *Angew. Chem. Int. Ed. Engl.* **30**, 854 (1991).
29. F. M. Menger: *Acc. Chem. Res.* **18**, 128 (1985) and references therein.
30. J. Polster and H. Lachmann: *Spectrometric Titrations – Analysis of Chemical Equilibria*, VCH Publishers Weinheim (1989).
31. K. Araki, H. Shimizu, and S. Shinkai: *Chem. Lett.* 205 (1993).
32. 'Chem3D$^+$' Molecular Modelling Software, Cambridge Scientific Computing Inc. (1989).
33. H.-J. Schneider, R. Kramer, and J. Rammo: *J. Am. Chem. Soc.* **115**, 8980 (1993).

Calixarenes: Structure of an Acetonitrile Inclusion Complex and Some Transition Metal Rimmed Derivatives[*]

WEI XU and RICHARD J. PUDDEPHATT[**]
Department of Chemistry, University of Western Ontario, London, Canada N6A 5B7

and

LJUBICA MANOJLOVIC-MUIR, KENNETH W. MUIR[**] and CHRISTOPHER S. FRAMPTON
Department of Chemistry, University of Glasgow, Glasgow, Scotland G12 8QQ

(Received: 3 March 1994; in final form: 14 June 1994)

Abstract. The host t-butylcalix[4]arene, **1a**, forms a 1 : 1 inclusion compound with acetonitrile as guest. The inclusion compound has been isolated and characterised by X-ray analysis of a twinned crystal at 123 K. The acetonitrile guest lies on a crystallographic four-fold symmetry axis passing through the centre of the bowl of **1a** which adopts a regular cone conformation. A known tetradentate and a new tridentate phosphinitocalix[4]arene derivative, **2a** and **2c** respectively, have been synthesized from **1a** and Ph$_2$Cl. Both **2a** and **2c** show a strong ability to coordinate with late transition metals and new complexes of gold(I), palladium(II) and platinum(II) are reported.

Key words: Calix[4]arene, phosphinite, coordination, transition metal.

Supplementary Data relating to this paper have been deposited with the British Library as Supplementary Publication No. SUP 92171 (5 pp.).

1. Introduction

The calixarenes are a group of phenolic macrocyclic compounds that have become significant in supramolecular chemistry during the past decade [1]. The most useful property of calixarenes is their ability to function as molecular baskets which can engulf small neutral or ionic guests [2–3]. The calixarene **1**, which contains hydroxyl groups at the bottom rim, has been extensively studied and its transition metal complexes are attracting attention [4–8]. Both fully hydroxylated **1a** and partially hydroxylated *tert*-butylcalixarene **1b** have been reacted with chlorodiphenylphosphine to give the fully substituted [9] and partially substituted [10] diphenylphosphinocalix[4]arenes, **2a** and **2b**. The compound **2a** was used as a phosphorus donor ligand in forming copper(I) [9] and iron(0) carbonyl [11] complexes. This paper

---
[*] This paper is dedicated to the commemorative issue on the 50th anniversary of calixarenes.
[**] Author for correspondence.

reports the synthesis of the tridentate phosphinitocalix[4]arene derivative **2c** and a study of the coordination chemistry of **2a** and **2c** with late transition metals. The conformations of the calixarenes with transition metal substituents on the lower rim will be discussed, and the structure of a host–guest complex of **1a** with acetonitrile will be described.

1 (1a: R=H; 1b: R=Me)

2a: $R^1=R^2=R^3=R^4=$ PPh$_2$;
2b: $R^1=R^2=$ Me; $R^3=R^4=$ PPh$_2$
2c: $R^1=R^2=R^3=$ PPh$_2$; $R^4 =$ H

2. Experimental

[AuCl(SMe$_2$)] [12], [PtCl$_2$(SMe$_2$)$_2$] [13], [PdCl$_2$(PhCN)$_2$] [14], and *p-tert*-butylcalix[4]arene [15] were prepared by literature methods. NMR spectra were recorded by using a Varian Gemini 300 MHz spectrometer. ^{1}H- and ^{13}C-NMR chemical shifts were measured relative to partially deuterated solvent peaks, but are reported relative to tetramethylsilane. ^{31}P-NMR chemical shifts were determined relative to 85% H$_3$PO$_4$ as an external standard. IR spectra were recorded on a Bruker IFS32 FTIR spectrometer with Nujol mulls. Elemental analyses were performed by Galbraith Laboratories Inc., Knoxville, TN, U.S.A.

p-tert-Butylcalix[4]arene(OPPh$_2$)$_4$, **2a**, was prepared as described in the literature [9, 11]. The ^{1}H-NMR spectrum is in good agreement with the literature [11]. The ^{31}P-NMR spectrum gives a singlet at $\delta = 123$ (Lit. $\delta = 29.27$, note the large discrepancy [11]) in CD$_2$Cl$_2$.

p-tert-Butylcalix[4]arene(OPPh$_2$)$_3$(OH), **2c**: Bu^{n}Li (2.5 M, 2 mL) was added at room temperature to a THF (15 mL) suspension of *p-tert*-butylcalix[4]arene **1** (1 g, 1.54 mmol). An orange solution was formed. Then the mixture was cooled to 0° and ClPPh$_2$ (1.14 mL, 6.25 mmol) was added slowly. The mixture was kept at 0°C for 2 h and then warmed to room temperature and stirred for 12 h. The resulting brown solution was heated to 60°C and kept at this temperature for another 48 h.

Most of the solvent was then evaporated under vacuum. MeCN (20 mL) was added and the mixture was cooled to 0°C for 4 h, when **2c** (1.15 g, 62.1%) was precipitated as a white solid, which was washed with MeCN and dried under vacuum. *Anal. Calc.* for **2c** $C_{80}H_{83}O_4P_3 \cdot MeCN$: C, 79.3; H, 7.0; *Found*: C, 79.1; H, 6.8%. Spectroscopic data of **2c**: NMR in CD_2Cl_2: $\delta(^{31}P)$ = 121.8 [s, 2P] and 112.9 [s, 1P]. $\delta(^1H)$ = 1.11 [s, 18H, t-Bu], 1.21 [s, 9H, t-Bu], 1.15 [s, 9H, t-Bu], 2.78 [d, $^2J_{HH}$ = 13.2 Hz, 2H, CH_2], 3.78 [d, $^2J_{HH}$ = 15.6 Hz, 2H, CH_2], 4.25 [d, $^2J_{HH}$ = 14.9 Hz, 2H, CH_2], 4.20 [d, $^2J_{HH}$ = 13.4 Hz, 2H, CH_2], 6.6–7.84, [PC$_6H_5$, and CC$_6H_5$], 1.97 [s, 3H, CH_3CN]. IR: ν_{OH} = 3351 cm^{-1}.

p-tert-Butylcalix[4]arene(OPPh$_2 \cdot$AuCl)$_3$(OH), **3**. A mixture of [AuCl(SMe$_2$)]$^-$ (0.147 g, 0.50 mmol) and *p-tert*-butylcalix[4]arene(OPPh$_2$)$_3$(OH), **2c**, (0.20 g, 0.167 mmol) in CH_2Cl_2 was stirred at room temperature for 15 h. The CH_2Cl_2 and SMe$_2$ were removed completely under vacuum. The residue was redissolved in CH_2Cl_2 (4 mL) and crystallized by diffusion with MeCN (10 mL). A white solid (0.17 g, yield 53.6%) was obtained, washed with MeCN and dried under vacuum. *Anal. Calc.* for **3** $C_{80}H_{83}Au_3Cl_3O_4P_3 \cdot CH_2Cl_2$: C, 49.1; H, 4.3; *Found*: C, 48.7; H, 4.0%. Spectroscopic data of **3**: NMR in CD_2Cl_2: $\delta(^{31}P)$ = 117.8 [s, 2P], 114.8 [s, 1P]. $\delta(^1H)$ = 0.93 [s, 18H, t-Bu], 0.99 [s, 9H, t-Bu], 1.25 [s, 9H, t-Bu], 2.91 [d, $^2J_{HH}$ = 12 Hz, 2H, CH_2], 3.16 [d, $^2J_{HH}$ = 14.8 Hz, 2H], 3.38 [d, $^2J_{HH}$ = 14.8 Hz, 2H], 3.51 [s, 1H, OH], 3.66 [d, $^2J_{HH}$ = 13.6 Hz, 2H, CH_2], 6.59 [t, J_{HH} = 4.6 Hz, 4H, ArH], 6.95 [d, J_{HH} = 8.4H, 4H, ArH], 7.3–7.9 [m, 30H, PC$_6H_5$], 5.33 [s, 2H, CH$_2$Cl$_2$]. IR: ν_{OH} = 3366 cm^{-1}.

p-tert-Butylcalix[4]arene(OPPh$_2$AuCl)$_4$, **4a**. A mixture of [AuCl(SMe$_2$)] (0.17 g, 0.58 mmol) and *p-tert*-butylcalix[4]arene(OPPh$_2$)$_4$, **2a**, (0.20 g, 0.145 mmol) in CH_2Cl_2 (20 mL) was stirred at room temperature for 15 h. The CH_2Cl_2 and SMe$_2$ were removed completely under vacuum. The residue was redissolved in CH_2Cl_2 and crystallized by diffusion with MeCN. Platelike colourless crystals (0.25 g, yield 74.5%) were obtained, washed with MeCN and dried under vacuum. Spectroscopic data of **4a**: *Anal. Calc.* for **4a** $C_{92}H_{92}Au_4Cl_4O_4P_4 \cdot MeCN$: C, 47.9; H, 4.1; *Found*: C, 48.1; H, 4.2%. NMR in CD_2Cl_2: $\delta(^{31}P)$ = 115.9 [s]. $\delta(^1H)$ = 1.03 [s, 36H, t-Bu], 1.97 [s, 3H, CH_3CN], 2.61 [d, $^2J_{HH}$ = 13.2 Hz, 4H, CH_2], 4.44 [d, $^2J_{HH}$ = 13.2 Hz, 4H, CH_2], 6.69 [s, 8H, ArH], 7.15–7.20 [m, 16H, PC$_6H_5$, meta to P], 7.42–7.48 [m, 8H, PC$_6H_5$, para to P], 7.62–7.70 [m, 16H, PC$_6H_5$, ortho to P].

p-tert-Butylcalix[4]arene(OPPh$_2$AuCCPh)$_4$, **4b**. A mixture of [(AuCCPh)$_n$] (0.17 g, 0.58 mmol) and *p-tert*-butylcalix[4]arene(OPPh$_2$)$_4$, **2a**, (0.20 g, 0.145 mmol) in CH_2Cl_2 (20 mL) was stirred at room temperature for 15 h. The same procedure as above was used and a white solid (0.19 g, yield 49.9%) was obtained and washed with MeCN and dried under vacuum. Spectroscopic data of **4b**: *Anal. Calc.* for **4b** $C_{124}H_{112}Au_4O_4P_4 \cdot MeCN$: C, 56.4; H, 4.4; *Found*: C, 56.5; H, 4.2%.

NMR in CD_2Cl_2: $\delta(^{31}P) = 132.3$ [2]. $\delta(^1H) = 1.04$ [s, 36H, t-Bu], 1.97 [s, 3H, CH_3CN], 2.69 [d, $^2J_{HH} = 13.4$ Hz, 4H, CH_2], 4.57 [d, $^2J_{HH} = 13.4$ Hz, 4H, CH_2], 5.30 [s, 4H, CH_2Cl_2], 6.70 [s, 8H, ArH], 7.07–7.84 [m, 68H, PC_6H_5, and CC_6H_5].

p-tert-Butylcalix[4]arene(OPPh$_2$)$_4$(PtCl$_2$)$_2$, 5a. A mixture of [PtCl$_2$(SMe$_2$)$_2$] (0.12 g, 0.30 mmol) and *p-tert*-butylcalix[4]arene(OPPh$_2$)$_4$, **2a**, (0.20 g, 0.145 mmol) in CH_2Cl_2 (20 mL) was stirred at room temperature for 15 h. The CH_2Cl_2 and SMe_2 were removed completely under vacuum. The residue was redissolved in CH_2Cl_2 and crystallized by diffusion with MeCN. A white solid (0.18 g, yield 67.1%) was obtained, washed with MeCN and dried under vacuum. *Anal. Calc.* for **5a** $C_{92}H_{92}Cl_4P_4Pt_2O_4 \cdot MeCN$: C, 57.6; H, 4.9; *Found*: C, 57.9; H, 5.1%. Spectroscopic data of **5a**: NMR in CD_2Cl_2: $\delta(^{31}P) = 83.5$ [s, $^1J_{Pt-P} = 4364.5$ Hz]. $\delta(^1H) = 0.90$ [s, 36H, t-Bu], 2.01 [s, 3H, CH_3CN], 2.88 [d, 2H, $^2J_{HH} = 14$ Hz, CH_2, exo], 2.91 [d, 2H, $^2J_{HH} = 14$ Hz, CH_2, exo to P—Pt—P ten-membered ring], 4.49 [d, 2H, $^2J_{HH} = 14$ Hz, CH_2, endo], 5.85 [dt, 2H, $^2J_{HH} = 14$ Hz, $^5J_{PH} = 3.4$ Hz, CH_2, endo to P—Pt—P ten-membered ring], 6.30 [s, 4H, ArH, endo to P—Pt—P ten-membered ring], 6.47 [s, 4H, ArH, exo to P—Pt—P ten-membered ring], 7.33–7.46 [m, 16H, PC_6H_5, para to P], 7.60–7.70 [m, 8H, PC_6H_5, meto to P], 8.12–8.22 [m, 16H, PC_6H_5, ortho to P].

p-tert-Butylcalix[4]arene(OPPh$_2$)$_4$(PdCl$_2$)$_2$, 5b. A mixture of [PdCl$_2$(PhCN)$_2$] (0.08 g, 0.29 mmol) and *p-tert*-butylcalix[4]arene(OPPh$_2$)$_4$ **2a** (0.20 g, 145 mmol) in CH_2Cl_2 (20 mL) was stirred at room temperature for 15 h. The CH_2Cl_2 and SMe_2 were removed completely under vacuum. The residue was redissolved in CH_2Cl_2 and crystallized by diffusion with MeCN. A white solid (0.11 g, yield 43.7%) was obtained, washed with MeCN and dried under vacuum. *Anal. Calc.* for **5b** $C_{92}H_{92}Cl_4P_4Pd_2O_4 \cdot MeCN$: C, 63.4; H, 5.4; *Found*: C, 63.8; H, 5.4%. Spectroscopic data of **5b**: NMR in $CDCl_3$: $\delta(^{31}P) = 115.7$ [s]. $\delta(^1H) = 0.89$ [s, 36H, t-Bu], 2.07 [s, 3H, CH_3CN], 2.78 [d, 2H, $^2J_{HH} = 13.6$ Hz, CH_2, exo], 2.81 [d, 2H, $^2J_{HH} = 13.6$ Hz, CH_2, exo to P—Pt—P ten-membered ring], 4.38 [d, 2H, $^2J_{HH} = 14$ Hz, CH_2, endo], 5.90 [dt, 2H, $^2J_{HH} = 14$ Hz, $^5J_{PH} = 3.4$ Hz, CH_2, endo to P—Pt—P ten-membered ring], 6.28 [s, 4H, ArH, endo to P—Pt—P ten-membered ring], 6.48 [s, 4H, ArH, exo to P—Pt—P ten-membered ring], 7.33–7.47 [m, 16H, PC_6H_5, para to P], 7.55–7.70 [m, 8H, PC_6H_5, ortho to P].

CRYSTAL STRUCTURE ANALYSIS OF **1a**·MeCN

Crystal data. $C_{44}H_{56}O_4 \cdot CH_3CN$, M = 689.98, T = 123 K, tetragonal, space group $P4/n$ (origin at $\bar{1}$ $a = 12.7194(4)$, $c = 12.7668(3)$ Å, $V = 2065.5(1)$) Å^3, $Z = 2$, $D_{calc} = 1.109$ g cm^{-3}, $F(000) = 748$, MoK$_\alpha$ X-rays, $\lambda = 0.71073$ Å, $\mu = 0.65$ cm^{-1}.

Measurements. The unit cell dimensions were obtained by a least-squares treatment of the setting angles of 25 reflections with $11° < \theta(MoK_\alpha) < 13°$. The

TABLE I. Fractional coordinates and equivalent isotropic displacement parameters (Å^2) of nonhydrogen atoms.

	x/a	y/b	z/c	U^*
O(1)	0.1147(1)	0.1861(1)	-0.0561(1)	0.019
C(2)	0.0655(1)	0.1573(1)	0.0363(1)	0.015
C(4)	-0.0038(1)	0.2298(1)	0.0814(1)	0.01
C(5)	0.1607(1)	-0.0190(1)	0.0362(1)	0.015
C(6)	0.0828(1)	0.0593(1)	0.0819(1)	0.014
C(7)	0.0282(1)	0.0344(1)	0.1740(1)	0.016
C(8)	-0.0422(1)	0.1035(1)	0.2207(1)	0.016
C(9)	-0.0563(1)	0.2012(1)	0.1726(1)	0.015
C(11)	-0.1059(2)	0.0762(2)	0.3189(1)	0.021
C(12)	-0.2233(2)	0.0757(2)	0.2911(2)	0.029
C(16)	-0.0861(2)	0.1598(2)	0.4039(2)	0.031
C(20)	-0.0773(2)	-0.0361(2)	0.3632(2)	0.034
C(28)	1/4	1/4	0.3688(7)	0.080
C(29)	1/4	1/4	0.2488(4)	0.038
N(27)	1/4	1/4	0.4677(7)	0.124

*U is one third of the trace of the orthogonalised anisotropic displacement tensor.

intensities of 4860 reflections with $1° < \theta MoK_\alpha < 27°$, h 0–16, k 0–16, l 6–16, were determined from $\omega/2\theta$ scans. Those of two standard reflections, remeasured every 2 h, showed random fluctuations of < 3.2% about their mean values. All measurements were made at 123 K on an Enraf–Nonius diffractometer. Averaging, assuming $4/m$ Laue symmetry, gave R_{int} = 0.037 for 2155 unique intensities measured more than once and yielded 2257 independent structure amplitudes of which 1846 with $I > 3\sigma(I)$ were used in the subsequent analysis.

Structure Analysis. The positions of the nonhydrogen atoms were obtained by direct methods [16]. Further progress was only possible after it was realised that the specimen crystal was twinned. Using the program CRYSTALS [17] the function $\sum w(|F_{obs}| - |F_{calc}|)^2$ was minimised with $|F_{calc}| = [pF^2(hkl) + (1 - p)F^2(khl)]^{1/2}$; p defines the relative amounts of the two twin components and $F(hkl)$ is the calculated structure factor for an untwinned crystal (with $p = 0$). Refinement of 176 parameters converged at R = 0.052 and R_ω = 0.059, with p = 0.715(2) and a maximum shifted/esd ratio < 0.001. Anisotropic displacement parameters were refined for nonhydrogen atoms. The hydrogen atoms of the calixarene host were observed in difference syntheses and their positional and isotropic displacement parameters were included in the refinement. The acetonitrile H-atoms were not located and are presumed to be disordered. Final parameters for nonhydrogen atoms are presented in Tables I and II.

TABLE II. Selected interatomic distances (Å) and angles(°).

(a) *Bond lengths:*

C(2)—C(4)	1.400(2)	C(2)—C(6)	1.393(2)
C(4)—C(5′)	1.520(2)	C(4)—C(9)	1.391(2)
C(5)—C(6)	1.521(2)	C(6)—C(7)	1.402(2)
C(7)—C(8)	1.389(2)	C(8)—C(9)	1.398(2)
C(8)—C(11)	1.533(3)	C(11)—C(12)	1.535(4)
C(11)—C(16)	1.540(4)	C(11)—C(20)	1.527(4)
C(28)—C(29)	1.532(11)	O(1)—C(2)	1.385(2)
C(28)—N(27)	1.263(13)		

(b) *Bond angles:*

C(29)—C(28)—N(27)	180.0	O(1)—C(2)—C(4)	117.4(2)
O(1)—C(2)—C(6)	121.4(2)	C(4)—C(2)—C(6)	121.1(2)
C(2)—C(4)—C(5′)	121.8(2)	C(2)—C(4)—C(9)	118.3(2)
C(5′)—C(4)—C(9)	119.8(2)	C(4″)—C(5)—C(6)	111.6(2)
C(2)—C(6)—C(5)	121.9(2)	C(2)—C(6)—C(7)	118.3(2)
C(5)—C(6)—C(7)	119.8(2)	C(6)—C(7)—C(8)	122.4(2)
C(7)—C(8)—C(9)	117.2(2)	C(7)—C(8)—C(11)	123.3(2)
C(9)—C(8)—C(11)	119.5(2)	C(4)—C(9)—C(8)	122.6(2)
C(8)—C(11)—C(12)	109.0(2)	C(8)—C(11)—C(16)	109.5(2)
C(8)—C(11)—C(20)	112.4(2)	C(12)—C(11)—C(16)	109.0(2)
C(12)—C(11)—C(20)	108.3(3)	C(16)—C(11)—C(20)	108.7(2)

(c) *Torsion angles:*

O(1)—C(2)—C(6)—C(5)	3.3(2)	C(9″)—C(4″)—C(5)—C(6)	88.6(2)
C(4″)—C(5)—C(6)—C(2)	89.6(2)	C(4″)—C(5)—C(6)—C(7)	-88.7(2)
C(7)—C(8)—C(11)—C(12)	-116.0(3)	C(7)—C(8)—C(11)—C(16)	124.9(3)
C(7)—C(8)—C(11)—C(20)	4.1(2)		

3. Results and Discussion

The calixarene system is known to be a clathrating agent for organic molecules [2, 18]. The determination of the crystal structure of **1a**·MeCN affords yet another illustration of this property. The calixarene host is centred around a crystallographic C_4 axis in the expected symmetrical cone conformation. The linear guest molecule occupies the resulting bowl-shaped cavity; its C and N atoms lie on the central C_4 axis of the calix with the methyl group directed towards the lower rim formed by the four hydroxyl groups (see Figure 1). The acetonitrile C—C and C—N bond lengths [1.53(1) and 1.26(l) Å] are sufficiently different from accepted values [1.47 and 1.14 Å, see Ref. 19] to suggest that the positions of atoms C(29), C(28) and N(27) are subject to systematic error which may arise from the complications caused by twinning (*vide supra*) or from the failure to include the disordered methyl

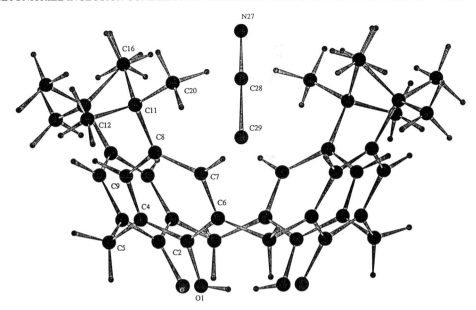

Fig. 1. A view of the *t*-butylcalix[4]arene host and its acetonitrile guest.

group hydrogen atoms attached to C(29) in the structure analysis. Some caution is therefore necessary in the interpretation of the shortest contacts between the host and the guest atoms C(29), C(28) and N(27), namely C(29)···C(2) 3.776(4) and C(28)···C(16) 4.449(3) Å; N(27) is more than 5.0 Å from any C or O atom of its host calixarene. These distances suggest that the acetonitrile methyl group is wedged into the bowl of **1a**, whereas the environments of C(28) and N(27) are more open. The progressive increase in the mean displacement parameters, U, along the sequence C(29)—C(28)—N(27) (see Table I) is consistent with this view.

The symmetrical cone conformation of **1a** found here may be characterised by the dihedral angle of 123.1° between the phenyl ring plane and that of the four hydroxyl oxygen atoms; similar values of 122.6°–125.8° have been found for the corresponding angles in **1a**·toluene [20], **1a**·dimethyl sulphoxide [21], (**1a**)$_2$·anisole [22] and (**1a**-H)·Cs(MeCN) [23]. In all of these complexes the **1a** host has exact C_4 symmetry and the disordered guest straddles the central C_4 symmetry axis of the calix cavity. Alternatively, the conformation of **1a** may be specified, as recently suggested by Ugozzoli and Andreetti [23], in terms of the point symmetry of the calixarene and of the torsion angles, ϕ and χ, about each of the independent C(aromatic)—CH$_2$ bonds [in Table II ϕ = C(9″)—C(4″)—C(5)—C(6) and χ = C(4″)—C(5)—C(6)—C(7)]; for **1a**·MeCN the conformation is specified as C4 +88.6(2) -88.7(2)°; for comparison, the **1a**·toluene complex is described by Ugozzoli and Andreetti as C4 +88.9 -89.4° and the similarity of the conformations of the **1a** hosts in the MeCN and toluene complexes is thereby strikingly revealed. As has been found in the structures of other calixarene

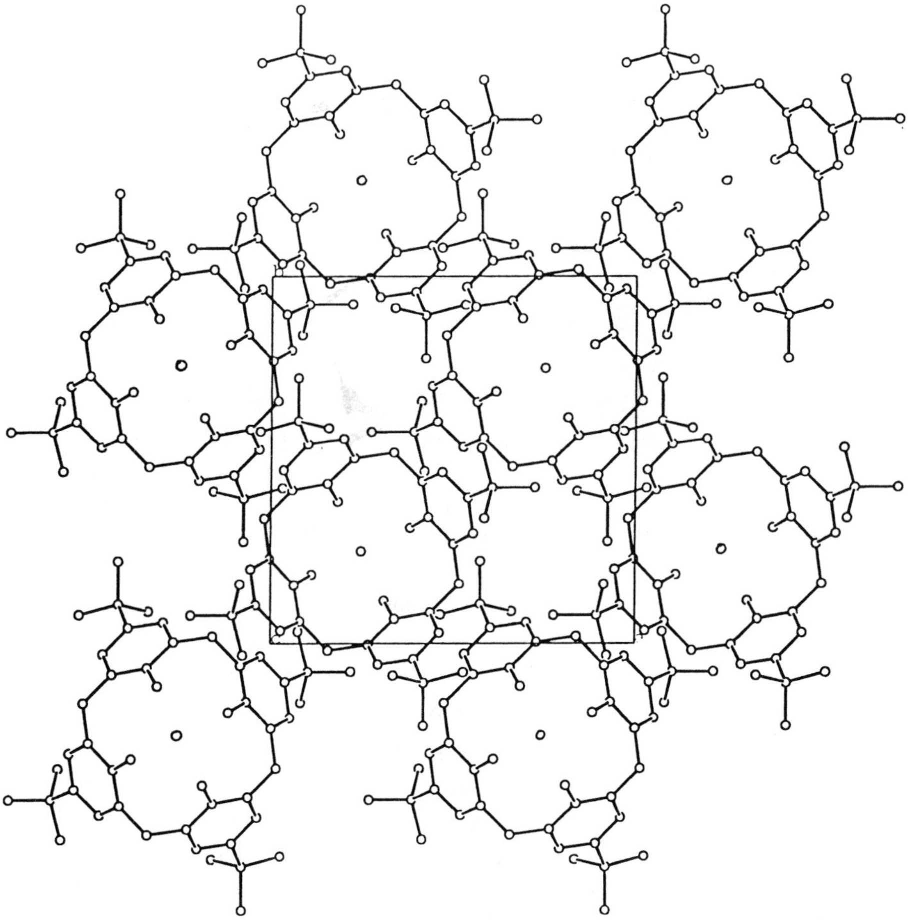

Fig. 2. The unit cell contents of **1a**·MeCN crystals viewed in projection down c. The cell origin is at the lower left corner, a is horizontal and b is vertical.

complexes the hydroxyl rim of the calix cavity in **1a**·MeCN is supported by O—H···O hydrogen bonds, although the O(1)···O(1) separation of 2.692(2) Å is slightly greater than corresponding values in related complexes [e.g. 2.670(9) Å in **1a**·toluene and 2.652(4) and 2.654(5) Å in (**1a**)$_2$·anisole].

Bond lengths and angles within **1a** (Table II) are unexceptional. The phenyl carbon atoms are coplanar to within 0.005(2) Å and the internal ring angles at C-substituted C(4), C(6) and C(8) are all ca. 2° less than 120°. O(1) and C(11) are displaced by 0.04–0.06 Å from the phenyl ring away from the C_4 axis and C(5) is displaced towards the axis by 0.04 Å. The t-butyl group has staggered conformations across its C—CH$_3$ bonds and its orientation brings C(20) close to the plane of the phenyl ring.

The crystal packing in **1a**·MeCN (Figure 2) contains two types of channel running parallel to the c-axis: the wider channels are centred on the four-fold symmetry axes and contain the guest MeCN molecules; in addition, there are channels centred on the bar-4 axes at $(3/4, 1/4, z)$ and $(1/4, 3/4, z)$ which are evidently too narrow to accommodate even small guest molecules. There are only three **1**···**1a** contacts involving nonhydrogen atoms which are 0.1 Å less than the sum of the appropriate van der Waals radii: C(2)···C(5i), 3.497(2), C(5)···C(6i) 3.483(2) and C(12)···C(20ii) 3.789(4) Å (i: $-x, -y, -z$; ii: $-1/2 - y, x, z$). The arrangement shown in Figure 2 closely resembles that found in **1a**·toluene [20], **1a**·dimethyl sulphoxide [21] and (**1a**-H)·Cs(MeCN) [23]. These compounds are approximately isomorphous: they all crystallize in the tetragonal space group $P4/n$ with unit cells containing two **1a** residues and of similar dimensions (a = 12.7–13.1, c = 12.5–13.8 Å). For tetragonal crystals a condition for twinning is that the square of the $c : a$ axial ratio must approach a rational number [24]; this condition is nearly satisfied for most of the structurally characterised **1a** complexes (e.g. **1a**·MeCN $c : a$ = 1.0037; **1a**·dimethyl sulphoxide $c : a$ = 1.02 [21]) and it may well be that twinning is a general feature of the crystal structures of the complexes of **1a** and related calix[4]arenes with small organic molecules.

In solution, the ^{1}H-NMR spectra of **1a**·MeCN and **1a**·2MeCN did not give separate resonances for included MeCN and free MeCN. Therefore, a rapid equilibrium between included MeCN and free MeCN in solution is expected. Even in the solid state, the included MeCN is only loosely bound in the calixarene cavity and it can be completely removed *in vacuo* (0.1 mm Hg) in one day. However, the structure is important as a model for the calixarene part of the more complex molecules described below which, though they were often obtained in crystalline form, always failed to diffract sufficiently to allow X-ray structure determination. Instead it is necessary to rely on spectroscopic techniques, comparing the NMR spectra with that of the parent calixarene whose structure is known (Figure 1).

Since polydentate ligands play an important role in coordination chemistry, the preparation of new polydentate ligands with unusual geometry, which may then coordinate to transition metals in unusual ways, is highly desirable. Although the ligand **2a** has been reported previously [9, 11], this ligand is unstable to hydrolysis by traces of water. It also decomposes quickly when dissolved in the NMR solvent CDCl$_3$. However, by changing reaction conditions, the ligand **2c** with three PPh$_2$ donors at the bottom rim of the calixarene can be obtained in high yield. Its ^{1}H-NMR spectrum shows three Bu^t groups at δ 1.2, 1.1 and 1.15 in 1 : 1 : 2 ratio, corresponding to three different Bu^t environments. The methylene groups also show four sets of doublet resonances [2]. The ^{31}P{^{1}H}-NMR spectrum contained only two singlet resonances in a 2 : 1 ratio, as expected for **2c** in the cone conformation. Less symmetrical conformations can be discounted on this basis. The presence of the hydroxy group also was confirmed by the IR spectrum, which gives ν(OH) at 3351 cm^{-1}. The higher stability of **2c** than **2a** is attributed to steric relief when there is one less bulky PPh$_2$ group present.

It is expected that **2c** and **2a** will show different coordination to transition metals. Thus, treatment of **2c** with [AuCl(SMe$_2$)] gave a conformer of **3**. The three PAuX

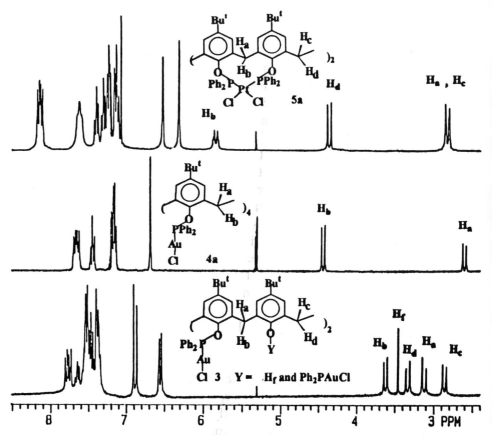

Fig. 3. The ^{1}H-NMR spectra of **3**, **4a** and **5a** (the *t*-butyl groups are omitted for clarity).

units are not identical. The ^{31}P{^{1}H}-NMR spectrum shows two singlets at $\delta = 114.8$ (s) and $\delta = 117.8$ (s). The observation of four pairs of methylene resonances and three singlet Bu^t resonances clearly supports the proposed structure with three PPh$_2$ substituents (see Figure 3a). The FTIR spectrum also confirmed that the ArOH group was still present from the absorption band at ν(OH) = 3366 cm^{-1}.

In order to compare the behavioral difference between the tridentate phosphintocalix[4]arene **2c** and the tetradentate **2a**, compound **2a** has also been coordinated with some later transition metals. Thus, treatment of **2a** with [AuCl(SMe$_2$)] or [(AuCCPh)$_n$] gave the air and moisture stable tetrametal complexes, **4** (**4a**, X = Cl, **4b**, X = CCPh), containing one gold(I) per phosphorus atom. The four PAuX units are now identical as shown by the ^{31}P-NMR spectra [$\delta = 105$ (s), X = Cl; $\delta = 135$ (s), X = CCPh]. The complexes are highly symmetric and contain a golden rim at the bottom of the calix[4]arene. Since the bridging CH$_2$ groups may serve as spectroscopic probes [2], the structures and conformations of **4** are clearly indicated by their ^{1}H-NMR spectra (see Figure 3b). There is only one pair of methylene signals, indicating that the calixarene bowls are still in the cone conformation. The inclusion behaviour of **4** is not as obvious as with other calixarenes [2]. Nevertheless, the complexes **4** crystallize with MeCN, which may be present inside the bowl, as in the precursor **1** (Figure 1) or outside the cavity forming a clathrate-like compound. The experimental data do not distinguish between these possible structures.

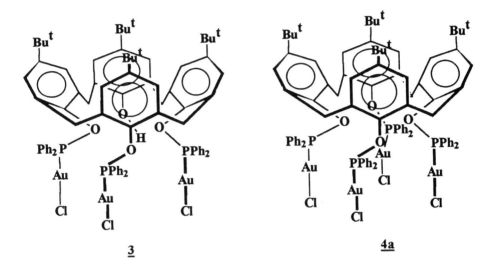

As expected, **2a** can either act as a monodentate ligand or as a chelating ligand. Reaction of [PtCl$_2$(SMe$_2$)$_2$] or [PdCl$_2$(PhCN)$_2$], with **2a** occurred with displacement of SMe$_2$ or PhCN to give the dimetal complexes, **5** (**5a**, M = Pt; **5b**, M = Pd) in which each metal is chelated by two phosphorus atoms. Though the four phosphorus atoms are still identical, as indicated by their ^{31}P-NMR spectra, there are two different pairs of methylene (ArCH_2Ar) resonances in the ^{1}H-NMR spectrum, indicating that the molecules no longer have four-fold symmetry (see Figure 3c). One of these pairs of resonances appears as two doublets at δ = 2.9 and 4.45, while the second pair appears as a doublet at δ = 2.9 and a doublet of triplets at δ = 5.9. The resonance with the extra triplet coupling is assigned to the protons on the CH$_2$ groups located between the chelating phosphorus atoms, and particularly to the two protons which are directed towards the metal atoms. These protons, embraced by the P—Pt—P ten-membered ring, couple not only to the geminal hydrogen but also to the two adjacent phosphorus atoms through space. This unusual ^{1}H—^{31}P through space coupling was confirmed by excluding all other proton couplings by the homonuclear decoupling technique and by recording the COSY spectrum (see Figure 4). Molecular mechanics calculations [25] also indicate that the distance between H$_a$ and the near ^{31}P atom is only 2.79 Å, and so it is reasonable to expect this through space coupling.

5 (5a: M=Pt; 5b: M=Pd)

Although most of these transition metal complexes are crystalline compounds, none was suitable for X-ray analysis. Hence the question of the ability of the complexes to act as hosts for small molecules such as MeCN could not be answered with certainty.

4. Conclusions

When phosphinite groups are introduced into the lower rim of the calix[4]arene, the trisubstituted derivatives are more stable to hydrolysis than the tetrasubstituted derivatives. This is proposed to be due to lower steric strain in the trisubstituted compound. The introduction of bulky substituents, such as the Ph_2P groups described here, on the lower rim of the calixarene is also expected to affect the geometry of the bowl such that the bulky Bu^t groups on the upper rim will be forced closer together and so reduce the size of the entrance to the bowl. The introduction of additional metal substituents accentuates this effect. Such relatively minor conformational changes may, of course, have a considerable effect on subsequent host–guest chemistry.

Acknowledgements

We wish to express our thanks to the N.S.E.R.C. (Canada) and S.E.R.C. (U.K.) for financial support and Dr. David J. Watkin, Chemical Crystallography Laboratory, University of Oxford, for carrying out the refinement calculation with the CRYSTALS program.

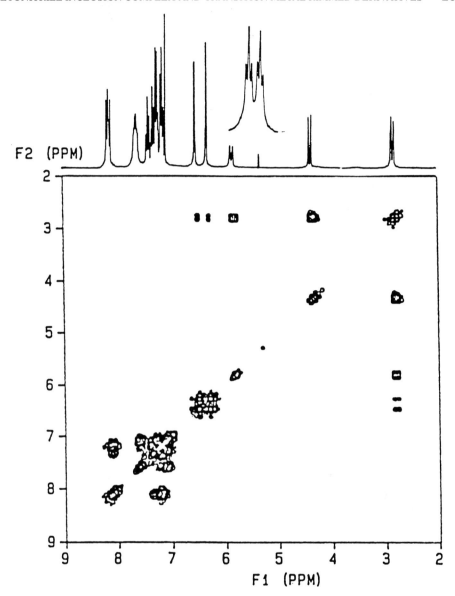

Fig. 4. The two-dimensional ^{1}H—^{1}H correlated NMR spectrum (COSY) of complex **5a**. The triplet coupling in the resonance at $\delta = 5.9$ is shown in the expansion. Note that two resonances overlap accidentally at $\delta = 2.9$ and the correlations with resonances at $\delta = 5.9$ and 4.45 are clear.

References

1. L. F. Lindoy: *The Chemistry of Macrocyclic Ligand Complexes*, Cambridge University Press (1989).
2. (a) C. D. Gutsche: *Calixarenes*, Royal Society Monographs in Supramolecular Chemistry, Cambridge (1989). (b) J. Vicens and V. Böhmer (Eds.): *Calixarenes: A Versatile Class of Macrocyclic Compounds*, Kluwer Academic Publishers, Dordrecht, the Netherlands (1990).
3. D. J. Cram: *Angew. Chem. Int. Ed. Engl.* **27**, 1009 (1988).
4. (a) F. Corazza, C. Floriani, A. Chiesi-Villa, and C. Guastini: *J. Chem. Soc. Chem. Commun.* 640 (1990). (b) F. Corazza, C. Floriani, A. Chiesi-Villa, and C. Rizzoli: *Inorg. Chem.* **30**, 4465 (1991). (c) F. Corazza, C. Floriani, A. Chiesi-Villa, and C. Guastini: *J. Chem. Soc. Chem. Commun.* 1083 (1990). (d) G. Calestani, F. Ugozzoli, A. Arduini, E. Ghidini, and R. Ungaro: *J. Chem. Soc. Chem. Commun.* 344 (1987).
5. J. L. Atwood, G. W. Orr, N. C. Means, F. Hamada, H. Zhang, S. G. Bott, and K. D. Robinson: *Inorg. Chem.* **31**, 657 (1992).
6. J. L. Atwood, S. G. Bott, C. Jones, and C. L. Raston: *J. Chem. Soc. Chem. Commun.* 1349 (1992).
7. Z. Asfari, J. M. Harrowfield, M. I. Ogden, J. Vicens, and A. H. White: *Angew. Chem. Int. Ed. Engl.* **30**, 854 (1991).
8. Wei Xu, J. P. Rourke, J. J. Vittal, and R. J. Puddephatt: *J. Chem. Soc. Chem. Commun.* 145 (1989).
9. C. Floriani, D. Jacoby, A. Chiesi-Villa, and C. Guastini: *Angew. Chem. Int. Ed. Engl.* **28**, 1376 (1989).
10. (a) J. K. Moran and D. M. Roundhill: *Inorg. Chem.* **31**, 4213 (1992). (b) D. Matt, C. Loeber, J. Vicens, and Z. Asfari: *J. Chem. Soc. Chem. Commun.* 604 (1993).
11. D. Jacoby, C. Floriani, A. Chiesi-Villa, and C. Rizzoli: *J. Chem. Soc. Dalton Trans.* 813 (1993).
12. A. Tamaki and J. K. Kochi: *J. Organomet. Chem.* **64**, 411 (1974).
13. P. L. Goggin, R. J. Goodfellow, R. S. Haddock, F. J. S. Reed, J. G. Smith, and K. M. Thomas: *J. Chem. Soc. Dalton Trans.* 1904 (1972).
14. G. K. Anderson and M. Lin: *Inorg. Syn.* **28**, 60 (1990).
15. C. D. Gutsche, M. Iqbal, and D. Stewart: *J. Org. Chem.* **51**, 742 (1986).
16. G. M. Sheldrick: *SHELXS* (Program for Solution of Crystal Structures), University of Göttingen, Germany (1986).
17. J. R. Carruthers: *CRYSTALS User Manual*, Oxford University Computing Laboratory (1975).
18. (a) W. J. Evans, S. C. Engerer, P. A. Piliero, and A. L. Wayda: *J. Chem. Soc. Chem. Commun.* 1005 (1979). (b) M. A. McKervey, E. M. Seward, G. Ferguson, and B. L. Ruhl: *J. Org. Chem.* **51**, 3581 (1986).
19. International Union of Crystallography: *International Tables for Crystallography, Volume C*, Kluwer Academic Publishers, Dordrecht, the Netherlands, pp. 691–706 (1992).
20. G. D. Andreetti, R. Ungaro, and A. Pochini: *J. Chem. Soc. Chem. Commun.* 1005 (1979).
21. B. M. Furphy, J. M. Harrowfield, M. I. Ogden, B. W. Skelton, A. H. White, and F. R. Wilner: *J. Chem. Soc. Dalton Trans.* 2217 (1989).
22. R. Ungaro, A. Pochini, G. D. Andreetti, and P. Domiano: *J. Chem. Soc. Perkin Trans.* 2 197 (1985).
23. F. Ugozzoli and G. D. Andreetti: *J. Incl. Phenom.* **13**, 337 (1992).
24. International Union of Crystallography: *International Tables for X-Ray Crystallography*, Volume II, Kynoch Press, Birmingham, England, p. 109 (1959). Distr. Kluwer Academic Publishers, Dordrecht, the Netherlands.
25. *PCMODEL Molecular Modeling Software*, Version 4.2, Serena Software (1992).

Synthesis, Characterization, and X-Ray Structure of 1,2-Bis-crown-5-calix[4]arene. Modeling of Metal Complexation*

ZOUHAIR ASFARI[a], JEAN-PIERRE ASTIER[b], CHRISTOPHE BRESSOT[a], JACQUES ESTIENNE[c**], GERARD PEPE[b] and JACQUES VICENS[a]
[a] *E.H.I.C.S., URA 405 du C.N.R.S., 1 rue Blaise Pascal, F-67008 Strasbourg, France*
[b] *C.R.M.C.2 – C.N.R.S., Campus de Luminy, Case 913, F-13288 Marseille, France*
[c] *Université de Provence, Centre Saint Jérôme, Laboratoire de Spectrométries et Dynamique Moléculaire, Case 542, F-13397 Marseille, France*

(Received: 11 May 1994; in final form: 12 September 1994)

Abstract. 1,2-*bis*-crown-5-calix[4]arene (**3**) was prepared by reacting calix[4]arene (**1**) with tetraethylene glycol di-*p*-toluenesulphonate (**2**) in the presence of cesium carbonate in 9% yield. The X-ray structure of (**3**) was determined. Crystal data for $C_{44}H_{52}O_{10}$ are as follows: monoclinic, space group $P_{2/c}$ with $a = 18.006(8)$ Å, $b = 10.680(4)$ Å, $c = 22.359(6)$ Å, $\beta = 112.93(3)°$, $V = 3958(6)$ Å^3, $Z = 4$, $D_{calc} = 1.2$ g cm^{-3}, the final R value is 0.11 for the 1851 observed reflections ($I > 3\sigma(I)$). The single crystal included two similar but slightly different molecules immobilized in a pinched-cone conformation with C_2 symmetry. The two enantiomorphous molecules were analyzed by molecular mechanics using the GenMol program to model the selective alkali-metal complexation.

Key words: Doubly crowned calix[4]arene, X-ray structure, proximal functionalization, alkali cations, modeling by molecular mechanics.

1. Introduction

The term calixcrown refers to a family of synthetic macrocyclic receptors presenting a hybrid structure [1] combining in their molecular frame calix[4]arene units and crown ether elements. The first member of this family was reported in 1983 by Alfieri *et al.* [2], who reacted *p-tert*-butylcalix[4]arene with pentaethylene glycol ditosylate to produce a *distal* or 1,3-capped or *mono*-crown-6 derivative in a cone conformation. In a subsequent paper the related 1,3-*p-tert*-butylcalix[4]arene-*bis*-crown-5 was isolated as a by-product during the production of the *mono*-crown-5 compound by a similar reaction [3]. The 1,3-*p-tert*-butylcalix[4]arene-*bis*-crown-5 was shown to be in the 1,3-alternate conformation due to the introduction of a second bridging in the calixarene macroring [3]. In order to prepare a new type of metal cation receptor the stepwise synthesis of a *proximal* or 1,2-*p-tert*-butylcalix[4]arene-*bis*-crown-5 in a cone conformation was reported [4]. The syn-

---

* This paper is dedicated to the commemorative issue on the 50th anniversary of calixarenes.
** Author for correspondence.

thesis began by the regioselective 1,2-*bis* demethylation of the tetramethoxy-*p-tert*-butylcalix[4]arene with $TiBr_4$ [4]. This regioselective Ti(IV)-assisted demethylation led to the formation of 1,2-*p-tert*-butylcalix[4]arene-*bis*-crown-5 [4].

As a part of our work on the synthesis of double calixcrown ethers [5], we reacted calix[4]arene with tetraethylene glycol ditosylate in the presence of cesium carbonate instead of potassium carbonate. Surprisingly, we isolated the 1,2-calix[4]arene-*bis*-crown-5 (**3**). Calixarene (**3**) was deduced to be in the cone conformation from spectroscopic data. This conformation was ascertained by X-ray diffractometry. Preliminary metal binding properties of ligand (**3**) were anticipated by modeling with molecular mechanics.

2. Experimental

2.1. MATERIAL FOR SYNTHESIS

Calix[4]arene (**1**) was prepared as described in the literature [6]. Tetraethylene glycol di-*p*-toluene sulphonate (**2**), cesium carbonate, and the solvents were commercial reagents and used without further purification.

2.2. ANALYTICAL PROCEDURES

The melting point was taken on a Büchi 500 apparatus in a capillary sealed under nitrogen. The Silica column was prepared with Kieselgel Merck (Art. 9385). The eluent is specified in the experimental procedure. The ^{1}H-NMR spectrum was recorded at 200 MHz on a Bruker SY200 spectrometer. The FAB mass spectrum was obtained on a VG-Analytical ZAB HF apparatus.

2.3. PREPARATION OF 1,2-*bis*-CROWN-5-CALIX[4]ARENE (**3**)

Into a 500 mL round-bottomed flask were added calix[4]arene (**1**) (2.00 g, 4.73 mmol) and cesium carbonate (15.42 g, 47.33 mmol) and acetonitrile (230 mL). The mixture was stirred magnetically for 20 min. Then tetraethylene glycol di-*p*-toluenesulphonate (2.37 g, 4.73 mmol) dissolved in acetonitrile (25 mL) was added. After 4 days of refluxing, the same quantities of tetraethylene glycol di-*p*-toluenesulphonate (**2**) and cesium carbonate were added. The reflux was maintained for 4 additional days. After cooling to room temperature, the mixture was filtered by suction and washed with dichloromethane. The filtrate was concentrated under reduced pressure to yield an oily residue which was dissolved in dichloromethane and washed with 1 N HCl. The organic layer was dried over sodium. After filtration the solvents were evaporated to dryness to give a transparent yellow oil which was chromatographed on silica with 80 : 20 dichloromethane : acetone mixture as eluent. 1,2-*bis*-crown-5-calix[4]arene (**3**) was eluted first and recrystallized from methanol. M.p. 206–207°. ^{1}H-NMR 6.69–6.57 (m, 12H, Ar*H*, *meta* and *para*), AB system 4.59 and 3.18 (d, 4H, J_{H-H} = 13.3 Hz, Ar—CH_2—Ar), A′B′ system

4.39 and 3.14 (d, 4H, J_{H-H} = 13.3 Hz, Ar'—CH_2—-Ar'), 3.6–3.1 (m, 32H, O—CH_2CH_2O). FAB m/z: 741.3 (for $C_{44}H_{52}O_{10}$). Yield 9%.

2.4. X-RAY CRYSTAL DATA

In order to avoid solvent molecule inclusion no recrystallisation was attempted and a suitable fragment (0.30 × 0.30 × 0.40 mm^3) was cut with a razor blade into a crystalline block obtained during the synthesis.

Data collection on an Enraf Nonius CAD-4 diffractometer (7796 measured reflections, T = 298 K, 1851 observed reflections, with $I_{net} > 3.0\sigma(I_{net})$). Program used to solve the structure: MULTAN 80 [7]. Program to refine the structure by full matrix least squares: SHELX [8]. Molecular graphics: ORTEP II [9], GenMol [10].

2.5. MODELING OF CALIXARENE

Modeling of compound (**3**) with and without cations has been performed with the GenMol Program. GenMol [10] is a molecular mechanics program using an original force field designed to obtain accurate geometries, well adapted to modeling molecules such as calixarenes in order to understand their ability to complex alkaline cations.

3. Results and Discussion

The synthesis of (**3**) was conducted according to Scheme 1. Calix[4]arene (**1**) was refluxed for 8 days under nitrogen with 2 equivalents of tetraethylene glycol di-*p*-toluenesulphonate (**2**) (added in two equal crops) in acetonitrile in the presence of cesium carbonate in large excess. The reaction mixture was observed to contain numerous products from which (**3**) could be isolated by chromatography probably because it was eluted first. Analytical data were in agreement with the proposed structure in which two glycolic chains are attached to the calix[4]arene (**1**) (FAB mass 741.3). The presence of two well-resolved AB systems at 4.59 and 3.18 and 4.39 and 3.14 for the methylene protons in the macroring were indicative that (**3**) is in the cone conformation with the two crown units linked in a *proximal* manner. This geometry should be compared to the one obtained during the formation of doubly crowned calix[4]arene in the 1,3-alternate conformation in which the glycolic chains are attached in a *distal* fashion [4]. The difference is probably due to the use of cesium carbonate instead of potassium carbonate. A recent publication reports the regioselective synthesis of calixcrowns derived from *p-tert*-butylcalix[5]arene [11]. The authors noticed the formation of a *proximal* derivative in a 20% yield in the presence of cesium fluoride [11]. These observations lead us to conclude that the cesium cation is effective in inducing the *proximal* dialkylation of the glycolic chains. It is assumed that after the first *O*-substitution had occurred the cesium

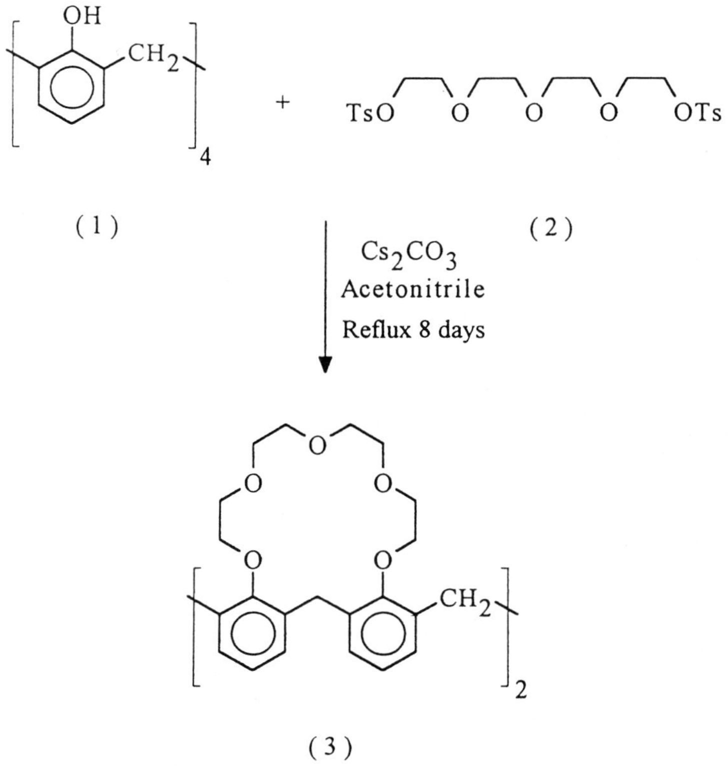

Scheme 1. Synthesis of 1,2-*bis*-crown-5-calix[4]arene (**3**).

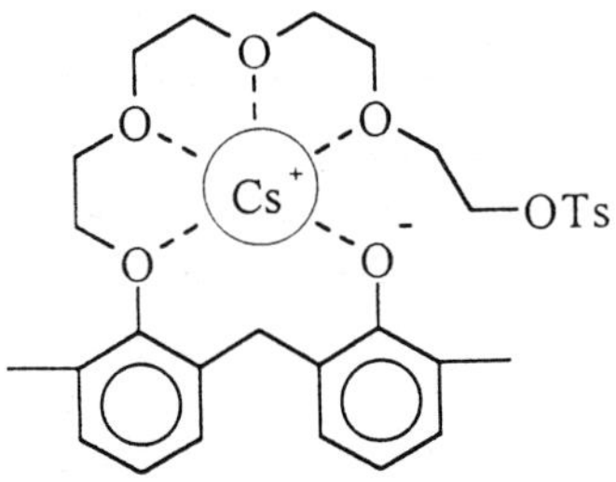

Scheme 2. Cesium template of the polyether chain.

cation templates the polyether chain to bring the unreacted *p*-toluenesulphonate leaving group in proximity to the adjacent phenolate ion (see Scheme 2).

The molecular arrangement of (**3**) has been ascertained by X-ray crystallographic analysis. Crystal data for $C_{44}H_{52}O_{10}$ (**3**) are: monoclinic space group $P_{2/c}$ with $a = 18.006(8)$ Å, $b = 10.680(4)$ Å, $c = 22.359(6)$ Å, $\beta = 112.03(3)°$, $V = 3958(6)$ Å^3, $Z = 4$, $D_{calc} = 1.24$ g cm^{-3}. The single crystal included two similar but slightly different molecules immobilized in a pinched-cone conformation with C_2 symmetry.

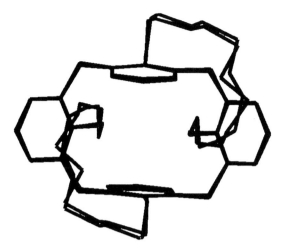

Fig. 1. The first crystalline molecule superimposed with the mirrored second one. The root mean square value between the two geometries is 0.04 Å.

In Figure 1, the two independent molecules have been superimposed after performing a best molecular fit between one molecule with the mirrored second molecule. Slight differences can be observed. The root mean square value between the two geometries is 0.04 Å.

In Figures 2a and 2b, which are projections of one crystalline molecule (along the binary axis, and orthogonal to this axis) it is seen that the molecular conformation can be characterized by two parameter families. One giving the orientation of the phenyl rings versus the methylene plane, and the other measuring the aperture of the oxygened chains which form an opened 'mouth'. The angles of the phenyl rings with the methylene plane are 140(2)° and 76(2)° for one molecule and 140(2)° and 80(2)° for the second molecule which indicate a great similarity in this part of the molecule. The 'mouth' aperture can be measured by the angle of the mean planes passing through the oxygen atoms, 133(2)° for one molecule and 110(2)° for the other one. The slight geometric differences come from this part of the molecules, which expresses the flexibility of the oxygened chains.

Remark. In solution only one molecular conformation is observed, the cone one, which is the intermediate form between the two enantiomorphous forms observed in the crystal. The two enantiomorphous forms observed in the crystalline state can be exchanged in the solution by a concerted motion of the phenyl rings and the glycolic chains, which can be related to the anisotropic thermal parameters of the corresponding atoms, see Figure 3.

In order to test the program we compared the calculated geometry obtained from GenMol to one geometry observed in the crystalline state. A best molecular fit performed on these molecules gives a root mean square value of 0.04 Å, analogous

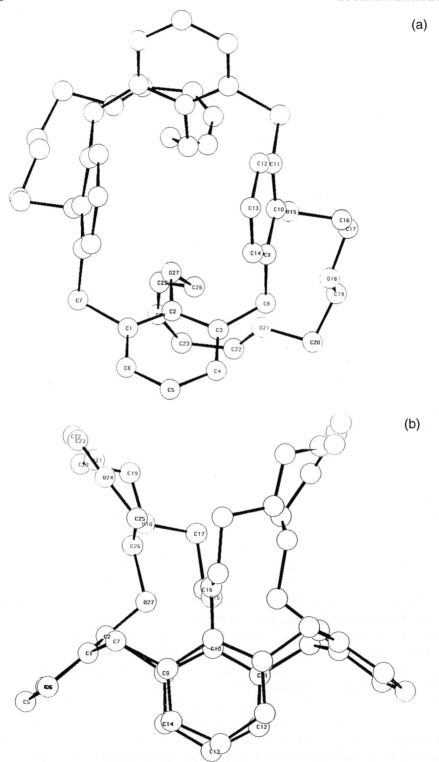

Fig. 2. Projection along (2a) and perpendicular to (2b) the crystallographic binary axis displaying the pinched-cone conformation of the molecule and the aperture of the glycolic chains.

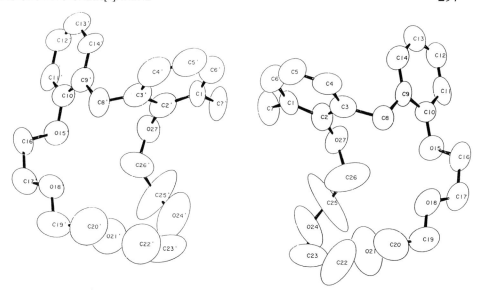

Fig. 3. ORTEP II drawing of the two crystalline independent parts of the molecules displaying the great thermal ellipsoid of the atoms, expressing exchange ability of one conformation to the other one.

to the value obtained for the fit between the two X-ray molecules, which expresses the quality of the modeled geometry.

The characteristic parameters of the calculated geometry are 143° and 79° for the angles between the phenyl rings and the methylene plane which correspond to the observed values (140(2)°, 140(2)°) and (80(2)°, 76(2)°) and the angle characteristic of the 'mouth' aperture is 136° instead of 133(2)° and 110(2)° for the experimental values which seems to indicate that *the larger, opened form is the one existing in solution*.

As in solution only the cone conformation occurs, which is the intermediate form, between the two enantiomers observed in the crystal, calculations were performed on the different molecular forms in order to understand the molecular behaviour. The strain energy of the cone form is 128 instead of 125 kcal/mole for both enantiomorphous pinched-cone forms. The pinched-cone forms are the most stable forms, which explains why they are observed in the solid state (the state always corresponding to an energy minimum).

In solution the phenyl rings are agitated, because of the absence of an energy barrier between the different forms. The observed cone form is in fact an average form of the molecule, while the enantiomorphous pinched-cone forms observed in the crystalline state are limit forms of the molecule.

Some alkali-cation extraction results are known on the related molecule to (**3**) bearing *tert*-butyl groups on the *para* position of the phenyl rings [4]. We decided to model the binding properties of calixarene (**3**) toward these cations in order to see, first, the geometric modification when adding cations and, second, to know if

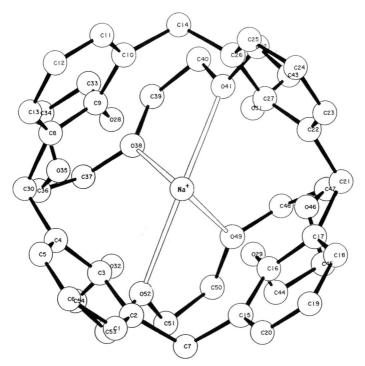

Fig. 4. Complex of calixarene **3** with Na$^+$. Only four bonds are formed with this cation.

the program is able to understand the cation extraction properties of ligand (**3**) and related molecules.

Computations of (**3**)-complexations with the different alkaline cations Na$^+$, K$^+$, Rb$^+$, Cs$^+$ (with increasing ionic radii values of 0.97, 1.33, 1.47 and 1.67 Å, respectively) were performed. Calculations indicate that the number of bonds between the cation and the oxygen atoms regularly increases from Na$^+$ (4 bonds) to Cs$^+$ (10 bonds) meaning that (**3**) will selectively extract Cs$^+$ cation, as shown from literature data.

Figures 4 and 5 display the geometries of the Na$^+$ and Cs$^+$ complexes. It is interesting to notice the geometric evolution of the molecule at the phenyl ring level: in both cases the symmetry increases from C_2 in the empty molecule to C_4 when complexed with the cation. The mean angle between the phenyl rings and the methylene plane is 114° in both complexes, which is a mean value between 140° and 80° of the free molecule. The angle between the mean oxygen atom plane, characteristic of the 'mouth' aperture is 82° instead of 133° and 110° in both molecules, which corresponds to the lower value of this parameter.

It is also interesting to remark on the much higher stability of the complexed molecule than of the uncomplexed one. The strain energy changes from 125 kcal/mole to 85 kcal/mole (value obtained without taking account of the counteranion).

1,2-BIS-CROWN-5-CALIX[4]ARENE 299

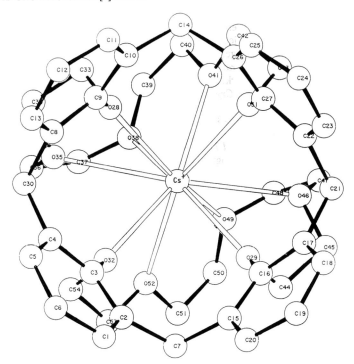

Fig. 5. Complex of calixarene **3** with Cs$^+$. Ten bonds are formed with this cation leading to a considerably great affinity of this calixarene for Cs$^+$.

4. Conclusion

In the present paper we have described an improved synthesis of 1,2-*bis*-crown-5-calix[4]arene (**3**). The synthesis was achieved by using cesium carbonate. The X-ray structure of (**3**) indicated the presence of two enantiomorphous molecules in the crystal. These two enantiomorphous molecules were analyzed by molecular mechanics using the GenMol program. It was concluded that while the conformation of (**3**) in solution is an average of several forms, in the solid state the more stable conformations are the pinched-cone forms which are the limit forms of (**3**). Calculations on supermolecules (**3**) complexed to alkali cations indicated that ligand (**3**) preferentially binds Cs$^+$ compared to the other alkali cations.

References

1. W. G. Gokel and A. Nakano: in *Crown Compounds toward Future Applications*, S. R. Cooper (Ed.), VCH, New York (1992).
2. C. Alfieri, E. Dradi, A. Pochini, R. Ungaro, and G. D. Andreetti: *J. Chem. Soc., Chem. Commun.* 1075 (1983).
3. E. Ghidini, F. Ugozzoli, R. Ungaro, S. Harkema, A. A. El-Fald, and D. N. Reinhoudt: *J. Am. Chem. Soc.* **112**, 6979 (1990).
4. A. Arduini, A. Casnati, L. Dodi, A. Pochini, and R. Ungaro: *J. Chem. Soc., Chem. Commun.* 1597 (1990).

5. Z. Asfari, R. Abidi, F. Arnaud, and J. Vicens: *J. Incl. Phenom.* **13**, 163 (1992).
6. C. D. Gutsche, J. A. Levine, and P. K. Sujeeth: *J. Org. Chem.* **50**, 5802 (1985).
7. P. Main, S. J. Fiske, S. E. Hull, L. Lessinger, G. Germain, J. P. Declercq, and M. M. Woolfson: *MULTAN 80. A System of Computer Programs for Automatic Solution of Crystal structures from X-Ray Diffraction Data*, University of York, England, and Louvain, Belgium (1980).
8. G. M. Sheldrick: *SHELX. Program from Crystal Structure Determination*, University of Cambridge, England (1976).
9. C. K. Johnson: *ORTEP II. Report ORNL-3974*, Oak Ridge National Laboratory, Tennessee (1976).
10. G. Pèpe and D. Siri: *Studies in Physical and Theoretical Chemistry* **71**, 93 (1990).
11. D. Kraft, R. Arnecke, V. Böhmer, and W. Vogt: *Tetrahedron* **49**, 6019 (1993).

Complexation Properties and Characterization of Four Conformers of a [2.1.2.1]Metacyclophane*

TSUYOSHI SAWADA[1], AKIHIKO TSUGE[2], THIES THIEMANN[1], SHUNTARO MATAKA[1] and MASASHI TASHIRO[1]**
[1]*Institute of Advanced Material Study, Kyushu University, 6-1, Kasuga-kohen, Kasuga-shi, Fukuoka 816, Japan*
[2]*Department of Chemistry, Kyushu Institute of Technology, Tobata-ku, Kitakyushu 804, Japan*

(Received: 3 March 1994; in final form: 26 October 1994)

Abstract. Cone, partial cone, 1,2-alternate, and 1,4-alternate conformers of tetrakis[(ethoxycarbonyl)-methoxy] [2.1.2.1]metacyclophane (MCP, **2**) were isolated and characterized by ^{1}H-NMR and an X-ray crystal structure analysis of the 1,4-alternate conformer. Of the four conformers, only the cone conformer **2d** forms a complex with alkali metals. The stability constants of **2d** with alkali metal ions were determined by the direct ^{1}H-NMR method and UV spectra, and the order observed as potassium $\ll$ cesium, sodium ions.

Key words: Metacyclophane, conformational isomers, ion-selectivity, calixarenes.

1. Introduction

Calix[n]arenes have attracted great attention as ionophoric receptors [1–4] and potential enzyme mimics [5] in host–guest chemistry. Shinkai *et al.* have reported the preparation and ionophoric properties of four conformers of tetra-*tert*-butyl-tetrakis[(ethoxycarbonyl)methoxy]calix[4]arene [6–8]. Cone and partial-cone conformers are obtained by the metal template effect using sodium and cesium ions, respectively, but the 1,2-alternate and 1,3-alternate conformers were synthesized by the protection–deprotection method [8–10]. They also found that the cone conformer shows a selectivity for sodium [6] and the other conformers show a selectivity for potassium [8].

In our laboratory we have investigated the preparation and reactivities of [2,2]metacyclophanes for two decades. We have developed the sulfur method using the *tert*-butyl group as a positional-protecting group. Using this method we have also prepared [2.n.2.n]metacyclophanes [11, 12]. These metacyclophanes can be regarded as calixarene homologues and are expected to have larger cavities. Their framework should be more flexible than that of the calix[4]arenes [13–15].

---

* This paper is dedicated to the commemorative issue on the 50th anniversary of calixarenes.
** Author for correspondence.

Here we describe the isolation and the ionophoric properties of the conformers of tetra-*tert*-butyl-tetrakis[(ethoxycarbonyl)methoxy][2.1.2.1]metacyclophane (MCP), a homologue of calix[4]arene.

2. Results and Discussion

2.1. Preparation and Conformational Properties

The tetrakis (ethoxycarbonyl)methoxy MCP derivative **2** was obtained as a mixture of conformers in the reaction of tetrahydroxy[2.1.2.1]MCP (**1**) [12] and ethyl bromoacetate in the presence of alkali carbonates in acetone or a mixture of DMF and THF. The ratio of the conformers, **2a**, **2b**, **2c**, and **2d** was determined by HPLC (Megapak Sil-10: 7.5 mm × 250 mm, a 8.5 : 1.5 mixture of hexane : ethyl acetate as an eluent) (Table I).

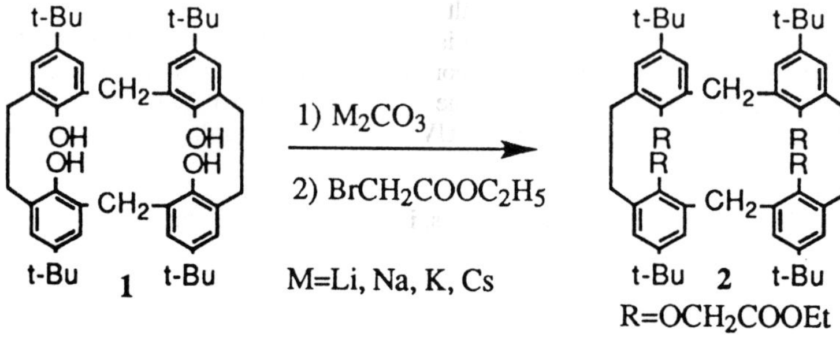

Scheme 1.

TABLE I. Reaction conditions and conformer distribution of **2**.

Solvent	Base	Yield of 2(%)	Distribution[a] of 2 (2a : 2b : 2c : 2d)	Other (%)
DMF–THF	Li_2CO_3	43	(30 : 28 : 39 : 3)	**3**[b] 40
	Na_2CO_3	92	(33 : 34 : 17 : 16)	
	K_2CO_3	83	(43 : 20 : 18 : 19)	
	Cs_2CO_3	80	(36 : 35 : 14 : 15)	
Acetone	Li_2CO_3	0		**1** 80
	Na_2CO_3	92	(25 : <1 : <1 : 74)	
	K_2CO_3	83	(36 : 11 : 13 : 40)	
	Cs_2CO_3	80	(35 : 24 : 13 : 28)	

[a] Determined by HPLC (Megapak Sil-10: 7.5 mm × 250 mm, a 8.5 : 1.5 mixture of hexane : ethyl acetate as eluent).
[b] Bis(ethoxycarbonylmethoxy)dihydroxy[2.1.2.1]MCP **3** was obtained as a by-product, but its region structure cannot be classified. **3** was separated by column chromatography (Wako gel C300, a 9 : 1 mixture of hexane : ethyl acetate as eluent).

In most cases four conformers were obtained but interestingly, when sodium carbonate in acetone was used, only the partial-cone and the cone conformer, **2a** and **2d**, were obtained in 23 and 68% yield. Each conformer could be isolated by careful column chromatography on silica gel (Wako C-300) using an 8.5 : 1.5 mixture of hexane : ethyl acetate as an eluent (R_f of **2a**: 0.25, **2b**: 0.33, **2c**: 0.37, and **2d**: 0.07).

If we suppose that there is no flipping but some flexibility of the benzene rings, we can expect five conformers of **2**, which can be expressed in terms of calix[4]arene nomenclature. As for the calix[4]arenes, for **2** the cone, partial-cone and 1,3-alternate conformations are possible. Two types of 1,2-alternate conformation, which we call the 1,2-alternate and 1,4-alternate, can be expected. Compound **2** has two alternate bridges with different lengths, an ethylene and a methylene bridge. In the 1,2-alternate conformation, the aromatic rings are reversed at the ethylene bridges, while in the 1,4-alternate, they are reversed at the methylene bridges (Figure 1).

The X-ray crystal structure of **2b** showed it to be the 1,4-alternate conformer (Figure 2); its benzene rings are reversed at the methylene bridges. In the cases of 1,2-alternate conformers of calix[4]arene derivatives, the ^{1}H-NMR spectra show the methylene bridges as one singlet and two doublets [8]. The singlet was identified as that of the alternate bridge protons and the two doublets as due to the other bridge protons. Just as for the calix[4]arenes, the methylene protons of **2b** were observed as a broad singlet in the ^{1}H-NMR spectrum (Figure 3). These protons are magnetically equivalent and do not show a geminal coupling.

Each of the ^{1}H-NMR spectra of **2c** and **2d** showed only one singlet for the *tert*-butyl groups. This result suggests that **2c** and **2d** are either the 1,2-alternate, 1,3-

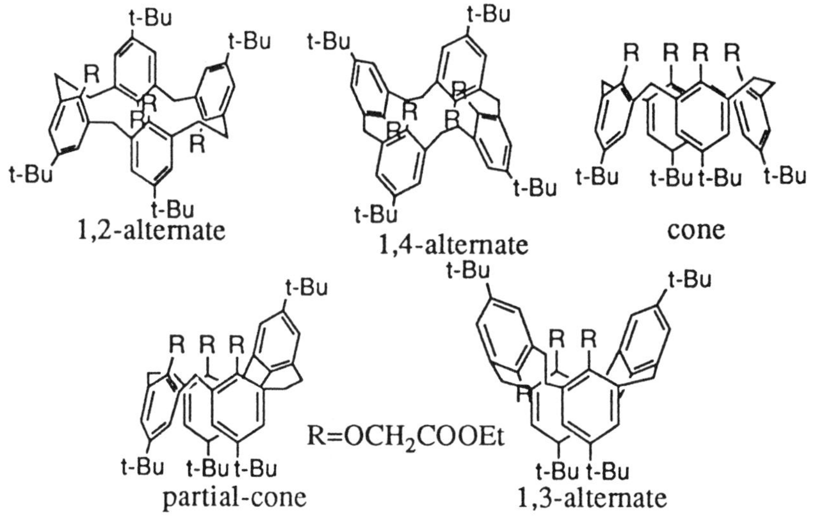

R=OCH$_2$COOEt

Fig. 1. Conformation of [2.1.2.1]MPC **2**.

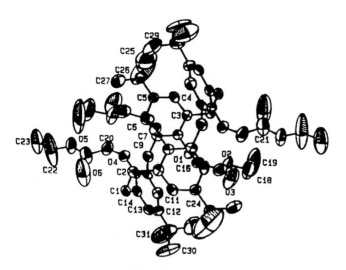

Fig. 2. Ortep view of **2b**.

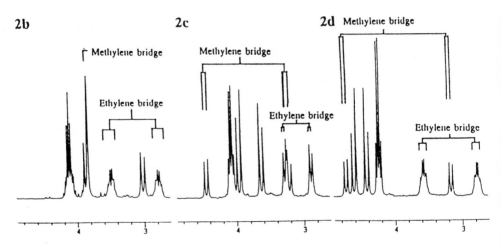

Fig. 3. Partial ^{1}H-NMR spectra of **2b**, **2c**, and **2d** (270 MHz, CDCl$_3$, 20°C).

alternate, or cone conformers, but neither the partial-cone nor any other conformer such as, for example, the flattened-cone conformer was observed. The partial ^{1}H-NMR spectra of **2b**, **2c** and **2d** are shown in Figure 3. ^{1}H-NMR spectra of **2c** and **2d** shows their methylene protons as two sets of doublets. This suggests that **2c** and **2d** may adopt the cone or 1,2-alternate conformation of the five conformations expected. Since the shape of the peaks due to the ethylene bridges of **2d** is very similar to that of **2b**, it can be assumed that the two ethylene bridges in **2d** are on the same side of the [2.1.2.1]MCP ring and those of **2c** are on opposite sides. Thus, **2c** adopts the 1,2-alternate conformation and **2d** takes the cone conformation.

The ^{1}H-NMR spectrum of **2a** shows the *tert*-butyl protons as four singlet peaks at δ 1.14, 1.19, 1.27, and 1.32 ppm, indicating that **2a** adopts an asymmetrical conformation. Of the five conformations we expected, only the partial-cone conformation is asymmetrical. Therefore, we can conclude that **2a** takes the partial-cone conformation.

Of the five conformers expected, only the 1,3-alternate type conformer of **2** was not obtained. Presumably, the 1,3-conformer of **2** is less stable than the other conformers, so that we could not obtain the conformer in detectable quantities.

2.2. COMPLEXATION PROPERTIES

The complexation properties of the four conformers of **2** were established using ^{1}H-NMR spectroscopy [1] and UV spectroscopy [6]. ^{1}H-NMR spectra were measured for **2** (4.9×10^{-3} M) in CD$_3$OD in the presence of alkali thiocyanate (0–4.9×10^{-2} M), and UV spectra were measured for alkali picrate solutions in THF (5.0×10^{-5} M) in the presence of **2** (0–1.0×10^{-3} M).

In the case of **2a**, **2b** and **2c**, we could find no difference in the spectra of **2** with or without alkali ions, but in the case of **2d**, a remarkable difference was observed (Figure 4 and Table II). The stability constants for **2d** were determined by ^{1}H-NMR and UV spectroscopy.

The molar ratio plots indicate the formation of a 1 : 1 complex. A typical example is given in Figure 5.

We shall not discuss the $\Delta\delta$ values of bridge protons, since some of them could not be determined due to overlapping of peaks. For the methoxy protons (OCH_2CO), high field shifts are observed in the case of potassium and cesium ion, but in the case of sodium ion a low field shift is observed. We assumed that

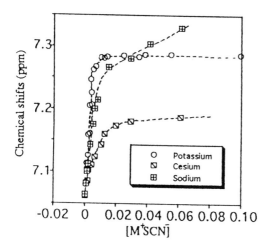

Fig. 4. The observed chemical shifts of aromatic protons of **2d** in the presence of alkali thiocyanate (270 MHz, 25°C, CD$_3$OD).

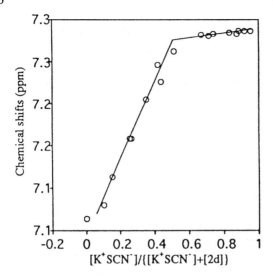

Fig. 5. Molar ratio plot of [K$^+$SCN$^-$]/([K$^+$SCN$^-$] + [**2d**]) and chemical shift of aromatic protons (270 MHz, 25°C, CD$_3$OD).

TABLE II. $\Delta\delta$ Values of **2d** in the presence of alkali cation in CD$_3$OD.

Alkali ion	$\Delta\delta$ of methoxy peaks (OCCH$_2$CO)	$\Delta\delta$ of bridge peaks (ArCH$_2$)	$\Delta\delta$ of aromatic peaks (ArH)
Na$^+$	-0.01^b, 0.10	-- ^{c}, 0.14	0.23, 0.14
K$^+$	-0.07, 0.03	-0.46, 0.18	0.21, 0.13
Cs$^+$	-0.11, 0.02	-0.20, -- ^{c}	0.11, 0.07

^{a} $\Delta\delta$ Values are the difference of the chemical shift between **2d** and **2d** in the presence of M$^+$SCN$^-$ ([**2d**] : [M$^+$SCN$^-$] = 1 : 6) in CD$_3$OD at 25°C.
^{b} The minus values indicate a the high field shift.
^{c} These chemical shifts could not be determined due to overlap with other peaks.

potassium and cesium ion coordinates at the carbonyl oxygen and sodium ion coordinates at the phenolic oxygen, since it is expected that the coordination at phenolic oxygen decreases the electron density of methoxy protons.

In all cases of aromatic protons, low field shifts are observed and $\Delta\delta$ values of the cesium complex are much smaller than those of potassium and sodium complexes. We consider that the aromatic groups come closer to each other on complexation, resulting in the low field shift of the aromatic proton. Therefore, sodium and potassium ion seem to enter into the cavity more deeply than the cesium ion.

Schneider et al. [6] have reported the direct ^{1}H-NMR method used for determining association constants. In the case of a 1 : 1 complex, the stability constant K can be calculated from Equation 2. The difference of the observed chemical shift δ_{obs} and the chemical shift δ_{free} of the uncomplexed **2d** is proportional to the ratio [HG] : [H₀], and the proportionally constant is the difference of the chemical shift of the pure complex δ_c and δ_{free} (Equation 3). As we can obtain Equation 4 from Equations 3 and 2, we can thus calculate the stability constant K by a curve fitting calculation (the nonlinear least-squares method) [17].

$$H + G \overset{K}{\rightleftarrows} HG \tag{1}$$

$$K\ (1/\text{mol}) = \frac{[HG]}{[H][G]} = \frac{[HG]}{([H_0] - [HG])([G_0] - [HG])} \tag{2}$$

where [H] = concentration of free host (**2d**); [H₀] = total concentration of host (**2d**); [G] = concentration of free guest (M⁺R⁻); [G₀] = total concentration of guest [M⁺R⁻]; and [HG] = concentration of complex (**2d** and M⁺R⁻).

$$[HG] = a[H_0] = \frac{\delta_{obs} - \delta_{free}}{\delta_c - \delta_{free}}[H_0] \tag{3}$$

$$[G_0] = \frac{a}{K(1-a)} + a[H_0] : a = \frac{\delta_{obs} - \delta_{free}}{\delta_c - \delta_{free}} \tag{4}$$

where a = ratio of complex HG to host (**2d**); δ_{obs} = observed chemical shift of **2d** in the presence of ion; δ_{free} = chemical shift of free **2d**; and δ_c = chemical shift of **2d** coordinating with alkali ions.

In UV spectroscopy, the spectra of sodium, potassium and cesium picrate in the presence of **2d** show a red shift of the absorption peak, this being dependent on the separation of the alkali ion from the picrate anion (Figure 6).

The molar ratio plots of **2d** and alkali picrate indicate the formation of a 1 : 1 complex in THF solution (Figure 7), and the stability constants K in THF are given by a curve fitting calculation (Equations 5, 6).

$$[HG] = b[G_0] = \frac{Abs_{obs} - Abs_{free}}{Abs_c - Abs_{free}}[G_0] \tag{5}$$

$$[H_0] = \frac{b}{K(1-b)} + a[G_0] : b = \frac{Abs_{obs} - Abs_{free}}{Abs_c - Abs_{free}} \tag{6}$$

where b = ratio of complex HG to guest (alkali picrate); Abs_{obs} = absorbance of alkali picrate in the presence of **2d**; Abs_{free} = absorbance of free alkali picrate; and Abs_c = absorbance of alkali picrate coordinating with alkali ions.

The log K of the complexes of **2d** and alkali cations in THF and CD₃OD solution are shown in Table III. The correlation factors of curve fitting calculation are high (> 0.99) in all cases. From these results, **2d** prefers potassium ion to sodium or cesium ion in CD₃OD solution. The difference of log K is small in THF solution, although the same order of complexation ability is observed.

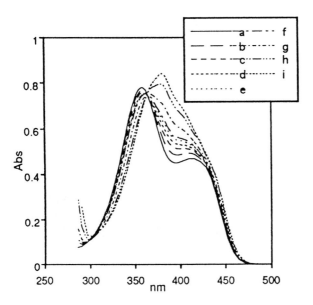

Fig. 6. UV spectra of 5.0×10^{-5} M of potassium picrate in the presence of **2d** in THF. (Potassium picrate: **2d** = a; 1 : 0, b; 1 : 0.2, c; 1 : 0.4; d, 1 : 0.6, e; 1 : 1.08; f, 1 : 1; g, 1 : 2, h; 1 : 5, i; 1 : 10).

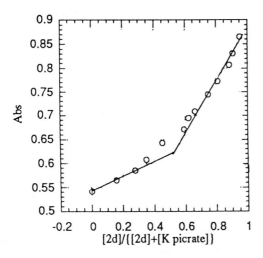

Fig. 7. Molar ratio plot of [**2d**]/([**2d**] + [K picrate]) and Abs at 380 nm.

Compared with the cone derivatives of calix[4]arene **4** [6], **2d** shows a high affinity for potassium, as the cavity size of **2d** is larger than that of calix[4]arene derivatives. This selectivity seems to be contradictory to the effect of sodium carbonate found in the selective preparation of **2d**, but we presume this difference to be dependent on the coordination site. In the preparation of **2d**, the sodium ion

TABLE III. The stability constants K of complexes of **2d** and tetra-*tert*-butyl-tetrakis[(ethoxycarbonyl)methoxy]calix[4]arene **4**[6] with sodium, potassium, and cesium ion.

	log K^a of **2d** (in MeOH-d_4)	log K^b of **2d** (in THF)	log K^c of **4** (in THF)
Na	2.42 ± 0.05	3.31 ± 0.13	3.95
K	3.86 ± 0.14	3.92 ± 0.04	3.08
Cs	2.47 ± 0.10	3.72 ± 0.17	1.60

[a] Determined by curve-fitting method of chemical shifts at 25°C.
[b] Determined by curve-fitting method of electron spectra at 23°C.
[c] Determined by Benesi–Hildebrand method of UV spectra in THF at 30°C.

will coordinate to the phenolic oxygen. On the other hand, in the case of complex-formation, the cavity of the lower ethoxycarbonyl groups seems to fit the potassium ion, being too large for the sodium ion.

3. Conclusion

As shown above, we could obtain the tetra-*tert*-butyl-tetrakis[(ethoxycarbonyl)methoxy] [2.1.2.1]MCP **2** as four conformers, the partial-cone **2a**, 1,4-alternate **2b**, 1,2-alternate **2c** and cone conformer **2d**. The conformational characterization was determined by ^{1}H-NMR spectra and an X-ray crystal structure analysis of **2b**. By using sodium carbonate in acetone, we could obtain the cone conformer **2d** selectively and in high yield.

The investigation of the complexation properties of **2** gave some interesting results. In the case of calix[4]arenes, the 1,2-alternate, 1,3-alternate, and partial-cone conformers show an affinity and a selectivity for potassium. The 1,2-alternate 1,4-alternate, and partial-cone conformers of **2**, a homologue of calix[4]arene, do not show any such ionophoric properties. It is assumed that the repulsion of the *tert*-butyl groups in **2** will affect the stability of the complexes to a larger extent than in the calix[4]arenes, since the [2.1.2.1]MCP skeleton is more flexible than that of calix[4]arene derivatives.

In the case of **2d**, the high affinities to alkali cations were observed in CD$_3$OD and THF solution. In contrast to the cone conformer of calix[4]arene derivatives **4**, **2d** prefers potassium ion to sodium ion. This difference of selectivity seems to be dependent on the size of cavity of host molecule.

Appendix: Numerical Data

X-RAY CRYSTALLOGRAPHIC DATA OF **2b**

Crystal data: $C_{62}H_{84}O_{12}$, Formula weight. 1020, triclinic; $a = 12.05(3)$, $b = 13.40(3)$, $c = 10.82(2)$ Å, $\alpha = 112.69(2)°$, $\beta = 111.06(2)°$, $\gamma = 92.60(3)°$, $V = 1469.7(7)$ Å^3, $Z = 1$ (the special position), $D_c = 1.153$ g cm^{-3}, space group $P_{\bar{1}}$, MoK_α radiation $\lambda = 0.71073$ Å, colorless prisms. Data were collected on an Enraf-Nonius CAD4 diffractometer, ω–2θ scan type, graphic-monochromatic MoK_a radiation, $\lambda = 0.71073$ Å. Of 5465 independent reflections collected in the range $1 < \theta < 35°$ 2520 with $I_0 > 3\sigma(I_0)$ were taken as observed. The crystal did not show any significant decay during the data collection. Positional parameters were determined by direct methods using SIR 88 [18], and were reflected by full-matrix least-squares calculations with all nonhydrogen atoms treated anisotropically and hydrogen atoms treated isotropically using the scheme $\omega = 4F_0^2/\sigma^2(F_0^2)^2$ to give the final residuals: $R = 8.88$, $R_w = 10.5$.

TABLE IV. Bond distances of **2b** (Å).

Atom 1	Atom 2	Distance	Atom 1	Atom 2	Distance
O(1)	C(8)	1.402(5)	C(7)	C(9)	1.501(9)
O(1)	C(6)	1.433(7)	C(9)	C(10)	1.516(7)
O(2)	C(17)	1.34(1)	C(10)	C(11)	1.412(7)
O(2)	C(18)	1.45(1)	C(10)	C(15)	1.389(9)
O(3)	C(17)	1.186(8)	C(11)	C(12)	1.359(7)
O(4)	C(15)	1.390(6)	C(12)	C(13)	1.38(1)
O(4)	C(20)	1.408(8)	C(12)	C(28)	1.517(8)
O(5)	C(21)	1.285(7)	C(13)	C(14)	1.412(7)
O(5)	C(22)	1.33(1)	C(14)	C(15)	1.392(7)
O(6)	C(21)	1.211(9)	C(16)	C(17)	1.476(9)
C(1)	C(14)	1.492(9)	C(18)	C(19)	1.28(2)
C(2)	C(3)	1.511(9)	C(20)	C(21)	1.45(1)
C(3)	C(4)	1.398(6)	C(22)	C(23)	1.33(1)
C(3)	C(8)	1.373(9)	C(24)	C(25)	1.49(2)
C(4)	C(5)	1.398(9)	C(24)	C(26)	1.43(2)
C(5)	C(6)	1.409(9)	C(24)	C(27)	1.44(1)
C(5)	C(24)	1.492(6)	C(28)	C(29)	1.50(2)
C(6)	C(7)	1.381(6)	C(28)	C(30)	1.57(1)
C(7)	C(8)	1.409(9)	C(28)	C(31)	1.46(2)

Numbers in parentheses are estimated standard deviations in the least significant digits.

TABLE V. Fractional atomic coordinate for **2b**.

Atom	x	y	z	$B(A2)$
O(1)	0.9662(3)	0.2036(3)	0.4687(4)	3.5(1)
O(2)	0.9594(5)	0.3662(4)	0.6869(5)	6.6(2)
O(3)	1.0737(7)	0.3181(5)	0.8557(5)	10.2(2)
O(4)	1.1398(4)	-0.0758(3)	0.2848(4)	3.7(1)
O(5)	1.1619(4)	-0.2365(4)	-0.0512(4)	6.0(1)
O(6)	1.3034(7)	-0.1656(9)	0.1721(8)	20.3(4)
C(1)	1.3563(5)	-0.0502(5)	0.5221(6)	3.3(2)
C(2)	0.7328(5)	0.1227(5)	0.4570(5)	3.1(2)
C(3)	0.7790(5)	0.0639(4)	0.3413(5)	2.9(1)
C(4)	0.7026(5)	-0.0270(5)	0.2115(6)	3.3(2)
C(5)	0.7326(5)	-0.0776(5)	0.0917(6)	3.3(2)
C(6)	0.8479(5)	-0.0305(5)	0.1107(5)	3.3(2)
C(7)	0.9273(5)	0.0569(4)	0.3255(5)	2.7(1)
C(8)	0.89001(5)	0.1046(4)	0.3515(5)	2.8(1)
C(9)	1.0417(5)	0.1134(5)	0.2403(5)	3.4(2)
C(10)	1.622(5)	0.1213(5)	0.3586(5)	3.0(1)
C(11)	1.2382(5)	0.2265(4)	0.4585(6)	3.5(2)
C(12)	1.3478(5)	0.2485(5)	0.5700(6)	3.4(2)
C(13)	1.3813(5)	0.1543(5)	0.5891(6)	3.4(2)
C(14)	1.3133(5)	0.0451(4)	0.4951(5)	2.9(1)
C(15)	1.2067(5)	0.0318(4)	0.3768(5)	2.5(1)
C(16)	1.0312(5)	0.2025(5)	0.6078(6)	4.0(2)
C(17)	1.0262(6)	0.2993(6)	0.7299(6)	5.3(2)
C(18)	0.9381(9)	0.4600(8)	0.794(1)	10.4(4)
C(19)	0.862(1)	0.510(1)	0.736(2)	22.8(6)
C(20)	1.1137(6)	-0.1191(6)	0.1341(7)	5.2(2)
C(21)	1.1986(6)	-0.1810(7)	0.0881(7)	6.6(2)
C(22)	1.227(1)	-0.292(1)	-0.120(1)	21.8(5)
C(23)	1.1923(9)	-0.3363(8)	0.264(1)	11.2(3)
C(24)	0.6494(5)	-0.1726(5)	-0.0481(6)	4.0(2)
C(25)	0.574(1)	-0.247(1)	-0.021(1)	19.5(6)
C(26)	0.566(2)	-0.136(1)	-0.145(2)	21.6(8)
C(27)	0.7125(8)	-0.2420(9)	-0.126(1)	11.8(4)
C(28)	1.4224(6)	0.3627(5)	0.6700(7)	5.1(2)
C(29)	1.355(1)	0.4281(9)	0.754(2)	16.5(5)
C(30)	1.5460(9)	0.3651(7)	0.7886(9)	10.0(3)
C(31)	1.435(1)	0.4195(8)	0.528(1)	13.2(5)

Anisotropically refined atoms are given in the form of the isotropic equivalent displacement parameter defined as: $(4/3) * [a^{2*}B(1,1) + b^{2*}B(2,2) + c^{2*}B(3,3) + ab(\cos\gamma)*B(1,2) + ac(\cos\beta)*B(1,3) + bc(\cos\alpha)*B(2,3)]$.

TABLE VI. Bond angles of **2b**.

Atom 1	Atom 2	Atom 3	Angle	Atom 1	Atom 2	Atom 3	Angle
C(8)	O(1)	C(16)	118.0(5)	C(1)	C(14)	C(15)	122.6(4)
C(17)	O(2)	C(18)	120.1(6)	C(13)	C(14)	C(15)	116.7(6)
C(15)	O(4)	C(20)	119.3(5)	O(4)	C(15)	C(10)	121.0(4)
C(21)	O(5)	C(22)	125.9(6)	O(4)	C(15)	C(14)	116.5(5)
C(2)	C(3)	C(4)	120.1(5)	C(10)	C(15)	C(14)	122.3(4)
C(2)	C(3)	C(8)	121.7(4)	O(1)	C(16)	C(17)	112.0(6)
C(4)	C(3)	C(8)	117.7(6)	O(2)	C(17)	O(3)	120.9(7)
C(3)	C(4)	C(5)	123.8(6)	O(2)	C(17)	C(16)	113.9(5)
C(4)	C(5)	C(6)	114.9(4)	O(3)	C(17)	C(16)	125.2(8)
C(4)	C(5)	C(24)	123.6(6)	O(2)	C(18)	C(19)	113.6(9)
C(6)	C(5)	C(24)	121.5(6)	O(4)	C(20)	C(21)	116.1(6)
C(5)	C(6)	C(7)	124.3(6)	O(5)	C(21)	O(6)	122.0(8)
C(6)	C(7)	C(8)	117.0(5)	O(5)	C(21)	C(20)	114.9(6)
C(6)	C(7)	C(9)	120.9(5)	O(6)	C(21)	C(20)	121.8(7)
C(8)	C(7)	C(9)	121.8(4)	O(5)	C(22)	C(23)	124.(1)
O(1)	C(8)	C(3)	121.8(5)	C(5)	C(24)	C(25)	111.5(7)
O(1)	C(8)	C(7)	115.2(5)	C(5)	C(24)	C(26)	111.4(7)
C(3)	C(8)	C(7)	122.3(4)	C(5)	C(24)	C(27)	113.5(6)
C(7)	C(9)	C(10)	118.0(6)	C(25)	C(24)	C(26)	107(1)
C(9)	C(10)	C(11)	119.0(6)	C(25)	C(24)	C(27)	105.8(9)
C(9)	C(10)	C(15)	125.1(4)	C(26)	C(24)	C(27)	108.(1)
C(11)	C(10)	C(15)	115.9(4)	C(12)	C(28)	C(29)	109.4(7)
C(10)	C(11)	C(12)	125.2(6)	C(12)	C(28)	C(30)	111.9(6)
C(11)	C(12)	C(13)	115.7(5)	C(12)	C(28)	C(31)	110.1(6)
C(11)	C(12)	C(28)	120.6(6)	C(29)	C(28)	C(30)	105.7(7)
C(13)	C(12)	C(28)	123.6(5)	C(29)	C(28)	C(31)	104.9(9)
C(12)	C(13)	C(14)	123.8(5)	C(30)	C(28)	C(31)	114.5(8)
C(1)	C(14)	C(13)	120.7(4)				

Numbers in parentheses are estimated standard deviations in the least significant digits.

^{1}H-NMR SPECTRUM OF **2a**

δ_H (270 MHz; CDCl$_3$; Me$_4$Si) 1.14 (s, 9H), 1.19 (s, 9H), 1.27 (s, 9H), 1.32 (s, 9H), 1.15–1.31 (m, 12H), 2.31–4.23 (m, 28H), 6.93 (d, 1H, J = 2.2 Hz), 6.94 (d, 1H, J = 2.2 Hz), 7.00 (d, 2H, J = 2.2 Hz), 7.07 (d, 1H, J = 22 Hz), 7.15 (d, 1H, J = 2.2. Hz), 7.27 (d, 1H, J = 2.2 Hz), 7.39 (d, 1H, J = 2.2 Hz).

^{1}H-NMR SPECTRUM OF **2b**

δ_H (270 MHz; CDCl$_3$; Me$_4$Si) 1.22 (t, 12H, J = 7.3 Hz), 1.26 (s, 36H), 2.68–2.77 (m, 4H), 3.01 (d, 4H, J = 14.7 Hz), 3.41–3.50 (m, 4H), 3.82 (s, 4H), 3.85 (d, 4H,

J = 14.7 Hz), 4.00–4.18 (m, 8H), 7.08 (d, J = 2.6 Hz, 4H), 7.13 (d, J = 2.6 Hz, 4H).

^{1}H-NMR SPECTRUM OF 2c

δ_H (270 MHz; CDCl$_3$; Me$_4$Si) 1.19 (12H, t, J = 7.1 Hz), 1.27 (36H, s), 2.88 (4H, br.d, J = 11.5 Hz), 3.21 (2H, d, J = 14.7 Hz), 3.27 (4H, br.d, J = 11.5 Hz), 3.62 (4H, d, J = 15.6 hz), 3.95 (4H, d, J = 15.6 Hz), 4.01–4.14 (8H, m), 7.08 (4H, d, J = 2.5 Hz), 7.20 (d, 4H, J = 2.5 Hz).

^{1}H-NMR SPECTRUM OF 2d

δ_H (270 MHz; CDCl$_3$; Me$_4$Si) 1.11 (s, 36H), 1.31 (12H, t, J = 7.1 Hz), 2.72–2.80 (4H, m), 3.15 (2H, d, J = 13.0 Hz), 3.52–3.61 (4H, m), 4.36 (4H, d, J = 15.0 Hz), 4.60 (4H, d, J = 15.0 Hz), 4.73 (2H, d, J = 13.0 Hz), 4.23 (8H, q, J = 7.1 Hz), 6.85 (4H, d, J = 2.3 Hz), 6.94 (4H, d, J = 2.3 Hz).

References

1. (a) G. D. Andreetti, G. Calestani, F. Ugozzoli, A. Auduni, E. Chidini, A. Pochini, and R. Ungaro: *J. Incl. Phenom.* **5**, 123 (1987). (b) A. Arduini, A. Pochini, S. Reverberi, R. Ungaro, G. D. Andreetti, and F. Ugozzoli: *Tetrahedron* **42**, 2089 (1986).
2. S. K. Chang and I. Cho: *J. Chem. Soc., Perkin Trans. 1* 211 (1989).
3. M. A. McKervey, E. M. Seward, G. Ferguson, B. Ruhl, and S. J. Harris: *J. Chem. Soc., Chem. Commun.* 388 (1985).
4. F. Arnaud-Neu, E. M. Collins, M. Deasy, G. Ferguson, S. J. Harris, B. Kaitner, A. J. Lough, M. A. McKervey, E. Marques, B. L. Ruhl, M. J. Schwing-Weill, and E. M. Seward: *J. Am. Chem. Soc.* **111**, 8681 (1989).
5. S. Shinkai: *Tetrahedron* **49**, 8933 (1993).
6. T. Arimura, M. Kubota, T. Matsuda, O. Manabe, and S. Shinkai: *Bull. Chem. Soc. Jpn.* **62**, 1674 (1989).
7. S. Shinkai, K. Fujimoto, T. Otsuka, and H. L. Ammon: *J. Org. Chem.* **57**, 1516 (1992).
8. K. Iwamoto and S. Shinkai: *J. Org. Chem.* **57**, 7066 (1992).
9. K. Iwamoto, K. Araki, and S. Shinkai: *J. Chem. Soc., Perkin Trans. 1* 1611 (1991).
10. C. D. Gutsche and P. A. Reddy: *J. Org. Chem.* **56**, 4783 (1991).
11. M. Tashiro, T. Watanabe, A. Tsuge, T. Sawada, and S. Mataka: *J. Org. Chem.* **54**, 2632 (1992).
12. M. Tashiro, A. Tsuge, T. Sawada, T. Makishima, S. Horie, T. Arimura, S. Mataka, and T. Yamato: *J. Org. Chem.* **55**, 2404 (1990).
13. J. Schmitz, F. Vögtle, M. Nieger, K. Gloe, H. Stephan, O. Heitzsch, H.-J. Buschmann, W. Hasse, and K. Cammann: *Chem. Ber.* **126**, 2483 (1993).
14. T. Yamato, Y. Saruwatari, L. K. Doamekpor, K. Hasegawa, and M. Koike: *Chem. Ber.* **126**, 2501 (1993).
15. D. H. Burns, J. D. Miller, and J. Santana: *J. Org. Chem.* **58**, 6526 (1993).
16. H.-J. Schneider, R. Kramer, S. Simova, and U. Schneider: *J. Am. Chem. Soc.* **110**, 6442 (1988).
17. The curve fitting calculation was done by 'KaleidaGraph' from Abelbeck software.
18. M. C. Burla, M. Camalli, G. Cascarano, C. Giacovazzo, G. Polidori, R. Spagna, and D. Viterbo: *J. Appl. Crystallogr.* **22**, 389 (1989).

Synthesis and Ion Selectivity of Macrocyclic Metacyclophanes Analogous to Spherand-Type Calixarenes[*]

TAKEHIKO YAMATO,[**] MASASHI YASUMATSU,
YOSHIYUKI SARUWATARI, and LOUIS KORBLA DOAMEKPOR
Department of Applied Chemistry, Faculty of Science and Engineering, Saga University, Honjo-machi 1, Saga-shi, Saga 840, Japan

(Received: 3 March 1994; in final form: 19 July 1994)

Abstract. Novel macrocyclic compounds, hexahydroxy[1.0.1.0.1.0]- (**2b**) and octahydroxy[1.0.1.0.-1.0.1.0]metacyclophane (**2c**) have been prepared in 50–70% yield by base-catalyzed condensation of 5,5'-di-*tert*-butyl-2,2'-dihydroxybiphenyl (**1**) with formaldehyde in refluxing xylene. An attempted alkylation of the flexible macrocycles **2b** and **2c** with ethyl bromoacetate in the presence of Cs_2CO_3 under acetonitrile reflux gave only one pure stereoisomer **3** and **4**, respectively, while other possible isomers were not observed. The structural characterization of these products is also discussed. The two-phase solvent extraction data indicated that hexaethyl ester **3** and octaethyl ester **4** show strong metal affinity, comparable with that of the corresponding calix[n]arenes, and a high K^+ selectivity was observed for octaethyl ester **4**. ^{1}H-NMR titration of hexaethyl ester **3** and octaethyl ester **4** with KSCN clearly demonstrate that a 1 : 1 complex is formed which is stable on the NMR time scale.

Key words: Spherand-type calixarene, ion selectivity, diastereoselective functionalization.

1. Introduction

Due to their importance in supramolecular chemistry [1] a large variety of macrocyclic compounds such as crown ethers [2], cryptands [3], cyclophanes [4], and spherands [5] have been synthesized and their properties investigated. Our present work is directly related to the studies of Cram and coworkers [5] in designing spherands possessing 1,1'-biarene units which contain enforced, spherical cavities lined with electron pairs of heteroatoms so that no conformational rearrangement is possible upon complexation with metal ions.

Closely related to the spherands are the 'calixarenes', [1_n]metacyclophanes, prepared by base-catalyzed condensation of *p*-substituted phenols with formaldehyde [6]. These are attractive building blocks, their phenolic hydroxyl groups being ordered in well-shaped cyclic arrays [6–8] which can be functionalized [7–14] to give novel guest inclusion blocks. The combination of structural elements of both

---

[*] This paper is dedicated to the commemorative issue on the 50th anniversary of calixarenes.
[**] Author for correspondence.

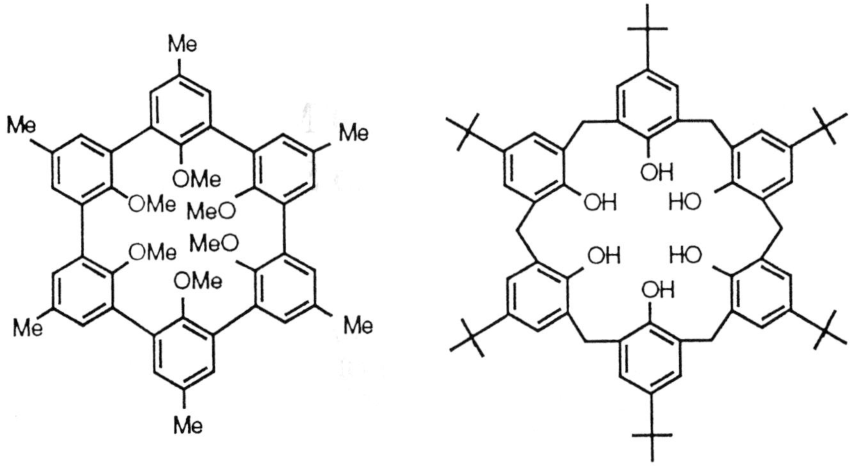

Chart 1.

spherands and calixarenes is expected to lead to novel macrocyclic compounds, which may greatly stimulate future work in the field of host–guest chemistry.

We have recently demonstrated a convenient and selective synthesis of a series of hydroxy[1_n]metabiphenylophanes with three and four biarene units **2** involving base-catalyzed condensation of 5,5'-di-*tert*-butyl-2,2'-dihydroxybiphenyl (**1**) with formaldehyde under refluxing xylene, and we have reported their unique properties [15]. The cyclic trimer **2b** was obtained as a major product with NaOH as base in xylene, while the action of the larger Cs$^+$ led to the larger macrocyclic tetramer **2c**. These results seem to indicate that the template effect of an alkali metal cation plays an important role in this condensation reaction as previously observed in the preparation of calixarenes [6].

In comparison to the structural characteristics of a calixarene family, spherand-type calixarene analogous metacyclophanes **2** attracted our further interest for the following reasons: (i) macrocycles **2b** with three methylene bridges and three biarene linkages may be regarded as a combination of half structural elements of both spherands and calix[6]arenes; (ii) since the MeO group is bulky enough to inhibit the oxygen-through-the-annulus rotation of macrocycles **2** [15], interconversion between the conformers derived from introduction of the higher alkyl group on the phenolic oxygens of macrocycles **2a** and **2b** should not take place; (iii) the substituent OR group at the 2-position of a biarene unit should be quite removed from the 2'-position in such a configuration, so avoiding steric hindrance at the Ar—Ar σ-bond; (iv) the restricted ring conformation due to the biarene units may increase the selectivity in the binding of metal ions as compared to the same functionalized phenolic units in calix[6]arene and calix[8]arene. We report here the first example of *O*-alkyl derivatives **3** and **4** with three and four biarene units from the reaction of macrocycles **2a** and **2b** with ethyl bromoacetate, their metal selectivity, and the binding-mode for metals. These compounds are expected to have combined properties of both the spherands and the calixarenes.

2. Experimental

All melting and boiling points are uncorrected. IR (KBr or NaCl): Nippon Denshi JIR-AQ2OM. ^{1}H-NMR: Nippon Denshi JEOL FT-270 in CDCl$_3$, TMS as reference. MS: Nippon Denshi JMS-01SA–2. Elemental analysis: Yanaco MT-5.

2.1. SYNTHESIS

2.1.1. *Preparation of 4,10,17,23,30,36-hexa-tert-butyl-7,13,20,26,33,39-hexahydroxy[1.0.1.0.1.0]metacyclophane* (**2b**). To a mixture of 5,5'-di-*tert*-butyl-2,2'-dihydroxybiphenyl (**1**) (5 g, 16.75 mmol) and paraformaldehyde (1.1 g, 35.75 mmol) in xylene (85 mL) was added aqueous 5 N NaOH (1.5 mL) under nitrogen with vigorous stirring. After the reaction mixture had been refluxed for 16 h, it was cooled to room temperature, acidified with 1 N HCl (50 mL) and extracted with CH$_2$Cl$_2$. The CH$_2$Cl$_2$ extract was washed with water, dried (Na$_2$SO$_4$) and the solvent was evaporated under reduced pressure. The residue was chromatographed over silica gel (Wako, C-300; 500 g) with hexane and benzene as eluent to give 3.20 g of crude **2b** and a trace amount of **2c**, respectively. Recrystallization from hexane gave the *title compound* **2b** (2.72 g, 52%) as colourless prisms; m.p. 253–256°C; ν_{max} (KBr)/cm^{-1} 3297 (OH); δ_H (CDCl$_3$, 20°C) 1.35 (54H, s, C(*CH$_3$*)$_3$), 4.02 (6H, Ar*CH$_2$*Ar), 7.20 (6H, d, J = 2.44, Ar*H*), 7.39 (6H, d, J = 2.44, Ar*H*), 8.48 (6H, broad s, *OH*); δ_H (CDCl$_3$, -50°C), 1.35 (54H, s, C(*CH$_3$*)$_3$), 3.79 (3H, broad s, Ar*CH$_2$*Ar), 4.21 (3H, broad s, Ar*CH$_2$*Ar), 7.21 (6H, d, J = 2.44, Ar*H*), 7.44 (6H, d, J = 2.44, Ar*H*), 8.48 (6H, broad s, *OH*); m/z: 931 (M$^+$); *Found*: C, 80.90; H, 8.60. *Calcd.* for C$_{63}$H$_{78}$O$_6$: C, 81.25; H, 8.44%.

2.1.2. *Preparation of 4,10,17,23,30,36,43,49-octa-tert-butyl-7,13,20,26,33,39, 46,52-octahydroxy[1.0.1.0.1.0]metacyclophane* (**2c**). To a mixture of **1** (5 g, 16.75 mmol) and paraformaldehyde (1.1 g, 35.75 mmol) in xylene (85 mL) was added aqueous 5 N CsOH (1.5 mL) under nitrogen with vigorous stirring. After the reaction mixture had been refluxed for 16 h, it was cooled to room temperature, acidified with 1 N HCl (50 mL) and extracted with CH$_2$Cl$_2$. The CH$_2$Cl$_2$ extract was washed with water, dried (Na$_2$SO$_4$) and the solvent was evaporated under reduced pressure. The residue was chromatographed over silica gel (Wako, C-300; 500 g) with hexane and benzene as eluent to give 204 mg (4%) of **2b** and 3.80 g of **2c**, respectively. Recrystallization from benzene gave the *title compound* **2c** (3.44 g, 66%) as colourless prisms; m.p. > 300°C; ν_{max} (KBr)/cm^{-1} 3245 (OH); δ_H (CDCl$_3$, 20°C) 1.30 (72H, s, C(*CH$_3$*)$_3$), 4.02 (8H, Ar*CH$_2$*Ar), 7.08 (8H, d, J = 2.44, Ar*H*), 7.43 (8H, d, J = 2.44, Ar*H*), 8.34 (8H, broad s, *OH*); δ_H (CDCl$_3$, -50°C) 1.29 (36H, s, C(*CH$_3$*)$_3$), 1.32 (36H, s, C(*CH$_3$*)$_3$), 3.63 (2H, d, J = 14.2,

ArCH$_2$Ar), 3.88 (2H, d, J = 14.2, ArCH$_2$Ar), 4.15 (2H, d, J = 14.2, ArCH$_2$Ar), 4.70 (2H, d, J = 14.2, ArCH$_2$Ar), 7.05 (4H, d, J = 2.44, ArH), 7.12 (4H, d, J = 2.44, ArH), 7.43 (4H, d, J = 2.44, ArH), 7.54 (4H, d, J = 2.44, ArH), 8.34 (4H, broad s, OH), 8.60 (4H, broad s, OH); m/z: 1241 (M$^+$); *Found*: C, 81.05; H, 8.40. *Calcd.* for C$_{84}$H$_{104}$O$_8$: C, 81.25; H, 8.44%.

2.1.3. *Preparation of 4,10,17,23,30,36-hexa-tert-butyl-7,13,20,26,33,39-hexa-[(ethoxycarbonyl)methoxy][1.0.1.0.1.0]metacyclophane* (**3**). A mixture of 4,10, 17,23,30,36-hexa-*tert*-butyl-7,13,20,26,33,39-hexahydroxy[1.0.1.0.1.0]metacyclophane (**2b**) (1.5 g, 1.61 mmol) and cesium carbonate (3.46 g, 10.63 mmol) in 100 mL of dry acetonitrile was heated at reflux for 3 h under nitrogen. Then ethyl bromoacetate (3.2 mL, 29 mmol) was added and the mixture heated at reflux for 16 h. After cooling the reaction mixture to room temperature it was filtered. The filtrate was concentrated to give a yellow oil, which was then distilled under reduced pressure to remove the excess of unreacted ethyl bromoacetate using a Kugelrohr apparatus. The residue was chromatographed over silica gel (Wako, C-300; 300 g) with benzene as eluent to give a solid, which was recrystallized from hexane to afford the *title compound* **3** (2.15 g, 92%) as colourless prisms; m.p. 191–192°C; ν_{max} (KBr)/cm^{-1} 1733, 1762 (C=O); δ_H (CDCl$_3$) 0.82 (6H, t, J = 7.1, OCH$_2$CH$_3$), 1.01 (6H, t, J = 7.1, OCH$_2$CH$_3$), 1.09 (6H, t, J = 7.1, OCH$_2$CH$_3$), 1.28 (18H, s, C(CH$_3$)$_3$), 1.30 (18H, s, C(CH$_3$)$_3$), 1.34 (18H, s, C(CH$_3$)$_3$), 3.31 (2H, d, J = 17.1, OCH$_2$CO), 3.39 (2H, J = 13.2, ArCH$_2$Ar), 3.68 (2H, d, J = 17.1, OCH$_2$CO), 3.47–4.02 (18H, overlapped signals due to OCH$_2$CO and OCH$_2$CH$_3$), 4.14 (2H, s, ArCH$_2$Ar), 4.28 (2H, d, J = 17.1, OCH$_2$CO), 5.24 (2H, d, J = 13.2, ArCH$_2$Ar), 6.89 (2H, d, J = 2.4, ArH), 7.04 (2H, d, J = 2.4, ArH), 7.12 (2H, d, J = 2.1, ArH), 7.31 (2H, d, J = 2.4, ArH), 7.34 (2H, d, J = 2.4, ArH), 7.44 (2H, d, J = 2.4, ArH); *Found*: C, 67.70; H, 7.76. *Calcd.* for C$_{87}$H$_{114}$O$_{18}$·CHCl$_3$: C, 67.44; H, 7.4%.

2.1.4. *Preparation of 4,10,17,23,30,36,43,49-octa-tert-butyl-7,13,20,26,33,39,46, 52-octa[(ethoxycarbonyl)methoxy][1.0.1.0.1.0.1.0]metacyclophane* (**4**). A mixture of 4,10,17,23,30,36,43,49-octa-*tert*-butyl-7,13,20,26,33,39,46,52-octahydroxy[1.0.1.0.1.0.1.0]metacyclophane (**2c**) (400 mg, 0.32 mmol) and cesium carbonate (1.89 g, 5.79 mmol) in 30 mL of dry acetonitrile were heated at reflux for 1 h under nitrogen. Then ethyl bromoacetate (1.4 mL, 15.1 mmol) was added and the mixture heated at reflux for 16 h. After cooling the reaction mixture to room temperature it was filtered. The filtrate was concentrated to give a yellow oil, which was then distilled under reduced pressure to remove the excess unreacted ethyl bromoacetate using a Kugelrohr apparatus. The residue was chromatographed over silica gel (Wako, C-300; 300 g) with chloroform–ethanol (1 : 1 v/v) as eluent to give a solid, which was recrystallized from chloroform–methanol (1 : 1 v/v) to afford the *title compound* **4** (479 mg, 82%) as colourless prisms; m.p. > 300°C; ν_{max} (KBr)/cm^{-1} 1736, 1762 (C=O); δ_H (CDCl$_3$) 1.02 (24H, t, J = 7.3; OCH$_2$CH$_3$), 1.28 (72H, s, C(CH$_3$)$_3$), 3.40 (16H, q, J = 7.3, OCH$_2$CH$_3$), 3.59 (4H, d, J = 15.1,

ArCH$_2$Ar), 3.72 (8H, d, J = 17.1, ArOCH$_2$), 3.99 (8H, d, J = 17.1, ArOCH$_2$), 5.13 (4H, d, J = 15.1, ArCH$_2$Ar), 7.02 (8H, d, J = 2.5, ArH), 7.21 (8H, d, J = 2.5, ArH); *Found*: C, 68.74; H, 7.64. *Calcd.* for C$_{116}$H$_{152}$O$_{24}$·CHCl$_3$: C, 68.56; H, 7.52%.

2.1.5. *Preparation of 4,10,17,23,30,36-hexa-tert-butyl-7,13,20,26,33,39-hexa-[(methoxycarbonyl)methoxy][1.0.1.0.1.0]metacyclophane* (**5**). A solution of hexaethyl ester (**3**) (100 mg, 0.069 mmol) and *p*-toluenesulfonic acid (20 mg, 10.63 mmol) in 50 mL of methanol was heated at reflux for a week under nitrogen. After cooling the reaction mixture to room temperature it was concentrated under reduced pressure. The residue was extracted with CH$_2$Cl$_2$, washed with water, dried (Na$_2$SO$_4$) and the solvent was evaporated under reduced pressure to give a solid which was recrystallized from hexane to afford the *title compound* **5** (80 mg, 85%) as colourless prisms; m.p. 235–236°C; ν_{max} (KBr)/cm^{-1} 1735, 1764 (C=O); δ_H (CDCl$_3$) 1.25 (18H, s, C(CH$_3$)$_3$), 1.28 (18H, s, C(CH$_3$)$_3$), 1.31 (18H, s, C(CH$_3$)$_3$), 3.07 (6H, s, OCH$_3$), 3.20 (6H, s, OCH$_3$), 3.39 (2H, d, J = 13.2, ArCH$_2$Ar), 3.40 (2H, d, J = 17.1, OCH$_2$CO), 3.24 (6H, s, OCH$_3$), 3.66 (2H, d, J = 17.1, OCH$_2$CO), 3.67 (2H, d, J = 17.1, OCH$_2$CO), 3.80 (2H, d, J = 17.1, OCH$_2$CO), 3.97 (2H, d, J = 17.1, OCH$_2$CO), 4.04 (2H, s, ArCH$_2$Ar), 4.16 (2H, d, J = 17.1, OCH$_2$CO), 5.10 (2H, d, J = 13.2, ArCH$_2$Ar), 6.94 (2H, d, J = 2.4, ArH), 7.02 (2H, d, J = 2.4, ArH), 7.10 (2H, d, J = 2.4, ArH), 7.31 (2H, d, J = 2.4, ArH), 7.36 (2H, d, J = 2.4, ArH), 7.45 (2H, d, J = 2.4, ArH); *Found*: C, 71.10; H, 7.86. *Calcd.* for C$_{81}$H$_{102}$O$_{18}$: C, 71.34; H, 7.54.

2.2. PICRATE EXTRACTION MEASUREMENTS

Metal picrates (2.5 × 10^{-4} M) were prepared *in situ* by dissolving the metal hydroxide (0.01 mol) in 2.5 × 10^{-4}M picric acid (100 mL); triply distilled water was used for all aqueous solutions. Two phase solvent extraction was carried out between water (5 mL, [alkali picrate] = 2.5 × 10^{-4} M) and CH$_2$Cl$_2$ (5 mL, [ionophore] = 2.5 × 10^{-4}M). The two-phase mixture was shaken in a stoppered flask for 2 h at 25°C. We confirmed that this period is sufficient to attain the distribution equilibrium. This was repeated three times, and the solutions were left standing until phase separation was complete. The extractability was determined spectrophotometrically from the decrease in the absorbance of the picrate ion in the aqueous phase as described by Pedersen [31].

3. Results and Discussion

3.1. BASE-CATALYZED CONDENSATION OF 5,5'-DI-*tert*-BUTYL-2,2'-DIHYDROXYBIPHENYL WITH PARAFORMALDEHYDE

The starting compound, 5,5'-di-*tert*-butyl-2,2'-dihydroxybiphenyl (**1**) [16] was prepared in two steps from 2,4-di-*tert*-butylphenol by using the *tert*-butyl group as a positional protective group on the aromatic ring [16–21].

When **1** was treated with paraformaldehyde in refluxing xylene under basic conditions, the expected products 4,10,17,23,30,36-hexa-*tert*-butyl-7,13,20,26,33,39-hexahydroxy[1.0.1.0.1.0]- (**2b**) and 4,10,17,23,30,36,43,49-octa-*tert*-butyl-7,13, 20,26,33,39,46,52-octahydroxy[1.0.1.0.1.0.1.0]metacyclophane (**2c**) were obtained [21]. These compounds were easily separated from the crude reaction mixture by column chromatography.

However, under these conditions the formation of the dimer, 4,10,17,23-tetra-*tert*-butyl-7,13,20,26-tetrahydroxy[1.0.1.0]metacyclophane **2a** has not been observed. This finding seems to support the strained nature of **2a** compared to **2b** and **2c**, which have larger rings. The structures of **2b** and **2c** were elucidated from their elemental analyses and spectral data.

The conformations of trimer **2b** and tetramer **2c** have been evaluated from their dynamic ^{1}H-NMR spectra [15]. The conformational ring inversion of macrocycles **2b** and **2c** was observed at room temperature in the same way as Gutsche's hydroxy[1_n]metacyclophanes (calix[n]arenes).

3.2. INTRODUCTION OF (ETHOXYCARBONYL)METHOXY GROUP TO PHENOLIC HYDROXY GROUP

Introduction of larger alkyl groups on the phenolic oxygens should lead to a situation where the OR groups within a biarene unit cannot pass each other. That means the rotation around the Ar—Ar σ-bond is hindered and each biarene unit is fixed in a certain configuration. For example, in the case of the tetramer **2c** four different diastereomers should be expected, two of which are chiral: (a) a compound in which all biarene units have the same configuration (*RRRR* or *SSSS*, D_4 symmetry);

D_4-symmmetry

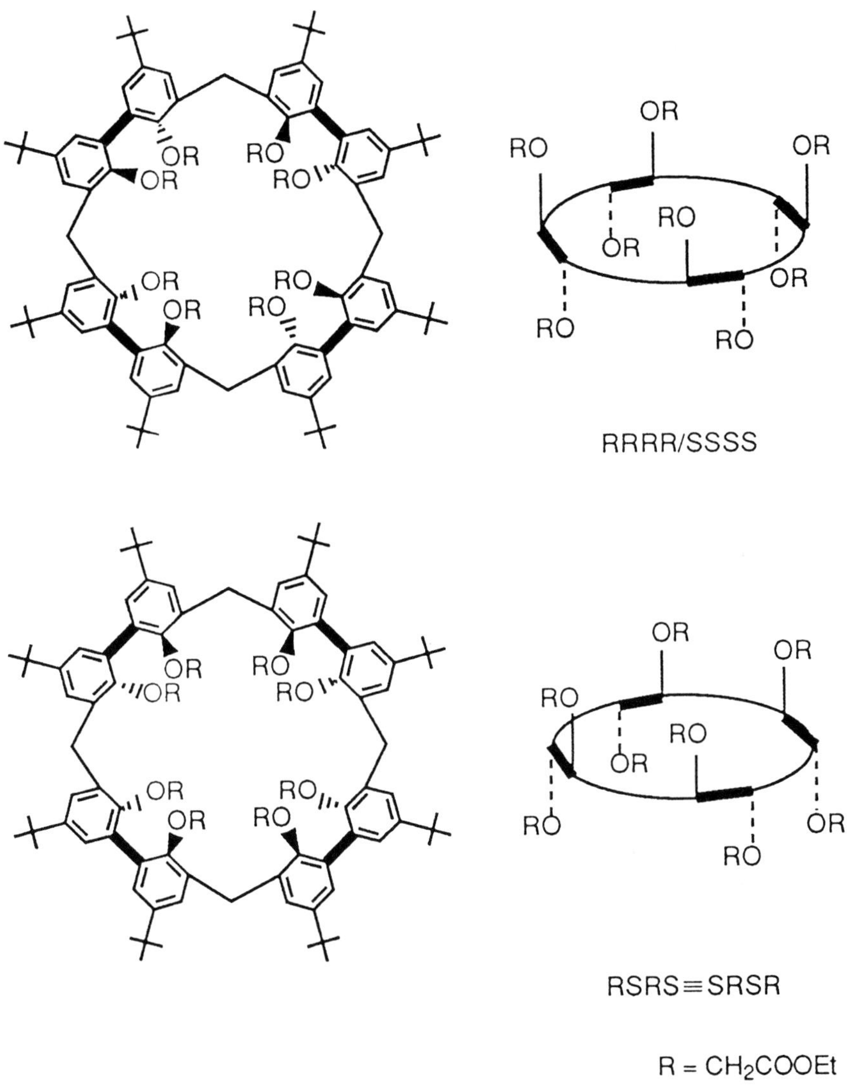

Fig. 1. Stereoisomers of the tetramer **2c**.

RRRR/SSSS

RSRS≡SRSR

R = CH$_2$COOEt

(b) a compound in which the configuration of one biphenyl unit differs from that of the other three (*RRRS* or *SSSR*); (c) (*RSRS* ≡ *SRSR*); or (d) *RRSS* ≡ *SSRR* (see Figure 1). Similarly, in the case of trimer **2b** there are two possible stereoisomers (*RRR/SSS* and *RRS/SSR*; D_3 symmetry and C_2 symmetry) (see Figure 2).

Alkylation of the flexible macrocycles **2b** and **2c** with ethyl bromoacetate in the presence of Cs$_2$CO$_3$ under acetonitrile reflux gave only one pure stereoisomer **3** and **4**, respectively. No other possible isomers were observed.

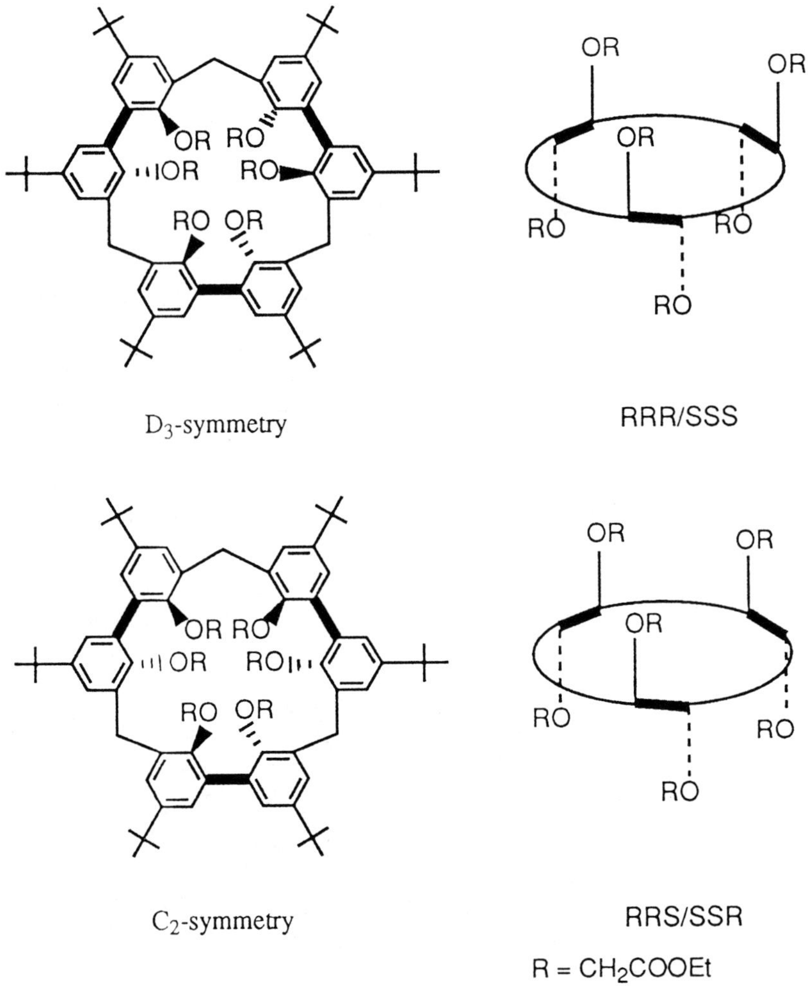

D₃-symmetry   RRR/SSS

C₂-symmetry   RRS/SSR

R = CH₂COOEt

Fig. 2. Stereoisomers of the trimer **2b**.

The ^{1}H-NMR spectrum of **3** shows three signals for the *tert*-butyl protons, three triplets of equal intensity for the methyl protons of the ethyl group, and six doublets of equal intensity for the aromatic protons. Furthermore, the resonance for the Ar*CH₂*Ar methylene protons appeared as a pair of doublets (δ 3.39 and 5.24, J_{AB} = 13.2 Hz) and a singlet (δ 4.14) (relative intensity 1 : 1 : 1). These signals correspond to an asymmetric structure (C_2 symmetry) but not a symmetric one (D_3 symmetry).

2
b; n = 3
c; n = 4

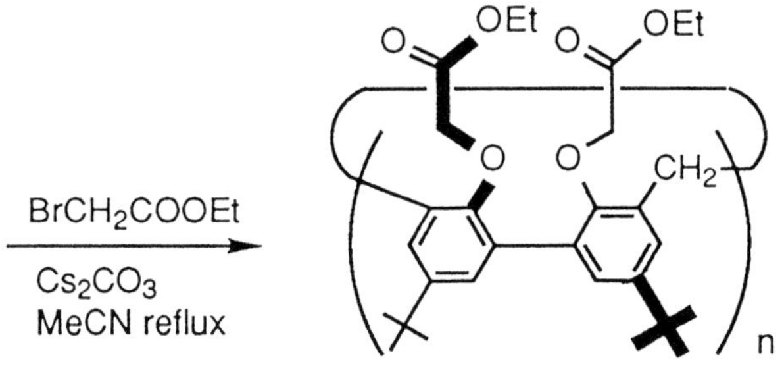

→ BrCH$_2$COOEt, Cs$_2$CO$_3$, MeCN reflux →

3; n = 3 (92%)
4; n = 4 (82%)

The resonances of the OCH$_2$CO methylene protons have not been assigned because of the overlapped signals due to OCH$_2$CO and OCH$_2$CH$_3$ methylene protons. It was therefore of interest to prepare the hexamethyl ester **5** whose ^{1}H-NMR spectrum might show much more clear patterns due to the ArCH$_2$Ar and OCH$_2$CO methylene protons. Transesterification of hexaethyl ester **3**, performed in refluxing methanol in the presence of p-toluenesulfonic acid, led to the desired hexamethyl ester **5** in 85% yield.

3 → p-toluenesulfonic acid, MeOH reflux for 1 week (85%) → **5**

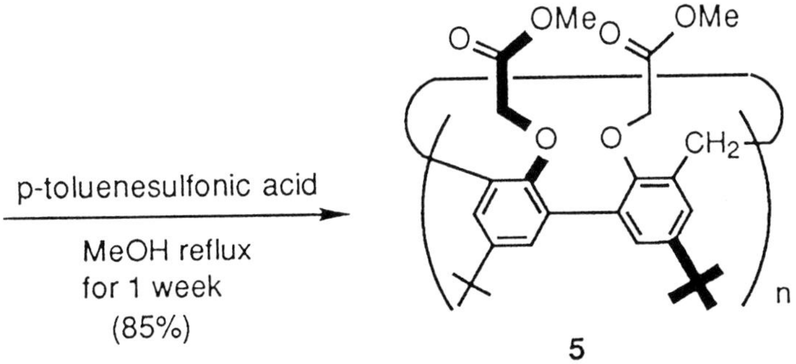

In addition to the three singlets of equal intensity for the OCH$_3$ methyl protons, three pairs of doublets (relative intensity 1 : 1 : 1) for the diastereotopic OCH$_2$CO methylene protons and a pair of doublets (δ 3.39 and 5.10, J_{AB} = 13.2 Hz) and a singlet (δ 4.04) (relative intensity 1 : 1 : 1) are observed. These signals are consistent with **3** and **5** having asymmetric structures (C_2 symmetry).

In contrast, the ^{1}H-NMR spectrum of **4** shows resonances for the *tert*-butyl protons at δ 1.28, for the methylene protons at δ 3.59, 5.13 (J_{AB} = 15.1 Hz), and for the aromatic protons at δ 7.02, 7.21, indicating a symmetrical structure. These signals correspond to symmetric structures, *RRRR/SSSS* or *RSRS*, described above. Recently, chiral calixarenes have been reported by many groups [22]. Böhmer and coworkers [23] demonstrated the chirality of dissymmetric calix[4]arenes with

C_2 and C_4 symmetry by interaction with Pirkle's reagent [(S)-(+)-1-(9-anthryl)-2,2,2-trifluoroethanol]. In fact, in the ^{1}H-NMR spectrum of compound **4** all peaks remained unchanged on addition of this shift reagent. This indicates that macrocycle **4** prepared here might adopt an *RSRS* configuration but not a chiral *RRRR/SSSS* one. Although the structure of **4** has not yet been completely assigned using the available data, one might also assume the more favorable *RSRS* stereoisomer for compound **4** instead of the *RRRR/SSSS* one because the intramolecular hydrogen bonding among OH groups between the diarylmethane units of the starting compound **2c** is stronger than that of biarene units. Unequivocal assignment of the structure of compound **4** must await X-ray analysis.

3.3. TWO-PHASE SOLVENT EXTRACTION OF ALKALI METALS

Ungaro et al. [10, 24], McKervey et al. [12, 25], Chang et al. [11], and Shinkai et al. [13, 14, 26–28] discovered that calix[n]arenes can be converted to neutral ligands by the introduction of ester groups into the OH groups. They demonstrated that metal selectivity depends on the calix[n]arene ring size and, in particular, calix[4]arylacetates and acetamides with a cone conformation show remarkably high Na^+ selectivity. Calix[4]arene and spherand-type calixarene analogous metacyclophanes have different ring sizes and ring flexibilities. It is thus interesting to assess what kind of ionophoric cavity the hexaethyl ester **3** and the octaethyl ester **4** provide. To the best of our knowledge, however, no precedent exists for the molecular design of such spherand-type calixarene analogous metacyclophane-based ionophores. We estimated this through two-phase solvent extraction of alkali metal picrates and compared these data with those for calix[n]arene aryl acetates. The results are summarized in Figure 3.

It is already known that the cone-conformer of a calix[4]arene tetraester shows Na^+ selectivity, whereas the partial-cone-conformer of calix[4]arene tetraester shows K^+ selectivity. The two-phase solvent extraction data indicated that hexaethyl ester **3** and octaethyl ester **4** show strong metal affinity, comparable with that of the corresponding calix[n]arenes, and a high K^+ selectivity was observed for octaethyl ester **4**. However, no significant high ion selectivity was observed in hexaethyl ester **3**. In contrast, octaethyl ester **4** shows higher K^+ selectivity than the corresponding calix[8]arene octatester, although the percentage extraction is somewhat lower than that for the partial-cone-conformer of the calix[4]arene tetraester. The effect of the restricted ring conformation due to the biarene units in spite of the larger ring size than that of a calix[8]arene clearly appears in the ion selectivity; i.e., octaethyl ester **4** extracted large ions like K^+ and Rb^+ more efficiently than small ones like Li^+ and Na^+. This behaviour forms a remarkable contrast to that of the partial-cone-conformer of calix[4]arene tetraester which has a rather larger ionophoric cavity than the cone-conformer of calix[4]arene tetraester, which shows notable Na^+ selectivity.

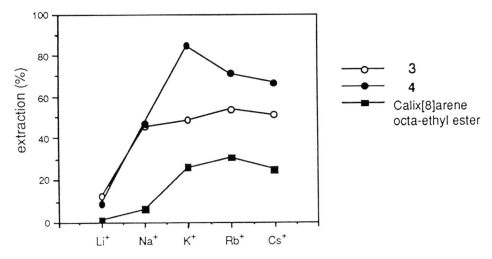

Fig. 3. Extraction (%) of alkali metal picrates by ionophores **3** and **4** in CH_2Cl_2. Extraction conditions: 2.5×10^{-4} M of ionophore in CH_2Cl_2; 2.5×10^{-4} M of picric acid in 0.1 M of alkaline hydroxide at 25°C. Ionophore solution (5.0 mL) was shaken for 2 h with picrate solution (5.0 mL) and % extraction was measured by the absorbance of picrate in CH_2Cl_2. Experimental error was ±2%.

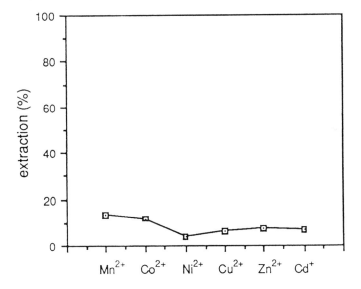

Fig. 4. Extraction (%) of transition metal picrates by ionophore **3** in CH_2Cl_2. Extraction conditions: 2.5×10^{-4} M of ionophore in CH_2Cl_2; 5.0×10^{-4} M of picric acid in 0.1 M of metallic nitrate at 25°C. Ionophore solution (5.0 mL) was shaken for 2 h with picrate solution (5.0 mL) and % extraction was measured by the absorbance of picrate in CH_2Cl_2. Experimental error was ±2%.

Based on these findings, a further investigation was performed on the extraction of transition metals by using ionophore **3**. The results are shown in Figure 4.

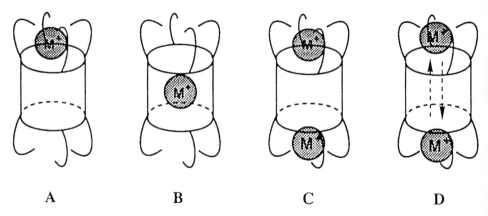

Fig. 5. Possible metal complexations for spherand-type calixarene.

Ionophore **3** exhibits low extractabilities and ion selectivities for transition metals. These results suggest that the effect of the restricted ring conformational flexibility due to the biarene units, despite the larger ring size than that of calix[8]arene, clearly suppress the contribution of the OCH_2CO units to the metal binding due to the conformational changes.

3.4. ^{1}H-NMR SPECTRA OF METAL COMPLEXES

Recently, Shinkai *et al.* reported that the 1,3-alternate conformer of calix[4]arene tetraester can form both a 1 : 1 and a 2 : 1 metal/calixarene complex and the two metal-binding sites display negative allostericity by ^{1}H-NMR titration [26]. In the present systems, due to the existence of two metal-binding sites there are several possibilities for metal complexation modes, as shown in Figure 5. Thus, a 1 : 1 and a 2 : 1 metal complexation of both hexaethyl ester **3** and octaethyl ester **4** might be possible.

In fact, the chemical shifts of the Ar*CH*$_2$Ar methylene protons of hexaethyl ester **3** were altered by titration with KSCN in $CDCl_3$–CD_3OD (1 : 1 v/v): i.e., a 1 : 1 mixture of **3** and KSCN showed a completely different ^{1}H-NMR spectrum, with sharp lines becoming evident for these protons. Their methylene proton peaks were concentrated at δ 3.60 and 4.74 (J_{AB} = 13.1 Hz) as a pair of doublets and δ 4.24 as a singlet in comparison to those in the metal free spectrum, as shown in Figure 6. In addition to this observation, the signals for the aromatic protons slightly downfield shifted (Figure 7) and *tert*-butyl, phenoxy methylene, and ethyl protons also showed different chemical shifts, respectively.

These results strongly suggest that the original C_2 symmetry might remain after the complete metal cation complexation, as shown in Figure 8. Thus the K^+ ion might exist in either the complexation mode **B** or **D**. In the case of the latter mode the rate of an intramolecular hopping between two possible metal-binding sites might be faster than the NMR time scale at room temperature. Despite lowering the

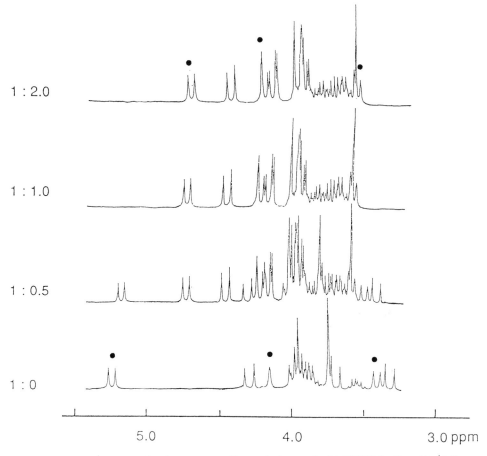

Fig. 6. Partial ^{1}H-NMR titration spectra of hexaethyl ester **3** with KSCN in (5×10^{-4} M), CDCl$_3$: CD$_3$OD = 1 : 1 v/v, 270 MHz. Ar*CH$_2$*Ar (filled circles). From the top to the bottom, molar ratios of **3** to KSCN of 2.0, 1.0 and 0.5 and in the absence of KSCN.

temperature to -80°C, no clear evidence for the intramolecular hopping behaviour was obtained.

Figure 9 shows the partial ^{1}H-NMR spectrum of octaethyl ester **4** of the free ligand and of its KSCN complex. By adding variable amounts of KSCN in CD$_3$OD to a CDCl$_3$ solution of the ionophore **4** the ^{1}H-NMR spectrum of the latter also greatly changes in all signals, as is observed with hexaethyl ester **3**. The ^{1}H-NMR titration experiment clearly indicates a 1 : 1 stoichiometry for the KSCN complex with **4**, since all signals remain essentially unchanged after the octaethyl ester **4**/KSCN ratio has reached a value of unity (Figure 9). A pair of doublets for the aromatic protons became split into two pairs of doublets and was shifted downfield. In addition to this observation, the signals for *tert*-butyl, and ethyl protons also showed two different chemical shift, respectively.

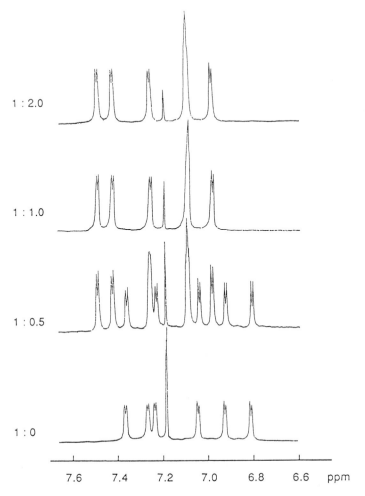

Fig. 7. Partial ^{1}H-NMR titration spectra of hexaethyl ester **3** with KSCN (5×10^{-4} M), CDCl$_3$: CD$_3$OD = 1 : 1 v/v, 270 MHz. From the top to the bottom molar ratios of **3** to KSCN of 2.0, 1.0 and 0.5 and in the absence of KSCN.

These phenomena may be attributed to the formation of four sets of nonequivalent aromatic protons and two sets of nonequivalent *tert*-butyl and ethyl protons due to the contribution of the asymmetric metal cation complexation on the one side of octaethyl ester **4** (Figure 8). However, no intramolecular hopping between two binding positions could be observed as with biscalix[4]arenes [29, 30]. This result might be attributable to the conformational changes in the other side of the binding site in the process of metal complexation. Further experiments on these metal complexations are currently in progress in our laboratory.

Hexaethyl ester

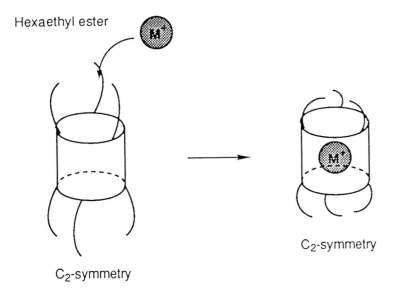

C$_2$-symmetry

C$_2$-symmetry

Octaethyl ester

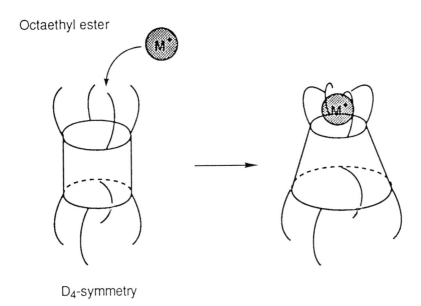

D$_4$-symmetry

Fig. 8.

4. Conclusion

We have demonstrated for the first time that the derivatives of the spherand-type calixarenes formed by alkylation with ethyl bromoacetate give ionophores with promising complexation properties and interesting stereochemistry. While to date

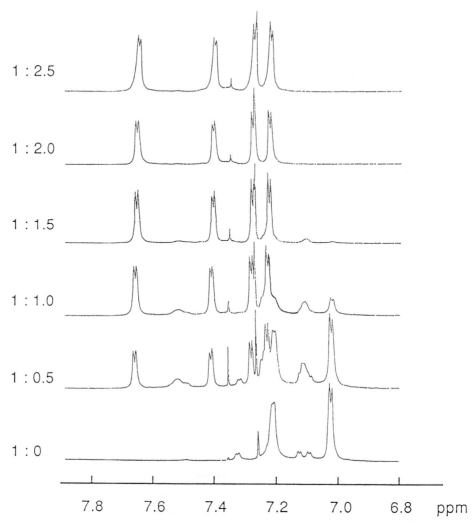

Fig. 9. Partial ^{1}H-NMR titration spectra of octaethyl ester **4** with KSCN in (5×10^{-4} M), CDCl$_3$: CD$_3$OD = 1 : 1 v/v, 270 MHz. From the top to the bottom, molar ratios of **4** to KSCN of 2.5, 2.0, 1.5, 1.0 and 0.5 and in the absence of KSCN.

only two stereoisomers have been obtained, variation of the alkylation conditions and reagents could lead to the derivatives with D_3 and D_4 symmetry, which will serve as interesting building blocks for larger potential host molecules.

References

1. J. M. Lehn: *Angew. Chem. Int. Ed. Engl.* **100**, 91 (1988).
2. D. J. Cram: *Angew. Chem. Int. Ed. Engl.* **100**, 1041 (1988).
3. A. Collet: *Tetrahedron* **43**, 5725 (1987).
4. P. M. Keehn and S. M. Rosenfeld: in *Cyclophanes I and II*, Academic Press, New York (1983).

5. D. J. Cram: *Angew. Chem. Int. Ed. Engl.* **25**, 1039 (1986).
6. For a comprehensive review of all aspects of calixarene chemistry, see: C. D. Gutsche: *Calixarenes*, Royal Society of Chemistry, Cambridge (1989).
7. J. D. van Loon, L. C. Groenen, S. S. Wijmenga, V. Verboom, and D. N. Reinhoudt: *J. Am. Chem. Soc.* **113**, 2378 (1991).
8. L. C. Groenen, J. D. van Loon, W. Verboom, S. Harkema, A. Casnati, R. Ungaro, A. Pochini, F. Ugozzoli, and D. N. Reinhoudt: *J. Am. Chem. Soc.* **113**, 2385 (1991).
9. C. D. Gutsche, B. Dahawan, J. A. Levine, K. H. No, and L. J. Bauer: *Tetrahedron* **39**, 409 (1983).
10. A. Arduini, A. Pochini, S. Reverberi, and R. Ungaro: *Tetrahedron* **42**, 2089 (1986).
11. S. K. Chang and I. Cho: *J. Chem. Soc., Perkin Trans. 1* 211 (1986).
12. M. A. McKervey, E. M. Seward, G. Ferguson, B. Ruhl, and S. Harris: *J. Chem. Soc., Chem. Commun.* 388 (1985).
13. T. Arimura, M. Kubota, T. Matsuda, O. Manabe, and S. Shinkai: *Bull. Chem. Soc. Jpn.* **62**, 1674 (1989).
14. S. Shinkai, K. Fujimoto, T. Otsuka, and H. L. Ammon: *J. Org. Chem.* **57**, 1516 (1992).
15. T. Yamamoto, K. Hasegawa, Y. Surawatari, and L. K. Doamekpor: *Chem. Ber.* **126**, 1435 (1993).
16. M. Tashiro, H. Watanabe, and O. Tsuge: *Org. Prep. Proceed. Int.* **6**, 117 (1974).
17. M. Tashiro: *Synthesis* 921 (1979).
18. M. Tashiro and T. Yamamoto: *J. Org. Chem.* **50**, 2939 (1985).
19. M. Tashiro, T. Yamamoto, K. Kobayashi, and T. Arimura: *J. Org. Chem.* **52**, 3196 (1987).
20. T. Yamato, T. Arimura, and M. Tashiro: *J. Chem. Soc., Perkin Trans. 1* 1 (1987).
21. T. Yamato, Y. Saruwatari, S. Nagayama, K. Maeda, and M. Tashiro: *J. Chem. Soc., Chem. Commun.* 861 (1992).
22. S. Shinkai, T. Arimura, H. Kawabata, H. Murakami, and K. Iwamoto: *J. Chem. Soc., Perkin Trans. 1* 2429 (1991).
23. A. Wolff, V. Böhmer, W. Fogt, F. Ugozzoli, and C. D. Andreetti: *J. Org. Chem.* **55**, 5665 (1990).
24. G. D. Andreetti, G. Calestani, F. Ugozzoli, A. Arduini, E. Ghidini, A. Pochini, and R. Ungaro: *J. Incl. Phenom.* **5**, 123 (1987).
25. F. Arnaud-Neu, E. M. Collins, M. Deasy, G. Ferguson, S. J. Harris, B. Kaitner, A. J. Lough, M. A. McKervey, E. Marques, B. Ruhl, M. J. Schwing-Weill, and E. M. Seward: *J. Am. Chem. Soc.* **111**, 8681 (1989).
26. K. Iwamoto and S. Shinkai: *J. Org. Chem.* **57**, 7066 (1992).
27. K. Araki, N. Hashimoto, H. Otsuka, and S. Shinkai: *J. Org. Chem.* **58**, 5958 (1993).
28. N. Sato and S. Shinkai: *J. Chem. Soc., Perkin Trans. 2* 621 (1993).
29. Z. Asfari, R. Abidi, F. Arnaud, and J. Vicens: *J. Incl. Phenom.* **13**, 163 (1992).
30. F. Ohseto and S. Shinkai: *Chem. Lett.* 2045 (1993).
31. C. J. Pedersen: *J. Am. Chem. Soc.* **89**, 2495 (1967).

Polymeric Membrane Sodium-Selective Electrodes Based on Calix[4]arene Ionophores*

Y. SHIBUTANI**, H. YOSHINAGA, K. YAKABE and T. SHONO
Department of Applied Chemistry, Osaka Institute of Technology, 5-16-1, Omiya Asahi-ku, Osaka 535, Japan

and

M. TANAKA
Department of Applied Chemistry, Faculty of Engineering, Osaka University, Yamada-oka, Suita, Osaka 565, Japan

(Received: 3 March 1994; in final form: 12 July 1994)

Abstract. Six kinds of tetra alkylester type calix[4]arene derivatives, ($R_1 = R_2 = CH_3$: **1**, C_2H_5: **2**, C_3H_7: **3**, n-C_4H_9: **4**, t-C_4H_9: **5**, n-$C_{10}H_{21}$: **6**), a diethyl-didecyl mixed ester type ($R_1 = C_2H_5$, $R_2 = C_{10}H_{21}$: **7**), and three kinds of lower rim bridged types ($R_1 = C_2H_5$, R_2—$R_2 = $ —$(CH_2)_{10}$—: **8**, —$(CH_2)_{12}$—: **9**, —$(CH_2)_2(OCH_2CH_2)_3$—: **10**) were characterized by electrochemical measurement to elucidate the effect of the length of the alkyl group of alkoxycarbonyl substituents on Na^+ selectivity. To obtain excellent Na^+ selective ionophores, introduction of short chain alkyl groups rather than long chain ones, such as a decyl group, and maintenance of sufficient solubility of the calix[4]arene derivatives in the membrane solvent are required concurrently. Among the calix[4]arenes tested, 25,26,27,28-tetrakis[(ethoxycarbonyl)methoxy]-p-tert-butylcalix[4]arene **2**, and the diethyl-didecyl mixed ester type derivative **7** are the best ionophores for a Na^+ selective electrode. On the other hand, sodium selectivity of the bridged type derivative **9** is comparable or even superior to that of the known bis(12-crown-4).

Key words: Calix[4]arene, Na^+ selective ionophore, lower rim bridged calix[4]arene.

1. Introduction

The cyclic tetramer of p-tert-butylphenol-formaldehyde condensates which maintains a cone-type conformation called a calix[4]arene from its cuplike shape, similar to that of a Greek crater vase, tends to complex with small organic compounds or metal ions. Further, incorporation of carbonyl groups such as ester, amide and ketone linkages into the phenolic oxygen atoms of the calixarene makes it function as an ionophore [1].

Previously, we have reported that the bulkier decyl ester type calix[4]arene derivative **6** [2] is available a sodium selective ionophore for polymeric membrane electrodes, independently of Diamond's work [3]. Selectivity coefficients for Na^+ with respect to other alkali and alkaline-earth metal ions, NH_4^+, and H^+ on the

---

* This paper is dedicated to the commemorative issue on the 50th anniversary of calixarenes.
** Author for correspondence.

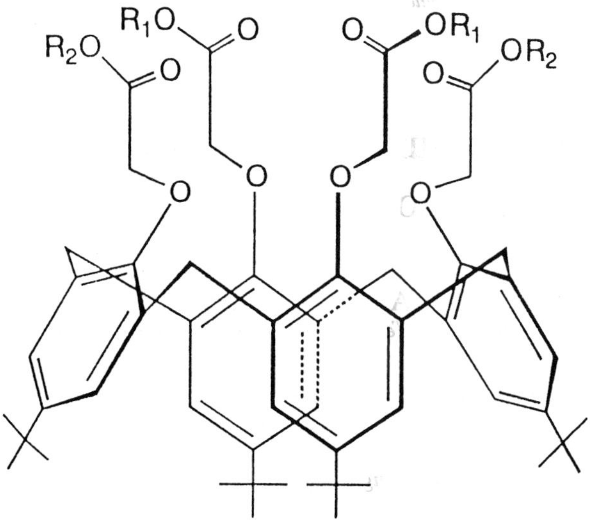

Fig. 1. Structural scheme of derived calix[4]arenes.

Tetra ester type
(1) $R_1 = R_2$: -CH_3
(2) $R_1 = R_2$: -C_2H_5
(3) $R_1 = R_2$: -C_3H_7
(4) $R_1 = R_2$: -n-C_4H_9
(5) $R_1 = R_2$: -t-C_4H_9
(6) $R_1 = R_2$: -$C_{10}H_{21}$

Diester-diester type
(7) R_1 : -C_2H_5, R_2 : -$C_{10}H_{21}$

Bridged type
R_1 : -C_2H_5
(8) [R_2 — R_2] : -$(CH_2)_{10}$-
(9) [R_2 — R_2] : -$(CH_2)_{12}$-
(10) [R_2 — R_2] : -$(CH_2)_2$-$(OCH_2CH_2)_3$ -

electrode based on derivative **6** are superior to Na^+ selective electrodes based on the known bis(12-crown-4) compound [4]. On the other hand, valuable membrane electrodes based on various calix[4]arene derivatives have also been reported [5–10]. In these cases, the sodium ion is encapsulated in the cavity of the calix[4]arene, interacting with the four carbonyl and four phenolic oxygen atoms as shown in Figure 1. The most important factor governing the selectivity characteristics of a calix[4]arene derivative is the spatial arrangement of the polar binding groups in a rigid conformation with optimum dimensions for sodium ions as the target ion. It is worthwhile to make a judicious choice of the substituents, R_1 or R_2, introduced onto the phenolic oxygen residues of the calix[4]arene with the intention of obtaining an excellent sodium ion selective ionophore.

In the present work, the effect of the length of the alkyl group of the alkoxycarbonyl substituents on sodium ion-selectivity is established by comparison of the electrochemical properties of the PVC membrane electrodes based on six kinds of tetra ester types, ($R_1 = R_2 = CH_3$; **1**, C_2H_5; **2**, C_3H_7; **3**, n-C_4H_9; **4**, t-C_4H_9; **5**, n-$C_{10}H_{21}$; **6**), one diester-diester mixed type ($R_1 = C_2H_5$, $R_2 = C_{10}H_{21}$; **7**) and three kinds of bridged type ($R_1 = C_2H_5$; R_2—R_2 = —$(CH_2)_{10}$—; **8**, —$(CH_2)_{12}$—; **9**, —$(CH_2)_2$—$(OCH2CH_2)_3$—; **10**) calix[4]arene derivatives.

2. Experimental

2.1. SYNTHESIS OF CALIX[4]ARENE DERIVATIVES

p-tert-Butylcalix[4]arene was prepared according to the Gutsche method [11]. The tetra ester type derivatives, **1–6**, and the diethyl-didecyl mixed ester type, **7** [12], were prepared by the Williamson reaction with the corresponding alkyl bromoacetates according to the literature methods [2, 3, 5, 13, 14]. Decyl α-bromoacetate was obtained by esterification of bromoacetic acid with decylalcohol in benzene in the presence of H_2SO_4. Methyl, ethyl, propyl and *tert*-butyl-α-bromoacetate were reagent grade and were used without further purification. Bridged type derivatives **8–10**, were synthesized by the subsequent reaction of calix[4]aryl diethyl ester [15, 16] with the corresponding alkyl bisbromoacetate according to the literature methods [17–19].

To *p-tert*-butylcalix[4]arene diethyl ester (1.36 mmol) dissolved in dry acetone (100 mL) was added K_2CO_3 (2.76 mmol) and a small amount of KI followed by alkyl bisbromoacetate (1.08 mmol) in dry acetone (50 mL). The mixture was refluxed for 75 h. After the reaction, the solvent was evaporated off and then water was added to the residue. The mixture was extracted with chloroform and the organic layer was dried over $MgSO_4$. The chloroform and excess alkyl bromoacetate were removed under reduced pressure. The resulting residue was dissolved in chloroform and methanol was added to obtain needle-like crystals or colorless oil as crude product. The crude product was dissolved in chloroform and chromatographed on a 2×25 cm column of ODS, using chloroform : acetonitrile (30 : 55) as an eluent. Identification of the synthesized calix[4]arene derivatives has been done by MS (Nippon Denshi JMW-AX506W), NMR (Varian unity 300) and elemental analysis.

2.1.1. *Mixed Ester Type Calix[4]arene Derivatives*

Diethyl didecyl ester derivative **7**. Colorless oil; MS (FAB): 1216 (M^+), 1239 (M^++Na); ^{1}H-NMR (CDCl$_3$) δ 0.86 (6H, s, $CH_3(CH_2)_7$), 1.05 and 1.08 (18H each, s each, $(CH_3)_3C$), 1.25 (28H, m, $(CH_2)_7CH_3$), 1.27 (6H, t, CH$_2$$CH_3$), 1.61 (4H, m, CH$_2$$CH_2$(CH$_2$)$_7$), 3.15 and 4.82 (8H each, d each, ArCH_2Ar), 4.10 and 4.18 (4H each, q each, COCH_2), 4.76 and 4.81 (4H each, s each, OCH_2CO), 6.74 and 6.77 (4H each, s each, *ArH*); Anal. Calcd. for $C_{76}H_{112}O_{12}$: C, 74.96; H, 9.27. *Found*: C, 74.91; H, 9.13.

Bridged type derivative **8**. Colorless oil; MS (FAB): 1076 (M^+), 1099 (M^++Na); ^{1}H-NMR (CDCl$_3$): δ 0.91 and 1.23 (18H each, s each, $(CH_3)_3C$), 1.04 (12H, m, $(CH_2)_6$), 1.27 (6H, t, CH$_2$$CH_3$), 1.64 (4H, m, CH_2(CH$_2$)$_6$$CH_2$), 3.17 and 4.84 (4H each, d each, ArCH_2Ar), 4.11 (4H, s, OCH_2CH$_3$), 4.20 (4H, s, OCH_2CH$_2$), 4.61 (4H, s, OCH_2CO), 5.02 (4H, s, (CH$_2$)$_{10}$OCH_2CO), 6.57 and 6.98 (4H each, s each, Ar*H*). *Anal. Calcd.* for C$_{66}$H$_{90}$O$_{12}$: C, 73.15; H, 8.44. *Found*: C, 73.71; H, 8.41.

Bridged type derivative **9**. Pale yellow crystals; m.p. 103–105°C, MS (FAB): 1102; ^{1}H-NMR (CDCl$_3$) δ 1.07 (16H, m, (*CH$_2$*)$_8$), 1.14 and 1.29 (18H each, s each, (*CH$_3$*)$_3$C), 1.41 (6H, t, CH$_2$*CH$_3$*), 1.80 (4H, m, *CH$_2$*(CH$_2$)$_8$*CH$_2$*), 3.38 and 4.82 (4H each, d each, Ar*CH$_2$*Ar), 4.13 and 4.27 (4H each, q each, CO*CH$_2$*), 4.47 and 4.49 (4H each, s each, O*CH$_2$*CO), 7.12 (s, 8H, Ar*H*). *Anal. Calcd.* for C$_{68}$H$_{94}$O$_{12}$: C, 74.01; H, 8.59. *Found*: C, 73.95; H, 8.48.

Bridged type derivative **10**. White crystals; m.p. 127–129°C, MS (FAB): 1094; ^{1}H-NMR (CDCl$_3$) δ 0.84 and 1.31 (18H each, s each, (*CH$_3$*)$_3$C), 1.33 (t, 6H, CH$_2$*CH$_3$*), 3.18 and 4.87 (8H each, d each, Ar*CH$_2$*Ar), 3.77 and 4.25 (18H each, m each, O*CH$_2$*), 4.51 and 5.15 (4H each, s each, O*CH$_2$*CO), 6.47 and 7.09 (4H each, s each, Ar*H*). *Anal. Calcd.* for C$_{64}$H$_{86}$O$_{15}$: C, 70.17; H, 7.91. *Found*: C, 70.88; H, 8.05.

2.2. OTHER CHEMICALS

Poly(vinyl chloride) (PVC) with an average polymerization degree of 1100 was purified by reprecipitation from THF in methanol. The plasticizers or membrane solvents, di-*n*-octylphthalate (DOP), *o*-nitrophenyl octyl ether (NPOE), and 2-fluorophenyl-2-nitrophenyl ether (FPNPE) were obtained from Dojindo Laboratories. The lipophilic salt, sodium tetrakis[3,5-bis(trifluoromethyl)phenyl]borate (NaTFPB), was obtained from Tokyo Chemical Industry Co., Ltd. and Dojindo Laboratories.

2.3. ELECTRODE FABRICATION AND EMF MEASUREMENT

The preparation of the electrode membranes was as described previously [2], i.e., PVC (25 mg), a plasticizer or membrane solvent (67.5 mg), the calix[4]arene neutral carrier (10–30 mg) and lipophilic salt (0.5 mg) were dissolved in THF. The solution was poured into a Petri dish (21 mm i.d.). THF was evaporated slowly to yield the membrane with 0.1–0.2 mm thickness.

The emf measurements were carried out at 25 ± 0.1°C using a pH/mV meter (Tokyo Chemical Laboratories Co., Ltd., TP-1000GP) equipped with a double-junction type Ag–AgCl reference electrode. The representative electrochemical cell for emf measurements was an Ag–AgCl | internal solution (1×10^{-3} M NaCl) | PVC membrane | measured solution | 0.1 M NH$_4$NO$_3$ | 4 M NaCl | AgCl–Ag. The selective coefficients, k_{NaM}^{Pot}, were determined using a fixed interference method (FIM) with background concentrations of 0.05 M for alkali metal and H$^+$ ion and 0.5 M for alkaline-earth metal ions and NH$_4^+$ ion.

TABLE I. Effect of membrane solvent on Na$^+$ selectivity of the **2**-based PVC membrane electrodes.

Cation	$-\log k_{\text{NaM}}^{\text{Pot}}$		
	DOP	NPOE	FPNPE
K$^+$	2.56	2.51	2.57
Li$^+$	3.44	2.98	3.39
H$^+$	3.49	3.10	2.49
NH$_4^+$	4.42	4.03	4.18
Mg^{2+}	4.64	4.39	4.62
Ca^{2+}	4.09	3.98	4.11

3. Results and Discussion

3.1. TETRAESTER TYPE CALIX[4]ARENE

The capability of neutral-carrier-type ion selective electrodes is considered to be dependent on its functional characteristics as an ionophore of the neutral carrier itself and on the compatibility of the carrier into the PVC membrane containing membrane solvent and lipophilic salt. At first, effects of membrane solvents on the sensitivity and selectivity of the Na$^+$ selective electrodes based on six of the calix[4]arene derivatives **1–6**, were determined. The membrane solvents tested were NPOE and FPNPE as the phenyl ether type solvents and DOP as the diester type one. Selectivity coefficients for Na$^+$ ion with respect to other alkali and alkaline-earth metal ions, NH$_4^+$ and H$^+$ on the electrodes based on derivative **2** are summarized in Table I.

Each membrane solvent, DOP and NPOE, gave good results with a Nernstian response, 58–59 mV/decade, in accordance with the wide Na$^+$ activity change in the activity range of $1.0 \times 10^{-5} - 1$ M and the Na$^+$ selectivity with respect to K$^+$, $-\log k_{\text{NaK}}^{\text{Pot}}$ are 2.56 and 2.51, respectively. Although the FPNPE solvent leads to a value of 2.57, it allowed not only a poor response of 50 mV/decade to the Na$^+$ activity changes but also a poor Na$^+$ selectivity against hydrogen ion ($-\log k_{\text{NaH}}^{\text{Pot}}$ = 2.49). Such reduced sensitivity and Na$^+$ selectivity tendencies derived from the use of FPNPE as a membrane solvent were also observed for the other tetra ester type derivatives, **1** and **3** to **6**. That is, as a membrane solvent, a low dielectric constant solvent such as DOP gave better results compared to the high dielectric solvents such as NPOE and FPNPE.

A comparison of the selectivity coefficients for the tetra ester type calix[4]arene derivatives–DOP system electrodes is depicted in Table II.

As summarized in Table II, although the **6**-based electrode presents a better Na$^+$ selectivity with respect to K$^+$, the selectivity coefficient $-\log k_{\text{NaK}}^{\text{Pot}}$ being 2.17 compared with the bis(12-crown-4) based electrode, the selectivity is inferior to the electrodes based on the other tetra ester type derivatives having a shorter alkyl

TABLE II. Selectivity comparison among Na^+ selective electrodes based on tetra ester type calix[4]arenes **1** to **6** and known bis(12-crown-4) type PVC membrane Na^+ electrodes.

Cation	$-\log k^{Pot}_{NaM}$						
	1	**2**	**3**	**4**	**5**	**6**	A–NPOE
K^+	2.25	2.56	2.51	2.49	1.70	2.17	2.05
Li^+		3.44	3.40	3.36		2.81	3.50
H^+	3.18	3.49	3.75	3.55	3.40	3.53	4.44
NH_4^+		4.42	4.26	4.20		3.34	3.55
Mg^{2+}		4.64	4.62	4.69		4.39	4.00
Ca^{2+}		4.09	4.10	4.06		3.94	3.93

Membrane solvent for derivative **1** to **6**: DOP.
A: dibenzyl-bis(12-crown-4) [4].

chain except for derivative **5**. On the other hand, derivative **1** and **5** based electrodes produced unsatisfactory results such that the neutral carriers were separated out as crystals in the membrane. In the derivative **5** based electrodes, the reduced Na^+ selectivity, the selectivity coefficient with respect to K^+, $-\log k^{Pot}_{NaK}$ being 1.70 is probably a consequence of the steric hindrance provided by the bulky *tert*-butyl groups around the cavity.

These facts led us to the following concept. That is, to obtain excellent Na^+ selective carriers, it is necessary to introduce shorter alkyl chains rather than longer alkyl ones such as a decyl group and to retain sufficient solubility of the resulting calix[4]arene derivatives into the membrane phase. Therefore, we designed and synthesized a diethyl-didecyl mixed ester derivative, **7**, and three kinds of lower rim bridged type calix[4]arenes, **8** to **10**. The other objects of capping the lower rim of the calix[4]arene with a bridge of a different length methylene groups are to fix the calixarene in a rigid cone conformation and to change the cavity size.

3.2. LIPOPHILICITY OF CALIX[4]ARENE DERIVATIVES

The chromatographic retention character of these calix[4]arenes were determined by means of HPLC using an ODS (4.6 × 500 mm) column and chloroform : acetonitrile (30 : 55) mixed solvent as an eluent. The observed retention times for calix[4]arene derivatives are summarized in Table III. Although the retention time does not necessarily reflect the compatibility of the calix[4]arenes with an electrode membrane, its values may have a correlation to the lipophilicity and/or the solubility of the calix[4]arenes into the membranes solvent.

Derivative **6** shows the most remarkable lipophilicity among these derivatives. On the other hand, the retention times of derivatives **7**, **9**, and **10** are 16.28, 14.10 and 22.13 min, respectively. The lipophilicities of these derivatives are not as good as derivative **6** but remarkably improved in comparison with derivatives **1** to **5**.

TABLE III. Comparison of retention times of calix[4]arene derivatives in HPLC.

Derivative	Retention time (min)
1	7.05
2	7.65
3	8.78
4	10.12
5	10.10
6	48.48
7	16.28
8	9.37
9	14.10
10	22.13

Column: 5-ODS-H 4.6 × 500
Eluent: chloroform : acetonitrile (30 : 55)
Flow rate: 1.0 ml/min.

3.3. MIXED ESTER TYPE CALIX[4]ARENES

3.3.1. *Diethyl-Didecyl Mixed Ester Derivative*

The selectivity coefficients of the mixed ester type derivative, **7**-based Na^+ selective electrodes are evaluated. Table IV shows the selectivity comparison among the neutral-carrier-type Na^+ selective electrodes based on derivative **7** and our previously reported Na^+ neutral carriers which possess one of the best Na^+ selectivities against K^+. In the **7**-based electrode, DOP and NPOE gave good results, allowing high Na^+ selectivity with respect to K^+, NH_4^+, H^+, Mg^{2+}, and Ca^{2+}. Particularly, the **7**–DOP system electrode is superior to the known bis(12-crown-4) and *tert*-octyl and de-*tert*-butylated calix[4]aryl tetradecyl ester based PVC membrane for Na^+ selectivity coefficient with respect to K^+, Li^+, NH_4^+ and H^+. Furthermore, another favorable characteristic of derivative **7** is its good solubility in a low dielectric constant solvent such as DOP which effectively avoids contamination phenomena of the membrane electrode by adhering proteins in human blood sera. Also, the lifetime of the **7**–DOP system electrode was studied by periodically recalibrating it in standard solution and calculating the response slope. During the 6 months or more of the observation period, the electrodes showed superior performance. These results must depend on the well-balanced lipophilicity of derivative **7**. Accordingly, derivative **7** is a very promising neutral carrier for sodium ion assay.

TABLE IV. Selectivity comparison among Na^+ selective electrodes based on derivative **7** and previously reported calix[4]arenes.

Cation	$-\log k_{NaM}^{Pot}$				
	7–DOP	7–NPOE	7–FPNPE	B–DOP	C–FPNPE
K^+	2.57	2.38	2.07	2.34	2.20
Li^+	3.49	3.40	3.02	3.21	2.95
H^+	4.00	3.18	3.64	3.31	2.39
NH_4^+	4.27	4.40	4.06	4.09	3.61
Mg^{2+}	4.96	4.35	4.57	4.25	4.69
Ca^{2+}	4.14	3.78	4.11	3.81	3.29

B: di-*tert*-butylated calix[4]aryl decyl ester [5].
C: *p-t*-octyl calix[4]aryl decyl ester [6, 21].

TABLE V. Selectivity comparison among Na^+ selective electrodes based on bridged type calix[4]arenes, **8** to **10**.

Cation	$-\log k_{NaM}^{Pot}$				
	8–DOP	8–NPOE	8–FPNPE	9–DOP	10–DOP
K^+	1.92	1.85	1.76	2.20	1.67
Li^+	3.30	3.08	2.79	3.15	3.29
H^+	3.49	2.94	2.03	2.58	2.76
NH_4^+	3.93	3.36	3.06	3.63	
Mg^{2+}	4.76	4.10	3.49	3.82	
Ca^{2+}	4.09	3.67	3.43	3.24	3.67

3.3.2. Bridged Type Derivative

Table V displays the sodium ion selectivity for three kinds of bridged type calix[4]arene derivatives. Since the capping of the lower rim of calix[4]arene with a bridge of different length methylene groups results in the fixation of a rigid cone conformation and the change of the cavity size, the Na^+ selectivity is expected to be improved.

Unfortunately, as seen in Table V, in the **8** and **10**-based electrodes, the Na^+ selectivity coefficient with respect to K^+, $-\log k_{NaK}^{Pot}$ were about 1.9 and 1.7, respectively. The reduced Na^+ selectivity against K^+ in the latter case may be caused by the three active oxygen atoms derived from the $-(CH_2)_2-(CH_2CH_2O)_3-$ group, because these oxygen atoms and the carbonyl ones form the crown ether ring which is favorable for K^+ complex formation. On the contrary, the **9**-based electrode was superior to the **8** and **10**-based electrodes; the Na^+ selectivity coefficient with respect to K^+, $-\log k_{NaK}^{Pot}$ being 2.2, and selectivity comparable to that of the commonly used bis(12-crown-4) ionophore was observed. These results suggest that the length of the methylene bridge moiety is effective for the improvement of Na^+ selectivity by calixarene, while oxygen atoms containing the bridge moiety are irrelevant.

Recently, Shinkai et al. [20] studied PVC membrane electrodes based on the calix[4]arene derivative and reported that the *p-tert*-octyl-calix[4]aryl tetraethyl ester–FPNPE system electrode allowed high Na^+ selectivity. The Na^+ selectivity coefficient being 3.1 [20], though its value was not evaluated by a fixed interference method (FIM). Its excellent Na^+ selectivity compared with those of our similar derivative, *p-tert*-octylcalix[4]aryl tetradecyl ester [6, 21], is probably derived from the introduction of shorter alkyl groups such as the ethyl ester and *tert*-octyl group into the upper rim moiety to produce lipophilicity, as already mentioned.

In conclusion, the diester-diester mixed type derivative **7** and diester-bridged type **9** are promising neutral carriers for sodium ion assay. Further investigation on the electrode membrane condition such as membrane solvents and lipophilic salts for the calix[4]arene derivatives, **7–10**, are currently being done in our laboratory.

Acknowledgement

This research was supported by a Grant-in Aid from the Ministry of Education, Science and Culture.

References

1. F. Arnaud-Neu, G. Barrett, S. J. Harris, M. Owens, M. A. McKervey, M. Schwing-Weill, and P. Schwinte: *Inorg. Chem.* **32**, 2644 (1993).
2. K. Kimura, M. Matsuo, and T. Shono: *Chem. Lett.* 615 (1988).
3. D. Diamond, G. Svehla, E. M. Seward, and M. A. McKervey: *Anal. Chim. Acta* **204**, 223 (1988).
4. M. Tanaka, H. Mizufune, and T. Shono: *Chem. Lett.* 1419 (1990).
5. K. Kimura, T. Miura, M. Matsuo, and T. Shono: *Anal. Chem.* **62**, 1510 (1990).
6. M. Tanaka, T. Kobayashi, Y. Yamashoji, Y. Shibutani, K. Yakabe, and T. Shono: *Anal. Sci.* **7**, 817 (1991).
7. A. M. Cadogan, D. Diamond, M. R. Smyth, M. Deasy, M. A. McKervey, and S. J. Harris: *Analyst* **114**, 1551 (1989).
8. R. J. Forster, A. Cadogan, M. T. Diaz, D. Diamond, S. J. Harris, and M. A. McKervey: *Sensors and Actuators* **B4**, 325 (1991).
9. J. A. J. Brunink, J. R. Haak, J. G. Bomer, D. N. Reinhoudt, M. A. McKervey, and S. J. Harris: *Anal. Chim. Acta* **254**, 75 (1991).
10. K. Cunningham, G. Svehla, S. J. Harris, and M. A. McKervey: *Analyst* **118**, 341 (1993).
11. C. D. Gutsche: *Acc. Chem. Res.* **16**, 161 (1983).
12. Y. Shibutani, H. Yoshinaga, K. Yakabe, and T. Shono: *Bunseki kagaku* **43**, 333 (1994).
13. M. Coruzzi, G. D. Andreetti, V. Bocchi, A. Pochini, and R. Ungaro: *J. Chem. Soc. Perkin Trans 2*, 1133 (1982).
14. T. Arimura, M. Kubota, T. Matsuda, O. Manabe, and S. Shinkai: *Bull. Chem. Soc. Jpn.* **62**, 1674 (1989).
15. F. Arnaud-Neu, E. M. Collins, M. Deasy, G. Ferguson, S. J. Harris, B. Kainter, A. J. Lough, M. A. McKervey, E. Marques, B. L. Ruhl, M. J. Schwing-Weill, and E. M. Seward: *J. Am. Chem. Soc.* **111**, 8681 (1989).
16. E. M. Collins, M. A. McKervey, and S. J. Harris: *J. Chem. Soc. Perkin Trans. 1*, 372 (1989).
17. J.-D. van Loon, D. Kraft, M. J. K. Ankone, W. Verboom, S. Harkema, W. Vogt, V. Böhmer, and D. N. Reinhoudt: *J. Org. Chem.* **55**, 5176 (1990).

18. E. Ghidini, F. Ugozzoli, R. Ungaro, S. Harkema, A. A. Ei-Fadl and D. N. Reinhoudt: *J. Am. Chem. Soc.* **112**, 6979 (1990).
19. P. J. Dijkstra, J. A. J. Brunik, K.-E. Bugge, D. N. Reinhoudt, S. Harkema, R. Ungaro, F. Ugozzoli, and E. Ghidini: *J. Am. Chem. Soc.* **111**, 7567 (1989).
20. T. Sakaki, T. Harada, G. Deng, H. Kawabata, Y. Kawahara, and S. Shinkai: *J. Incl. Phenom.* **14**, 285 (1992).
21. Y. Shibutani, K. Yakabe, T. Shono, Y. Yamashoji and M. Tanaka: *Anal. Sci.* **7** (supplement II), 1671 (1991).

Cation Recognition by New Diester- and Diamide-Calix[4]arenediquinones and a Diamide-Benzo-15-Crown-5-Calix[4]arene*

PAUL D. BEER,** ZHENG CHEN, PHILIP A. GALE, JENNIFER A. HEATH, RACHEL J. KNUBLEY and MARK I. OGDEN
Inorganic Chemistry Laboratory, University of Oxford, South Parks Road, Oxford, OX1 3QR, U.K.

and

MICHAEL G. B. DREW
Department of Chemistry, University of Reading, Whiteknights, Reading, RG2 6AD, U.K.

(Received: 3 March 1994; in final form: 3 June 1994)

Abstract. The synthesis, metal, ammonium and alkyl ammonium cation coordination chemistry and electrochemical recognition studies of new diester- and diamide-calix[4]arenediquinone receptors are described. In addition the synthesis and coordination properties of a novel diamide benzo-15-crown-5-calix[4]arene molecule is reported.

Key words: Calix[4]arenediquinones, cation coordination, electrochemistry, benzo-15-crown-5-calix[4]arene.

1. Introduction

In pursuit of advancing chemical sensor technology considerable interest is being shown in the incorporation of organic and transition metal redox-active centres into various crown ether, cryptand and calixarene macrocyclic structural frameworks [1]. With the cyclic polyethers, these systems can be designed to selectively electrochemically recognise the binding of metal [2] and ammonium guest cations [3] either by through-space interactions and/or via various bond linkages between the receptor site and the redox centre. In addition, redox responsive anion receptors based on cobalticinium [4], ruthenium (II) bipyridyl derivatives [5] and functionalised calix[4]arenes [6] have also been described. Although the synthesis [7, 8] and electrochemical properties [9, 10] of calix[4]arenequinones have recently been reported, their potential use and application to amperometric sensor technology has not, to our knowledge, been exploited. We report here the first examples of such an application in which simple new lower rim diester and diamide modi-

---

* This paper is dedicated to the commemorative issue on the 50th anniversary of calixarenes.
** Author for correspondence.

fied calix[4]arenediquinones bind and electrochemically recognise Group 1 and 2, ammonium and alkyl ammonium guest cations. In addition the synthesis and coordination properties of a novel diamide benzo-15-crown-5-calix[4]arene receptor are also described.

2. Experimental

2.1. SOLVENT AND REAGENT PRETREATMENT

Where necessary solvents were purified and dried prior to use using standard procedures. Unless otherwise stated commercial grade chemicals were used without further purification.

p-tert-Butylcalix[4]arene bis ethyl ester [11], *p-tert*-butylcalix[4]arene bis diethyl amide [11], *p-tert*-butylcalix[4]arene bis acid chloride [11], and 4-aminobenzo-15-crown-[5] [12] were synthesised according to literature methods.

2.2. INSTRUMENTAL METHODS

Melting points were recorded on a Gallenkamp melting point apparatus in open capillaries and are uncorrected.

NMR spectra were recorded on a Brüker AM 300 instrument. IR spectra were recorded on a Perkin Elmer 1710 IR FT spectrometer. UV spectra were recorded on a Perkin-Elmer Lambda 6 spectrometer. Electrochemical measurements were conducted on a Princeton Applied Research Potentiostat/Galvanostat Model 273. Fast atom bombardment mass spectra were performed at University College, Swansea by the S.E.R.C. mass spectrometry service. All elemental analyses were carried out by the Inorganic Chemistry Laboratory, Oxford.

2.3. SYNTHESIS

5,17-Di-tert-butyl-26,28,-bis-(carboethoxymethyl)-calix[4]-25,27-diquinone (**3**)

p-tert-Butylcalix[4]arene bis ethyl ester (0.82 g, 1 mmol) was stirred in 0.88 M thallium trifluoroacetate (TTFA) in trifluoroacetic acid (TFA) solution (6.8 mL, 6 mmol), for 2 h in the dark. The TFA was then removed *in vacuo* and the residue poured onto ice/water (50 mL). The product was then extracted into chloroform (100 mL), separated and the organic layer washed with water (100 mL). The organic layer was then dried over $MgSO_4$ and reduced *in vacuo*. The residue was purified by column chromatography on silica gel eluting with chloroform/methanol 90 : 10 (v/v). The product was collected and triturated with methanol to give a yellow powder (0.48 g, 65%).

^{1}H-NMR (CD_2Cl_2, 300 MHz) 1.10 (s, 18H, (CH_3)$_3$C), 1.28 (t, J = 7.1 Hz, 6H, CH$_2$CH_3), 1.58 (s, 2H, H$_2$O), 3.27 (d, J = 13.3Hz, 4H, ArCH$_2$Ar), 4.02 (d, J = 13.3Hz, 4H, ArCH$_2$Ar), 4.23 (q, J = 7.1Hz, 4H, CH_2CH$_3$), 4.39 (s, 4H, OCH$_2$), 6.71 (s, 4H, C=CH), 6.85 (s, 4H, ArH).

^{13}C-NMR (CD$_2$Cl$_2$, 75.42 MHz) 14.10 (CH$_3$CH$_2$), 31.27 ((CH$_3$)$_3$C), 32.71 (ArCH$_2$Qu), 33.88 ((CH$_3$)$_3$C), 61.35 (CH$_2$O), 71.29 (CH$_2$O), 126.81 (CH=C), 128.36, 132.88 (ArH), 146.61, 147.86, 169.26 (C=O), 185.40 (C=O), 188.57 (C=O).
IR: 1760.0 cm^{-1}, 1733.9 cm^{-1}, 1654.8 cm^{-1} (carbonyl stretches). M.P. 235°C (dec.). Microanalysis: (**3**) + H$_2$O. *Calcd*. C, 70.0; H, 6.7. *Found*: C, 70.0; H, 6.7. FAB MS: m/z 738 ((M + H)$^+$ 737).

5,17-Di-tert-butyl-26,28,-bis-(diethylcarbamoylmethoxy)-calix[4]-25,27-diquinone (**4**)

5,11,17,23-Tetra-*tert*-butyl-25,27-bis(diethylcarbamoylmethoxy)-26,28-dihydroxycalix[4]arene (1.0 g, 1.18 mmol) was stirred in 0.88 M TTFA/TFA solution (8.1 mL, 7.1 mmol) for 2 h in the dark. The TFA was then removed *in vacuo* and the residue poured onto ice/water (50 mL). The product was then extracted with chloroform (100 mL) and then washed with 100 mL water. The organic layer was then reduced *in vacuo*. The residue was purified by column chromatography on silica gel eluting with chloroform/methanol 90 : 10. (v/v) The product was collected and reduced *in vacuo*. The solid was collected and dried *in vacuo*. The product was a yellow powder (0.12 g, 15%).
^{1}H-NMR (CDCl$_3$, 300 MHz) 1.13 (s, 18H, (CH$_3$)$_3$C, 1.28 (m, 12H, NCH$_2$CH$_3$), 3.32 (d, J = 13.3Hz, 4H, ArCH$_2$Ar), 3.44 (m, 4H, NCH$_2$CH$_3$), 3.66 (d, J = 13.3 Hz, 4H, ArCH$_2$Ar), 4.35 (s, 4H, OCH$_2$), 5.48 (CH$_2$Cl$_2$), 6.62 (s, 4H, C=CH), 6.96 (s, 4H, ArH).
^{13}C-NMR (CDCl$_3$, 75.42 MHz) 12.61 (CH$_3$CH$_2$), 14.44 (CH$_3$CH$_2$), 31.34 ((CH$_3$)$_3$C), 33.18 (ArCH$_2$Qu), 34.04 ((CH$_3$)$_3$C), 40.29 (CH$_3$CH$_2$), 41.97 (CH$_3$CH$_2$), 61.35 (CH$_2$O), 73.57 (CH$_2$O), 127.21 (CH=C), 128.89, 132,53 (ArH), 146.62, 147.66, 167.17 (C=O), 185.47 (C=O), 188.59 (C=O). M.P. 195°C (dec.). Microanalysis: (**4**) + CH$_2$Cl$_2$. *Calcd*. C, 67.19; H, 6.90; N, 3.20. *Found*: C, 67.33; H, 6.86, N, 3.18. FAB MS: (M + Na)$^+$ @813 (M + Cs)$^+$ @923. Sodium salts were used to clean the FAB spectrometer and CsI to calibrate it.

Diamide-benzo-15-crown-5-calix[4]arene (**7**)

A solution of 4-aminobenzo-15-crown-5 (1.20 g, 4.23 mmol) and triethylamine (0.6 g, 6.0 mmol) in dry dichloromethane (50 mL), was added to the *p-tert*-butylcalix[4]arene bis acid chloride (**5**) (1.54 g, 1.92 mmol). The resulting solution was stirred for 12 h, and then washed with 1 M HCl (50 mL) and then water (50 mL). The organic layer was dried (MgSO$_4$), filtered and the solvent removed under reduced pressure leaving a pale brown solid. Recrystallisation from dichloromethane/hexane gave the product as a white powder (1.54 g, 60%).
^{1}H-NMR (CDCl$_3$, 300 MHz) 1.10 (s, 18H, (CH$_3$)$_3$C), 1.28 (s 18H, (CH$_3$)$_3$C, 3.52 (d, J = 13.3 Hz, 4H, ArCH$_2$Ar), 3.74–3.99 (m, 32H, OCH$_2$), 4.21 (d, J =

13.3 Hz, 4H, ArCH$_2$Ar)), 4.59 (s, 4H, ArOCH$_2$), 6.76 (d, J = 8.6 Hz, 2H, ArH), 6.84 (d, J = 8.6, 1.8 Hz, 2H, ArH), 6.97 (s, 4H, ArH), 7.06 (s, 4H, ArH), 7.30 (d, J = 1.8 Hz, 2H, ArH), 8.16 (s, 2H, OH), 10.15 (s, 2H, NH).

^{13}C-NMR (CDCl$_3$, 75.42 MHz) 31.00, 31.62 ((CH$_3$)$_3$C); 32.37 (ArCH$_2$Ar); 33.98, 34.23 ((CH$_3$)$_3$$C$); 68.56, 69.69, 69.80, 70.60, 70.73, 71.14 (OCH$_2$$CH_2$O); 74.79 (CH$_2$O); 106.21, 111.27, 114.88, 125.80, 126.44 (ArH); 127.22, 132.00, 132.26, 143.61, 145.61, 148.44, 148.87, 149.23, 149.37 (Ar); 164.92 (C=O).

M.P. 155–159°C. Microanalysis: *Calcd*. C, 70.45; H, 7.62; N, 2.16. *Found*: C, 70.32; H, 7.62; N, 2.26. FAB MS (in the presence of K$^+$): MK$^+$ @1336.

2.4. UV/VIS SPECTRAL EXPERIMENTS FOR DETERMINING ASSOCIATION CONSTANTS

The titrations were conducted by progressively adding a cation solution (0.1 M in CH$_3$CN) using a 10 μL syringe, to the cuvette containing 3.0 mL of the compound solution (5 × 10^{-5} M in CH$_2$Cl$_2$/CH$_3$CN 4 : 21 (v/v) for (**3**) and 5 × 10^{-5} M in CH$_3$CN for (**4**) and (**7**)). The maximum addition for all the cations was less than 150 μL to minimise the change of the solution volume. The spectrum was recorded after each addition. The added equivalents of the cation were then plotted against the absorption intensity change at a certain wave length around the absorption peak of the spectrum (310–370 nm). This was repeated at three different wave lengths. A locally-written non-linear curve fitting computer program was used to fit the experimental titration curves using the equation for a 1 : 1 complex:

$$\alpha = (I - I_f^0)/(I_c^0 - I_f^0) + (I - I_f^0)/[C_f^0 K(I_c^0 - I)] \quad (1)$$

where α is the added equivalents of cations, I_f^0, I and I_c^0 are absorption intensities of the free ligand, the ligand plus α equivalents of cations and the ligand plus large excess cation solutions. C_f^0 is the concentration of the free ligand and K is the association constant. The fitting procedure was repeated for each cation for all the three titration curves at different wave lengths and the averaged value of K is reported.

2.5. CRYSTALLOGRAPHY

Crystal data are given in Table I, together with refinement details. Data were collected with MoK_α radiation using the MAR research Image Plate System. The crystal was positioned at 70 mm from the image plate: 120 frames were measured at 2° intervals with a counting time of 5 min. Data analysis was carried out with the XDS program [13]. The structure was solved using direct methods with the SHELX86 program [14]. All non-hydrogen atoms were refined anisotropically and hydrogen atoms isotropically in calculated positions using the SHELXL program [15]. All calculations were carried out on a Silicon Graphics R4000 Workstation at the University of Reading. Positional parameters are listed in Table II, molecular dimensions in Table III.

TABLE I. Crystal data and structure refinement for (**3**).

Empirical formula	$C_{44}H_{48}O_{10}$
Formula weight	736.82
Temperature	293(2) K
Wave length	0.71071 Å
Crystal system	orthorhombic
Space group	$C222_1$
Unit cell dimensions	$a = 14.158(10)$ Å, $b = 11.993(10)$ Å, $c = 22.029(10)$ Å
Volume	3741(4) Å^3
Z	4
Density (calculated)	1.31 Mg/m^3
Absorption coefficient	0.09 mm^{-1}
$F(000)$	1568
Crystal size	$0.3 \times 0.3 \times 0.2$ mm
Theta range for data collection	2.23–24.05 deg.
Index ranges	$0 \leq h \leq 16, -13 \leq k \leq 13, -23 \leq l \leq 23$
Reflections collected	4956
Independent reflection	2698 [R(int) = 0.0441]
Refinement method	Full-matrix least-squares on F^2
Data/restraints/parameters	2698/0/256
Goodness-of-fit F^2	0.616
Final R indices $[I > 2\sigma(I)]$	$R1 = 0.0693, wR2 = 0.2255$
Largest diff. peak and hole	0.283 and -0.223 e Å^3

3. Results and Discussion

3.1. Synthesis and Conformational Studies

Using the methodology of McKillop [16] which was applied to calixarenes by Gutsche and coworkers [8], the oxidation of *p-tert*-butylcalix[4]arene bis ethyl ester (**1**) [11] and *p-tert*-butylcalix[4]arene bis ethyl amide (2) [11] with thallium trifluoroacetate in trifluoroacetic acid for 2 h at room temperature produced the new diester- and diamide-calix[4]arenediquinones (**3**) and (**4**) as yellow powders in yields of 65% and 15% respectively (Scheme 1). The room temperature ^{1}H-NMR spectrum of both receptors exhibit the typical AB splitting patterns of the methylene protons consistent with a cone conformation for the calix. Low temperature ^{1}H-NMR spectra, however, in CD$_2$Cl$_2$, revealed the coexistence of cone and partial cone conformers. For example with (**3**) Figure 1 shows that on cooling the proton resonances broaden and it is noteworthy that the tertiary butyl resonances split into two resonances at low temperatures suggesting the presence of two species in solution. Casnati and coworkers [10] have recently investigated the conformational properties of alkyl calixarenediquinones and report a similar dynamic process.

TABLE II. Atomic coordinates ($\times 10^4$) and equivalent isotropic displacement parameters ($\text{Å}^2 \times 10^3$) for (**3**). $U_{(eq)}$ is defined as one third of the trace of the orthogonalised U_{ij} tensor.

	x	y	z	$U_{(eq)}$
C(11)	4325(4)	-528(4)	4014(2)	42(1)
C(12)	3873(4)	-1147(4)	4460(2)	48(1)
C(13)	2957(4)	-1488(4)	4406(2)	48(1)
C(14)	2521(4)	-1265(4)	3862(2)	48(1)
C(15)	2933(4)	-651(4)	3406(2)	45(1)
C(16)	3810(3)	-239(4)	3502(2)	40(1)
C(17)	2494(4)	-584(5)	2791(2)	49(1)
C(21)	3088(3)	-1153(4)	2328(2)	42(1)
C(22)	3323(4)	-2224(5)	2379(3)	49(1)
C(23)	3942(4)	-2747(4)	1958(3)	52(1)
C(24)	4394(4)	-2076(4)	1503(3)	51(1)
C(25)	4167(3)	-1010(4)	1421(2)	41(1)
C(26)	3450(4)	-520(4)	1805(2)	41(1)
C(27)	5339(4)	−294(5)	4036(2)	51(1)
O(1)	3133(3)	386(4)	1691(2)	65(1)
O(2)	4095(4)	-3741(3)	1981(3)	93(2)
O(30)	4250(3)	364(3)	3061(2)	47(1)
C(31)	3887(4)	1451(4)	2970(3)	52(1)
C(32)	4473(4)	2345(5)	3228(3)	56(1)
O(33)	5210(4)	2233(4)	3443(3)	96(2)
O(34)	4086(3)	3298(3)	3142(3)	82(2)
C(35)	4577(7)	4290(7)	3312(6)	102(3)
C(36)	4321(11)	4676(12)	3871(7)	177(7)
C(100)	2480(5)	-2164(5)	4893(3)	60(2)
C(101)	2860(9)	-1940(10)	5501(4)	132(5)
C(102)	1451(8)	−1918(18)	4911(7)	219(10)
C(103)	2530(13)	-3348(8)	4744(6)	182(8)

The mobility of the quinone moieties rotating through the calix cavity whilst the two aromatic rings remain fixed relative to one another is a possible explanation for the variable temperature ^{1}H-NMR results. At low temperatures this dynamic process is slowed on the NMR timescale and the quinone groups adopt fixed cone and partial cone conformations. Respective relative integrations of the methylene proton resonances suggest both (**3**) and (**4**) at low temperatures exist in CD_2Cl_2 solution at 50 : 50 mixtures of cone and partial cone conformers. The condensation of the 1,3-distally substituted chlorocarbonyl calix[4]arene (**5**) [11] with two moles of 4-aminobenzo-15-crown-5 (**6**) in the presence of triethylamine gave the new diamide-benzo-15-crown-5-calix[4]arene (**7**) as a white solid in 60% yield (Scheme 2). The room temperature ^{1}H-NMR spectrum suggests (**7**) adopts a rigid

TABLE III. Bond lengths [Å] and angles [deg] for (**3**).

C(11)—C(16)	1.387(7)	C(13)—C(14)—C(15)	123.4(5)
C(11)—C(12)	1.387(8)	C(16)—C(15)—C(14)	118.0(5)
C(11)—C(27)	1.464(8)	C(16)—C(15)—C(17)	120.3(5)
C(12)—C(13)	1.365(8)	C(14)—C(15)—C(17)	121.0(5)
C(13)—C(14)	1.374(8)	C(15)—C(16)—O(30)	120.1(4)
C(13)—C(100)	1.503(8)	C(15)—C(16)—C(11)	121.2(4)
C(14)—C(15)	1.377(7)	O(30)—C(16)—C(11)	118.2(4)
C(15)—C(16)	1.353(7)	C(21)—C(17)—C(15)	111.2(4)
C(15)—C(17)	1.492(7)	C(22)—C(21)—C(17)	121.8(5)
C(17)—C(21)	1.487(7)	C(26)—C(21)—C(17)	119.8(4)
C(21)—C(22)	1.332(8)	C(21)—C(22)—C(23)	121.7(5)
C(21)—C(26)	1.472(7)	O(2)—C(23)—C(22)	121.1(5)
C(22)—C(23)	1.421(8)	O(2)—C(23)—C(24)	120.0(5)
C(23)—O(2)	1.213(7)	C(22)—C(23)—C(24)	118.9(5)
C(23)—C(24)	1.435(8)	C(25)—C(24)—C(23)	121.7(5)
C(24)—C(25)	1.332(7)	C(24)—C(25)—C(26)	118.7(5)
C(25)—C(26)	1.445(7)	C(24)—C(25)—C(27*)	122.0(5)
C(26)—O(1)	1.203(6)	C(26)—C(25)—C(27*)	119.3(4)
C(27*)—C(25)	1.496(7)	O(1)—C(26)—C(25)	120.5(5)
O(30)—C(31)	1.415(6)	O(1)—C(26)—C(21)	120.0(5)
C(31)—C(32)	1.469(9)	C(25)—C(26)—C(21)	119.6(4)
C(32)—O(33)	1.153(8)	C(11*)—C(27*)—C(25)	109.1(4)
C(32)—O(34)	1.281(7)	C(16)—O(30)—C(31)	115.0(4)
O(34)—C(35)	1.428(9)	C(16)—O(30)—C(31)	115.0(4)
C(35)—C(36)	1.36(2)	O(30)—C(31)—C(32)	114.4(5)
C(100)—C(103)	1.460(12)	O(33)—C(32)—O(34)	123.4(6)
C(100)—C(101)	1.468(10)	O(33)—C(32)—C(31)	125.8(6)
C(100)—C(102)	1.487(13)	O(34)—C(32)—C(31)	110.7(5)
		C(32)—O(34)—C(35)	119.8(6)
C(16)—C(11)—C(12)	117.9(5)	C(36)—C(35)—O(34)	112.9(9)
C(16)—C(11)—C(27)	119.6(4)	C(103)—C(100)—C(101)	111.4(9)
C(12)—C(11)—C(27)	122.1(5)	C(103)—C(100)—C(102)	104.3(12)
C(13)—C(12)—C(11)	122.5(5)	C(101)—C(100)—C(102)	107.3(10)
C(12)—C(13)—C(14)	116.3(5)	C(103)—C(100)—C(13)	110.0(6)
C(12)—C(13)—C(100)	121.8(5)	C(101)—C(100)—C(13)	112.8(5)
C(14)—C(13)—C(100)	121.6(5)	C(102)—C(100)—C(13)	110.6(7)

*Symmetry element $1 - x, y, 0.5 - z$.

cone conformation in solution, which in contrast to (**3**) and (**4**) does not broaden at lower temperatures.

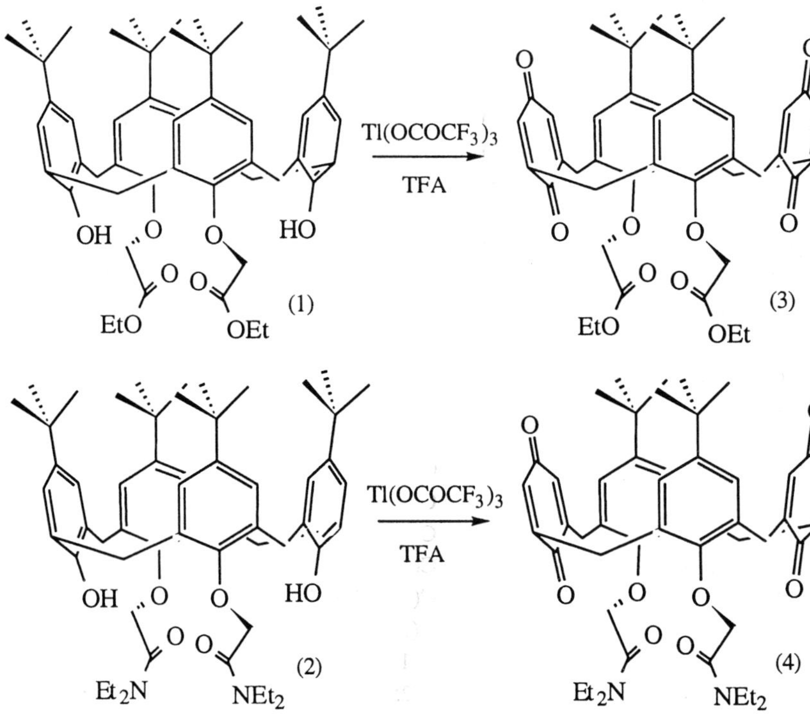

Scheme 1. The synthesis of the calixarenediquinones (**3**) and (**4**).

3.2. STRUCTURE DETERMINATION OF (**3**)

Crystals of (**3**) suitable for X-ray crystallographic determination were obtained by slow evaporation of a CH_2Cl_2/MeOH solution of (**3**). The molecule has crystallographically imposed C_2 symmetry. The most striking feature of the structure is the difference in orientation of the rings and this is illustrated in Figures 2 and 3. The two phenyl rings are spread out considerably by comparison with the calix[4]arene structure in that they intersect at an angle of 114.2° (see Figure 2). The distance between the two oxygen atoms O(30) and O(30*) is 3.26 Å. By contrast, the two quinone rings intersect at 31.1°, but unlike the calix[4]arene structures, the rings are oriented such that they are closer at the top of the cone than at the bottom (see Figure 3). The distance between the two oxygens at the top of the cone (O(2) and O(2*)) is 3.43 Å and at the bottom of the cone is 6.38 Å. This gives rise to a cramped cavity and it is perhaps not surprising that no solvent was located in the crystal.

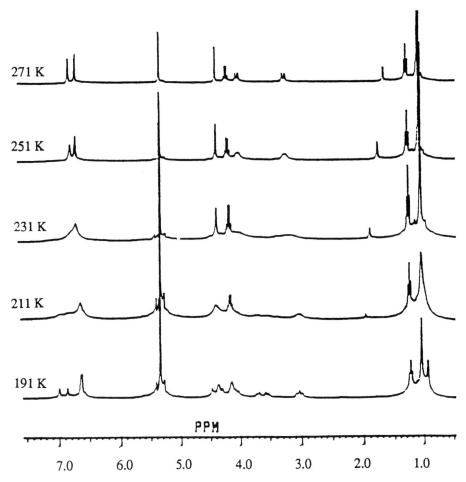

Fig. 1. The NMR spectrum of (3) from 271–191 K.

3.3. COORDINATION STUDIES

3.3.1. ^{1}H-NMR Titrations of (3) and (4) with Cationic Guest Species

Proton NMR solution complexation investigations of Group 1 and 2 metal cations, ammonium and butyl ammonium cations were carried out with (3) and (4), and for comparison purposes with the recently reported dimethylether calix[4]arenediquinone (8) (Figure 4).

In a typical titration experiment the stepwise addition of a concentrated solution of a cation salt in 40 : 60 (v/v) deuterated chloroform/acetonitrile to a dilute solution of (3) in the same deuterated solvent mixture of 100% deuterated acetonitrile for ligand (4) resulted in either significant downfield shifts of the receptor protons or the evolution of a new set of resonances corresponding to a solution complexed species. This most notably occurred with Ba^{2+} with (3), and Na^+ with (4). In all

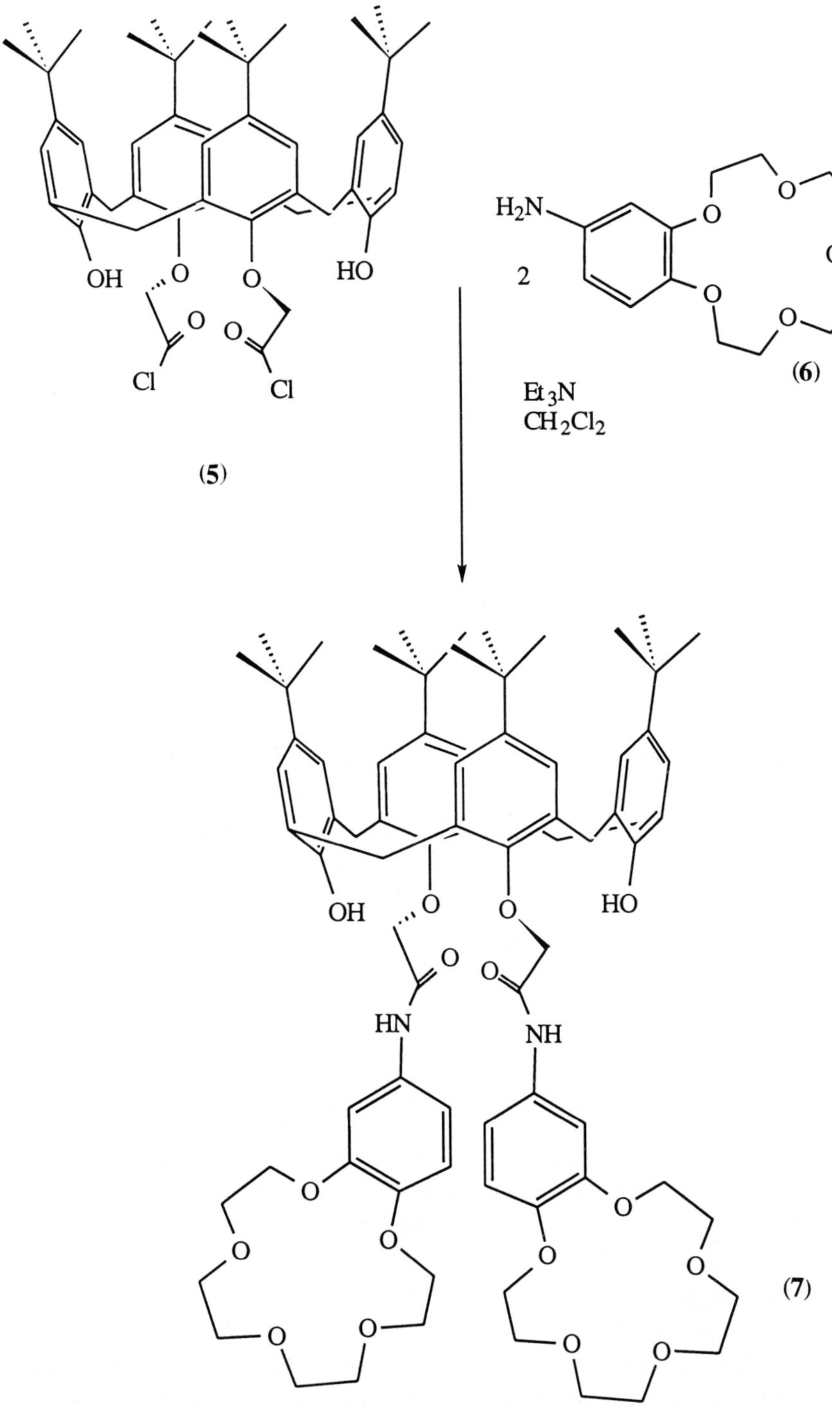

Scheme 2. Synthesis of diamide-benzo-15-crown-5-calix[4]arene.

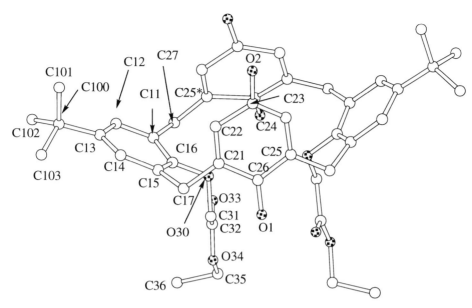

Fig. 2. The crystal structure of (**3**).

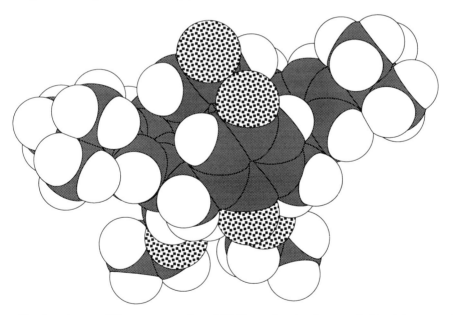

Fig. 3. A space filling representation of (**3**) illustrating the close proximity of the upper rim oxygens.

cases the resulting titration curves of $\Delta\delta$ ppm proton chemical shift perturbation of respective ligand versus equivalents of metal cation suggested 1 : 1 solution complex stoichiometries (Figure 5) in which the metal cationic guests are bound via favourable electrostatic interactions with the ester or amide carbonyl oxygen donor

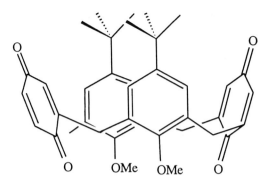

Fig. 4. The dimethylether calix[4]arenediquinone (**8**).

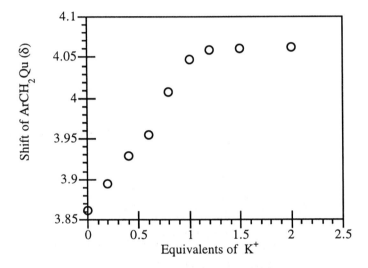

Fig. 5. Shift of one ArCH$_2$Qu proton (**3**) upon addition of potassium cations.

atoms and the quinone carbonyl moieties. This mode of metal cation coordination would have the effect of inhibiting the quinone calix-ring inversion process and consequently rigidifying the receptor into a cone conformation. In support of this hypothesis a low temperature ^{1}H-NMR spectrum (**3**) in the presence of Ba^{2+} exhibited exclusively one pair of doublets for the methylene protons.

Interestingly, the ammonium and alkyl ammonium cations also formed 1 : 1 stoichiometric complexes with (**3**) and (**4**) but did not bind to (**8**) highlighting the importance of additional favourable hydrogen bonding interactions with the respective ester or amide carbonyl moieties of these receptors.

Analogous ^{1}H-NMR titration experiments with (**7**) in deuterated acetonitrile gave the titration curves suggesting 1 : 1 stoichiometries with K$^+$ (Figure 6), NH$_4^+$, Ba^{2+}, and a 2 : 1 metal-to-ligand stoichiometry with Na$^+$.

CATION RECOGNITION BY DIESTER- AND DIAMIDE-CALIX[4]ARENEDIQUINONES 355

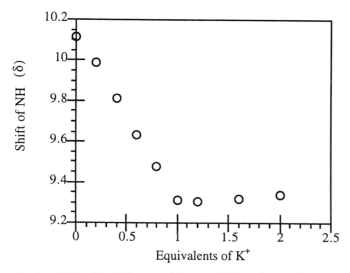

Fig. 6. Shift of the NH proton (**7**) upon addition of potassium cations.

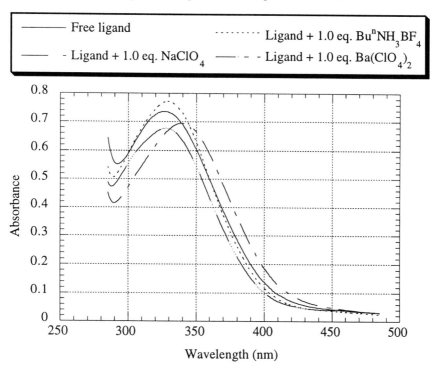

Fig. 7. The UV/vis spectrum of ligand (**4**) in acetonitrile (2.7×10^{-4} M) in the absence and presence of guest cations.

3.3.2. *UV/Vis Spectroscopic Investigations – Association Constant Determinations*

For the calix[4]arenediquinone receptors the respective electronic absorption spectrum exhibits an absorbance ca. 330 nm corresponding to the n to π^* transition of

TABLE IV. Association-constants calculated for (**3**), (**4**) and (**8**) (maximum error ±15%).

Cation	(**3**)	(**4**)	(**8**)
Na^+	a	1.58×10^5	b
K^+	7.2×10^4	4.9×10^4	1.9×10^4
Ba^{2+}	4.8×10^5	1.8×10^5	5.3×10^4
NH_4^+	1.1×10^4	1.2×10^3	b
$Bu^nNH_3^+$	6.6×10^3	1.0×10^4	b

[a] Association constant too large to be reliably calculated using curve fitting method.
[b] No evidence of binding was seen.

the quinone groups. The addition of cationic guests to acetonitrile solutions of, for example (**4**) (Figure 7), resulted in significant red shifts and intensity perturbations of this absorption band. Association constant data for (**3**), (**4**) and (**8**) with various guest cations were determined using a non-linear curve fitting computer program (see Experimental) and the results are presented in Table IV.

For both (**3**) and (**4**) Na^+ and Ba^{2+} cations form very stable complexes. It is noteworthy that the diamide calix[4]arenediquinone (**4**) exhibits a larger association constant for $Bu^nNH_3^+$ than for NH_4^+ whereas with the diester derivative (**3**) the reverse trend is observed. Preliminary electronic absorption titration data of (**7**) with K^+ and Ba^{2+} cations gave association constants of 5×10^4 and 3×10^3 respectively. Interestingly, this K^+ over Ba^{2+} selectivity preference is similar to that exhibited by the dimethyl ether calix[4]diquinone ligand (**8**) but is the reverse shown by (**3**) and (**4**).

3.3.3. Electrochemical Studies

The electrochemical properties of (**3**), (**4**) and (**8**) were investigated using cyclic and square wave voltammetries. The redox behaviour of all these ligands were very similar. For example the cyclic and square wave voltammograms of (**4**) are shown in Figure 8. Potentials refer to a Ag/Ag^+ (1.0 mM $AgNO_3$ in CH_3CN) reference electrode whose potential is 330 ± 10 mV versus SCE. Both of the cyclic and square wave voltammograms of the free receptor (**4**) exhibit reversible/quasi-reversible wave couples (1/1′ and 2/2′), a broad irreversible wave (3) and prewaves (4 and 5). Except for the prewaves, the multiwave feature has also been observed by others [9, 10]. Detailed studies [9, 10] have revealed that waves 1 and 2 correspond to the consecutive one-electron transfer to each of the quinone moieties, leading to the formation of a diradical structure. The addition of one or more equivalents of $NaClO_4$ or KPF_6 to electrochemical solutions of (**4**) resulted in the disappearance of waves 1 and 2 and the evolution of reversible new wave couples at more anodic potentials. Anodic potential perturbations were generally observed with (**3**), (**4**), (**8**) and all cationic guests, and the results are summarised in Table V. Interestingly

TABLE V. A summary of voltammetric properties of (3), (4) and (8) in acetonitrile, supporting electrolyte: 0.10 M TBABF$_4$.

	No. of redox couples	E_{pc} (V) of each redox couple
Free (3)	4	-0.75 (s); -0.85 (r); -1.10 (r); -1.24 (s)
(3) + Na$^+$	2	-0.70 (r); -0.80 (s)
(3) + K$^+$	3	-0.75 (s); -0.85 (r); -0.93 (r)
(3) + Ba^{2+}	1	-0.52 (ec, a)
(3) + NH$_4^+$	1	-0/67 (ec)
(3) + Bu^{n}NH$_3^+$	1	-0.68 (ec)
Free (4)	4	-0.81 (r); -0.91 (r); -1.11 (r); -1.26 (q)
(4) + Na$^+$	2	-0.81 (r); -0.89 (r)
(4) + K$^+$	2	-0.83 (r); -0.92 (r)
(4) + Ba^{2+}	1	-0.64 (ec, a)
(4) + NH$_4^+$	1	-0.69 (ec)
(4) + Bu^{n}NH$_3^+$	1	-0.78 (ec)
Free (8)	2	-1.05 (r); -1.09 (r)
(8) + Na$^+$	2	-0.90 (r); -0.99 (s)
(8) + K$^+$	2	-1.02 (r); -1.09 (r)
(8) + Ba^{2+}	1	-0.52 (ec, a)
(8) + NH$_4^+$	1	-0.85 (ec)
(8) + Bu^{n}NH$_3^+$	1	-0.94 (ec)

E_{pc} The potential of the reduction current peak; r: reversible; q: quasi-reversible; s: single reduction peak without corresponding reoxidation peak; ec: electron transfer followed by a chemical reaction; ec, a: electron transfer followed by a chemical reaction with insoluble product which adsorbs onto the electrode surface.

addition of one or more equivalents of NH$_4$PF$_6$ or Bu^{n}NH$_3$BF$_4$ to electrochemical solutions of (4) resulted in EC mechanistic behaviour (Figure 8) which was not affected by subsequent addition of equivalent amounts of Na$^+$ or K$^+$ cations. This finding is contrary to the respective association constant data calculated from UV/vis titration results in which the ammonium association constants are comparatively smaller in magnitude to those of the alkali metal cations. The relatively strong interactions of these ammonium cations with the radical anions formed by the reduction of the respective quinone moieties of (4) may be responsible for this electrochemical observation and EC mechanistic behaviour. In the presence of more than one equivalents of Ba(ClO$_4$)$_2$, both of the cyclic and square wave voltammograms of (4) showed typical adsorption characteristics.

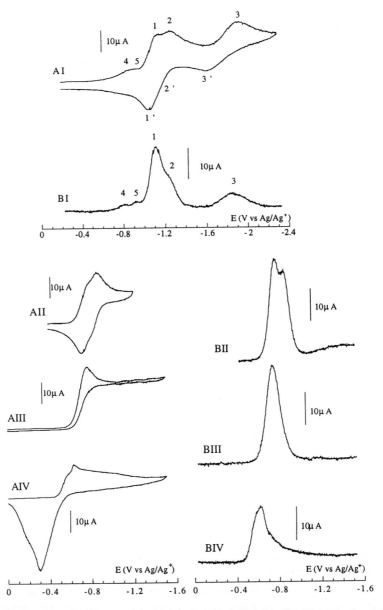

Fig. 8. Cyclic (A, scan rate: 100 mV s^{-1}) and square wave (B, frequency: 10 Hz for BI and 60 Hz for the others) voltammograms of ligand (**4**) (5×10^{-4} M) in acetonitrile in the absence (I) and presence of two equivalents of NaClO$_4$ (II), Bu^{n}NH$_3$BF$_4$ (III) and Ba(ClO$_4$)$_2$ (IV). The units of current (vertical) and potential (horizontal) are Ampère and Volt (vs. Ag/Ag$^+$), respectively.

4. Conclusions

New diester/diamide-calix[4]arenediquinones and a diamide-benzo-15-crown-5-calix[4]arene receptor molecule have been prepared and shown to complex Group 1 and 2 metal, ammonium and alkylammonium cations with association constants up to 4.8×10^5 with Ba^{2+} and (3) in acetonitrile solution. The redox-active quinone containing receptors electrochemically recognise these cationic guest species, including for the first time, the amperometric detection of an alkyl ammonium cationic guest species by a redox-active ionophore.

Acknowledgements

We thank the E.P.S.R.C. for a studentship to P.A.G. and for the use of the Mass Spectrometry service at University College, Swansea.

References

1. (a) P. D. Beer: *Adv. Inorg. Chem.* **39**, 79 (1992). (b) P. D. Beer: *Chem. Soc. Rev.* **18**, 409 (1989).
2. (a) P. D. Beer, H. Sikanyika, C. Blackburn, and J. F. McAleer: *J. Chem. Soc. Chem. Commun.* 1831 (1989). (b) C. D. Hall, N. W. Sharpe, I. P. Danks, and Y. P. Sang: *J. Chem. Soc. Chem. Commun.* 419 (1989). (c) M. L. H. Green, W. B. Heiser, and G. C. Saunder: *J. Chem. Soc. Dalton Trans.* 3789 (1990). (d) J. C. Medina, T. T. Goudnow, M. T. Rodgers, J. L. Atwood, B. C. Lynn, A. E. Kaifer, and G. W. Gokel: *J. Am. Chem. Soc.* **114**, 10583 (1992).
3. P. D. Beer, D. B. Crowe, M. I. Ogden, M. G. B. Drew, and B. Main: *J. Chem. Soc. Dalton Trans.* 2107 (1993).
4. P. D. Beer, D. Hesek, J. Hodacova, and S. E. Stokes: *J. Chem. Soc. Chem. Commun.* 270 (1992).
5. P. D. Beer, C. A. P. Dickson, N. Fletcher, A. J. Goulden, A. Grieve, J. Hodacova, and T. Wear: *J. Chem. Soc. Chem. Commun.* 828 (1993).
6. P. D. Beer, M. G. B. Drew, C. Hazelwood, D. Hesek, and S. E. Stokes: *J. Chem. Soc. Chem. Commun.* 229 (1993).
7. (a) Y. Morita, T. Agawa, Y. Kai, N. Kanehisa, N. Kasai, E. Nomura, and H. Taniguchi: *Chem. Lett.* 1349 (1989). (b) Y. Morita, T. Agawa, E. Normura, and H. Taniguchi: *J. Org. Chem.* **57**, 3658 (1992).
8. P. A. Reddy, R. K. Kashyap, W. M. Watson, and C. D. Gutsche: *Isr. J. Chem.* **32**, 89 (1992).
9. K. Suga, M. Fujihira, Y. Monta, and T. Agawa: *J. Chem. Soc. Faraday Trans.* **87**, 1575 (1991).
10. A. Casnati, E. Comelli, M. Fabbi, V. Bucchi, G. Mori, F. Ugozzoli, A. M. M. Lanfredi, A. Puchini, and R. Ungaro: *Recl. Trav. Chim. Pays-Bas* **112**, 384 (1993).
11. E. M. Collins, M. A. McKervey, E. Madigan, M. B. Moran, M. Owens, G. Ferguson, and S. J. Harris, *J. Chem. Soc. Perkin Trans. 1* **12**, 3137 (1991).
12. R. Ungaro, B. El Haj, and J. Smid: *J. Am. Chem. Soc.* **98**, 5198 (1976).
13. W. Kabsch: *J. Appl. Crystallogr.* **21**, 916 (1989).
14. G. M. Sheldrick: *Acta Crystallogr.* **A46**, 467 (1990).
15. SHELXL Program for Structure Refinement: G. M. Sheldrick, University of Göttingen (1993).
16. A. McKillop, B. P. Swann, and E. C. Taylor: *Tetrahedron* **26**, 4031 (1970).

The Complexation of Ferrocene Derivatives by a Water-Soluble Calix[6]arene*

LITAO ZHANG[a,b], ALBA MACIAS[a], RAHIMAH ISNIN[a], TIANBAO LU[a,b], GEORGE W. GOKEL,[a,b] and ANGEL E. KAIFER[a,**]
[a]*Chemistry Department, University of Miami, Coral Gables, FL 33124, U.S.A.;* [b]*Department of Molecular Biology and Pharmacology, Washington University Medical School, St. Louis, MO 63110, U.S.A.*

(Received in final form: 3 March 1994)

Abstract. The complexation of several ferrocene derivatives by the water-soluble host p-sulfonatocalix[6]arene was investigated using electrochemical and ^{1}H-NMR spectroscopic techniques. The electrochemical results indicate that both oxidation states of the guests are bound to the calixarene host, although the oxidized (ferrocenium) forms are complexed more strongly than the reduced (ferrocene) species. ^{1}H-NMR spectroscopic data indicate that the complexation phenomena involves the inclusion of the guest's ferrocene moiety into the flexible calixarene cavity.

Key words: Inclusion complexation, sulfonated calixarenes, redox-active guests, voltammetry, apolar binding.

1. Introduction

The calixarenes [1–3] constitute a class of host compounds that is attracting substantial attention in the field of supramolecular chemistry. Calixarenes are macrocyclic oligomers formed by the condensation of a p-alkylphenol and formaldehyde. Cyclic tetramers, hexamers, and octamers are readily produced in these reactions, serving as the core nuclei for further synthetic elaboration. While the seminal work of Gutsche and coworkers on calixarenes [4] was published more than ten years ago, the literature does not contain many examples of well-characterized solution complexes of calixarenes [5–12]. In this regard, Shinkai's method for the preparation of sulfonated calixarenes [5] opened the possibility of studying calixarenes as hosts for complexation in aqueous media. Sulfonated calixarenes have become one of the most promising type of molecular hosts, since they can be viewed as conformationally flexible cyclodextrin analogs.

Our continuous interest in molecular recognition phenomena involving redoxactive hosts or guests led us to explore the binding of ferrocene derivatives by sulfonated calixarenes in aqueous media. Therefore, we have investigated the binding of a series of ferrocene-containing guests (see Chart 1) by the water-soluble

---

* This paper is dedicated to the commemorative issue on the 50th anniversary of calixarenes.
** Author for correspondence.

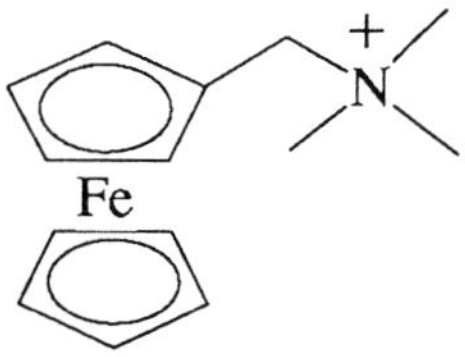

Chart 1.

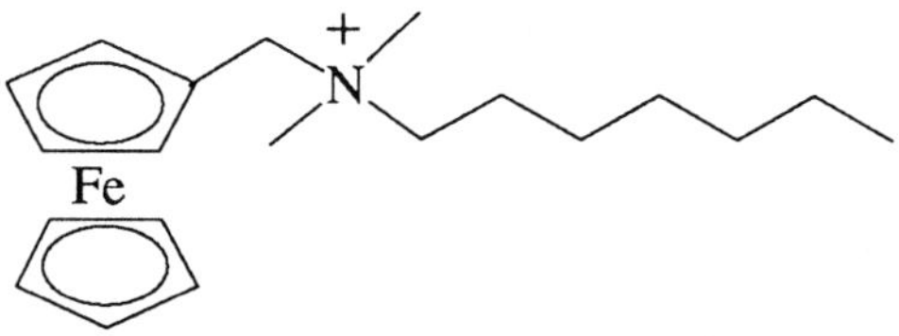

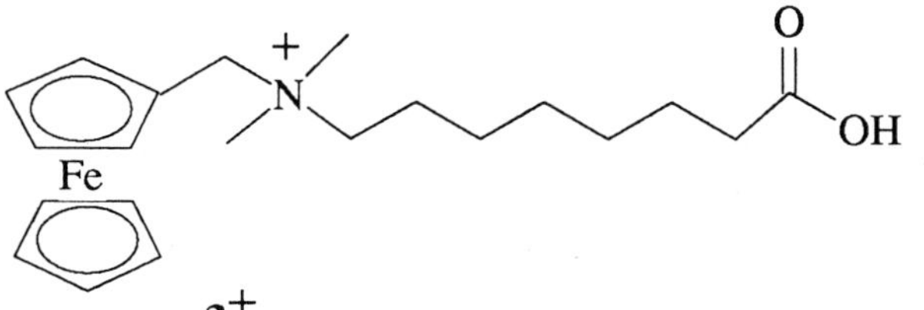

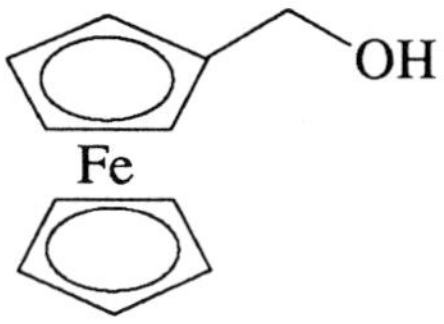

host p-sulfonatocalix[6]arene. We describe here the results of this investigation. A preliminary report has been published elsewhere [13].

2. Experimental

2.1. MATERIALS AND REAGENTS

Guest **4** was purchased from Aldrich and recrystallized from ethanol before use. The cationic guests **1**$^+$ and **2**$^+$ were prepared as described previously [14]. Guests **1**$^+$ and **2**$^+$ were used as their hexafluorophosphate and bromide salts, respectively. The bromide salt of guest **3**$^+$ was synthesized by a similar procedure involving the quarternization of ((dimethylamino)methyl)ferrocene (Aldrich) with 8-bromooctanoic acid (Aldrich). The yellow product was first recrystallized from acetone/ether and then washed with CH_2Cl_2 to remove impurities. The remaining solid was filtered off and dried under vacuum at 60°C to yield pure **3**·Br. Aqueous titration of **3**·Br with a solution of NaOH yields a pK_a value of 4.44 and a neutralization equivalent of 452.3 (calculated value: 466.2, 3% error). ^{1}H-NMR (400 MHz, DMSO-d_6): δ 4.50 (s, 2H) 4.38 (d, 4H), 4.25 (s, 5H), 3.10 (br, 2H), 2.83 (s, 6H), 2.20 (t, 2H), 1.70 (br, 2H), 1.5 (t, unresolved, 2H), 1.3 (br, 6H). DCI-MS (NH$_3$ as carrier gas): [M$^+$ = 386]. *Anal. Calcd.* for $C_{21}H_{32}O_2NFeBr$: C, 53.96%; H, 6.87%. *Found*: C, 54.10%; H, 6.92%.

The sodium salt of the anionic host p-sulfonatocalix[6]arene (Na$_8$·**5**) was synthesized by the procedure of Shinkai and coworkers [5]. All other reagents and solvents were of the best commercial quality available. D$_2$O was purchased from

Aldrich. Distilled water was further purified by passage through a four-cartridge Barnstead Nanopure water purification system.

2.2. Equipment

^{1}H-NMR spectra were recorded on a Varian VXR-400-S spectrometer operating at a frequency of 399.99 MHz. Mass spectra were obtained in VG-trio-2 spectrometer. The electrochemical instrumentation has been described elsewhere [15].

2.3. Procedures

The voltammetric experiments were conducted in a single-cell compartment. The working electrodes were either a Bioanalytical Systems glassy carbon disk (0.080 cm^2) for cyclic voltammetry or a Pine Instruments glassy carbon disk of larger area (0.196 cm^2) for rotating disk electrode voltammetry. The auxiliary (Pt flag) and reference (sodium chloride saturated calomel) electrodes were homemade. Typically, the voltammetric behavior of a 1.0 mM solution of the ferrocene derivative in 50 mM NaCl was recorded at the beginning of each experiment. Immediately afterwards, aliquots from a stock solution containing 10 mM 5^{8-}, 1.0 mM ferrocene derivative, and 50 mM NaCl were added to adjust the concentration of the host in the test solution without diluting the redox-active guest or the supporting electrolyte. No effort was made to keep the test solutions oxygen-free as control experiments demonstrated that the presence of oxygen did not affect the voltammetric results.

Rotating disk electrode voltammetry was used to determine the apparent diffusion coefficients (D_{app}) of the ferrocene derivatives. The D_{app} values depend on the added concentration of host and reflect the position of the host–guest association equilibrium because the inclusion complexes diffuse substantially more slowly than the corresponding free guests. We utilized this concentration dependence to determine the equilibrium constants for the formation of each of the host–guest complexes surveyed here. The details of this method have been already published by us [14].

3. Results and Discussion

3.1. Voltammetric studies

As is widely known, the ferrocene subunit undergoes rapid monoelectronic oxidation to yield the corresponding cationic ferrocenium species. Thus, the ferrocenium/ferrocene redox couple is electrochemically reversible and provides a convenient handle to study the complexation of the ferrocene-containing guests 1^+–4^+ by the anionic calixarene host. Our experiments clearly indicate that all guests are strongly complexed by the calixarene host. For instance, Figure 1 shows the anodic voltammetric behavior of **4** in the absence and in the presence of host

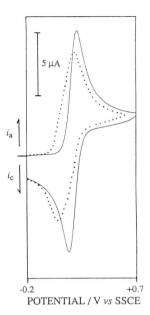

Fig. 1. Voltammetric response (100 mV s^{-1}) on glassy carbon of a 0.5 mmol dm^{-3} solution of **4** in 50 mmol dm^{-3} NaCl in the absence (continuous line) and in the presence (dotted line) of 1.0 equiv. of host **5**$^{-8}$.

5^{8-}. It is clearly evident that the voltammetric response is strongly affected by the presence of the calixarene host as evidenced by the shift of the voltammetric waves to less positive potentials and the marked decrease in the current levels. The electrochemical conversions taking place at the electrode surface are represented by the following equation:

Complexation of **4** in either or both oxidation states is expected to shift the couple's half-wave potential ($E_{1/2}$). As seen in Figure 1, the $E_{1/2}$ value undergoes a substantial displacement in the negative direction, revealing that the oxidation process takes place more easily in the presence of the calixarene. This observation, in turn, reveals the stabilization of the ferrocenium form by complexation with the anionic host. However, the decreased current levels observed throughout the voltammogram indicate that the reduced form of **4** is also complexed by **5**$^{8-}$. Similar voltammetric results were obtained with all the remaining guests. It is interesting to compare these data with those previously reported by us for the complexation of guests **1**$^+$ and **2**$^+$ by β-CD [14]. In contrast to the anionic calixarene host, the

TABLE I. Voltammetric data at 25°C for the oxidation of ferrocene-based guests as a function of the concentration of added calixarene host. Medium: 50 mmol dm^{-3} NaCl.

Guest	[Calix.]/[Guest]	$E_{1/2}$/V^a	ΔE_p/mV^b	K/M$^{-1\,c}$
1$^+$	0:1	0.400	60	
	1:1	0.295	90	10,900
	2:1	0.280	80	
	3:1	0.280	80	
2$^+$	0:1	0.400	60	
	1:1	0.290	100	7,600
	2:1	0.270	80	
	4:1	0.270	80	
4	0:1	0.180	60	
	1:1	0.130	110	3,700
	2:1	0.110	90	
	4:1	0.090	80	

^{a} Measured against a sodium chloride saturated calomel electrode.
^{b} Potential difference between the cathodic and anodic peaks determined at 100 mV/s.
^{c} Binding constants measured using the method given in reference[14]. See text for details.

presence of β-CD shifts the half-wave potential for the oxidation of either guest to more positive values, reflecting the preferential complexation of the reduced form of the guest. In fact, in our voltammetric studies with β-CD [14], we did not detect any complexation of the oxidized (ferrocenium) form of the guests. Thus, the anionic nature of the calixarene host investigated in this work substantially alters the complexation processes of these guests as compared to those observed with the uncharged cyclodextrin receptor. While both types of hosts interact with the reduced guests, only the calixarene seems capable of strongly complexing the oxidized species, a finding which can be easily rationalized by taking into account the electrostatic attraction between the positively charged (oxidized) guest and the negatively charged host.

A summary of relevant voltammetric data with guests 1^+, 2^+, and **4** is provided in Table I. The binding constant values correspond to the association equilibria between the reduced guest and the calixarene host 5^{8-} and were obtained from the concentration dependence of the calixarene-induced decrease in the apparent diffusion coefficients as measured by rotating disk electrode voltammetry (see Experimental Section). Interestingly, all three guests exhibit binding constants in the range 10^3–10^4 M^{-1}. The cationic guests 1^+ and 2^+ show larger values than the neutral guest **4**. This is probably due to the favorable electrostatic interactions

TABLE II. Calixarene-induced effects on the half-wave potential[a] of guest 3^+ at 25°C as a function of medium pH. $[3]^+ = 1.0$ mmol dm^{-3}.

pH	$E_{1/2}$ at [Calix] = 0	$E_{1/2}$ at [Calix] = 2.0 mmol dm^{-3}
2	0.360	0.360
4	0.380	0.290
8	0.360	0.250

[a] Volts vs. SSCE.

between the cationic guests and the anionic calixarene. However, the binding constant measured with **4** is only a factor of 2–3 lower than those measured with the cationic guests, suggesting that the electrostatic component in the binding interactions is not the predominant binding force, i.e., a large fraction of the free energy released in the complexation processes is due to nonelectrostatic interactions. This point will be revisited later.

As discussed before, the negative direction of the calixarene-induced shift in the half-wave potentials leads to the conclusion that the *oxidized (ferrocenium) form of the guests is more strongly bound than the reduced (ferrocene) form.* This is true for all three guests in Table I, regardless of their cationic or neutral nature before electrochemical oxidation. Another important observation can be made from the voltammetric data: The peak-to-peak potential splittings (ΔE_p values) of the voltammetric set of waves are strongly affected by the presence of the calixarene host. This is also observed for all three guests. This finding clearly suggests that the complexation reactions interfere with the electrochemical conversions of the ferrocene subunit. In our studies with β-CD [14], we noted that the electrochemical oxidation of 1^+ (or 2^+) only occurs after dissociation of the corresponding cyclodextrin complex. The voltammetric data obtained with β-CD indicated that the oxidized guest is never included by the cyclodextrin host. In this work, the situation is more complex since both oxidation states of the ferrocene guests are included by the anionic calixarene host. Further electrochemical studies are necessary to obtain a detailed mechanistic picture of the oxidation of the calixarene complexes surveyed in this work.

The aliphatic chain of guest 3^+ ends up in an ionizable carboxylic acid group. This guest was included in this work with the idea of studying its binding to the calixarene host as a function of pH, because the overall charge of the reduced guest may be altered by appropriate pH changes in the reaction medium. However, the host is also pH-sensitive, which complicates considerably the interpretation of the results. Table II provides some preliminary data on the pH dependence of the interactions of this guest with the anionic calixarene host. Upon the addition of 2 equiv. of host, the largest change in the half-wave potential of 3^+ takes place at pH 8. A slightly lower change takes place at pH 4, and no significant shift is observed at pH 2. As reported in the Experimental Section, the pK_a of the carboxylic acid group of 3^+ is 4.44 so that, at pH 2, this group is fully protonated (uncharged) and, at pH 8, is fully ionized (negatively charged). The fact that the largest $E_{1/2}$ shift is observed at the highest pH is probably the result of changes in the structure of the calixarene. At pH 8, the calixarene is present in aqueous solution as an octaanion. According to Atwood and coworkers [16], this octaanion has a solution structure that can be described as a double partial cone, with four negative charges (one phenolate and three sulfonate groups) pointing to each opening of the cavity. As the pH of the medium decreases, it is anticipated that the phenolate groups will undergo protonation [16], thus losing their negative charges and reducing the overall anionic character of the calixarene framework. These protonation-driven changes in the number and distribution of the negative charges of the host may affect its predominant solution conformation, perhaps, leading to less organized conformers which may not exhibit a clearly defined cavity for molecular complexation. Increased conformational flexibility associated to removal of negative charge might be responsible for the results that we obtained at pH 2, where no evidence for complexation was detected using electrochemical techniques.

3.2. ^{1}H-NMR SPECTROSCOPIC STUDIES

Although the substantial calixarene-induced shifts observed in the half-wave potential for the oxidation of all the guests strongly suggest that their main site of interaction with the host is the ferrocene subunit, such detailed information cannot be obtained only from electrochemical data. As we have done before [14, 15], we decided to obtain ^{1}H-NMR spectroscopic data to address this point in more detail. The results are compiled in Table III.

In general, the NMR data are in excellent agreement with the electrochemical results. The proton spectra of all three ferrocene-based guests are strongly affected by the addition of 1 equiv. of host 5^{8-}. The resonances for all guest protons shift upfield, which is consistent with guest inclusion in a cavity formed by aromatic rings. The magnitude of this displacement is largest for the aromatic protons on the ferrocene subunit or for neighboring protons. For instance, for guest 1^+ all proton resonances undergo substantial upfield shifts upon addition of 1 equiv. of calixarene. In contrast, under identical conditions, the protons labeled 3–7 on guest 2^+, do not exhibit appreciable shifts, while all the protons in the vicinity of the ferrocene moiety (as well as the ferrocene protons) experience upfield shifts of magnitude similar to those observed with guest 1^+. These observations clearly indicate that *the calixarene host interacts with the ferrocene residue of the guests*. No significant interaction is observed between the calixarene and the pendant heptyl chain of 2^+. The binding constant between 5^{8-} and 2^+ is lower than the corresponding K value between 5^{8-} and 1^+ (see Table I). This difference might be due to steric hindrance imposed on the complexation process by the heptyl chain of 2^+.

TABLE III. 400 MHz ^{1}H-NMR chemical shifts (δ) for guests **1$^+$**, **2$^+$**, and **4** in D$_2$O solution. Values for the corresponding complexes were recorded after addition of 1 equiv. of **5^{8-}**.

Position	δ (free guest)	δ (complex)
α'	4.17	4.01
α	4.41	3.83
β	4.33	4.01
1'	4.28	3.52
1	2.84	1.67
α'	4.16	3.97
α	4.37	3.83
β	4.31	3.83
1'	4.25	3.62
1"	2.76	1.66
1	2.97	2.30
2	1.61	1.14
3/6	1.17	1.12
7	0.72	0.72
α'	4.13	3.95
α	4.30	4.22
β	4.20	4.05
1	4.13	3.95

4. Conclusions

We have clearly demonstrated the complexation in aqueous media of several ferrocene-based guests by the anionic calixarene host **5^{8-}**. Control experiments with a monomeric 'analog' of the host, 4-hydroxybenzenesulfonate, showed no effect on the electrochemical or spectroscopic parameters of the guests, even after addition of large excesses of this anion. Therefore, our data indicate that *the anionic calixarene host forms inclusion complexes with all the guests surveyed.* The binding constants measured for the association of **5^{8-}** with guest **1$^+$**, **2$^+$**, and **4** are all in the range 10^3–10^4 M^{-1}. However, the voltammetric data reveal that in all cases the oxidized forms of the guests are more tightly bound than the reduced forms. This is probably a reflection of the stronger electrostatic attraction between the host and the oxidized guests.

Although attractive host–guest coulombic forces play an important role in the binding of these guests to the anionic calixarene, other interactions must be responsible for most of the binding free energy as evidenced by the strong binding of neutral **4**. At this point, it seems reasonable to view these host–guest interactions as another example of *apolar binding*, as defined by Diederich and coworkers [17]. We are currently setting up calorimetric measurements to determine the relative enthalpic and entropic contributions to the free energy of complexation in these systems.

Since host 5^{8-} seems capable of apolar binding in aqueous media, it is of some interest to compare briefly its binding properties with those of cyclodextrin hosts. Host 5^{8-} exhibits a cavity, capable of molecular inclusion, which is similarly sized to that of β-CD. Although both 5^{8-} and β-CD exhibit roughly similar binding affinities for ferrocene-containing guests, the calixarene shows an even higher affinity for the oxidized (ferrocenium) guests, while the cyclodextrin does not bind the oxidized forms at all. This is probably due to the dissimilar charges on these receptors, i.e., the anionic nature of the calixarene and the cyclodextrin's lack of charge. Furthermore, while the cyclodextrin has a fairly rigid structure and maintains its well-defined cavity and binding properties throughout the pH 2–10 range, the same is not true for the calixarene. Our data suggest that, as the pH is lowered, a combination of negative charge removal (by protonation) and molecular flexibility tends to decrease the binding properties of the calixarene host. Thus, the anionic calixarene surveyed here shows some similarities in its host properties to β-CD, but its pH-dependent charge and its inherent conformational flexibility give rise to binding properties differing from those found in the substantially more rigid cyclodextrins.

Acknowledgments

The authors gratefully acknowledge the support of this research by the N.S.F. (to A.E.K., CHE–9304262) and the N.I.H. (to G.W.G., GM–36262 and AI–27179). L.Z. thanks the University of Miami for a Maytag graduate fellowship. R.I. gratefully acknowledges the Malaysian Petroleum Company (PETRONAS) for a graduate fellowship.

References

1. C.D. Gutsche: *Calixarenes*, Royal Society of Chemistry, Cambridge, England (1989).
2. V. Böhmer and J. Vicens (Eds.): *Calixarenes: A Versatile Class of Macrocyclic Compounds*, Kluwer, Dordrecht, the Netherlands (1990).
3. G. D. Andreeti, F. Ugozzoli, R. Ungaro, and A. Pochini: in *Inclusion Compounds*, Eds. J. L. Atwood, J. E. D. Davies, and D. D. MacNicol, Oxford University Press, Oxford, Vol. 4 (1991).
4. C. D. Gutsche: *Acc. Chem. Res.* **16**, 161 (1983).
5. S. Shinkai, S. Mori, H. Kiroshi, T. Tsubaki, and O. Manabe: *J. Am. Chem. Soc.* **108**, 2409 (1986).
6. G. E. Hofmeister, F. E. Hahn, and S. F. Pedersen: *J. Am. Chem. Soc.* **111**, 2318 (1989).
7. Y. Aoyama, Y. Nonaka, Y. Tanaka, H. Toi, and H. Ogoshi: *J. Chem. Soc. Perkin Trans.* 2, 1025 (1989).
8. S. Shinkai: *J. Incl. Phenom.* **7**, 193 (1989).
9. T. Arimura, T. Nagasaki, S. Shinkai, and T. Matsuda: *J. Org. Chem.* **54**, 3766 (1989).

10. H. Kawabata and S. Shinkai: *Kagaku Kogyo* **42**, 278 (1991).
11. J. D. Van Loon, J. F. Heida, W. Verboom, and D. N. Reinhoudt: *Recl. Trav. Chim. Pays-Bas* **111**, 353 (1992).
12. S. K. Chang, M. J. Jang, S. Y. Han, Y. H. Lee, M. H. Kang, and K. T. No: *Chem. Lett.* 1937 (1992).
13. L. Zhang, A. Macias, T. Lu, J. I. Gordon, G. W. Gokel, and A. E. Kaifer: *J. Chem. Soc., Chem. Commun.* 1017 (1993).
14. R. Isnin, C. Salam, and A. E. Kaifer: *J. Org. Chem.* **56**, 35 (1991).
15. A. R. Bernardo, J. F. Stoddart, and A. E. Kaifer, *J. Am. Chem. Soc.* **114**, 10624 (1992).
16. J. L. Atwood, D. L. Clark, R. K. Juneja, G. W. Orr, K. D. Robinson, and R. L. Vincent: *J. Am. Chem. Soc.* **114**, 7558 (1992).
17. D. B. Smithrud, T. B. Wyman, and F. Diederich: *J. Am. Chem. Soc.* **113**, 5420 (1991).

Thermodynamic and Electrochemical Aspects of *p-tert*-Butylcalix[*n*]arenes (*n* = 4, 6, 8) and Their Interactions with Amines*

A. F. DANIL DE NAMOR,**, J. WANG, I. GOMEZ ORELLANA
F. J. SUEROS VELARDE and D. A. PACHECO TANAKA
Laboratory of Thermochemistry, Department of Chemistry, University of Surrey, Guildford, Surrey GU2 5XH, U.K.

(Received in final form: 3 March 1994)

Abstract. Attention is drawn to the need of detailed thermodynamics in calixarene chemistry. The reasons for increasing efforts in this area are underlined and suggestions for new issues to be addressed are given. The solution thermodynamics of *p-tert*-butylcalix[*n*]arenes (*n* = 4, 6, 8) is discussed with particular reference to transfer Gibbs energies which reflect the selective solvation that the tetramer and the octamer undergo in the various solvents. This is followed by recent solution studies on amine-*p-tert*-butylcalix[*n*]arene (*n* = 4, 6, 8) in nitrobenzene and in benzonitrile at 298.15 K which indicate the lower acidic character of the tetramer relative to the hexamer and the octamer in these solvents. As an implication of these results, very low conductivities are observed in studies involving the interaction of the former with amines. Thus, thermodynamic studies suggest that *p-tert*-butylcalix[4]arene interacts with triethylamine in benzonitrile and in nitrobenzene through hydrogen bonding or ion-pair formation. A thermodynamic cycle is used to investigate the effect associated with the interaction of the amine with the tetramer in these solvents.

Key words: *p-tert*-Butylcalix[*n*]arenes, amines, thermodynamics, electrochemistry.

1. Introduction

Synthetic developments in the area of Supramolecular Chemistry are currently leading to a massive production of new macrocycles. The driving force for this continuous growth is the search for selective hosts to target a particular neutral or ionic species. There is no doubt that the impact produced by the discovery of macrocyclic ligands such as the crown ethers [1] and the cryptands [2] resulted from their cation complexation and this prompted us to consider the thermodynamic characterisation of these systems (mainly cryptands) which has been extensively reported [3–5]. Calixarenes, an important class of macrocyclic compounds, are products of the base-catalysed condensation reaction of *p*-substituted phenols and formaldehyde [6, 7]. These compounds are characterised by their low solubilities in most solvents, although until recently [8], no quantitative data has been reported. Functionalisation of the lower or upper rim of parent calix[*n*]arenes has

---

* This paper is dedicated to the commemorative issue on the 50th anniversary of calixarenes.
** Author for correspondence.

led to a number of interesting derivatives which are known to interact with metal cations in solution and stability constants for various calixarene derivatives and metal cations have been reported [9, 10]. As far as thermochemical studies are concerned, there are two communications based on research carried out mainly at the Thermochemistry Laboratory (Surrey) in which the first calorimetric (titration macrocalorimetry) data were reported on (i) calixarene esters and metal cations in two solvents (methanol and acetonitrile) [11] and (ii) parent calixarenes and amines in benzonitrile [12]. More recently, these two aspects of calixarene chemistry have been further considered [13–16]. Although the contributions so far made on the solution thermodynamics of calixarene chemistry have provided some useful insights, there are some aspects of the existing work that need to be examined and new issues should be addressed. In order to do so it is important to present a brief account on: (a) what can thermodynamicists contribute to the understanding of solution calixarene chemistry; (b) how far these issues have been addressed: and (c) what aspects of thermodynamics need to be considered.

In order to answer these question it should be stressed that in our opinion, a detailed thermodynamic analysis should involve: (a) the solution thermodynamics of the host, the guest and if possible, the resulting complex (adduct) in the appropriate solvents; (b) the thermodynamic characterisation of the binding process in the same solvents. In both cases, the relevant parameters are Gibbs energy, enthalpy, entropy and heat capacity. As far as (a) is concerned, for a solid compound, solution data involve the contributions of the crystal lattice and that of solvation. The former can be eliminated by the calculation of transfer data from a reference solvent to another. These data reflect the difference in solvation of a solute between two solvents. Availability of these data are extremely useful, particularly in calixarene chemistry where the presence of hydrophobic regions gives unusual properties to these molecules to the extent that solute–solvent interactions often occur. These are likely to have implications on many processes, such as binding, choice of reaction media used for recrystallisation, solvent extraction technology, design of ion-selective electrodes, etc. Within this context, heat capacity data provide a thermodynamic probe of conformational changes taking place as a result of solute–solvent interaction.

Attention has been focussed on the possible applications of calixarene derivatives to mimic biological carriers. Although investigations carried out at the standard temperature are most encouraging since these compounds appear to fulfil the requirements for efficient carrier modelling in the sense that these are insoluble in water and their metal-cation complex stability is not very high, there are several aspects to be considered in order to place this interesting hypothesis on firmer grounds. The main ones are the kinetics and thermodynamics associated with these processes at the biological temperature.

From calixarene derivatives, through complexation with metal cation salts, new electrolytes are derived $[M^+Calix]X^-$. Despite the fact that these new entities may find important applications in many aspects of research, not many efforts have been devoted to their characterisation. In fact, these comments can be extended to the general area of supramolecular chemistry. A representative example of the applications of macrocyclic electrolytes is found in recent work carried out at Surrey [17] which demonstrate that the conductivities of some lithium coronate salts are greater than those observed for common lithium salts in dipolar aprotic media, a development likely to lead to a major contribution in the area of high energy lithium batteries and for which a detailed thermodynamic study made it possible to interpret the nature of these findings. As far as (b) is concerned, one of the most important questions regarding the interactions of calixarene derivatives with different guests in solution concerns selectivity. Therefore, if the factors controlling it are to be understood and rationalised, thermodynamics should be considered a priority area in calixarene chemistry. A direct implication of this statement is the need for accurate thermodynamic data, the derivation of which is largely dependent on the suitability of the methodology employed and this will be discussed in detail elsewhere [18]. However, it is important to mention that as far as stability constants are concerned, attempts should be made to check these data by different techniques since this information provides the guide for the isolation of metal-ion calixarene salts and the basis to establish a selectivity index in calixarene chemistry. This is particularly relevant for high stability complexes ($\log K_s > 5$) where new methodology needs to be developed. As far as enthalpies are concerned, calorimetry is by far the most suitable technique to derive these data particularly in view of the recent advances in microcalorimetry [19]. There are two reasons by which titration microcalorimetry offers advantages relative to classical macrocalorimetry; particularly for processes involving calixarenes. The first one is related to the lack of heat found when using the latter technique when some cations interact with calixarene derivatives in nonaqueous media [20]. Preliminary studies using titration microcalorimetry have shown that the failure to characterise thermodynamically the binding process between metal cations with these ligands by macrocalorimetry is not due to the fact that calorimetry is not a suitable reporter for molecular events but to kinetically slow processes being involved. It is well established that classical calorimetry is unsuitable to deal with these processes. The second reason is that in terms of accuracy, classical calorimetry cannot compete with the most sensitive calorimetric devices recently designed [21]. From the above discussion, it seems appropriate to state that the time has now come to characterise thermodynamically:

(a) the effect of temperature on the solution properties of parent calixarenes and their derivatives in a large variety of solvents.
(b) the factors governing selectivity as a function of temperature across a wide variety of calixarene derivatives and cations in different media. Particular emphasis should be placed on heat capacity measurements of the host, guest and the resulting complex.

These studies should be accompanied by ^{1}H- and ^{13}C-NMR evidence [22] and conductance studies. The latter should be carried out particularly in low permittivity media where the extent of ion-pair formation for these new electrolytes need to be investigated. No such studies have been undertaken in these media. We are carrying out extensive research on calixarene derivatives which are currently synthesised at Surrey using a new methodology developed by us [23] which increases their yields with respect to existing methods.

Although considerable progress has been made by our group on structural, thermodynamic and electrochemical aspects of these derivatives and metal cations in a wide variety of solvents, in this special commemorative volume on calixarenes, we consider it appropriate to discuss the 'parent compounds' and their interactions with amines in nitrobenzene and in benzonitrile. The background leading to this work has been described elsewhere by Danil de Namor and coworkers [13, 14] together with various aspects of calixarene-amine interactions in benzonitrile. Although a phase separation between nitrobenzene (or benzonitrile) in water can be achieved and therefore, this solvent can be also used in extraction processes, nitrobenzene (ε = 34.82 at 298.15 K) offers a higher permittivity medium than benzonitrile (ε = 25.2 at 298.15 K) [24] and this may have interesting implications on amine-calixarene interactions. Since the solution properties of *p-tert*-butylcalix[n]arene (n = 4, 6, 8) in nitrobenzene are unknown, these are discussed first.

2. Experimental

2.1. CHEMICALS

All chemicals were used as described previously [14]. The water content of nitrobenzene determined by Karl-Fischer titration was found to be less than 0.01% and its conductivity at 298.15 K was 7.98×10^{-8} S cm^{-1}.

2.2. SOLUBILITY AND HEAT OF SOLUTION MEASUREMENTS

These were carried out as described elsewhere [25].

2.3. CONDUCTIMETRIC, POTENTIOMETRIC AND CALORIMETRIC TITRATIONS

Except for microcalorimetric titrations, readers are referred to the literature [14]. Microcalorimetric experiments at 298.15 K were carried out using the titration vessel of the 2277 Thermal Activity Monitor. The vessel was filled with 2.8 mL^3 of a solution of *p-tert*-butylcalix[4]arene in benzonitrile (5×10^{-4} mol dm^{-3}) or in nitrobenzene (5×10^{-3} or 9×10^{-3} mol dm^{-3}). Triethylamine solutions in the appropriate solvent (concentration range; 0.07–0.95 mol dm^{-3}) were injected (16 injections; 0.015–0.025 mL for each run) from a 0.5 mL gas-tight Hamilton syringe, attached to a computer-operated syringe drive at 5 minute intervals. *p-tert*-Butylphenol solutions were 0.04 mol dm^{-3} in both solvents. A dynamic correction

TABLE I. Thermodynamic parameters of solutions of *p-tert*-butylcalix[n]arenes (n = 4, 6, 8) in nitrobenzene at 298.15 K. Derived parameters of transfer from benzonitrile.

	n = 4	n = 6	n = 8
Solubility/mol dm^{-3} (PhNO$_2$)	$(1.83 \pm 0.04) \times 10^{-2}$	$(2.26 \pm 0.03) \times 10^{-2}$	$(2.57 \pm 0.06) \times 10^{-3}$
$\Delta_s G^0$/kJ mol^{-1} (PhNO$_2$)	9.92 ± 0.05	9.40 ± 0.03	14.78 ± 0.06
$\Delta_s G^0$/kJ mol^{-1} (PhCN)	17.26^a	12.88^a	11.09^a
$\Delta_t G^0$/kJ mol^{-1} (PhCN $\to$ PhNO$_2$)	-7.34	-3.48	3.69
$\Delta_s H^0$/kJ mol^{-1} (PhNO$_2$)	-14.67 ± 2.88	-34.85 ± 1.96	-29.09 ± 1.50
$\Delta_s H^0$/kJ mol^{-1} (PhCN)	-14.20^a	-23.67^a	-45.90^a
$\Delta_t H^0$/kJ mol^{-1} (PhCN $\to$ PhNO$_2$)	-0.47	-11.18	16.80
$\Delta_s S^0$/J K^{-1} mol^{-1} (PhNO$_2$)	-82.5 ± 9.7	-148.4 ± 6.6	-147.1 ± 5.0
$\Delta_s S^0$/J K^{-1} mol^{-1} (PhCN)	-105.5^a	-122.6^a	-191.1^a
$\Delta_t S^0$/J K^{-1} mol^{-1} (PhCN $\to$ PhNO$_2$)	23.0	-25.8	44.0

^{a}From Reference [14].

based on Tian's equation [26, 27] was used to calculate the integrals from the microcalorimetric curve. Corrections for the thermal effect arising from dilution were applied. The reliability of the equipment was checked by carrying out the standard reactions suggested in the literature [28].

3. Results and Discussion

3.1. THERMODYANMICS OF SOLUTIONS OF *p-tert*-BUTYLCALIX[n]- ARENES (n = 4, 6, 8)

Table I lists the solubility data, derived Gibbs energies, $\Delta_s G^0$, enthalpies, $\Delta_s H^0$ and entropies, $\Delta_s S^0$ of solutions of *p-tert*-butylcalix[n]arenes (n = 4, 6, 8) (calix n) in nitrobenzene (PhNO$_2$) at 298.15 K referred to the process

$$\text{calix}[n] \text{ (solid)} \to \text{calix } n \text{ (PhNO}_2\text{)}. \tag{1}$$

The standard deviation of the data are also included in Table I. Taking benzonitrile (PhCN) as the reference solvent, the thermodynamic parameters of transfer

($\Delta_t G^0$, $\Delta_t H^0$ and $\Delta_t S^0$) of these macrocycles to nitrobenzene (Equation 2) are calculated.

$$\text{calix}[n]\ (\text{PhCN}) \rightarrow \text{calix}[n]\ (\text{PhNO}_2). \tag{2}$$

and these are reported in Table I.

As far as transfer Gibbs energies are concerned, as the size of the calixarene increases $\Delta_t G^0$ becomes more positive to the extent that the tetramer and the hexamer are better solvated in nitrobenzene than in benzonitrile while the converse is true for the cyclic octamer. Availability of $\Delta_s G^0$ values for *p-tert*-butylcalix[n]arenes ($n = 4, 8$) in various solvents at the same temperature allows the calculation of transfer Gibbs energies of these macrocycles to other media. Thus, for the cyclic tetramer, $\Delta_t G^0$ values from benzonitrile to various solvents [methanol, (MeOH); 1.17 kJ mol^{-1}; ethanol (EtOH); 2.61 kJ mol^{-1}; *N,N*-dimethylformamide (DMF); -0.37 kJ mol^{-1}; acetonitrile (MeCN); 7.43 kJ mol^{-1}; *n*-hexane(*n*-Hex); 3.71 kJ mol^{-1}; chloroform (Cl$_3$CH); -3.78 kJ mol^{-1})] reflect the fact that *p-tert*-butylcalix[4]arene is more favourably transferred to nitrobenzene (see Table I) than to any of the solvents considered. The Gibbs energy term determines the equilibrium position associated with the transfer process and therefore, these results unambiguously demonstrate that this position is shifted considerably by changing the solvent to which *p-tert*-butylcalix[4]arene is transferred as to favour the medium which provides the macrocycle with the highest stability as a result of solvation. Consequently, the most striking feature of the transfer Gibbs energy data is the selective solvation that the tetramer undergoes with the various solvents to the extent that on the basis of $\Delta_t G^0$ values, a solvation selectivity index for *p-tert*-butylcalix[4]arene can be for the first time established. Thus, the capability of the medium to solvate *p-tert*-butylcalix[4]arene is reflected in the following sequence

$$\text{PhNO}_2 > \text{Cl}_3\text{CH} > \text{DMF} > \text{PhCN} > \text{MeOH} > \text{EtOH} > n\text{-Hex} > \text{MeCN}. \tag{3}$$

Selective solvation is also observed for the cyclic octamer Thus, $\Delta_t G^0$ values (kJ mol^{-1}) of 4.08, 16.17, 15.17 and 1.50 were found for the transfer of *p-tert*-butylcalix[8]arene from benzonitrile to DMF, MeCN, *n*-Hex and CHCl$_3$, respectively. These findings lead to a selectivity sequence for calix[8] in the order:

$$\text{PhCN} > \text{Cl}_3\text{CH} > \text{PhNO}_2 > \text{DMF} > n\text{-Hex} > \text{MeCN}. \tag{4}$$

Much significance should be attributed to these results which provide clear evidence that the transfer Gibbs energy is a suitable reporter of molecular events related to solvation. Based on these data it is concluded that in calixarene chemistry, the role of the solvent is such that it could affect significantly their reactivity or indeed their host properties.

Analysis of the enthalpy and entropy contributions to the transfer process reflects the different origins of the favourable (negative) Gibbs energies of the tetramer

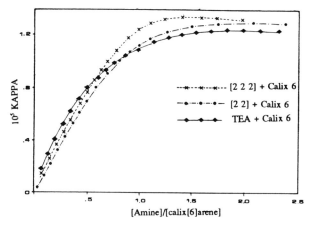

Fig. 1. Conductimetric titrations curves of *p-tert*-calix[6] and amines in nitrobenzene at 298.15 K.

(entropy controlled) relative to the hexamer (enthalpy controlled). For the octamer, the positive entropic contribution is not favourable enough to overcome the energy barrier (endothermic process) and therefore, the net result (positive $\Delta_t G^0$) is that the cyclic octamer is better solvated in benzonitrile and therefore, reluctant to transfer to nitrobenzene. However, an interesting feature of these results is that shown by the hexamer since its transfer to nitrobenzene occurs with heat release (exothermic process) accompanied by a considerable loss in entropy. This pattern is similar to that found by Danil de Namor and coworkers for the transfer of (a) cyclodextrins from water to N,N-dimethylformamide [29], (b) an ester derivative of *p-tert*-butylcalix[4]arene from methanol to acetonitrile, where in both cases the receiving media are believed to enter specific interactions with the ligand. These data suggest that the hexamer exhibits significant interactions with nitrobenzene and this is now being further investigated.

3.2. AMINE-CALIXARENE INTERACTIONS. CONDUCTANCE STUDIES

In order to establish whether or not ions are formed as a result of the interaction of *p-tert*-butylcalix[n]arene ($n = 6, 8$) and amines (A = triethylamine, cryptand 22 and cryptand 222) in nitrobenzene, conductance measurements were carried out at 298.15 K. A representative example is given in Figure 1 in which the conductimetric titration curves for the hexamer and the different amines in PhNO$_2$ at the standard temperature are shown.

Since the starting solution contains a nonelectrolyte (calixarene; nonconducting species), the increase in conductance by the addition of the amine results from the formation of a 1 : 1 electrolyte resulting from the transfer of the proton from the calixarene to the amine. The initial portion of the titration curve was used to calculate the thermodynamic equilibrium constants for the formation of the adduct;

K_s, (Equation 5) and for the ion-pair (association) process, K_a (Equation 6),

$$\text{A (PhNO}_2\text{)} + \text{calix}[n] \text{ (PhNO}_2\text{)} \xrightarrow{K_s} [\text{AH}^+\text{calix}[n]^-] \text{ (PhNO}_2\text{)} \quad (5)$$

$$\text{AH}^+ \text{ (PhNO}_2\text{)} + \text{calix}[n] \xrightarrow{K_a} [\text{AH}^+\text{calix}[n]^-] \text{ (PhNO}_2\text{)} \quad (6)$$

following the procedure described elsewhere [14]. Table II lists the limiting molar conductances, Λ^0, of the ammonium calixarenate salts; $\log K_s$ (Equation 5) and $\log K_a$ (Equation 6) of p-tert-butylcalix[n]arene ($n = 6, 8$) and amines in nitrobenzene at 298.15 K. Equilibria data (Equations 5 and 6) are combined to derive corresponding values for the proton transfer reaction (Equation 7).

$$\text{calix}[n] \text{ (PhNO}_2 + \text{A (PhNO}_2\text{)} \xrightarrow{K_p} \text{calix}[n]^- \text{ (PhNO}_2\text{)} + \text{AH}^+ \text{ (PhNO}_2\text{)} \quad (7)$$

and these data are also included in Table II.

As far as Λ^0 values are concerned, these are the results of the anion and cation contributions. Unfortunately, very few conductance studies for electrolyte in nitrobenzene have been reported and the data available are mostly referred to the tetra-n-alkylammonium salts [30]. Values for these salts in this solvent at the standard temperature vary from approximately 20 to 40 S cm^2 mol^{-1}. Interestingly, a Λ^0 value of 22.31 S cm^2 mol^{-1} in nitrobenzene at 298.15 K has been reported for Bu$_4$NPh$_4$B, an electrolyte constituted by a cation larger than Et$_3$NH$^+$ and an anion smaller than calix[6]$^-$ or indeed calix[8]$^-$. Therefore, the values given in Table II for the triethylammonium p-tert-butylcalix[n]arene ($n = 6, 8$) electrolytes in this solvent seem reasonable. To our knowledge no values have been reported for the ionic conductivities, λ^0+, of protonated amines in nitrobenzene as to derive the λ^{0-} for the p-tert-butylcalix[n]arenate ($n = 6, 8$) anions. However, the anion effect can be assessed by considering the differences in Λ^0 values for a given cation and the two different anions. Thus, in moving from calix[6]$^-$ to calix[8]$^-$, a decrease in the limiting conductance of about 0.3 S cm^2 mol^{-1} (average value) is observed which is likely to be attributed to the size increase of the anion. It should be emphasised that this difference is relatively small and may reflect the higher solvation of the hexamer relative to the octamer in nitrobenzene as discussed above. However, this is difficult to assess since the solvation of the conjugated base (anion) would differ from that of the parent calixarene.

Similarly, for a given calixarenate anion (common anion), the cation effect is reflected in the different Λ^0 values obtained for the various electrolytes. Thus, the limiting conductance decreases significantly (~ 6.5 S cm^2 mol^{-1}) from the triethylammonium containing electrolyte to the protonated cryptand salt. This drop in conductance for the latter relative to the former electrolyte is likely to be the result not only of the considerable size increase of the cation but also to the shielding effect of the cryptands on the proton which will undoubtedly reduce the electric field in the vicinity of the ion.

Availability of Λ^0 values for the same electrolytes at the same temperature in benzonitrile (see Table II, values in brackets) allows comparison between the

TABLE II. Limiting molar conductances, Λ^0 of p-tert-butylcalixarene salts and equilibria data of p-tert-butylcalix[n]arene ($n = 6, 8$) and amines in nitrobenzene at 298.15 K. Comparison with corresponding data in benzonitrile.

Amine	calix[6]				calix[8]			
	Λ^0/S cm^2 mol^{-1}	$\log K_s$	$\log K_a$	$\log K_p$	Λ^0/S cm^2 mol^{-1}	$\log K_s$	$\log K_a$	$\log K_p$
Triethylamine	22.79 ± 0.03 (31.07)[a]	1.69 (3.19)[a]	1.36 (2.44)[a]	0.33 (0.75)[a]	(30.66)[a]	(3.98)[a]	(3.23)[a]	(0.75)[a]
Cryptand 22	16.32 ± 0.06 (23.86)[a]	2.06 (3.04)[a]	1.13 (1.88)[a]	0.93 (1.16)[a]	15.82 ± 0.07 (23.03)[a]	2.30 (3.24)[a]	1.21 (2.62)[a]	1.09 (0.62)[a]
Cryptand 222	16.95 ± 0.04 (24.70)[a]	3.20 (3.18)[a]	2.13 (1.73)[a]	1.07 (1.45)[a]	16.79 ± 0.04 (24.37)[a]	2.74 (3.70)[a]	1.94 (2.52)[a]	0.80 (1.18)[a]

[a]From Reference [14].

two sets of data. Thus, for a given electrolyte, the differences in the limiting conductances in benzonitrile relative to nitrobenzene are within the experimental error, constant. The Λ^0 values in the former solvent are about 7.5 S cm^2 mol^{-1} higher than in the latter and this is attributed to the lower viscosity of benzonitrile ($\eta = 0.0124$ p at 298.15 K) [24] relative to nitrobenzene ($\eta = 0.0186$ p) [24]. This statement is based on the suggestion put forward by Walden [31] ($\Lambda_1^0, \eta_1, -\Lambda_2^0 \eta_2$) which enables the calculation of approximate values of limiting conductances for electrolytes constituted by large ions in a given solvent from known Λ^0 for the same electrolyte in another solvent. Despite the complexity of the systems involved, Walden's rule works reasonably well for the cryptands. Thus, using the Λ^0 for [22H$^+$ calix[6]$^-$] in benzonitrile, a corresponding value of 15.96 S cm^2 mol^{-1} is calculated for this electrolyte in nitrobenzene which does not differ significantly from the value of 16.32 S cm^2 mol^{-1} reported in Table II. Similar agreement is found for other cryptand containing electrolytes.

Equilibria data ($\log K_s$) for the process represented in Equation 5 result from the contribution of (i) the ion-pair formation between the protonated amine and the calixarenate anion ($\log K_a$) and (ii) the proton transfer reaction ($\log K_p$). The data shown in Table II for the overall process (Equation 5) shows that this is less favoured in nitrobenzene than in benzonitrile (values in brackets). Judging from $\log K_s$ values, the selectivity pattern observed for the octamer in nitrobenzene (cryptand 222 > cryptand 22 > Et$_3$N) differs from that found in benzonitrile (Et$_3$N > cryptand 222 > cryptand 22). Regarding $\log K_s$ values for the hexamer in benzonitrile, these data reflect that this ligand is unable to selectively recognise among these amines. In nitrobenzene, the selectivity sequence for the hexamer and the different amines follows the same pattern to that found for the octamer. However, the lower stability ($\log K_s$) in nitrobenzene is mainly attributed to the weaker acidic and basic behaviour of *p-tert*-butylcalix[n]arene ($n = 6, 8$) and these amines respectively; in this solvent relative to benzonitrile as assessed from potentiometric (qualitative) studies in nitrobenzene. However, an interesting aspect of these results is that in benzonitrile, the formation of ion-pairs is more favoured for the electrolyte containing the smaller cation. Thus, $\log K_a$ values decrease in going from R$_3$NH$^+$calix[n]$^-$ ($n = 6, 8$) to Cry·H$^+$calix[n]$^-$. Considering that the proton is more likely to be exposed in the former than in the latter, where ^{13}C-NMR studies [14] have revealed that the cation is likely to be sitting in the cavity (cryptand 222) or in the hole (cryptand 22), these results are somehow expected. However, this is not the pattern observed in nitrobenzene. It should be emphasized that transfer Gibbs energies from benzonitrile to nitrobenzene reflect the fact that triethylamine is slightly better solvated in nitrobenzene than in benzonitrile (as discussed below). In an attempt to gain further understanding on these systems we are now proceeding with ^{13}C-NMR studies and with the calculation of transfer data for the protonated amines in these two solvents [32].

In summary, as far as amine-*p-tert*-butylcalix[*n*]arene (*n* = 6, 8) interactions are concerned, conductance studies provided information regarding

(a) the nature of these interactions
(b) the stoichiometry of the reaction
(c) the behaviour of the calixarenate anion (delocalised charge) in these solvents [14]
(d) the extent of ion-pair formation
(e) the limiting conductances for these new electrolytes in these solvents
(f) the factors contributing to the overall interaction of calixarenes with amines including the solvent effect on the ability of *p-tert*-butylcalix[*n* = 6, 8]arene to selectively recognise amines [14].

Although we have now completed a thermodynamic study on amine-*p-tert*-butylcalix[*n*]arene (*n* = 6, 8) in nitrobenzene, in the remaining part of this paper we address new issues related to *p-tert*-butylcalix[4]arene–amine interactions in benzonitrile and in nitrobenzene.

3.2. AMINO *p-tert*-BUTYLCALIX[4]ARENE INTERACTIONS IN BENZONITRILE AND IN NITROBENZENE

Unlike calix[6] and calix[8], very low conductivities were obtained in the titration of *p-tert*-butylcalix[4]arene and these amines in benzonitrile and nitrobenzene. The relatively low conductances observed in the titration of the tetramer and amines in these solvents do not necessarily imply that these two species are unable to interact in solution. Essentially, these findings reflect the fact that very few ions are formed in solution. These could be attributed to an extensive formation of ion-pairs or to a process in which hydrogen bonding rather than a proton-transfer (ion) reaction predominates. These two possibilities are feasible since benzonitrile and, to a lesser extent nitrobenzene, are media of lower permittivity than acetonitrile (the solvent commonly used to study calixarene–amine interactions). Therefore, electrolytes will have a greater tendency to undergo ion-pair formation in PhCN or $PhNO_2$ than in MeCN. On the other hand, hydrogen bond formation is known to occur between the monomer (*p-tert*-butylphenol) and amines and the thermodynamics of phenol–amine interactions have been reported [33]. In order to identify which of these processes are responsible for *p-tert*-butylcalix[4]arene–amine interactions in solution it is necessary to have information regarding the acid and basic strengths of the species involved. It must be a condition that for a proton (ion formation) transfer reaction to occur, the amine should be basic enough to be able to remove the proton from the calixarene. The dissociation constants (pK_a) of various amines in benzonitrile at 298.15 K were previously reported [14].

Thus, Figure 2 shows the potentiometric curve (and its first derivative) for the titration of *p-tert*-butylcalix[4]arene in benzonitrile at 298.15 K. For comparison

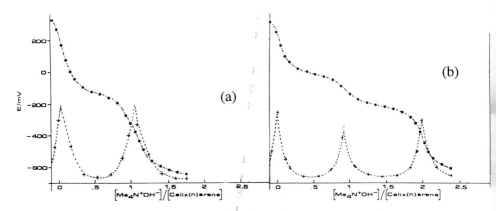

Fig. 2. Potentiometric titrations of (●) *p-tert*-butylcalix[*n*]arene (*n* = 4, 6) with tetramethylammonium hydroxide in benzonitrile at 298.15 K (a) calix[4], (b) calix[6]. (+) First derivative of titration curve.

purposes, the corresponding titration curve for calix[6] in the same solvent and at the same temperature is included. It should be noted that unlike the cyclic octamer (four inflection points) and the hexamer (two inflection points) [14] only one proton is removed for *p-tert*-butylcalix[4]arene in benzonitrile. The potentiometric titration data were analysed using a MINIQUAD program [34]. The dissociation constant (expressed as pK_{d_1}) for the process

$$\text{calix[4] (PhCN)} \xrightarrow{K_{d_1}} \text{calix[4]}^- \text{ (PhCN)} + \text{H}^+ \text{ (PhCN)} \tag{8}$$

in benzonitrile (pK_{d_1} = 19.33) is combined with pK_a values for the various amines (Equation 9).

$$\text{A (PhCN)} + \text{H}^+ \text{ (PhCN)} \xrightarrow{K_a} \text{AH}^+ \text{ (PhCN)} \tag{9}$$

so data for the proton transfer reaction (Equation 7) from *p-tert*-butylcalix[4]arene to the various amines in benzonitrile can be obtained and these are reported in Table III. Also included in this table are the derived Gibbs energies, $\Delta_p G^0$ for this process.

These data show that unlike calix[6] and calix[8] (negative $\Delta_p G^0$), the proton transfer reaction for *p-tert*-butylcalix[4]arene and these amines in benzonitrile is much less favoured (positive $\Delta_p G^0$) and these findings explain the low conductivities observed for these systems. Unfortunately, some experimental difficulties are found in measuring the pK_d values for *p-tert*-butylcalix[4]arene in nitrobenzene and we were unable to obtain quantitative data on this system.

However, in an attempt to gain further insight on these processes, we proceeded with the thermodynamics associated with the interactions of the cyclic tetramer and amines in benzonitrile and in nitrobenzene at the standard temperature. For these purposes, titration microcalorimetry was the technique used since inconsistent data were obtained by the use of classical macrocalorimetry mainly due

TABLE III. Equilibria data for the proton-transfer reaction from *p-tert*-butylcalix[4]arene and amines in benzonitrile at 298.15 K. Derived Gibbs energies (molar scale).

Amine	$\log K_p{}^a$	$\Delta_p G^0$/kJ mol^{-1}
atropine	-1.62	9.25
cryptand 22	-1.54	8.79
cryptand 222	-0.93	5.31
triethylamine	-1.35	7.71
tert-butylamine	-1.54	8.79

^{a} Obtained from a combination of pK_a values of *p-tert*-butylcalix[4]arene (see text) and amines (References [14]) in benzonitrile at 298.15 K.

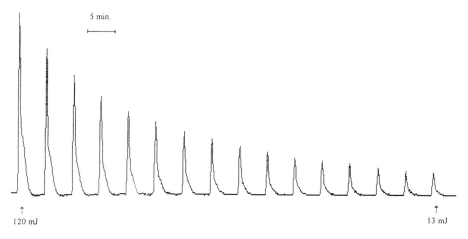

Fig. 3. Microcalorimetric recorded graph for the titration of *p-tert*-butylcalix[4]arene and triethylamine in nitrobenzene at 298.15 K.

to the low sensitivity of the latter relative to the former technique. Since there are no published data in nonaqueous media originated from the Thermal Monitor Analyser (microcalorimeter, see Experimental) we consider it appropriate to show a typical microcalorimetric recorded graph (Figure 3) for the titration of *p-tert*-butylcalix[4]arene and triethylamine in nitrobenzene at 298.15 K. In this figure, the integral values for the first (120 mJ) and the last (13 mJ) areas in the plot are indicated. Also included is the time interval (5 min) at which samples were introduced in the calorimetric vessel. Similar titration graphs were obtained in benzonitrile. From these data, equilibrium constants and enthalpy changes were calculated using a nonlinear least squares procedure based on Marquardt's method, including an algorithm in which linear parameters are eliminated [35]. Uncertainties for the K_s and $\Delta_c H$ values were obtained from the diagonal of the variance–covariance matrix [36]. Calorimetric data were consistent with a model assuming

1 : 1 amine–*p-tert*-butylcalix[4]arene interaction. It must be noted that despite the small formation of ions observed for this system in these solvents, this stoichiometry was also found in the conductance measurements. Table IV lists $\log K_s$ and derived Gibbs energies, enthalpies and entropies for the interaction of triethylamine and *p-tert*-butylcalix[4]arene in benzonitrile and in nitrobenzene at 298.15 K. For comparison purposes, corresponding values for *p-tert*-butylphenol (monomer) and the same amine in these solvents at the same temperature are also included in this table. The medium effect is clearly reflected in the thermodynamic data. Although the process involving the tetramer is enthalpically more stable ($\Delta_c H^0$ more negative) in nitrobenzene than in benzonitrile, the loss of entropy is greater in the former relative to the latter. These results seem to suggest that there is a greater loss of translational freedom upon interaction of these species in nitrobenzene than in benzonitrile. The outcome of enthalpy and entropy contributions makes the process more favoured ($\Delta_c G^0$ more negative) in PhCN than in PhNO$_2$. Note that the opposite is found for the monomer and triethylamine in these solvents.

Medium effects on adduct formation are best analysed by considering the solvation properties of the species involved. Thus, using transfer thermodynamic data for *p-tert*-butylcalix[4]arene (Table I), corresponding values for triethylamine from benzonitrile to nitrobenzene ($\Delta_t G^0 = -1.11$ kJ mol^{-1}, $\Delta_t H^0 = 1.67$ kJ mol^{-1} and $\Delta_t S^0 = 9.3$ J K^{-1} mol^{-1}) combined with the complexation data in these solvents given in Table IV, the thermodynamic parameters for the transfer of the triethylamine–*p-tert*-butylcalix[4]arene adduct [R$_3$N·calix[4]] between these two solvents are calculated by inserting the appropriate quantities in the following cycle, where P = G, H or S.

$$R_3N \text{ (PhCN)} + \text{calix[4] (PhCN)} \xrightarrow{\Delta_c P^0} R_3N \text{ calix[4] (PhCN)}$$

$$\Delta_t P^0 \qquad\qquad \Delta_t P^0 \qquad\qquad\qquad\qquad \Delta_t P^0 \qquad\qquad (10)$$

$$R_3N \text{ (PhNO}_2\text{)} + \text{calix[4] (PhNO}_2\text{)} \xrightarrow{\Delta_c P^0} R_3N \text{ calix[4] (PhNO}_2\text{)}$$

As far as Gibbs energies are concerned, the $\Delta_t G^0$ values reveal the differences in solvation of these species in these solvents. Thus, nitrobenzene while stabilising both *p-tert*-butylcalix[4]arene and triethylamine better than benzonitrile ($\Delta_t G^0$ negative) it weakens the acidic properties of the tetramer and the basic character of the amine. This statement is based on comparative potentiometric studies carried out with this system in these solvents. This is further corroborated by the decrease observed in the stability of the adduct in nitrobenzene relative to benzonitrile as shown in the $\Delta_c G^0$ values listed in Table V. Therefore, the role of the solvent is such that conclusions drawn for the same (or similar) process in the solid state are not necessarily applicable to solution processes [37]. It should be emphasised that the most interesting feature of the data shown in Table V are observed in the enthalpy and entropy contributions to the Gibbs energy of the process. Thus,

TABLE IV. Thermodynamic parameters for the interaction of *p-tert*-butylcalix[4]arene and *p-tert*-butylphenol with triethylamine in benzonitrile and in nitrobenzene at 298.15 K.

Compound	log K_s	$\Delta_c G^0$/kJ mol^{-1}	$\Delta_c H^0$/kJ mol^{-1}	$\Delta_c S^0$/J K mol^{-1}
		Benzonitrile		
calix[4]	2.39 ± 0.04	-13.65 ± 0.20	-27.33 ± 0.02	-45.9 ± 0.7
p-tert-butylphenol	0.70 ± 0.01	-3.99 ± 0.03	-21.72 ± 0.06	-59.5 ± 0.2
		Nitrobenzene		
calix[4]	1.57 ± 0.01	-8.97 ± 0.03	-40.25 ± 0.01	-105.0 ± 0.1
p-tert-butylphenol	1.18 ± 0.01	-6.73 ± 0.04	-29.09 ± 0.01	-75.0 ± 0.1

TABLE V. Thermodynamic parameters for the transfer of triethylamine–*p-tert*-butylcalix[4]arene adduct from benzonitrile to nitrobenzene at 298.15 K.

$\Delta_c G^0$ (PhCN)^a kJ mol^{-1}	$\Delta_c G^0$ (PhNO$_2$)^a kJ mol^{-1}	$\Delta_t G^0$ (PhCN $\rightarrow$ PhNO$_2$)/kJ mol^{-1}		
		Et$_3$N^b	calix[4]^c	R$_3$N·calix[4]^d
-13.65	-8.97	-1.11	-7.34	-3.77
$\Delta_c H^0$ (PhCN)^a kJ mol^{-1}	$\Delta_c H^0$ (PhNO$_2$)^a kJ mol^{-1}	$\Delta_t H^0$ (PhCN $\rightarrow$ PhNO$_2$)/kJ mol^{-1}		
		Et$_3$N^b	calix[4]^c	R$_3$N·calix[4]^d
-27.33	-40.25	1.67	-0.47	-11.72
$\Delta_c S^0$ (PhCN)^a J K^{-1} mol^{-1}	$\Delta_c S^0$ (PhNO$_2$)^a J K^{-1} mol^{-1}	$\Delta_t S^0$ (PhCN $\rightarrow$ PhNO$_2$)/J K^{-1} mol^{-1}		
		Et$_3$N^b	calix[4]^c	R$_3$N·calix[4]^d
-45.9	-105.0	9.3	23.0	-26.7

^{a}From Table IV. ^{b}See text. ^{c}From Table I. ^{d}Calculated via cycle (Equation 10).

for triethylamine and the tetramer, their favourable transfer to nitrobenzene is entropy driven, while for the adduct, the process is enthalpy controlled and takes place with a loss of entropy. Since the transfer enthalpy for triethylamine ($\Delta_t H^0$ = 0.67 kJ mol^{-1}), these findings unambiguously demonstrate that the difference observed in the $\Delta_c H^0$ values in these solvents (Table V) is mainly attributed to the higher enthalpic stability of the adduct in nitrobenzene relative to benzonitrile.

It is therefore concluded that the interaction of *p-tert*-butylcalix[4]arene and triethylamine in these solvents occurs predominantly through hydrogen bonding or ion-pair formation. So far for the tetramer, the thermodynamic data fits a one step process which do not distinguish between hydrogen bonding or ion-pair formation. However, due to the increased acidity of the hexamer and the octamer relative to the tetramer in these solvents, their interactions with amines result in the formation of ions and ion-pairs. Thus, for *p-tert*-butylcalix[4]arenes (n = 6, 8) and this amine thermodynamic data fit a two-step process where the formation of ions are considered and these will be shortly reported. We are now proceeding with heat capacity measurements in order to gain further information on these systems.

Finally, it must be said that in this paper discussion on previous work carried out with these (or similar) systems [22, 38–40] in different media has been deliberately omitted since the behaviour of amines and mainly calixarenes depends crucially upon the medium and for the purpose of useful comparison, potentiometric, conductimetric and thermodynamic data in the relevant solvents are required.

Acknowledgement

J.W., F.J.S.V. and D.A.P.T. thank the EC for financial support and I.G.O. the Ministry of External Relations, Spain, for a fellowship.

References

1. C. J. Pedersen: *J. Incl. Phenom.* **6**, 341 (1988).
2. J. M. Lehn: *J. Incl. Phenom.* **6**, 351 (1988).
3. A. F. Danil de Namor and L. Ghousseini: *J. Chem. Soc., Faraday Trans. 1* **80**, 2349 (1984); **81**, 781 (1985); **82**, 3275 (1986).
4. A. F. Danil de Namor: *J. Chem. Soc., Faraday Trans. 1* **84**, 2441 (1988).
5. A. F. Danil de Namor: *Pure Appl. Chem.* **61**, 2121 (1990).
6. A. Zinke and E. Ziegler: *Ber.* **77B**, 264 (1944).
7. C. D. Gutsche and R. Muthukrishnan: *J. Org. Chem.* **43**, 4905 (1978).
8. A. F. Danil de Namor: *Pure Appl. Chem.* **65**, 193 (1993).
9. R. Ungaro and A. Pochini: p. 127 and M.J. Schwing-Weill and M. A. McKervey: p. 149 in *Calixarenes: A Versatile Class of Macrocyclic Compounds*, J. Vicens and V. Böhmer (Eds.), Kluwer Academic Publishers, Dordrecht (1990).
10. G. Barrett, M. A. McKervey, J. F. Malone, A. Walker, F. Arnaud-Neu, L. Guerra, M. J. Schwing-Weill, C. D. Gutsche, and D. R. Stewart: *J. Chem. Soc., Perkin Trans. 2* 1475 (1993) and references therein.
11. A. F. Danil de Namor, N. Apaza de Sueros, M. A. McKervey, G. Barrett, F. Arnaud-Neu, and M. J. Schwing-Weill: *J. Chem. Soc., Chem. Commun.* 1546 (1991).
12. A. F. Danil de Namor, M. T. Garrido Pardo, L. Muñoz, D. A. Pacheco Tanaka, F. J. Sueros Velarde, and M. C. Cabaleiro: *J. Chem. Soc., Chem. Commun.* 855 (1992).
13. A. F. Danil de Namor, P. M. Blackett, M. T. Garrido Pardo, D. A. Pacheco Tanaka, F. J. Sueros Velarde, and M. C. Cabaleiro: *Pure Appl. Chem.* **65**, 415 (1993).
14. A. F. Danil de Namor, M. T. Garrido Pardo, D. A. Pacheco Tanaka, F. J. Sueros Velarde, J. D. Cárdenas García, M. C. Cabaleiro, and J. M. A. Al Rawi: *J. Chem. Soc., Faraday Trans.* (special issue in Biophysical Chemistry) **89**, 2727 (1993).
15. A. F. Danil de Namor, M. C. Cabaleiro, B. M. Vuano, M. Salomon, O. I. Pieroni, D. A. Pacheco Tanaka, C. Y. Ng, M. A. Llosa Tanco, N. M. Rodríguez, J. Cárdenas García, and A. R. Casal: *Pure Appl. Chem.* **66**, 435 (1994).
16. A. F. Danil de Namor, F. J. Sueros Velarde, L. Pulcha Salazar, (work in progress) (1993).
17. A. F. Danil de Namor, M. A. Llosa Tanco, M. Salomon, and C. Y. Ng: *J. Phys. Chem.* **98**, 11796 (1994).
18. Commission V.8, IUPAC: *Critical Evaluation on Thermodynamic Aspects of Macrocyclic Chemistry* (1993), work in progress.
19. J. Suurkuusk and I. Wadsö: *Chem. Scr.* **20**, 155 (1982).
20. A. F. Danil de Namor: unpublished results.
21. D. Hallen and I. Wadsö: *Pure Appl. Chem.* **61**, 123 (1989).
22. L. J. Bauer and C. D. Gutsche: *J. Am. Chem. Soc.* **107**, 6063 (1985).
23. A. F. Danil de Namor, F. J. Sueros Velarde and I. Gómez Orellana (unpublished work).

24. R. C. Weast (Ed.): *Handbook of Chemistry and Physics*, 62nd edn., CRC Press (1981).
25. A. F. Danil de Namor, R. Traboulssi, and M. Salomon: *J. Chem. Soc., Faraday Trans. 1* **86**, 2193 (1990).
26. M. Bastos, S. Hägg, P. Lönnbro, and I. Wadsö: *J. Biochem. Biophys. Methods* **23**, 255 (1991).
27. S. L. Randzio and J. Suurkuusk: in *Biological Microcalorimetry*, A. E. Beezer (Ed.), Academic Press, London (1980).
28. L. E. Briggner and I. Wadsö: *J. Biochem. Biophys. Methods* **22**, 101 (1991).
29. A. F. Danil de Namor, R. Traboulssi, and D. F. V. Lewis: *J. Am. Chem. Soc.* **112**, 8442 (1990).
30. R. Fernandez Prini: in *Physical Chemistry of Organic Solvents*, A. K. Covington and T. Dickinson (Eds.), Plenum Press (1973).
31. P. Walden: *Z. Phys. Chem.* **55**, 207 (1906); **78**, 257 (1912).
32. A. F. Danil de Namor and J. Wang: in progress (1993).
33. N. S. Isaacs: *Physical Organic Chemistry*, Longman (1987).
34. P. Gans, A. Sabatini, and A. Vacca: *Inorg. Chim. Acta* **18**, 237 (1976).
35. P. H. Bevington: *Data Reduction and Error Analysis in the Physical Sciences*, McGraw-Hill, New York (1969).
36. W. H. Lawton and E. A. Sylvestre: *Thermometrics* **13**, 461 (1971).
37. C. Bavoux and M. Perrin: *J. Incl. Phenom.* **14**, 247 (1992).
38. C. D. Gutsche, M. Iqbal, and I. Alam: *J. Am. Chem. Soc.* **109**, 4314 (1987).
39. V. Böhmer and J. Vicens: in *Calixarenes: A Versatile Class of Macrocyclic Compounds*, J. Vicens and V. Böhmer (Eds.), Kluwer Academic Publishers, Dordrecht (1990).
40. G. Görmar, K. Seiffarth, M. Schulz, and C. L. Chachimbombo: *J. Prakt. Chem.* **333**, 475 (1991).

Liquid Crystalline Calixarenes*

TIMOTHY M. SWAGER** and BING XU
Department of Chemistry, University of Pennsylvania, Philadelphia, PA 19104-6323, U.S.A.

(Received in final form: 3 March 1994)

Abstract. Bowl-shaped (bowlic) liquid crystals are reviewed and new bowlic materials containing rigid tungsten-oxo calix[4]arene based cores are discussed. Tungsten-oxo calix[4]arenes with 8 and 12 dodecyloxy sidechains have been investigated and exhibit bowlic columnar phases which are stable over approximately a 200° temperature range. The uncomplexed tetra-phenol ligands display only a transient liquid crystallinity on the first heating, and the conformational rigidity provided through tungsten-oxo complexation is necessary for well behaved mesomorphism. For the 8 sidechain analog the clearing point is at 320°C and the addition of four more sidechains results in a lower clearing point at 267°C. Polarized optical microscopy and DSC indicate that the 12 sidechain analog displays a phase with the columns packed in a hexagonal lattice which is conductive to the formation of polar phases. Both complexes exhibit a pronounced tendency to bind Lewis base guests in their cavities, and DMF forms very strong complexes which were spectroscopically characterized. The DMF guest produces large effects on the phase behavior by suppressing mesomorphism and lowering the isotropic points by 115°C and 84°C for the 8 and 12 sidechain compounds respectively. This extreme sensitivity to the DMF guest is conclusive proof that bowlic tungsten-oxo calix[4]arene liquid crystals organize in head-to-tail structures.

Key words: Mesomorphism, calixarenes, bowlic liquid crystals, nonlinear optics, ferroelectricity

1. Introduction

Much of the interest in calix[4]arenes has arisen from their ability to form bowl (cone) shapes. Bowl-shaped compounds have been key materials in the development of the field of molecular recognition since they exhibit cavities with high degrees of preorganization necessary for strong binding of guest molecules. The majority of the research in this vast effort has been fueled by those wishing to mimic catalysis and signaling processes in nature. As a result, less attention has been given to the application of these types of compounds in the design of new materials. Nevertheless there is growing interest in the integration of host–guest design principles to create novel materials [1]. While this area of research is still young, it is clear that bowl-shaped compounds present unique opportunities in the design of novel new materials. Notable studies include the demonstration that incorporating nonlinear optical (NLO) chromophores into calix[4]arenes can produce materials with greater transparency without loss of hyperpolarizability [2]. Hence calix[4]arenes offer an approach to a long-standing goal of the NLO com-

---

* This paper is dedicated to the commemorative issue on the 50th anniversary of calixarenes.
** Author for correspondence.

munity. Another notable example is the use of calix[6]arenes for the formation of porous materials for separation technologies [3]. The focus of this article will be on the use of calixarene-based liquid crystals for the formation of bowl-shaped (bowlic) liquid crystals [4].

2. Liquid Crystallinity and Bowlic Compounds

The liquid crystalline (mesomorphic) state is produced by a combination of dispersive and attractive forces [5]. In this state of matter, intermolecular interactions are sufficiently weakened to force a substance into a liquid state but attractive dipolar interaction hold the 'molten molecules' into a low dimensional lattice which lacks long range positional order. The weakening of the intermolecular interactions can be accomplished by heating the material which creates molecular motions or by the addition of a liquid component which partially solvates the material. Liquid crystalline phases formed by heating and solvation are termed thermotropics and lyotropics respectively and the remainder of our discussion will focus on thermotropic materials.

Molecular shape plays a large role in determining the structure of a liquid crystalline phase (mesophase). Typical liquid crystals contain a rigid polarizable core (mesogen) to which flexible sidechains are attached. The critical shape is a composite of the mesogen and the sidechains. Molecules with a rod-like (calamitic) shape generally display nematic and/or smectic phases. Nematics are the least ordered liquid crystal phases and the molecules have only directional order. In smectics the molecules have directional order but they also organize into two dimensional or lamellar superstructures [6]. Molecules which present a circular or disc (discotic) shape most often organize into phases in which the molecules stack in columns [7]. While there is only liquid-like order between molecules in the columns, the columns pack in a two dimensional superstructure which may have very long ($> 10^3$ Å) correlation lengths (Figure 1).

The bowlic liquid crystals investigated to date (Figure 2) are related to discotic materials since a projection of their mesogenic cores into the plane parallel to their greatest aspect ratio results in a circular shape. This circular shape is responsible for the fact that they all exhibit columnar structures in their liquid crystalline phases. There are however important differences since bowlic molecules lack the mirror symmetry through the mesogenic core common to most discotic molecules. In other words, the bowlic mesogen has inequivalent faces which are concave and convex. The first examples of bowlic crystals were based on the cyclotriveratrylene mesogen **1** [8] and these compounds have received the most extensive study [9]. A crystal structure on one derivative showed the bowls have head-to-tail or ferroelectric order within the column [8a]. However, this is not proof of head-to-tail order in the mesophase since the organization in the liquid crystal phase is not necessarily the same as that in the crystal phase. NMR investigations of the dynamic processes in these materials suggest that the motion of these compounds in the liquid crystal

LIQUID CRYSTALLINE CALIXARENES

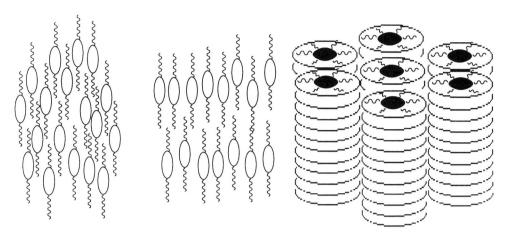

Fig. 1. Calamitic liquid crystals in nematic (left) and smectic (center) phases. Discotic liquid crystals in a columnar phase (right).

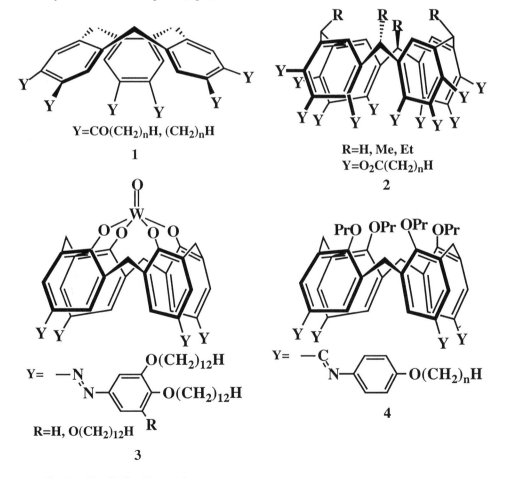

Fig. 2. Bowlic liquid crystals.

phase may be more restricted than most liquid crystals [9e]. Based upon these results Zimmerman suggested that these materials "are very similar to soft solids" [9e]. Indeed the entropy changes associated with the liquid crystal to isotropic phase transitions of these bowlics are higher than those observed for most discotics and indicate that the liquid crystal phase is more highly organized. However it is clear that dynamic processes are present in these materials, since by isolating chiral analogs of **1** and monitoring the racemization, it has been determined that the cores undergo inversion in the mesophase [9c]. There have also been a number of studies on pyrogallol based calixarenes **2** [10]. In this case single crystal X-ray studies and structural investigations of the mesophases reveal that there is head-to-head or antiferroelectric order between neighboring cores within the columns. We have investigated liquid crystals containing rigid tungsten-oxo calix[4]arene cores **3** [11]. These materials are the only metal containing bowlic liquid crystals and will be discussed in more detail in the following sections. More recently another class of bowlic calix[4]arenes **4** were reported to display a columnar mesophase [12]. The same group had previously reported that related calix[8]arene compounds are liquid crystalline [13].

3. Polar Order in Bowlic Liquid Crystals

Our interest in bowlic liquid crystals has arisen from the proposal that bowl shaped molecules may exhibit polar (noncentrosymmetric) organization in the liquid crystalline phases [4, 8, 9]. Indeed bowlic liquid crystals are natural noncentrosymmetric building blocks since a head-to-tail organization maximizes the interactions between bowlic cores. New methodologies for the creation of noncentrosymmetric structures in molecular solids and liquids are critical to the development of new materials with ferroelectric and second order nonlinear optical (NLO) properties [14, 15]. Liquid crystalline methods are particularly attractive since liquid crystalline materials are easily deposited for device construction and are readily aligned.

While the bowlic structure is conductive to the formation of columns with head-to-tail or ferroelectric order, the presence of polar columns does not necessarily generated a bulk polar structure. Indeed it is generally difficult to generate thermodynamically stable polar assemblies due to the tendency of materials to avoid internal electric fields and adopt antiferroelectric structures. To overcome this obstacle and find an optimal balance of interactions which will generate a ferroelectric liquid crystalline state, we have been interested in hexagonal columnar superstructures with axial polarity [16]. The importance of this structure is illustrated for polar bowlic mesogens in Figure 3 and as shown the triangular symmetry of a hexagonal lattice cannot accommodate bulk antiferroelectric order.

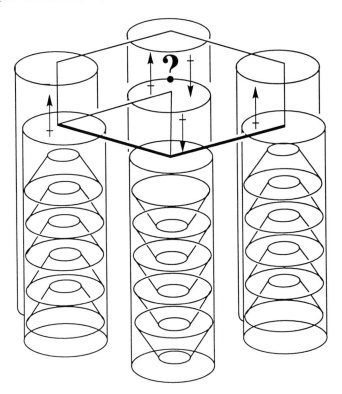

Fig. 3. Demonstration of frustration in a hexagonal columnar structure in which bowlic mesogens have a head-to-tail arrangement and thereby create dipoles within each column. The triangular symmetry of the lattice will not allow the dipoles to organize into an antiferroelectric state and forces either a paraelectric or a ferroelectric state.

4. Liquid Crystallinity in Tungsten-Oxo Calix[4]arenes

Our attention was drawn to calix[4]arenes due to the propensity for bowlic conformations and the extensive reaction chemistry of this class of macrocycles. In the synthesis of liquid crystalline materials we generally find that reactions subsequent to attaching the sidechains must be high yielding to easily obtain pure materials. We have been able to satisfy this requirement using diazonium addition reactions to calix[4]arene [17] which we have found to proceed in nearly quantitative yields. Hence we have synthesized azo-substituted calix[4]arene derivatives **5** with dodecycloxy sidechains as shown in Figure 4. Subsequent complexation with tungsten was performed under conditions similar to those reported for the parent t-butylcalix[4]arene and also proceeded in excellent yield [18].

^{1}H-NMR of the uncomplexed calix[4]arene **5** shows it to exhibit a bowl conformation in solution. However the bridging methylenes are broadened indicating dynamic behavior. As a result it is likely that recrystallization from solution produces a crystal phase which exhibits a bowl conformation. When heated the virgin crystals melt to give a fluid phase which is birefringent when viewed with a polar-

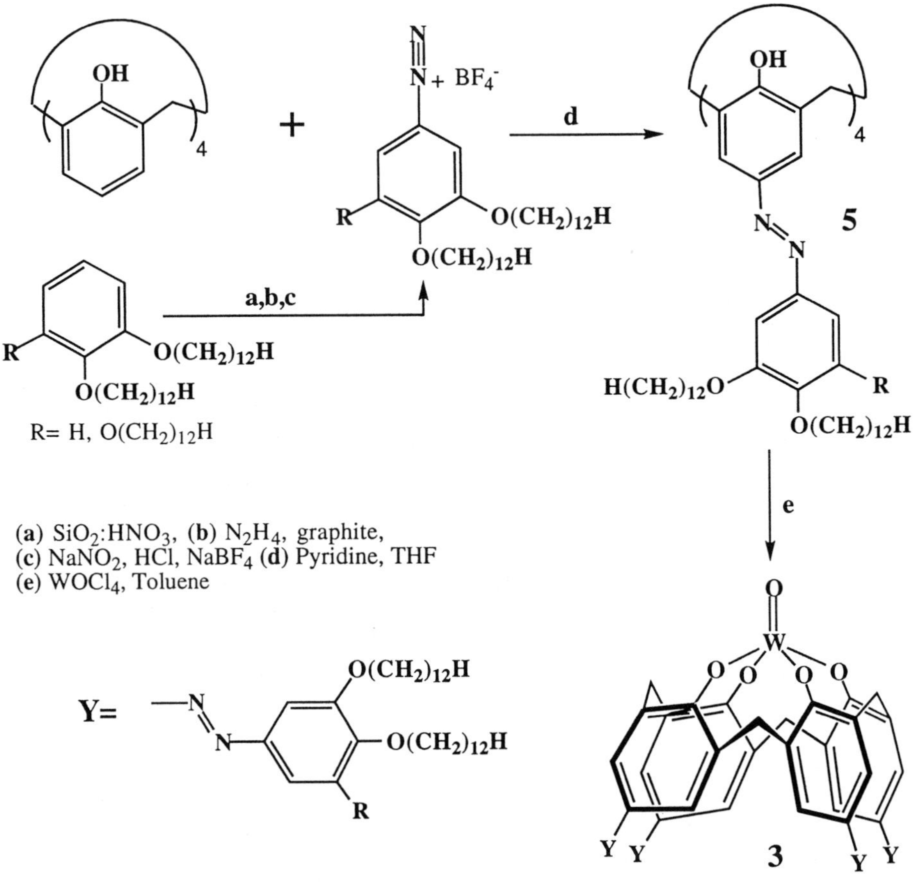

Fig. 4. Synthesis of liquid crystalline tungsten-oxo based calixarenes.

(a) SiO_2:HNO_3, (b) N_2H_4, graphite, (c) $NaNO_2$, HCl, $NaBF_4$ (d) Pyridine, THF (e) $WOCl_4$, Toluene

ized microscope. The fluidity and the birefringence of this phase suggest that it is liquid crystalline. However, this mesophase is transient since it is only observed on first heating. Once heated into the isotropic phase, these compounds loose their mesomorphism and cooling produces only nonbirefringent glasses. As illustrated in Figure 5 an explanation of this phenomena is that the initial crystal phase exhibits a bowl conformation which is compatible with liquid crystalline behavior, but when heated into the isotropic phase a mixture of conformations are produced which prevent a return to the liquid crystalline phase.

Capping **5** with a tungsten-oxo group to form **3** produces a mesogen with a rigid bowl conformation. These compounds exhibit well-behaved liquid crystallinity which is regained after heating into the isotropic phase. These results are also consistent with the assumption that the bowl conformation is necessary for liquid crystalline behavior. We note that others have also arrived at similar conclusions, since in the case of **4** the bulky groups on the lower rim were necessary to preserve the cone conformation and the liquid crystalline behavior [12].

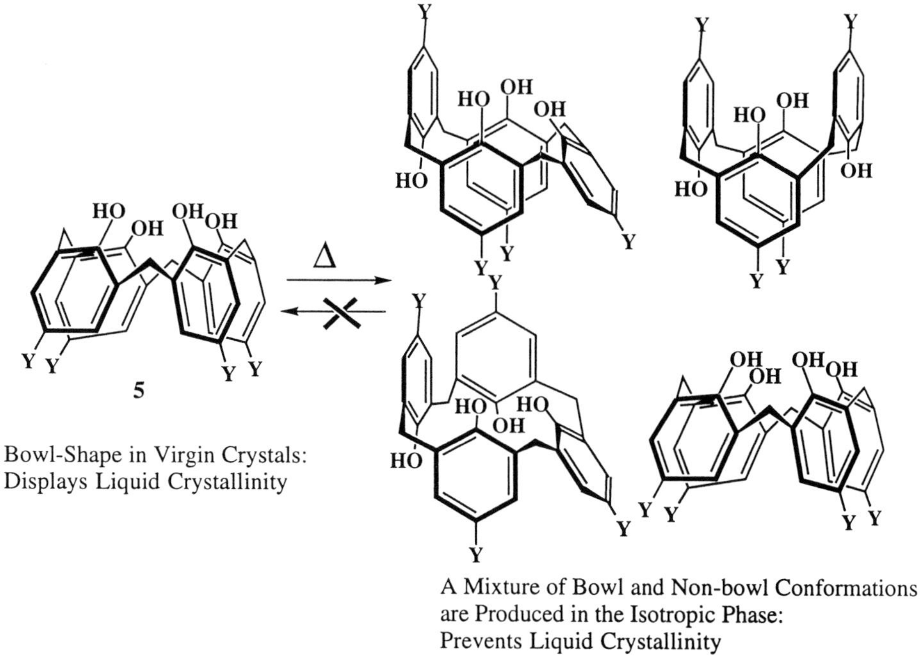

Fig. 5. Explanation of the transient liquid crystalline behavior of **5**.

Bowl-Shape in Virgin Crystals: Displays Liquid Crystallinity

A Mixture of Bowl and Non-bowl Conformations are Produced in the Isotropic Phase: Prevents Liquid Crystallinity

R=H $\quad$ K $\xrightarrow{\underset{19.8 \text{ kcal/mol}}{136°C}}$ M $\xrightarrow{\underset{1.5 \text{ kcal/mol}}{320°C}}$ I_d

R=OC$_{12}$H$_{25}$ $\quad$ K $\xrightarrow{\underset{7.9 \text{ kcal/mol}}{54°C}}$ M$_1$ $\xrightarrow{\underset{0.6 \text{ kcal/mol}}{77°C}}$ M$_2$ $\xrightarrow{\underset{5.1 \text{ kcal/mol}}{267°C}}$ I

Fig. 6. Transition temperatures and enthalpies of **3**.

Both derivatives of **3** display columnar mesophases which are stable over a very wide temperature range as shown in Figure 6. The nature of the mesophases has been investigated by DSC and optical textures were viewed with a polarizing microscope. The mesophase to isotropic transition enthalpies are fairly small (1.5 kcal/mol and 5.1 kcal/mol for the 8 and 12 sidechain analogs respectively) indicating that these mesophases are likely disordered phases with liquid-like correlation between mesogens. The structure of the mesophase of **3** (R=H), M could not be determined conclusively by an optical microscope, however the textures were indicative of a columnar structure. The R=OC$_{12}$H$_{25}$ derivative displays two columnar phases M$_1$ and M$_2$, and the additional four sidechains lowers the melting and isotropic transitions by 82° and 53° respectively. Miscibility studies determined that M$_1$ and M$_2$ have different structures. By slow cooling of the isotropic phase an optical texture of the M$_2$ phase developed which displays digitized contours, leaf patterns, and large regions of uniform extinction. The large areas of extinction are suggestive of a uniaxial structure, and the observation of digitized stars with six-fold symmetry and 120° facets confirms a hexagonal arrangement of the columns in M_2. We therefore characterize M_2 as a hexagonal bowlic phase B_h.

The combination of the rigid cavity and the Lewis acidic nature of the square pyramidal tungsten produces a pronounced tendency for 3 to display host–guest interactions with Lewis bases. Although a number of compounds may serve as guests, we have focused on DMF complexes due to their strong association and 1 : 1 complexes are obtained by recrystallization from anhydrous DMF. Infrared spectroscopy confirms that DMF forms a Lewis base complex. The W=O band undergoes a characteristic shift from 1074 cm^{-1} (^{18}O 1019 cm^{-1}) to 990 cm^{-1} (^{18}O 938 cm^{-1}) with DMF complexation. Likewise, the C=O band of the DMF of the host–guest complex in CH_2Cl_2 solution occurs at 1645 cm^{-1} which is 25 cm^{-1} lower in energy than uncomplexed DMF in CH_2Cl_2. NOESY NMR experiments conclusively show the DMF to occupy the cavity and all of the DMF protons experience a large up-field shift consistent with the shielding environment provided by the cavity.

The most dramatic consequence of host–guest complexation is its effect upon the mesomorphic behavior of the complexes. While the R=H analog exhibits a discotic mesophase from 135° to 330°C, its DMF complex melts directly to an isotropic phase at 115°C ($\Delta H = 12.3$ kcal/mol). Likewise, the R=OC$_{12}$H$_{25}$ DMF complex melts to form an isotropic phase at 84°C. Complexes with pyridine guests were found to exhibit isotropic transitions at the same temperatures, indicating that a filled cavity is more important than the nature of the guest. With further heating (200°C to 250°C), the DMF complex slowly dissociates to form the liquid crystalline phase.

The deleterious effect of a DMF or pyridine guest on the mesophase stability indicates that occupation of the cavity is critical. This fact, combined with the columnar structure of the mesophases, is proof that the bowlic cores exhibit a head-to-tail arrangement whereby tungsten-oxo groups protrude into the cavity of the neighboring mesogen (Figure 7). Hence, in 3 with R=OC$_{12}$H$_{25}$ we have an example of a material which exhibits ferroelectric order within the columns and a hexagonal lattice. As discussed earlier these features preclude bulk antiferroelectric order and most likely force the material into a frustrated paraelectric state. Bulk ferroelectric order may be introduced by formation of the mesophase in the presence of an electric field.

5. Conclusion

Calixarenes are attractive materials for the design of new materials. The established synthetic procedures for their functionalization, their bowl shapes, and their host–guest chemistry offers a wealth of opportunities for the design of novel materials. In this report we have focused upon the use of the bowlic shape to generate materials with polar order. However, there are many other aspects of these materials

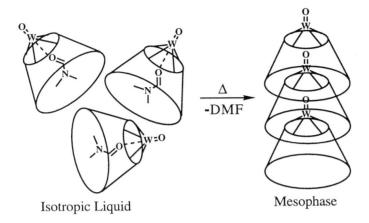

Fig. 7. Illustration of how blocking of the cavity by the DMF guest prevents mesophase formation. Heating to temperatures greater than 200°C liberates DMF and allows mesophase formation.

which may impart useful effects in materials design. For example, we found that the isotropic point of **3** is depressed dramatically through host–guest complexation. This type of process may be useful in processing materials, since it provides a route to a low viscosity state at reduced temperatures. As a result host–guest complexes of very high melting bowlic complexes may be poled at lower temperatures as host–guest melts and subsequent dissociation of the guest will produce noncentrosymmetric materials with high temporal stability.

Acknowledgment

We are grateful for financial support from the Office of Naval Research.

References

1. H.-J. Schneider and H. Dürr (Eds.): *Frontiers in Supramolecular Organic Chemistry and Photochemistry*, VCH (1991).
2. E. Kelderman, L. Derhaeg, G. J. T. Heesink, W. Verboom, J. R. J. Engbersen, N. R. van Hulst, A. Persoons, and D. N. Reinhoudt: *Angew. Chem. Int. Ed. Engl.* **31**, 1075 (1992).
3. M. D. Conner, V. Janout, I. Kudelka, P. Dedek, J. Zhu, and S. L. Regen: *Langmuir* **9**, 2389 (1993).
4. L. Lei: *Mol. Cryst. Liq. Cryst.* **146**, 41 (1987).
5. G. W. Gray (Ed.): *Thermotropic Liquid Crystals: Critical Reports on Applied Chemistry Vol. 22*, Society of Chemical Industry (1987).
6. G. W. Gray and J. W. G. Goodby: *Smectic Liquid Crystals: Textures and Structures*, Leonard Hill Pub., Glasgow (1984).
7. S. Chandrasekhar and G. S. Ranganath: *Rep. Prog. Phys.* **53**, 57 (1990).
8. These materials where reported nearly simultaneously by two groups. (a) J. Malthete and A. Collet: *Nouv. J. Chim.* **9**, 151 (1985). (b) H. Zimmerman, R. Poupko, Z. Luz, and J. Billard: *Z. Naturforsch., A: Phys., Phys. Chem. Kosmophys.* **40A**, 149 (1985).
9. (a) H. Zimmerman, R. Poupko, Z. Luz, and J. Billard: *Z. Naturforsch., A: Phys., Phys. Chem. Kosmophys.* **41A**, 1137 (1986). (b) A.-M. Levelut, J. Malthete, and A. Collet: *J. Physique* **47**,

351 (1986). (c) J. Malthete and A. Collet: *J. Am. Chem. Soc.* **109**, 7544 (1987). (d) J. M. Buisine, H. Zimmermann, R. Poupko, Z. Luz, and J. Billard: *Mol. Cryst. Liq. Cryst.* **151**, 391 (1987). (e) R. Poupko, Z. Luz, N. Spielberg, and H. Zimmermann: *J. Am. Chem. Soc.* **111**, 6094 (1989). (f) W. Kranig, H. W. Spiess, and H. Zimmermann: *Liq. Cryst.* **7**, 123 (1990). (g) L. Wang, Z. Sun, X. Pei, and Y. Zhu: *Chem. Phys.* **142**, 335 (1990).
10. (a) G. Cometti, E. Dalcanale, A. Du vosel, and A.-M. Levelut: *Chem. Commun.* 163 (1990). (b) S. Bonsignore, G. Commetti, E. Dalcanale, and A. Du vosel: *Liq. Cryst.* **8**, 639 (1990). (c) E. Dalcanale, A. Du vosel, A.-M. Levelut, and J. Malthete: *Liq. Cryst.* **10**, 185 (1991). (d) G. Cometti, E. Dalcanale, A. Du vosel, and A.-M. Levelut: *Liq. Cryst.* **11**, 93 (1992). (e) S. Bonsignore, A. Du vosel, G. Guglielmetti, E. Dalcanale, and F. Ugozzoli: *Liq. Cryst.* **13**, 471 (1993).
11. B. Xu and T. M. Swager: *J. Am. Chem. Soc.* **115**, 1159 (1993).
12. T. Komori and S. Shinkai: *Chem. Lett.* 1455 (1993).
13. T. Komori and S. Shinkai: *Chem. Lett.* 901 (1992).
14. J. W. Goodby, R. Blinc, N. A. Clark, S. T. Lagerwall, M. A. Osipov, S. A. Pikin, T. Sakurai, Y. Koshino, and B. Zeks: *Ferroelectric Liquid Crystals: Principles, Properties, and Applications*, Gordon and Breach Science Publishers, Amsterdam (1991).
15. S. R. Marder, J. E. Sohn, and G. D. Stucky (Eds.): *Materials for Nonlinear Optics: Chemical Perspectives*, ACS Symposium Series 455, Washington (1991).
16. (a) H. Zheng, P. J. Carroll, and T. M. Swager: *Liq. Cryst.* **14**, 1421 (1993). (b) A. G. Serrette and T. M. Swager: *J. Am. Chem. Soc.* **115**, 8879 (1993).
17. S. Shinkai, K. Araki, J. Shibata, D. Tsugawa, and O. Manabe: *J. Chem. Soc., Perkin Trans.* 3333 (1990).
18. F. Corazza, C. Floriani, A. Chiesi-Villa, and C. Rizzoli: *Inorg. Chem.* **30**, 4465 (1991).

Nuclear Waste Treatment by Means of Supported Liquid Membranes Containing Calixcrown Compounds*

C. HILL, J.-F. DOZOL**, V. LAMARE, H. ROUQUETTE, S. EYMARD and B. TOURNOIS
S.E.P./S.E.T.E.D. Centre d'Etudes de Cadarache, Commissariat à l'Energie Atomique (C.E.A.), 13108 Saint-Paul-lez-Durance, France

J. VICENS, Z. ASFARI and C. BRESSOT
Ecole Européenne des Hautes Etudes des Industries Chimiques de Strasbourg, France

and

R. UNGARO and A. CASNATI
Dipartimento di Chimica Organica e Industriale, Università degli Studi di Parma, Italy

(Received: 11 May 1994; in final form: 27 July 1994)

Abstract. Permeability variation with repeated caesium transport experiments has been chosen to measure the leaching of the supported liquid membrane by the contacting aqueous solutions. This allowed us to characterize the SLM stability. Whereas classical crown ethers such as the widely used 21C7 derivatives were revealed to be poorly efficient and poorly stable in SLMs, crown-6-calix[4]arene compounds in the so-called *1,3-alternate* configuration led to very stable (over 50 days), highly selective (concentration factor > 100) and efficient (decontamination factor = 20) SLMs, for the removal of caesium from high salinity and acidity media. These results were achieved by using proper organic diluents and introducing hydrophobic substituents in the frame of the calixarenes.

Key words: Caesium, calixcrown, transport, SLM stability.

1. Introduction

Nuclear fuel reprocessing operations produce both high and medium level activity liquid wastes (HLW/MLW). The major nuclides in these radioactive wastes are those with long half-lives, mainly β/γ emitters such as: Tc, I, Zr, Se, Cs, etc., or α emitters such as transuranics: Np, Pu, Am, Cm, etc. [1]. That is why great efforts have been devoted throughout the world to propose harmless storage of these wastes.

The burial of vitrified reprocessed HLWs (containing fission products and α emitters) has been considered as the safest method for their permanent disposal [2], whereas MLWs are treated by evaporation or other conventional techniques such as

---
* This paper is dedicated to the commemorative issue on the 50th anniversary of calixarenes.
** Author for correspondence.

chemical precipitation, ion exchange, etc., in order to concentrate their radioactivity into the smallest possible volume [3]. This treatment nevertheless leads to large volumes of concentrates composed of active and inactive salts (mainly: $NaNO_3$: 4 mol L^{-1} and HNO_3: 1 mol L^{-1} as the matrix). The greater part of these concentrates has to be disposed of in geological formations after embedding due to their activity in long-life radionuclides (actinides, strontium, caesium, etc.).

Therefore it would be desirable to remove these very long-life radionuclides from the contaminated liquid wastes before embedding, in both the above mentioned scenarios of treatment and storages. This would allow, on the one hand, the volume and the radiotoxicity of the wastes to be reduced, and, on the other hand, part of these decontaminated wastes to be directed to surface repositories. This goal could be achieved by separating the long-life radionuclides from the matrix and turning them into short-life radionuclides or at least into nonradioactive elements through transmutation.

One chemical separation process among others such as liquid–liquid extraction by specific extractants, or chemical precipitation to decontaminate HLWs and/or MLWs, could be *coupled transport* through supported liquid membranes (SLM) using specific carriers [4–5].

2. Process Description

A SLM consists of an organic liquid solution (mixture of an organic diluent and a carrier) absorbed by capillary forces onto a microporous support separating two aqueous solutions [6–8]: the first solution, referred to as the *feed solution* containing the permeating ions, is the solution of waste to be decontaminated, and the second solution, referred to as the *stripping solution*, initially free of ions (demineralized water) receives and concentrates the selectively transported radionuclides.

SLMs containing selective carriers (ionophores) give higher fluxes and selectivities than conventional semipermeable porous polymeric membranes, because diffusion is faster in liquids than in solids. The carrier dissolved in the liquid membrane favours the distribution of one species out of a mixture by specific complexation and extraction, if properly chosen [9].

Although they have not yet been industrially applied, SLMs are of great interest both for potential technological applications (in hollow fibers), and for basic research (in flat sheet SLMs) to determine the transport mechanisms. Because they require only small amounts of organic solutions, SLMs allow the study of very sophisticated tailor-made compounds. Unfortunately, one of their major drawbacks is their great instability. This instability is caused by the disfavourable volume ratio between the aqueous solutions and the organic membrane, which leads to substantial partitioning of the diluent and/or the carrier in the aqueous solutions [10].

That is why very lipophilic carriers and very hydrophobic diluents have to be used to avoid their partitioning. Membrane stability can be improved by choosing the proper diluent of high interfacial tension with water and low water solubility,

TABLE I. Physico-chemical characteristics of organic diluents.

Alkyl group	Mol. weight (g mol^{-1})	Density (g cm^{-1})	Dielectric Cst. (Debye)	Viscosity (centipoise)	Surface tension (dynes cm^{-1})
Hexyl	223.3	1.066	25.7	8.9	34.3
Octyl	251.3	1.036	31.8	12.4	34.3

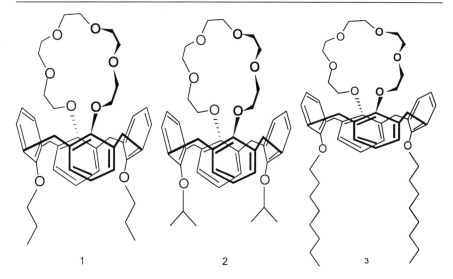

Fig. 1. 1,3-Dialkoxy-2,4-crown-6-calix[4]arenes (compounds **1, 2, 3**) in the *1,3-alternate* conformation.

lower surface tension than the solid microporous support (e.g.: 35 dynes cm^{-1} for polypropylene), high boiling-point, etc., such as Nitrophenylalkylethers (number of carbon atoms > 5) and modified carriers with hydrophobic alkyl or aryl groups [10–12].

3. Experimental

3.1. REAGENTS

The organic diluents: 1,2-nitrophenylhexyl/octylether (*o*-NPH/OE), were synthetized at the CHIMIE PLUS laboratory and used without further purification. Their physico-chemical characteristics are summarized in Table I. The tested mono-crown-6-calix[4]arene compounds **1, 2** and **3** (Figure 1) were synthesized by Professor R. Ungaro of the University of Parma, in the framework of a European Community program collaboration with the French Atomic Energy Commission (CEA). The bis-crown-6-calix[4]arene compounds **4, 5, 6, 7** and **8** (Figure 2) were synthesized by Drs. J. Vicens and Z. Asfari of the University of Strasbourg (E.H.I.C.S.) [13].

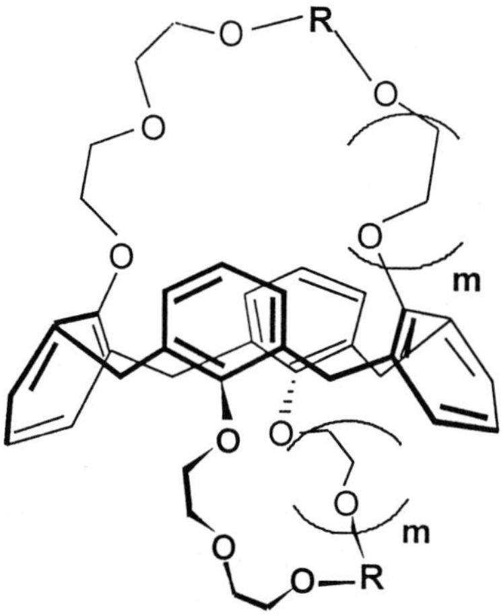

compound 6
m = 2 and R = [tolyl]

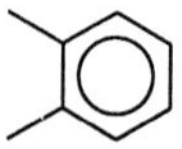

compound 7
m = 2 and R = [naphthyl]

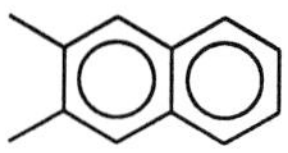

compound 4
R = CH₂—CH₂
and m = 1

compound 5
R = CH₂—CH₂
and m = 2

compound 8
R = CH₂—CH₂
and m = 3

Fig. 2. Bis-crown-calix[4]arenes (compounds **4, 5, 6, 7** and **8**).

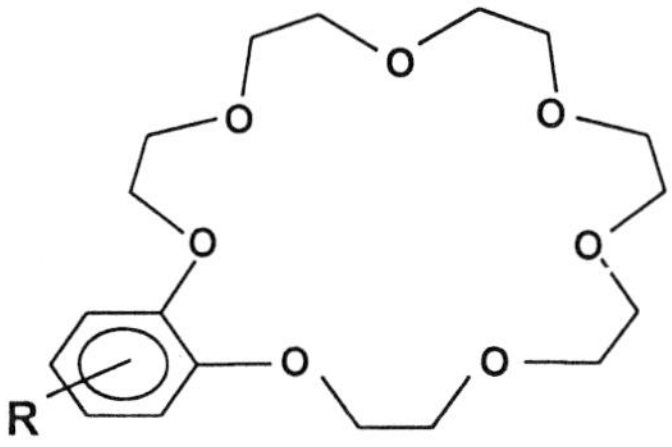

compound 9
R = $C_{10}H_{21}$

compound 10
R = t-Bu

Fig. 3. Compound **9**: *n*-decyl-benzo-21-crown-7. Compound **10**: *t*-butyl-benzo-21-crown-7.

The inorganic salts used to prepare the synthetic feed solutions ($NaNO_3$, HNO_3) were analytical-grade products from Prolabo and Aldrich. Radioactive ^{137}Caesium was provided by the Amersham company.

3.2. CAESIUM EXTRACTION

In the search for specific extractants able to achieve sufficiently high separation rates of the major radionuclides of interest (*vide supra*), substituted calixarenes are new macrocycles in addition to crown ethers or cryptands which promise selective complexation and high extraction yields.

For the removal of caesium from HLW to MLW, for instance, crown ethers (compounds **9** and **10**, Figure 3) are well known for their ability to complex and extract alkali cations from acidic media. The stability of the complexes is linked to the relative size of the cavity of the crown ether as compared to the complexed cation: the most strongly bound cation is the one which best fits into the crown cation, according to the *complementarity principle* [15–18].

Unfortunately, caesium (0.338 nm) which corresponds to the cavity of 21-crown-7 derivatives (0.34–0.43 nm), is not efficiently and selectively extracted from ML concentrates which contain large amounts of sodium salts and nitric acid as the inactive matrix [4].

On the other hand, crown-6-calix[4]arene compounds which consist of a calixarene frame (i.e., phenolic units *meta*-linked by methylene groups) in the so-called *1,3-alternate* conformation (where two opposite phenolic units are flipped upward and bridged with a polyethylene glycol chain of various length, Figure 1, compounds **1, 2, 3**), simultaneously offer the selectivity of crown compounds through the polyether-chain ring size, and the preorganization of calixarenes fixed in a more or less rigid conformation [19–21].

The two remaining phenolic units are either derived with long alkyl chains to enhance carrier lipophilicity or bridged with a second polyethylene glycol chain to rigidify the calixarene frame leading to bis-crown-calix[4]arene (Figure 2: compounds **4, 5, 6, 7** and **8**).

Although these latter compounds present two potential complexing sites, complex stoichiometry studies have proved that it was of the 1 : 1 type (calixarene : caesium cation). Moreover, introducing aryl substituents such as phenyl or naphthyl into the polyether ring(s) (**6, 7**) greatly enhances the *carrier lipophilicity* which favours SLM stability.

These two properties (complementarity and preorganization) lead to great efficiency and selectivity by favouring the distribution of caesium cations over that of sodium cations. Table II summarizes some extraction results obtained with calixcrown compounds and classical crown ethers on simulating synthetic MLW solutions. Extraction experiments were performed by mixing equal volumes (5 to 7 mL) of aqueous and organic solutions in sealed tubes for an hour. Aliquots of each phase were analyzed by γ spectrometry after centrifugation. The selectivity of the tested compounds towards caesium, in the presence of sodium, is expressed as the ratio of the distribution coefficients obtained separately for both cations:

$$\alpha_{Cs/Na} = \frac{D_{Cs}}{D_{Na}} \quad \text{with} \quad D_M = \frac{\sum[\bar{M}]}{\sum[M]}$$

where $\sum[\bar{M}]$ denotes the metal species total concentration in the organic phase and $\sum[M]$ the metal species total concentration in the aqueous phase.

As shown in Table II, calixcrown compounds containing six oxygen atoms in the ring(s) are much more selective than classical crown ethers. The complexes best formed with the least hydrated alkali cations (caesium) are probably stabilized by the π-bonding interactions arising from the phenyl rings of the *1,3-alternate* phenolic units of the calixcrown compounds [22].

TABLE II. Liquid–liquid extraction experiments: selectivity $\alpha_{(Cs/Na)}$ determination. Aqueous feed solution: $M^+(NO_3^-)$: 5×10^{-4} mol L^{-1} in HNO_3: 1 mol L^{-1}. Organic solution: extracting agent: 0.01 mol L^{-1} in o-NPHE.

No.	Extracting agent used	D_{Na}	D_{Cs}	$\alpha_{Cs/Na}$
1	1,3-dipropoxy-2,4-crown-6-calix[4]arene	1.8×10^{-3}	19.2	10 500
2	1,3-diisopropoxy-2,4-crown-6-calix[4]arene	$< 10^{-3}$	28.5	$> 28\,500$
3	1,3-di(n-octyloxy)-2,4-crown-6-calix[4]arene	$< 10^{-3}$	33	$> 33\,000$
4	bis-crown-5-calix[4]arene	2.3×10^{-3}	0.5	220
5	bis-crown-6-calix[4]arene	1.3×10^{-2}	19.5	1500
6	bis-(1,2-benzo-crown-6)-calix[4]arene	1.7×10^{-3}	32.5	19 000
7	bis-(1,2-naphtho-crown-6)-calix[4]arene	$< 10^{-3}$	29.5	$> 49\,000$
8	bis-crown-7-calix[4]arene	$< 10^{-3}$	0.3	> 300
9	n-decyl-benzo-21-crown-7	1.3×10^{-3}	0.3	250
10	t-butyl-benzo-21-crown-7	1.3×10^{-3}	0.3	250

3.3. TRANSPORT EXPERIMENTS

The transport of ions through the SLM occurs because a chemical gradient is established between the *feed solution* and the *stripping solution*. The use of neutral carriers such as calixcrown compounds leads to the coupled cotransport of cations and anions through the SLM [20]. When concentrates or fission product solutions are used as the *feed solution* and demineralized water as the *stripping* solution, the concentration gradient of the nitrate anions will force the transport of caesium cations against their own concentration gradient, thus leading to a concentration of radioactive caesium in the stripping solution. Nevertheless, the basicity as well as the polarity of the organic diluent that we have chosen to use in order to improve caesium extraction – by a better solvation of the complex paired anion in the membrane (nitrates are hydrated anions not easily extracted in major conventional organic solvents) – lead to substantial transport of nitric acid from the feed to the stripping solution. This reduces caesium permeation through the SLM by decreasing the nitrate concentration gradient.

3.3.1. *Permeability Determination*

The transport of caesium from synthetic aqueous solutions of $NaNO_3$: 4 mol L^{-1}/HNO_3: 1 mol L^{-1}, spiked with ^{137}Cs (# 2000 kBq L^{-1}) was followed by regular measurement of the decrease of radioactivity in the feed solution by γ spectrometry analysis, using a detection chain from Intertechnique, equipped with germanium detectors. The counting was always sufficiently long to insure a relative error in the activity measurements of less than 5%. This allowed graphical determination of the constant permeabilities P_M (cm h^{-1}) of caesium cation permeation through the SLM for 6–7 h by plotting the logarithm of the ratio C/C^0 versus time, as described in the model of mass transfer proposed by P. Danesi [5]:

$$\ln\left(\frac{C}{C^0}\right) = -\frac{\varepsilon S}{V} P_M t, \qquad (1)$$

where C = concentration of the cation in the feed solution at time t (mol L^{-1}); C^0 = initial concentration of the cation in the feed solution (mol L^{-1}); ε = volumic porosity of the SLM (%); S = membrane surface area (cm^2); V = volume of feed and stripping solution (mL); and t = time (h).

If experiments are performed under the same conditions (stirring rates, concentration gradients) determination of the permeabilities P_M allows the selectivity and efficiency of different tested carriers to be characterized and quantified.

With certain assumptions (limited diffusion and controlled process), the permeabilities P_M can be evaluated by the formula:

$$P_M = \frac{D_M}{D_M \Delta_a + \Delta_o}, \qquad (2)$$

where D_M is the distribution coefficient of the permeating cation; Δ_a is the ratio of the thickness of the aqueous diffusion boundary layer to the aqueous diffusion coefficient; Δ_o is the ratio of the thickness of the membrane to the organic diffusion coefficient.

Expression (2) shows that under certain conditions where Δ_a is constant throughout the experiments and $D_M \Delta_a$ negligible in front of Δ_o, P_M is proportional to D_M (and D_M directly depends on the actual organic carrier concentration in the SLM).

When repeated transport experiments are performed, in which both the aqueous feed and receiving solutions are renewed every day while the membrane remains the same as in the first run, daily partitioning of the carrier from the SLM to the renewed aqueous solutions causes a decrease of the carrier concentration in the membrane and thus a decrease of D_M and proportionally of P_M. The evolution of the permeability measurements versus the number of runs is a way to describe SLM leaching by the aqueous solutions and characterizes SLM stability in time.

3.3.2. Materials and Device

We used a thin flat sheet SLM device described by T. Stolwijk [23] and shown in Figure 4. The volume of both aqueous solutions was 50 mL. The membrane was a Celgard^{Ä} 2500 (of 25 μm thickness and 45% volume porosity) polypropylene microporous support soaked with a solution 10^{-2} mol L^{-1} of the tested calixarenes in 1,2-nitrophenyloctylether (O-NPOE). The surface area of the membrane was about 15–16 cm^2, depending on the device.

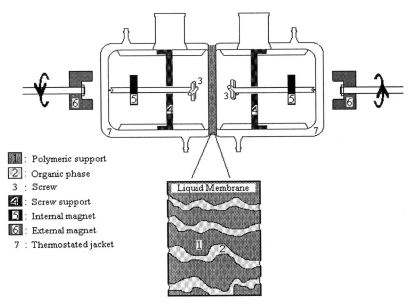

Fig. 4. Flat sheet supported liquid membrane device for transport experiments.

3.3.3. *Results and Discussion*

As summarized in the following diagram (Figure 5), repeated caesium transport experiments show that 1,3-di(*n*-octyloxy)-2,4-crown-6-calix[4]arene (compound **3**), is more efficient and above all more stable in an SLM than the classical crown ether 21-C-7 (which is 50 times more concentrated in *o*-NPOE). The pendant alkyl chains attached to the remaining phenolic units furthermore enhance the calixarene solubility in *o*-NPOE.

Whereas bis-crown-6-calix[4]arene (compound **5**) rapidly leaks out of the SLM ($P_{Cs} < 0.1$ cm h^{-1} after 15 runs) because of its low partition constants between the organic diluent and the aqueous solutions, very good stability and efficiency have been observed with the more lipophilic benzo (compound **6**) and naphtho (compound **7**) derivatives.

Both families of calixcrown compounds allowed selective removal of ^{137}caesium from sodium containing solutions: less than 100 mg of sodium is transported within 24 h for compound **7**, whereas more than 95% of trace level ^{137}caesium is concentrated in the stripping solution. Nitric acid transport, due to the basicity of both the organic diluent and the calixarene, could not be limited to less than 5% (0.05 mol L^{-1}) within 24 h, thus leading to concentration factors (ratio of initial waste concentration to final waste concentration) greater than *100 for a single step process*.

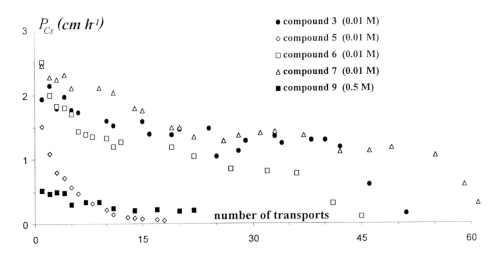

Fig. 5. Caesium permeability measurements for repeated transport experiments.

4. Conclusion

Calixcrown compounds appear to be a new promising family of carriers for the selective removal of caesium from high salinity and acidity media such as MLW through an SLM. By choosing a highly hydrophobic organic diluent, *o*-NPOE, and a lipophilic crown-6-calix[4]arene derivative in the *1,3-alternate* conformation suitable for caesium complexation over that of sodium, very selective and stable SLMs can be obtained (over a period of 50 days).

References

1. J. Lefevre: *Techniques de l'Ingénieurs – Mécanique et Chaleur* **B8 II**, B 36600-1 à B 3361–10 (1990).
2. *Revue Générale Nucléaire*, No. 5 September–October (1992).
3. P. Chauvet and T. Dippel: *Radioactive Waste Management*, Vol. 1, Harwood, London (1979).
4. J. F. Dozol, S. Eymard, R. Gambade, G. La Rosa, and J. Casas i Garcia: *Rapport EUR 13887 FR*.
5. P. R. Danesi: *Sep. Sci. Technol.* **19** (11 & 12), 857 (1984–1985).
6. J. D. Lamb, R. L. Bruening, R. M. Izatt, Y. Hirashima, P.-K. Tse, and J. J. Christensen: *J. Memb. Sci.* **37**, 13 (1988).
7. J. D. Way, R. D. Noble, and B. R. Bateman: *Advances in Chemistry Series*, No. 269 (1985).
8. D. Y. Takigawa: *Sep. Sci. Technol.* **27**(3), 325 (1992).
9. W. F. Van Straaten-Nijenhuis, F. De Jong, and D. Reinhoudt: *Recl. Trav. Chim. Pays-Bas* **112**, 317 (1993).
10. T. B. Stolwijk, E. J. R. Sudhölter, and D. N. Reinhoudt: *J. Am. Chem. Soc.* **111**, 6321 (1989).
11. J. F. Dozol, J. Casas, and A. Sastre: *J. Memb. Sci.* **82**, 237 (1993).
12. J. F. Dozol, H. Rouquette, and B. Tournois: *1993 Intern. Conf. Nuclear Waste Manag. Environ. Remediat.*, Prague, Czech Republic, March (1993).
13. Z. Asfari, S. Pappalardo, and J. Vicens: *J. Incl. Phenom.* **14**(2), 189 (1992).
14. P. R. Brown, J. L. Hallman, L. W. Whaley, D. H. Desai, M. J. Pugia, and R. A. Bartsch: *J. Memb. Sci.* **56**, 195 (1991).
15. I. H. Gerow and M. W. Davis, Jr.: *Sep. Sci. Technol.* **14**(5), 395 (1979).

16. I. H. Gerow, J. E. Smith, Jr., and M. W. Davis, Jr.: *Sep. Sci. Technol.* **16**(5), 519 (1981).
17. W. J. McDowell and G. N. Case: *Anal. Chem.* **64**, 3013 (1992).
18. E. Blasius and K. H. Nilles: *Radiochim. Acta*, Part I, **35**, 173 (1984).
19. E. Ghidini, F. Ugozzoli, R. Ungaro, S. Harkema, A. El-Fadl, and D. N. Reinhoudt: *J. Am. Chem. Soc.* **112**, 6979 (1990).
20. W. Nijenhuis, E. Buitenhuis, F. De Jong, E. Sudhölter, and D. N. Reinhoudt: *J. Am. Chem. Soc.* **113**, 7963 (1991).
21. P. J. Dijkstra, J. Brunink, K. Bugge, D. N. Reinhoudt, S. Harkema, R. Ungaro, F. Ugozzoli, and E. Ghidini: *J. Am. Chem. Soc.* **111**, 7567 (1989).
22. R. Ungaro, A. Casnati, F. Ugozzoli, A. Pochini, J. F. Dozol, and H. Rouquette: *Angew. Chem.* (in press, 1994).
23. T. B. Stolwijk, E. J. R. Sudhölter, and D. N. Reinhoudt: *J. Am. Chem. Soc.* **109**, 7042 (1987).

CONFERENCE DIARY

THIRD INTERNATIONAL CONFERENCE ON CALIXARENES AND RELATED COMPOUNDS
Fort Worth, USA, 21–25 May, 1995.
Further information from: Professor C.D. Gutsche, Department of Chemistry, Texas Christian University, Fort Worth, Texas 76129, USA. Fax: +817-921-7330.

EIGHTH INTERNATIONAL SYMPOSIUM ON INTERCALATION COMPOUNDS
Vancouver, Canada, 28 May–1 June, 1995.
Further information from: ISIC 8, Conference Services, Simon Fraser University, Halpern Centre, Burnaby, British Columbia, Canada V5A 1S6. Fax: +604-291-3420.

INTERNATIONAL SCHOOL OF CRYSTALLOGRAPHY. 22ND COURSE: CRYSTALLOGRAPHY OF SUPRAMOLECULAR COMPOUNDS
Erice, Sicily, 2–12 June, 1995.
Further information from: Professor G. Tsoucaris, CNRS UPR 180, Tour B, Centre Pharmaceutique, 92290 Chatenay-Malabry, France. Fax: +33-1-46-83-13-03.

THIRD GORDON RESEARCH CONFERENCE ON ZEOLITES AND LAYERED MATERIALS
Plymouth, USA, 18–23 June, 1995.
Further information from: Professor M.E. Davis, Chemical Engineering, California Institute of Technology, Pasadena, California 91125, USA. Fax: +818-568-8743.

TWENTIETH INTERNATIONAL SYMPOSIUM ON MACROCYCLIC CHEMISTRY
Jerusalem, Israel, 2–7 July, 1995.
Further information from: The Secretariat, XXth International Symposium on Macrocyclic Chemistry, PO Box 50006, Tel Aviv 61500, Israel. Fax: +972-3-5175674.

TWELFTH INTERNATIONAL CONFERENCE ON THE CHEMISTRY OF THE ORGANIC SOLID STATE
Matsuyama, Japan, 9–14 July, 1995.
Further information from: Professor F. Toda, Department of Applied Chemistry, Faculty of Engineering, Ehime University, Matsuyama, Ehime 790, Japan. Fax: +899-23-0672.

SIXTH INTERNATIONAL SEMINAR ON INCLUSION COMPOUNDS
Istanbul, Turkey, 27–31 August, 1995.
Further Information from: Professor Dr S. Akyüz, Physics Department, Istanbul University, Fen Fakültesi, Vezneciler, 34459 Istanbul, Turkey. Fax: +212-519-0834; e-mail: fen12@earn.triuvml1.

EIGHTH INTERNATIONAL SYMPOSIUM ON CYCLODEXTRINS
Budapest, Hungary, 1–3 April, 1996.
Further information from: Professor J. Szejtli, Cyclolab, H-1525, PO Box 435, Budapest, Hungary. Fax: +1-212-5020.

Author Index

Aleksiuk, O., Grynszpan, F., and Biali, S.E. – Preparation, Structure and Stereodynamics of Phosphorus-Bridged Calixarenes   237
Asfari, Z., Astier, J.-P., Bressot, C., Estienne, J., Pepe, G., and Vicens, J. – Synthesis, Characterization, and X-Ray Structure of 1,2-Bis-crown-5-calix[4]arene. Modeling of Metal Complexation   291
Asfari, Z., Wenger, S., and Vicens, J. – Calixcrowns and Related Molecules   137
Asfari, Z. – see Hill, C. et al.   399
Astier, J.-P. – see Asfari, Z. et al.   291

Beer, P.D., Chen, Z., Gale, P.A., Heath, J.A., Knubley, R.J., Ogden, M.I., and Drew, M.G.B. – Recognition by New Diester and Diamide-Calix[4]arene Diquininones and a Diamide-Benzo-15-Crown-5 Calix[4]arene   343
Biali, S.E., – see Aleksiuk, O. et al.   237
Boerrigter, H. – see Timmerman, P. et al.   167
Böhmer, V., Kraft, D., and Tabatabai, M. – Inherently Chiral Calixarenes   17
Bottino, F. and Pappalardo, S. – Synthesis and Properties of Pyridinocalixarenes   85
Bressot, C. – see Asfari, Z. et al.   291
Bressot, C. – see Hill, C. et al.   399
Brodesser, G. and Vögtle, F. – Homocalixarenes and Homocalixpyridines   111

Casnati, A. – see Hill, C. et al.   399
Chen, Z. – see Beer, P.D. et al.   343

Danil de Namor, A.F., Wang, J., Gomez Orellana, I., Sueros Velarde, F.J., and Pacheco Tanaka, D.A. – Thermodynamic and Electrochemical Aspects of p-$tert$-butylcalix(n)arenes (n = 4, 6, 8) and Their Interactions with Amines   371
Diamond, D. – Calixarene-Based Sensing Agents   149
Doamekpor, L.K. – see Yamato, T. et al.   315
Dozol, J.-F. – see Hill, C. et al.   399
Drew, M.G.B. – see Beer, P.D. et al.   343

Estienne, J. – see Asfari, Z. et al.   291
Eymard, S. – see Hill, C. et al.   399

Frampton, C.S. – see Xu, W. et al.   277

Gale, P.A. – see Beer, P.D. et al.   343
Georghiou, P.E. and Li, Z. – Conformational Properties of Calix[4]naphthalenes   55
Georgiev, E. – see Roundhill, D.M. et al.   101
Gokel, G.W. – see Zhang, L. et al.   361
Gomez Orellana, I. – see Danil de Namor, A.F. et al.   371
Grynszpan, F. – see Aleksiuk, O. et al.   237

Harkema, S. – see Timmerman, P. et al.   167
Harrowfield, J.M., Richmond, W.R., and Sobolev, A.N. – Inclusion of Quaternary Ammonium Compounds by Calixarenes   257
Heath, J.A. – see Beer, P.D. et al.   343
Hill, C., Dozol, J.-F., Lamare, V., Rouquette, H., Eymard, S., Tournois, B., Vicens, J., Asfari, Z., Bressot, C., Ungaro, R., and Casnati, A. – Nuclear Waste Treatment by Means of Supported

Liquid Membranes Containing Calixcrown Compounds 399

Iki, H., Kikuchi, T., Tsuzuki, H., and Shinkai, S. – X-Ray Crystallographic Studies of Tricarbonylchromium Complexes of Calix[4]arene Conformers on an Unusual Conformation which Appears in Cone Conformers 227
Isnin, R. – see Zhang, L. et al. 361

Kaifer, A.E. – see Zhang, L. et al. 361
Kappe, T. – The Early History of Calixarene Chemistry 3
Khan, I.U. – see Takemura, H. et al. 193
Kikuchi, T. – see Iki, H. et al. 227
Knubley, R.J. – see Beer, P.D. et al. 343
Kraft, D. – see Böhmer, V. et al. 17

Lamare, V. – see Hill, C. et al. 399
Li, Z. – see Georghiou, P.E. 55
Lu, T. – see Zhang, L. et al. 361

Macias, A. – see Zhang, L. et al. 361
Manojlovic-Muir, L. – see Xu, W. et al. 277
Mataka, S. – see Sawada, T. et al. 301
Miura, H. – see Takemura, H. et al. 193
Muir, K.W. – see Xu, W. et al. 277

Nishimura, J. – see Okada, Y. 41

Ogden, M.I. – see Beer, P.D. et al. 343
Okada, Y. and Nishimura, J. – The Design of Cone-Fixed Calix[4]arene Analogs by Taking *syn*-[2.*n*]Metacyclophanes as a Building Block 41

Pacheco Tanaka, D.A. – see Danil de Namor, A.F. et al. 371
Pappalardo, S. – see Bottino, F. 85
Pepe, G. – see Asfari, Z. et al. 291
Puddephatt, R.J. – see Xu, W. et al. 277

Reinhoudt, D.N. – see Timmerman, P. et al. 167
Richmond, W.R. – see Harrowfield, J.M. et al. 257
Roundhill, D.M., Georgiev, E., and Yordanov, A. – Calixarenes with Nitrogen or Phosphorus Substituents on the Lower Rim 101
Rouquette, H. – see Hill, C. et al. 399

Saruwatari, Y. – see Yamato, T. et al. 315
Sawada, T., Tsuge, A., Thiemann, T., Mataka, S., and Tashiro, M. – Complexation Properties and Characterization of Four Conformers of a [2.1.2.1]Metacyclophane 301
Schneider, H.-J. and Schneider, U. – The Host-Guest Chemistry of Resorcinarenes 67
Schneider, U. – see Schneider, H.-J. 67
Shibutani, Y., Yoshinaga, H., Yakabe, K., Shono, T., and Tanaka, M. – Polymeric Membrane Sodium-Selective Electrodes Based on Calix[4]arene Ionophores 333
Shinkai, S. – see Iki, H. et al. 227
Shinmyozu, T. – see Takemura, H. et al. 193
Shono, T. – see Shibutani, Y. et al. 333
Sobolev, A.N. – see Harrowfield, J.M. et al. 257
Sueros Velarde, F.J. – see Danil de Namor, A.F. et al. 371
Swager, T.M. and Xu, B. – Liquid Crystalline Calixarenes 389

AUTHOR INDEX

Tabatabai, M. – *see* Böhmer, V. *et al.*   17
Takemura, H., Shinmyozu, T., Miura, H., Khan, I.U., and Unazu, T. – Synthesis and Properties of *N*-Substituted Azacalix[*n*]arenes   193
Tanaka, M. – *see* Shibutani, Y. *et al.*   333
Tashiro, M. – *see* Sawada, T. *et al.*   301
Thiemann, T., – *see* Sawada, T. *et al.*   301
Timmerman, P., Boerrigter, H., Verboom, W., Van Hummel, G.J., Harkema, S., and Reinhoudt, D.N. – Proximally Functionalized Cavitands and Synthesis of a Flexible Hemicarcerand   167
Tournois, B. – *see* Hill, C. *et al.*   399
Tsuge, A. – *see* Sawada, T. *et al.*   301
Tsuzuki, H. – *see* Iki, H. *et al.*   227

Unazu, T. – *see* Takemura, H. *et al.*   193
Ungaro, R. – *see* Hill, C. *et al.*   399

Van Hummel, G.J. – *see* Timmerman, P. *et al.*   167
Verboom, W. – *see* Timmerman, P. *et al.*   167
Vicens, J. – *see* Asfari, Z. *et al.*   137, 291
Vicens, J. – *see* Hill, C. *et al.*   399
Vögtle, F. – *see* Brodesser, G.   111

Wang, J. – *see* Danil de Namor, A.F. *et al.*   371
Weiss, J. – *see* Wytko, J.A.,   207
Wenger, S. – *see* Asfari, Z. *et al.*   137
Wytko, J.A. and Weiss, J. – Arranging Coordination Sites around Cyclotriveratrylene   207

Xu, B. – *see* Swager, T.M.   389
Xu, W., Puddephatt, R.J., Manojlovic-Muir, L., Muir, K.W., and Frampton, C.S. – Calixarenes: Structure of an Acetonitrile Inclusion Complex and Some Metal Rimmed Derivatives   277

Yakabe, K. – *see* Shibutani, Y. *et al.*   333
Yamato, T., Yasumatsu, M., Saruwatari, Y., and Doamekpor, L.K. – Synthesis and Ion Selectivity of Macrocyclic Metacyclophanes Analogous to Spherand-Type Calixarenes   315
Yasumatsu, M. – *see* Yamato, T. *et al.*   315
Yordanov, A. – *see* Roundhill, D.M. *et al.*   101
Yoshinaga, H. – *see* Shibutani, Y. *et al.*   333

Zhang, L., Macias, A., Isnin, R., Lu, T., Gokel, G.W., and Kaifer, A.E. – The Complexation of Ferrocene Derivatives by a Water-Soluble Calix[6]arene   361

Subject Index

acetonitrile inclusion complex 277
acetylcholine 67, 257
azacalix[n]arenes 193

caesium 399
calix(aza)crowns 137
calixcrowns 137
calix[4]arenediquinones 343
calix[4]naphthalenes 55
catalytic activity 41
cation recognition 343
cavitands 167
choline-type compounds 67
coordination sites 207
cyclotriveratrylene 207

diastereoselective functionalization 315

early history 1
electrochemistry 371

ferrocene derivatives 361

hemicarcerand 167
homocalixarenes 111
homocalixpyridines 111

inherent chirality 17
interactions with amines 371

large functionalized cavities 111

liquid crystalline 389
liquid-liquid-extraction 111

syn-[2.*n*] Metacyclophanes 41
[2.1.2.1]metacyclophane 301
MM3 calculation 193
modeling of metal complexation 291

nitrogen or phosphorus substituents 101

phosphorus-bridged calixarenes 237
pyridinocalixarenes 85

quaternary ammonium 257

resorcarenes 17
resorcinarenes 67

sensing agents 149
sodium-selective electrodes 333
soft metal complexation 137
spherand-type 315
strong hydrogen bond 193
supported liquid membranes 399

thermodynamics 371
transition metal 277
tricarbonylchromium complexes 227
Tungsten - oxo Calix[4]arenes 393

water-soluble calix[6]arene 361

Radicals on Surfaces

edited by **Anders Lund, Christopher Rhodes**

Due to its commercial potential and intellectual challenge, catalysis has seen a great amount of activity from researchers into surface science. The bulk of this activity has previously been focused on the physical properties of catalytic materials, and little has been done on the mechanistic details by which molecular transformations occur on catalyst surfaces. Growing evidence, however, indicates that organic free radicals act as intermediates in a number of reactions, particularly those catalysed by zeolites and other metal oxides.

This book contains invited contributions by leading scientists and addresses the broad aspect of radicals on surfaces. It deals with the properties of paramagnetic surface sites, the structure and reactivity of radicals on surfaces, and those trends and developments in spectroscopic techniques which are revealing new horizons.

This book will be of interest to researchers in organic free radicals and reactive catlysts in physical, organic and biological chemistry, chemical physics, radiation physics and chemistry.

New Publication

Contents and Contributors
I: Properties of Catalytic Surfaces. I.1 Study of Catalytic Site Structure and Diffusion of Radicals in Porous Heterogenic Systems with ESR, ENDOR, and ESE; *R.L. Samoilova, Yu.D. Tsvetkov, A.D. Milov.* **I.2** Metal Ion Active Sites; *M. Che, C. Louis.* **I.3** Theoretical Investigations of XPS and XAS Spectra of Adsorbates on Metal Surfaces; *M. Ohno.* **II: Structure and Reactivity of Radicals on Surfaces. II.1** Aromatic Radicals on Metal-Oxide Surfaces; *R.B. Clarkson, K. Matson, W. Ski, W. Wang, R.L. Belford.* **II.2** Structure and Reactions of Radical Cations in Zeolites; *C. Rhodes.* **II.3** Reactions of Surface Trapped Electrons with Adsorbed Molecules; *D. Murphy, E. Giamello.* **II.4** Formation of Radicals on Surfaces by Ionising Radiation; *M. Shiotani, M. Lindgren.* **II.5** Photostimulated Formation of Radicals on Surfaces; *V.A. Bolshov, A.M. Volodin, T.A. Konovalova.* **III: Trends in Modern Techniques. III.1** FT-EPR Studies in Heterogeneous Systems; *P. Levstein, H. van Willigen.* **III.2** Muonium Spin Resonance of Radicals on Surfaces; *E. Roduner, M. Schwager, M. Shelley.* **III.3** Investigation of Radical Ions with Surface-Enhanced Raman Spectroscopy; *R.L. Birke, J.R. Lombardi.* Index.

1995, 260 pp.   ISBN 0-7923-3108-7
Hardbound USD 154.00/NLG 240.00/GBP 100.00

TOPICS IN MOLECULAR ORGANIZATION AND ENGINEERING 13

Reprinted from MOLECULAR ENGINEERING, 4:1-3

P.O. Box 322, 3300 AH Dordrecht, The Netherlands
P.O. Box 358, Accord Station, Hingham, MA 02018-0358, U.S.A.

KLUWER ACADEMIC PUBLISHERS

Transition Metals in Supramolecular Chemistry
Proceedings of the NATO Advanced Research Workshop, Santa Margherita Ligure (Genoa), Italy, April 14–16, 1994

edited by **Luigi Fabbrizzi**, **Antonio Poggi**, *Dipt. di Chimica Generale, Università di Pavia, Italy*

Since the pioneering publications on coordination chemistry by Lehn and Pedersen in the late 1960s, coupled with the more orthodox interest from the transition metal chemists on template reactions (Busch, 1964), the field of supramolecular chemistry has grown at an astonishing rate. The use of transition metals as essential constituents of multi-component assemblies has been especially sharp in recent years, since the metals are prone to quick and reversible redox changes, and there is a wide variety of metal–ligand interactions. Such properties make supramolecular complexes of transition metal ions suitable candidates for exploration as light–energy converters and signal processors.
Transition Metals in Supramolecular Chemistry focuses on the following main topics: (1) metal controlled organization of novel molecular assemblies and shapes; (2) design of molecular switches and devices operating through metal centres; (3) supramolecular catalysts that mimic metalloenzymes; (4) metal–containing sensory reagents and supramolecular recognition; and (5) molecular materials that display powerful electronic, optoelectronic and magnetic properties.

Contents
Preface. Supramolecular chemistry and photophysics. Energy conversion and information-processing devices based on transition metal complexes. Transition metal redox active ligand systems for recognising cationic and anionic guest species. Ligand design for enhanced molecular organization – selectivity and specific sequencing in multiple receptor ligands, and orderly molecular entanglements. Metal-ions: a self-assembly motif in supramolecular oligomers. Biomedical targeting: a role for supramolecular chemistry. Redox chemistry of metal ion complexes: preparation of new materials. pH and redox switches based on metal centres. Schiff base macrocycles and metallo-biosite modelling. Self-assembly of mono and dinuclear metal complexes; oxidation catalysis and metalloenzyme modelling. Artificial porphyrins containing cyclopropane units functioning as electron shuttles. Homo and heterobinuclear metal complexes with bis-macrocyclic ligands. Ferrocene as a building block for supramolecular systems. Macrocyclic polyamine complexes beyond metalloenzyme models. Towards molecular wires and switches: exploiting coordination chemistry for nonlinear optics and molecular electronics. Molecular interactions between metalloproteins involved in electron transfer processes: tetrahaeme cytochrome c_3 flavodoxin. ^{1}H NMR and molecular modelling studies. Supramolecular models of metalloproteins. The role of macrocyclic receptors in the organization of metal centres. Metallomacrocycles and clefts, receptors for neutral molecules and anions. Chiral recognition by functionalized cyclodextrin metal complexes. Transition metal-directed threading and knotting processes. Expanded porphyrins. Receptors for cationic, anionic and neutral substrates. Following the self-assembly process in solution. Author Index. Subject Index.

1994, 456 pp.   ISBN 0–7923–3196–6
Hardbound USD 195.00/NLG 305.00/GBP 128.00

NATO ADVANCED SCIENCE INSTITUTES SERIES
C: Mathematical and Physical Sciences 448

P.O. Box 322, 3300 AH Dordrecht, The Netherlands
P.O. Box 358, Accord Station, Hingham, MA 02018-0358, U.S.A.

KLUWER ACADEMIC PUBLISHERS